4

BEAUTIFUL EQUATIONS

Vadim Baines-Jones
Patrick Baird
Harish Cherukuri
Massimiliano Ferrucci
Dave Flack
Mike Goldsmith
Peter de Groot
Han Haitjema
Olaf Kievit
Richard Leach
Ian Lee-Bennett
Campbell McKee
Giuseppe Moschetti
Kang Ni
Robert Oates
Frank Roberts
Paul Rubert
Stuart Smith
Trevor Stolber
Antonio Torrisi

COPYRIGHT PAGE

Introduction

In 2015, an online community of scientists and mathematicians got together to share a love of equations, and, every day that year, a beautiful equation appeared on Facebook. The page has since attracted almost 6000 "likes", showing that lots of other people love equations as much as we do.

What do we mean by "beautiful" in the context of equations? Well, as John Keats put it almost exactly 200 years ago:

> "Beauty is truth, truth beauty – that is all
> Ye know on earth, and all ye need to know"

The equations in this book each encapsulate a truth. That truth may be drawn from physics, mathematics, biology, engineering, computing - or from economics, psychology, gaming, etc. - any topic can potentially generate an equation. Many of us can appreciate the beauty in a painting or even a sports car, but equations capture the beauty of simplicity. A beautiful equation is one where the bare essentials of a particular relationship between one aspect of reality and another have been captured in a mathematical statement. The beauty is not in the symbols on the paper (although this can be beautiful to some people), but in the almost magical capability of the equation to capture so much about reality. Equations have a profundity that can be breath-taking; stirring up almost every aspect of human emotion.

"Truth" is the key here, but there's more to mathematical beauty than truth. Unlike a line of poetry, we can recognise when an equation is as clear and concise as it can possibly be. If the equation is unnecessarily long or unwieldy, or limited in its scope, then it may lack the beauty for which it was destined. There are no such equations here!

In setting up the Facebook page, and in publishing this book, we had other aims than simply to revel in the beauty of mathematics. There are more popularisations of science and other technical topics on the market than ever before, for children and adults alike, enthusiastically covering all aspects of their chosen topics. All except one: publishers of popular science books, convinced of the terrors, inextricable complexities and deep unpopularity of mathematics, do their best to eradicate mathematical formulae from their publications. To do this is not only to present an unbalanced view of science, which necessarily focuses on its superficialities at the expense of its heart and soul, but also to deny access to a source of wonder and knowledge. It is like publishing an image-free book about the world's greatest paintings.

We hope that, in a small way, this book will help to redress that imbalance.

Colour versions of many of the figures can be found on the original Facebook page: https://www.facebook.com/Beautifulequations2015/

Mike Goldsmith and Richard Leach, April 2017

Equation 1: Newton's Second Law of Motion

This equation essentially captures Newton's Second Law of Motion, which states: "The rate of change of momentum of a body is equal to the resultant force acting on the body and is in the same direction". F is force, m is mass and a is acceleration. Notice the arrows above the F and a – these are needed (but often omitted) because F and a are vector quantities; meaning they have both magnitude and direction (i.e. in the equation, F and a are in the same direction). Despite its apparent simplicity, this equation is a second order differential equation and is very powerful for predicting the motion of classical bodies.

$$F = ma$$

RL

Equation 2: D'alembert's Principle

$$\sum_i (F_i - m_i a_i - \dot{m}_i v_i) \cdot \delta r_i = 0$$

where, for the i_{th} particle:

F_i is the total applied force (excluding constraint forces), m_i is the mass, $\dot{m}_i$ is the acceleration, a_i is the virtual displacement, consistent with the constraints, v_i is the velocity, and δr_i is the rate of change of mass with respect to time.

The vector nature of Newton's equations of motion makes them difficult to use with non-Cartesian coordinate systems. The n vector equations of an n-body system can be transformed into a single scalar equation by making use of the concept of 'virtual displacement'. This equation is the mathematical formulation of d'Alembert's principle: "The total virtual work of the impressed forces plus the inertial [*aka* 'fictitious' or 'pseudo'] forces disappears for displacements that are reversible."

This principle is the foundation of analytical mechanics; from it, Lagrange's equations can be derived. The latter are particularly used with non-inertial frames of reference.

FR

Equation 3: Snell's Law of Refraction

Although traditionally attributed to the Dutch scientist Snellius in the early 17th century, this was first formulated by a Persian scientist more than 1000 years ago, and was recorded in tabulated form by Ptolemy many centuries before that. A simple equation with no differential functions, but essential to all optics. It describes what happens to a light beam as it moves from one medium (or material) into another, for example air to liquid or glass. For a first medium 1 and second medium 2, n is the refractive index of that medium. This is equivalent to the ratio of the speed of light in a vacuum to the speed in the medium and is dependent on the wavelength (and is therefore also what causes dispersion in a prism); theta is the angle (relative to normal or perpendicular incidence) that the incident beam makes with the interface between the two media. The angle of refraction, as with the angle of reflection (equal to the angle of incidence) can be derived from Fermat's principle (path of least time). If the angle is zero (normal incidence) there is no change in the direction of propagation. If there is an angle between the direction of the light beam and the interface, the direction of propagation of the beam changes. There are two reasons for this: 1) the speed of the light beam is slowed down in a medium of higher refractive index (due to its interaction with matter - in a vacuum the speed is c, the of light, or any electromagnetic radiation, roughly 3 x 10^8 m/s, and in air it is very slightly less); 2) light propagates as a transverse wave. Due to the wave nature of light, successive crests of a wavefront will enter the optically denser medium at later times and faster speeds than previous ones, resulting in a new wavefront propagating at a different angle. This is a clear demonstration of light having wave properties, and was instrumental in the description of light as a wave phenomenon rather than a "corpuscular" one, as Newton had interpreted it. Much later experiments investigating quantum phenomena revealed that light has aspects of both. Unless there is absorption by the medium by electronic or molecular resonance (another story), the frequency (energy) of the light is not changed. The absorption component can be conveniently accounted for by defining a complex refractive index. In the case of white light, a combination of many frequencies, dispersion occurs (the light beam will separate at different angles). It is important to realise that in refraction the true speed of light does not change - the phase velocity, the speed of propagation of a particular amplitude (and hence the effective wavelength) changes. This applies to isotropic media, and there are exceptions such as birefringent crystals in which the direction of propagation depends on polarisation, and some materials that aren't transparent have a refractive index less than 1 (the phase velocity is faster than c - this is not a violation of the speed of light). In some cases, negative refractive index can occur in specifically designed nanostructures (metamaterials with units less than a wavelength). On the other end of the scale, there are some materials (Bose-Einstein condensates) in which the phase velocity of light becomes very slow. Generally, there will be some reflection in an optical material and the relative amounts of light reflected and refracted are described by the Fresnel equations.

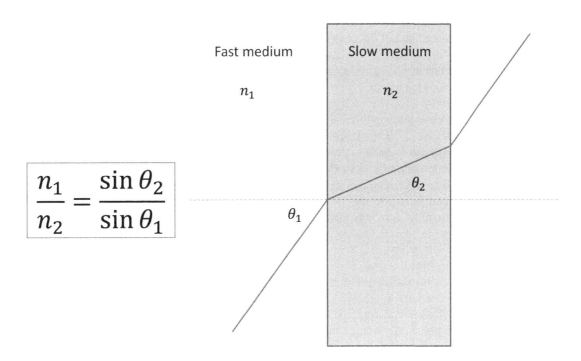

$$\frac{n_1}{n_2} = \frac{\sin \theta_2}{\sin \theta_1}$$

PB

Equation 4: The Schwarzschild Radius

$$R = \frac{2GM}{c^2}$$

The Schwarzschild radius, R. This is the radius at which an object will become a black hole. The equation holds for all objects, however massive. If a mass is over some critical value, between two and three solar masses and has no fusion process to keep it from collapsing, then gravitational forces alone make the collapse to a black hole inevitable. Note that the object does need to be spherical and non-rotating for this equation

11

to hold. When collapsing under its own gravity, an object will first compress its matter to very high density, then compress its atoms past what's called electron degeneracy, then past neutron degeneracy, and finally down to the Schwarzschild radius – the point of no return. The final phase of collapse results in a singularity and we can't really be exact about what that is (although it appears to have infinite density and zero spatial extent). In the equation, G is the gravitational constant (6.673×10^{-11} N(m/kg)^2) (sometimes called Newton's constant), M is the object's mass and c is the speed of light in a vacuum (299 792 458 m/s). The equation was found by Karl Schwarzschild (German) in 1916 as an exact solution to Einstein's field equations. This was one of the first equations that really made me wonder at how such a complex process can result in such a simple solution involving universal constants. Truly beautiful!

RL

Equation 5: Temperature-Corrected Length Measurements

The reference temperature for Geometric Product Specifications is 20 °C. The length of an object at this temperature is given by:

$$L_{20} = L \cdot \left(1 - \alpha \cdot (T - 20\,^{\circ}C)\right)$$

where: L_{20} is the object length at 20 °C, L is the measured length of an object, T is the temperature of the measured object and α is the thermal linear expansion coefficient of the object.

This may not look like a 'beautiful' equation, especially the 20 °C is a bit awkward and deliberate, but it is a very basic one with huge consequences.

ISO 1 states that the reference temperature for Geometrical Product Specification, i.e. the length of objects made in industrial production, is 20 °C. If the temperature is different, it should be corrected according to the equation above. In order to keep deviations and the corresponding implied uncertainties low (i.e. the term $\alpha \cdot (T - 20\,^{\circ}C)$) the temperature in workshops and dimensional metrology laboratories must be kept close to 20 °C. This lead to huge amounts of energy usage, especially for cooling and dehumidifying, in countries such as USA and Japan. In the 1990's an attempt was made to raise this temperature, but the consequences for all mechanical drawings made in the past could not be overseen.

HH

Equation 6: Euler-Lagrange Equation of Motion

Lagrangian mechanics allows one to describe the motion of a physical system by a formulation different to the one proposed by Newton. In Newtonian mechanics, the equations of motion for a given system can be determined using Newton's equation,

$$F = ma$$

and the relationship between force and potential energy,

$$F = \frac{-\mathrm{d}U}{\mathrm{d}x}$$

where x is a spatial coordinate.

On the other hand, Lagrangian mechanics make use of a new variable L, which is accordingly named the Lagrangian. L is given by the following equation:

$$L = T - V$$

where T is the total kinetic energy of the system and V is the total potential energy (there is conceptually no difference between U and V). Once L has been determined, the following relationship is used to produce the equations of motion:

$$\frac{d}{dt}\left(\frac{dL}{d\dot{x}}\right) = \frac{dL}{dx}$$

The equation above is known as the Euler-Lagrange equation of motion: x is a generalized spatial coordinate in the system and $\dot{x}$ is the first time derivative of that coordinate. The Euler-Lagrange equation is performed for each of the spatial coordinates of the system under question.

While this method for determining equations of motion might seem like a trivial re-formulation, the advantages of the method over the Newtonian formulation are more obvious for complex systems.

MF

Equation 7: Ohm's Law

Ohm's law states that the current through a conductor between two points is directly proportional to the potential difference across the two points. The usual mathematical equation that describes this relationship when including the constant of proportionality, the resistance, is today's beautiful equation namely:

$$I = \frac{V}{R},$$

where I is the current through the conductor in units of amperes, V is the potential difference measured across the conductor in units of volts, and R is the resistance of the conductor in units of ohms. More specifically, Ohm's law states that the R in this relation is constant, independent of the current.

It is often put into the Triangle formulation below to make it easy to understand. Reading the triangle gives $I = \frac{V}{R}$, $V = IR$ and $R = \frac{V}{I}$.

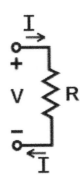

VB-J

Equation 8: Time Dilation

$$\Delta t' = \frac{\Delta t}{\sqrt{1 - \frac{v^2}{c^2}}}$$

Einstein's time dilation (or dilatation) equation (1905) captures the essence of his revolution, in which the traditional view of time as something that ticks on steadily at the same rate, forever and everywhere, was consigned to the dustbin of history. What Einstein realised was that the passage of time changes depending on who measures it - and in particular, how that observer is moving.

In the equation, $\Delta t'$ represents a period of time. This might be, for example, one minute as measured by the wrist-watch of an astronaut. Δt is the equivalent period of time that the astronaut might read on another clock - say, a clock face mounted on a space-station which (s)he is speeding past.

Let's call this passing-speed v. If the astronaut hovers near the station, v is 0. In the equation, c is the velocity of light (about 300,000 km per sec). This means that the fraction $\frac{v2}{c2}$ equals 0, and so the denominator (bottom half) of the equation is $\sqrt{1}$, which is 1, so that means that $\Delta t' = \Delta t$. The station clock and the wristwatch, in other words, tell the same times. But if the astronaut moves past the space station at half the velocity of light, then the denominator of the equation becomes $\left(1 - \frac{0.5^2}{1^2}\right)$ (the 0.5 refers to the half-light velocity of the astronaut). This then gives $\Delta t' \approx \frac{\Delta t}{0.866}$ which means that, by the time the wristwatch indicates that one minute has past $\Delta t'$, the space station clock has recorded the passing Δt of only about 52 seconds.

But time is stranger still. If someone on the space station, equipped with a telescope, zoomed in on the watch of the passing astronaut, one might think that once 52 seconds had passed on the station, (s)he would see the astronauts watch reach 1 minute. In fact, she would see it indicate just 45 seconds. In other words, from the point of view of the person in the space station, the astronaut's time has slowed down, and from the point of view of the astronaut the space-station time has slowed. And this is the other key discovery of Einstein: that there is no privileged frame of reference, which is to say that whether we say that the astronaut moves past the space-station or that the station moves past the astronauts makes no difference. The only thing that matters is that that they move relative to each other. Hence, "relativity."

MJG

14

Equation 9: Airy Disk

$$\sin \theta = 1.22 \frac{\lambda}{d}$$

Airy - Disk

where: λ is the wavelength of illuminating light, d is the diameter of the circular aperture, and θ is the angle at which the first pattern minimum occurs from the direction of the incoming light.

Sir George Airy (1801-1892), a British scientist first derived the equation that describes the diffraction pattern of light passing through a small circular aperture when the image is observed in the far field (from a distance). He published it in the 1835 paper "On the Diffraction of an Object-glass with Circular Aperture".

The Airy disk is also known as the theoretical limit of the resolution of an optical imaging system such as the eye, a microscope, telescope and camera. In microscopy and astronomy, the Airy disk can be stated as the Rayleigh Criterion for resolving details in the object plane, where the pattern of the first minimum of the Airy disk from a point object coincides with the maximum from an adjacent point.

KN

Equation 10: Maxwell–Faraday Equation

$$\nabla \times E = -\frac{\partial B}{\partial t}$$

where: E is the electric field, B is the magnetic field, and $\nabla \times$ is the curl operator.

The Maxwell–Faraday equation is a generalisation of Faraday's law that predicts how a magnetic field will interact with an electric circuit by way of a phenomenon known as electromagnetic induction. The curl operator represents the "vorticity" or the amount of rotation of electric charges (if the charges are free to move) in the plane through which the magnetic flux varies in time.

The equation is named after the Scottish physicist and mathematician James Clerk Maxwell, who formulated the full set of Maxwell equations between 1861 and 1862 and the English scientist Michael Faraday, who demonstrated electromagnetic induction in 1831.

In other words, the equation states that the induced potential difference in any closed circuit is equal to the negative time rate of change of the magnetic flux enclosed by the circuit. This principle is employed in many electric devices, such as generators, electric motors and solenoids.

When I first studied the theory behind this equation I liked how it describes the concept behind the electric power we receive in our houses. In a generator, mechanical energy is converted in electric energy by moving magnets to vary periodically the magnetic flux in an electric circuit. As a consequence, an alternating potential difference is induced in the electric circuit and therefore an alternating current (AC) we all use in our houses.

Further reading: *A Student's Guide to Maxwell's Equations* (2011), Daniel Fleisch

GM

Equation 11: The Pythagorean Theorem

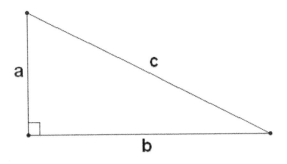

$$a^2 + b^2 = c^2$$

The Pythagorean theorem, is a relation in Euclidean geometry among the three sides of a right triangle. It states that the square of the hypotenuse (the side opposite the right angle) is equal to the sum of the squares of the other two sides. The theorem can be written as an equation relating the lengths of the sides a, b and c, often called the "Pythagorean equation". Often referred to as Pythagoras' theorem, however, because of the secretive nature of his school and the custom of its students to attribute everything to their teacher, there is no evidence that Pythagoras himself worked on or proved this theorem. For that matter, there is no evidence that he worked on any mathematical or meta-mathematical problems.

There is some evidence that Babylonian mathematicians understood the formula, although little of it indicates an application within a mathematical framework. Mesopotamian, Indian and Chinese mathematicians are all known to have discovered the theorem independently and, in some cases, provide proofs for special cases.

The formula can be proved in 370 (known) ways but the most common ways are by similar triangles, Euclid's proof, proof by rearrangement, algebraic proofs and proof by differentiation. More generally, in Euclidean n-space, the Euclidean distance between two points can be calculated using the equation (we do this in co-ordinate metrology).

The Pythagorean theorem is a special case of the more general theorem relating the lengths of sides in any triangle: the law of cosines.

The Scarecrow in the film The Wizard of Oz exhibits his "knowledge" by reciting a mangled and incorrect version of the theorem: "The sum of the square roots of any two sides of an isosceles triangle is equal to the square root of the remaining side. Oh, joy! Oh, rapture! I've got a brain!"

Further reading: *The Pythagorean Theorem: A 4,000-Year History* (2010), Eli Maor

DF

Equation 12: Fourier Series

$$f(t) = a_0 + \sum_{n=1}^{\infty} R_n \cos\left(\frac{2\pi n}{\tau} t - \varepsilon_n\right)$$

$$= a_0 + \sum_{n=1}^{\infty} A_n \cos(\omega_n t) + \sum_{n=1}^{\infty} B_n \sin(\omega_n t)$$

$$A_n = R_n \cos \varepsilon_n, \quad B_n = R_n \sin \varepsilon_n, \quad \omega_n = \frac{2\pi n}{\tau}$$

$$a_0 = \frac{1}{\tau} \int_0^\tau f(t)\, dt, \quad A_n = \frac{2}{\tau} \int_0^\tau f(t) \cos(\omega_n t)\, dt, \quad B_n = \frac{2}{\tau} \int_0^\tau f(t) \sin(\omega_n t)\, dt$$

$$\int_0^\tau \cos(\omega_n t) \cos(\omega_m t)\, dt = 0, m \neq n$$

$$= \frac{\tau}{2}, m = n$$

$$\int_0^\tau \sin(\omega_n t) \sin(\omega_m t)\, dt = 0, m \neq n$$

$$= \frac{\tau}{2}, m = n$$

$$\int_0^\tau \sin(\omega_n t) \cos(\omega_m t)\, dt = 0$$

In his 1822 treatise on the Analytic Theory of Heat, Jean Baptiste Joseph Fourier presented a general method for representation of a single-valued, continuous, periodic function. This series had appeared previously for solutions of physical problems, in particular, in the solution for the dynamics of strings presented by Daniel Bernoulli in the middle of the 18th century. At a first glance an equation comprising an infinite sum of harmonic functions (sine and cosine waves) looks a little complicated. However, the reasoning behind this equation is relatively straightforward. Before discussing this, it is important to understand the nature of all of the parameters in the equation.

Because this is a periodic function the shape will repeat after a fixed time period τ that, of course, is a constant. A major attribute of this series is that, as stated before, provided the function $f(t)$ is continuous, single-valued and periodic in the variable t, all other parameters in this series, a_0, R_n, ε_n, A_n & B_n are simply constants. A method to visualize this series is to study the sum shown in very first equation. Each term in the sum comprises a harmonic function (note that the sine and cosine waves are also often referred to as trigonometric, or circular, or are grouped within other transcendental functions), multiplied by an amplitude R_n and frequency ω_n. The first term in the sum is a harmonic wave that will return to the same value after the periodic time τ as is necessary. If this has to have a specific value at a time within this cycle, this can easily be arranged by suitable selection of the two constants R_1 and ε_1. Those familiar with phasor representations might visualize this as a vector of length R_1 rotating at speed ω_1 and starting with an offset angle provided by ε_1. However, if it is necessary to hit a second value within this period, a second phasor having a different frequency is needed. However, if a second phasor is used, it is necessary that it also returns to the exact same value after the period of the original signal, something that is only possible if the frequency is an exact integer multiple of the fundamental. Hence a second frequency of twice the fundamental can be used and so on until an infinitude of points are traced out by the trajectory of the sum.

Not mentioned above was how to determine the values of the constants. To do this, use is made of the orthogonal properties of harmonic functions shown in the second box. Using these properties (in what seems like a circular procedure), multiplying the original function in its Fourier series form by either $\cos(\omega_n t)$ or $\sin(w_n t)$ and integrating over the period of the function immediately produces the expressions in the last line of the Fourier series.

One might be tempted to ask what is so beautiful about such a long equation replacing a single function? The amazing thing about this is the original periodic function has an arbitrary shape and will therefore often require complicated equations in its description that, when used in other mathematical procedures, are often very difficult to solve. After converting to a Fourier series, the function can now be expressed in terms containing only sine and cosine waves of constant magnitude and frequency varying about a zero mean value (plus the average of the signal of course). For many physical models, particularly those comprising linear differential equations, analytic solutions for such harmonic inputs can be readily obtained. Another outcome of this analysis is the fact that any periodic function can contain only harmonics of integer multiples of a fundamental frequency. For example, functions containing irrational multiples of frequencies (such as $\cos(t) + \cos(\pi t)$) cannot be periodic!

Reference
Fourier J. B. J., 1878, *The analytical theory of heat*, The Cambridge University press, translated by A. Freeman.

SS

Equation 13: The Black-Scholes Equation

$$\frac{\partial V}{\partial t} + \frac{1}{2}\sigma^2 S^2 \frac{\partial^2 V}{\partial S^2} + rS\frac{\partial V}{\partial S} - rV = 0$$

where: V is the option price, t is the time, S is the stock price, r is the interest rate, and σ is the volatility of the underlying stock.

The Black-Scholes equation is a financial formula for pricing a European stock option. The intention of this formula is to effectively remove the risk element associated with volatile stocks and choose options which hedge a portfolio into a delta neutral position.

The delta neutral position basically creates a risk-less portfolio which will remain stable whether the underlying stocks in the portfolio go up or down in price. Options are purchased to hedge a portfolio if the value equates to a favourable valuation. Options prices fluctuate a lot with stock market volatility and time decay. The process of selecting options based on price fluctuation is referred to as dynamic hedging.

The equation is a partial differential equation which factors in time decay and effectively eliminates the uncertainty in the underlying instrument.

This is the model used by the majority of the financial world and is currently used to price options on the major exchanges around the world. It applies to both European put and call options. It is less commonly used with American style options which can be executed at any time whereas European style options are only able to be executed at expiry rather than at any point in time.

The inventors of the equation received the Nobel prize and were widely celebrated. They also started a commercial entity based on their dynamic hedging principle and made investors worldwide billions with annual returns of 30-40% for 3 years before spectacularly failing and accruing massive debt when the underlying market conditions changed outside of the parameters in which they had allowed for.

This is a beautiful equation in my mind because it effectively eliminate risk, something no other financial model or instrument can do. Note that this equation does not factor in any dividends or future changes in rates and is generally used over fairly short time periods.

Further reading: *Probability Theory in Finance: A Mathematical Guide to the Black-Scholes Formula* (2013), Seán Dineen

TS

Equation 14: Photosynthesis

$$6\,CO_2 + 6\,H_2O \rightarrow C_6H_{12}O_6 + 6\,O_2$$

Photosynthesis is the process whereby plants use energy from sunlight to convert carbon dioxide and water into molecules needed for growth, chiefly the sugar glucose.

The green chemical chlorophyll which is present in all plants, absorbs light energy, which in turn allows the production of glucose by the reaction between carbon dioxide and water. Oxygen is also produced as a waste product. The glucose made by the process of photosynthesis may be used in three ways: it can be converted into cellulose, required for growth of plant cells, or into starch, a storage molecule, and then back into glucose when the plant requires it. It can also be broken down during the process of respiration, releasing energy stored in the glucose molecules.

The usefulness of this simple chemical reaction for human life can hardly be overstated: at the macroscopic whole-earth level, forests help reduce the amount of carbon dioxide contributing to the greenhouse effect, and at the microscopic level plants can help improve stuffy atmospheres in confined spaces by converting exhaled carbon dioxide into oxygen.

Many individuals played a part in the development of the theory of photosynthesis, going back to Jan van Helmont's experiments in the mid-seventeenth century on changes to the mass of a plant and of its soil as the plant grew. Later, Joseph Priestley was able to show that air which had been "injured" by a burning candle or by a breathing mouse, could be "restored" by a plant. Jean Senebier and Nicolas de Saussure also made important discoveries.

The beauty of the equation is due to its simplicity: it shows how two ordinary compounds which are highly stable can react together under conditions of sunlight to form the basis for all life on earth.

PR

Equation 15: Heisenberg's Uncertainty Principle

Heisenberg's uncertainty principle sets a fundamental limit to the uncertainty with which certain pairs of physical properties of a particle can be known simultaneously. Δx is the uncertainty in the position, Δp is the uncertainty in the momentum and the h with a bar through it is Planck's constant ($6.62606957 \times 10^{-34}$ m^2 kg/s) divided by 2π. The principle was developed on the cusp of a revolution in physics around the turn of the 20th Century: the development of quantum mechanics – our best theory of how things happen on the scale of atoms and particles. It was first stated by Werner Heisenberg (German) in 1927 and formally written out in the form in the figure by Hesse Kennard later that year.

The certain pairs of properties that are referred to above are known as complementary variables. The pair in the figure is position and momentum. So the principle states that there is a finite limit to the uncertainty in which one can know both position and momentum, or to put it another way, the more precisely you know the momentum, the less precisely you know the position, and vice versa.

Other complimentary variables are energy and time, and phase and amplitude. The former is the reason that the vacuum is awash with particle-anti-particle pairs popping into existence by essentially borrowing energy for a very short amount of time.

The uncertainty principle is often misinterpreted. The principle tells us something fundamental about nature – it is not due to the presence of an observer in a physical system – that is the observer effect – basically, you cannot observe something without affecting it in some manner.

$$\Delta x \Delta p \geq \frac{\hbar}{2}$$

Further reading: *Uncertainty: Einstein, Heisenberg, Bohr, and the Struggle for the Soul of Science* (2008), David Lindley

RL

Equation 16: The Malthusian Law

$$P(t) = P_0 e^{rt}$$

where: $P(t)$ is the population size at time t, $P_0 = P(0)$ is the initial population size, r is the Malthusian parameter, aka population growth rate, i.e. replication rate plus immigration rate-death rate, and t is the time.

The Malthusian Law, *aka* the Malthusian growth model, is named after the Rev Thomas Malthus, FRS, prominent in the fields of political economy and demography.

In 1798 Malthus wrote the then controversial 'An Essay on the Principle of Population', a very influential book on population dynamics. In this book, he observed that in nature plants and animals produce far more offspring than can survive, given food supply constraints.

Both Darwin and Wallace, influenced by Malthus's Essay, independently arrived at similar theories of Natural Selection. They extended Malthus' logic by pointing out that a population producing more offspring than can survive establishes a competitive environment. The combination of the latter and genetic variation in offspring will result in an evolutionary progression of a species.

A Malthusian parameter range $r > 0$ implies positive population growth, i.e. increase; $r < 0$ implies negative population growth, i.e. decline. A negative parameter implies death rate in a population exceeds birth and net immigration rates.

Although in real life scenarios r may not be constant, the Malthusian growth model works well in early stages of population growth. It has provided a platform for the development of the several subsequent growth models such as the logistic model which produce a flattening out of what was an exponential curve into an 'S' shape.

The Malthusian Law / growth model is just the exponential law / growth model used in the context of population dynamics. In other contexts, the exponential law is used to describe: accumulation of savings with compound interest; spread of epidemics; radioactive decay; attenuation of photons travelling through a medium; and so on.

Further reading: *Malthus* (2014), Robert J. Mayhew

FR

Equation 17: The Schrödinger Equation

$$i\hbar \frac{\partial}{\partial t} \psi(x, t) = -\frac{\hbar}{2m} \nabla^2 \psi(x, t) + V(x)\, \psi(x, t)$$

The general form of the Schrödinger equation is

$$i\hbar \frac{\partial}{\partial t}\psi(\boldsymbol{r},t) = \hat{H}\,\psi(\boldsymbol{r},t).$$

where $\psi(\boldsymbol{r},t)$ is defined as a "wave function" in general spatial coordinate vector $\boldsymbol{r}$ and time t, $\hbar$ is Planck's constant h divided by 2π, $\hat{H}$ is the Hamiltonian operator (representing the total energy of a system) and i is the square root of -1. The wave function describes the quantum state of a system using partial derivatives and contains all the information about the system. It is not, in itself, a measurable quantity, and the results of measurements, due to their inherent statistical nature, are represented by a probability density (the complex-valued wave function multiplied by its complex conjugate generates a real number).

One of the most familiar forms, displayed at the top, is the form for a single particle of mass m in 3 dimensions, in which the two terms on the right are the parts of the Hamiltonian which represent the kinetic and potential energy of the particle in an electric field (this could be a simplistic description for the solitary electron in a hydrogen atom). The Laplacian operator ∇^2 is a spatial second derivative, so this is a second order partial differential equation (PDE). At first sight it looks like a diffusion type equation (a parabolic PDE), but the time-dependent form includes complex-valued functions and the imaginary term on the left ensures the solution is wave-like with certain properties. The classical wave equation (BE48, a typical hyperbolic PDE) has a second derivative with respect to time; the Schrödinger equation is different in that it has a first derivative with respect to time. This is a consequence of requiring a wave-like solution and satisfying the recently discovered de Broglie relations (BE102) for "matter waves" (the French physicist Louis de Broglie had recently shown that matter particles have a wavelength associated with them, related to the momentum and Planck's constant). It was later realised that the wave function itself has to be a complex function to satisfy some of the postulates of quantum theory used in calculations (the main reason for this is another branch of the story and quite a lengthy one, relating to the mathematical preservation of probability amplitudes).

The first true observations of quantum effects were observed back in the 19th century in the form of atomic emission and absorption lines. Atoms only emit or absorb light at specific wavelengths. The Balmer series for hydrogen (lines in the visible range of the spectrum) was first measured by Anders Ångström in 1853. These discrete emission/absorption energies were not understood at the time and it was not until after the work by Max Planck on black body radiation in 1900 that quantisation of energy was taken seriously (although something of that nature had been speculated on since the 1870s). Later on, in developing what is now called the "new" quantum theory, the spectroscopic lines of hydrogen became an important test of the new atomic theory and the development of quantum theory. Schrödinger formulated his wave mechanics in 1926, a year after Werner Heisenberg had developed his own method of matrix mechanics, an equivalent but different approach which is somewhat less intuitive. Both were used to solve the energy levels of the hydrogen atom, and were a significant improvement over Niels Bohr's previous model of the atom which represented the electron "shells" as circular, discrete orbits. Heisenberg's method was successfully applied by Wolfgang Pauli in the same year, and in Schrödinger's case (with a bit of help from the mathematician Hermann Weyl) it involved a time-independent form of the above equation (in which the left hand side is replaced with an $E\psi$ term). The equation has paired solutions of certain values of energy (eigenvalues) and wave functions (eigenvectors). The solution is rather involved and requires separation of variables in spherical polar coordinates, resulting in the first three quantum numbers (principal, orbital and magnetic) that we know of today. The fourth quantum number, intrinsic spin, was discovered around the same time by Pauli due to the fine structure of atomic spectra unaccounted for and applied later by Paul Dirac, who developed the equations further to include relativistic effects. Eventually quantum theory was developed further to account for hyperfine structure such as the Lamb shift (an additional effect resulting from the interaction of charged particles with the vacuum), and led the way to quantum electrodynamics and an alternative way of calculating quantum states (Feynman's path integrals).

However, Schrödinger's equation, and Dirac's extensions, provided the basis for understanding atomic structure including a proper understanding of the periodic table. It is one of the greatest achievements in physics in the 20th century. Schrödinger and Dirac shared the Nobel Prize for physics in 1933 for "the discovery of new productive forms of atomic theory".

The formalism can be extended and built up to calculate a wide range of phenomena including the energy states of larger atoms with many electrons, the lower energy levels associated with molecular vibrations (the quantum harmonic oscillator) and rotations (the quantum rigid rotor), and conduction bands in bulk solids. The equation is essentially the same in all cases but with the energy terms (Hamiltonian) selected according to the system.

Further reading:
The Schrödinger Equation (1991), F. A. Berezin & M.A. Shubin

PB

Equation 18: Hooke's Law

Hooke's law states that the force needed to extend or compress a spring by a certain distance is proportional to that distance. F is force, x is distance and k is the constant of proportionality, known as the spring constant or stiffness. The law was first stated in 1660 by Robert Hooke (English) – one of the big guns in the Royal Society, often unfairly quoted as an unsavoury character, but this was only towards the end of his life when he became ill.

Hooke's law applies to many materials that have elastic properties, e.g. a cantilever or a building blowing in the wind. It does have its limits and is only a first order approximation of the behaviour of a material. At some limit (the elastic limit), more complex behaviour will be exhibited, e.g. work hardening or plastic (permanent) deformation.

$$F = -kx$$

RL

Equation 19: Temperature-Corrected Length Measurements

The length of an object, when measured by a length reference, is given by

$$L_{obj,20} = L_{ref} \cdot \left(1 + \frac{\alpha_{ref} + \alpha_{obj}}{2} \cdot \left(T_{ref} - T_{obj} \right) + \left(\alpha_{ref} + \alpha_{obj} \right) \cdot \left(\frac{T_{ref} - T_{obj}}{2} - 20 \, ^\circ C \right) \right)$$

with: $L_{obj,20}$ is the object length at 20 °C, L_{ref} is the length indication of reference standard (e.g. line scale, calliper), T_{obj} is the temperature of the measured object, T_{ref} is the temperature of the used reference/measuring instrument, α_{obj} is the thermal linear expansion coefficient of the object, and α_{ref} is the thermal linear expansion coefficient of the reference.

This equation gives the temperature correction of the length of an object when it is measured using a reference standard or measuring instrument, both having a different temperature and expansion coefficient. The measuring instrument is supposed to be properly calibrated at 20 °C and having no further deviations. The items are supposed to be in thermal steady-state.

From the equation the following generals aspects of length measurements can be derived:

- If object and reference have the same expansion coefficient, the average deviation from 20 °C is not important, however they should have the same temperature

- If object and reference have the same temperature, the average expansion coefficient is not important, however they should have the same expansion coefficient
- If the average expansion coefficient is low, the temperature difference is less important
- If the average temperature is close to 20 °C, the difference in expansion coefficient is less important

Alternatively, in quasi-logical terms: (The temperature difference or the thermal expansion coefficients must be minimal) and (the expansion coefficients must be the same or the temperatures must be close to 20 °C).

Overall: A temperature close to 20 °C and a minimum temperature difference between object and reference gives the smallest deviations.

Side observation: Steel objects are best measured using steel references. If the reference is made of a low-expansion material this may look attractive, but in fact a proper measurement of a steel item will become more difficult.

HH

Equation 20: Ideal Gas Law ('Puvnert')

The Ideal Gas Law provides a relationship between various parameters describing the state of an ideal gas. According to the ideal gas law, a gas confined to a container of a fixed volume increases in pressure as the temperature inside the container is increased. Similarly, as the temperature inside a container is increased, the volume must also increase if the pressure is to be maintained constant. The gas constant is a constant of proportionality that relates the energy of a physical system to its temperature.

The Ideal Gas Law neglects molecular size and inter-molecular behaviors. Therefore, it is an approximation for the dynamics of many real gases. First stated by French physicist Benoit Clapeyron in 1834, the Ideal Gas Law was formulated from two earlier laws: Boyle's Law and Charles' Law. Boyle's Law states that the pressure of a gas is inversely proportional to its temperature. In addition to combining the two laws, the Ideal Gas Law provides a constant of proportionality and the relationship to the amount of gas in the system.

$$PV = nRT$$

where: P is the pressure of the gas, V is the volume of the gas, n is the amount of the gas (in moles), R is the ideal gas constant, and T is the temperature of the gas.

MF

Equation 21: The Quadratic Formula

A quadratic equation is a second-order polynomial equation in a single variable

$$ax^2 + bx^2 + c = 0 \qquad (if \ a \neq 0)$$

The fundamental theorem of algebra guarantees that it has two solutions because it is a second-order polynomial equation. These solutions may be both real or both complex.

The quadratic equation involving one unknown is named "univariate". The quadratic equation only contains powers of x that are non-negative integers, and therefore it is a polynomial equation, and is a second degree polynomial equation since the greatest power is 2.

The roots x can be found by completing the square,

$$x^2 + \frac{b}{a}x = -\frac{c}{a}$$

$$\left(x + \frac{b}{2a}\right)^2 = -\frac{c}{a} + \frac{b^2}{4a^2} = \frac{b^2 - 4ac}{4a^2}$$

$$x + \frac{b}{2a} = \frac{\pm\sqrt{b^2 - 4ac}}{2a}$$

Solving for x then gives

$$x = \frac{-b \pm \sqrt{b^2 - 4ac}}{2a}$$

This equation is known as the "Quadratic Formula". In the quadratic formula, the expression in the square root sign is called the discriminant of the quadratic equation, and is denoted using an upper case Greek delta:

$$\Delta = b^2 - 4ac.$$

The quadratic formula with discriminant notation:

$$x_{1,2} = \frac{-b \pm \sqrt{\Delta}}{2a}.$$

This expression is important because it can tell us about the solution

- When $\Delta > 0$, there are two real roots $x_1 = \frac{-b+\sqrt{\Delta}}{2a}$ and $x_2 = \frac{-b-\sqrt{\Delta}}{2a}$
- When $\Delta = 0$, there is one root $x_1 = x_2 = -\frac{b}{2a}$
- When $\Delta < 0$, there are no real roots, there are two complex roots $x_1 = \frac{-b+i\sqrt{\Delta}}{2a}$ and $x_2 = \frac{-b-i\sqrt{\Delta}}{2a}$.

Babylonian mathematicians as early as 2000 BC (displayed on Old Babylonian clay tablets) could solve problems relating the areas and sides of rectangles. There is evidence dating this algorithm as far back as the Third Dynasty of Ur. The first known solution of a quadratic equation is the one given in the Berlin papyrus from the Middle Kingdom (c. 2160-1700 BC) in Egypt. This problem reduces to solving

$$x^2 + y^2 = 100$$

$$y = \frac{3}{4}x$$

The Greeks were able to solve the quadratic equation by geometric methods, and Euclid's (c. 325-270 BC) data contains three problems involving quadratics. In his work Arithmetica, the Greek mathematician Diophantus (c. 210-290) solved the quadratic equation, but giving only one root, even when both roots were positive.

A number of Indian mathematicians gave rules equivalent to the quadratic formula. It is possible that certain altar constructions dating from c. 500 BC represent solutions of the equation, but even should this be the case, there is no record of the method of solution. The Hindu mathematician Aryabhata (75 or 76-550) gave a rule for the sum of a geometric series that shows knowledge of the quadratic equations with both solutions, while Brahmagupta (c. 628) appears to have considered only one of them. Similarly, Mahavira (c. 850) had substantially the modern rule for the positive root of a quadratic. Sridhara (c. 1025) gave the positive root of the quadratic formula, as stated by Bhaskara (c. 1150). The Persian mathematicians AI-Khwarimzi (c. 825) and Omar Khayyam (c. 1100) also gave rules for finding the positive root. Francois Viete, at the end of the 16th century, was among the first to replace geometric methods of solution with analytic ones, although he apparently did not grasp the idea of a general quadratic equation.

VB-J

Equation 22: Gauss's Law for Magnetism

$$\nabla \cdot B = 0$$

At first glance, Gauss's Law for Magnetism might seem rather pointless: in words, it says that the divergence of a magnetic field is equal to zero. Divergence here refers to the existence of sources or sinks, so the equation is stating that there are so such things for magnetic fields or, to put it another way, that magnetic field lines are loops, without beginnings or ends. This is in contrast to electric fields, which have sources/sinks in the forms of electrons, positrons and other charged particles. Electric field lines begin/end at these particles. One might think of the magnetic north pole as the magnetic equivalent of an electron, but the north pole is actually part of a dipole; if one attempted to remove the end of a magnet, one would just end up with two smaller magnets, each with its own north and south pole.

Gauss's Law is not trivial however, it is one of Maxwell's Laws of Electromagnetism, from which the whole of classical electromagnetic theory can be derived, and which many, including Einstein, regard as one of the most important sets of physical laws there are.

Yet, for such a key law, Gauss's Law for Magnetism rests on very shaky foundations - simply the fact that no free magnetic pole (monopole) has ever been found. It is, therefore, a very bold statement indeed - a bit like raising the observation that all birds have two eyes to the status of a law of nature.

Neither this law nor any other proves that monopoles do not exist. In fact, there is some reason to believe that if monopoles did exist in the early Universe they would now be hidden inside stars and other magnetic objects in space.

MJG

Equation 23: Lensmaker's Equation

$$P = \frac{1}{f} = (n-1)\left[\frac{1}{R_1} - \frac{1}{R_2} + \frac{(n-1)d}{nR_1R_2}\right]$$

where: P is the optical power of the lens, f Is the focal length of the lens, n is the refractive index of the lens material, R_1 is the radius of curvature (with sign) of the surface closest to the light source, R_2 is the radius of curvature (with sign) of the length surface farthest from the light source, and d is the thickness of the lens (the distance along the lens axis between two surface vertices).

The earliest written records of lenses date to Ancient Greece with Aristophanes' play "The Clouds" when a burning-glass was used to focus the Sun's rays to produce fire. Similarly, Archimedes built his heat ray to destroy Roman ships at the Siege of Syracuse (.c 214–212 BC). A parabolic reflector also indicates the optical power of a mirror (a lens with only one optical surface) as a fire maker.

In geometric optics, the optical power is the degree to which an optical system converges or diverges light. It is equal to the reciprocal of the focal length of the optical system. The focal length of a single lens is calculated by the lensmaker's equation, which can be derived by using geometry, trigonometry and Snell's law. The reason that it's called as "lensmaker's equation" is probably because one can calculate the focal length of a lens by knowing how the lens was made: the radius of curvature of lens surfaces, its thickness and the refractive index of the lens.

One of the widely used approximations of the lensmaker's equation is the so called "thin lens formula". This simplified equation determines the distances of an object and its image relative to the location of the lens, the understanding of which explains how human eyes and cameras function by adjusting focus automatically.

KN

Equation 24: Fermat's Last Theorem

$$a^n + b^n = c^n$$

The Fermat's Last Theorem is based on the Pythagorean theorem (see equation 11). However, the Pythagorean equation has an infinite number of positive integer solutions for a, b, and c; these solutions are known as Pythagorean triples. A French mathematician, Pierre de Fermat stated that the more general equation $a^n + b^n = c^n$ has no positive integer solutions for a, b and c if n is an integer greater than 2.

Pierre de Fermat conjectured the Theorem in 1637 in the margin of a copy of Arithmetica where he claimed he had a proof that was too large to fit in the margin.

Cubum autem in duos cubos, aut quadratoquadratum in duos quadratoquadratos, et generaliter nullam in infinitum ultra quadratum potestatem in duos eiusdem nominis fas est dividere cuius rei demonstrationem mirabilem sane detexi. Hanc marginis exiguitas non caperet.

Which translates to English as:

It is impossible to separate a cube into two cubes, or a fourth power into two fourth powers, or in general, any power higher than the second, into two like powers. I have discovered a truly marvellous proof of this, which this margin is too narrow to contain.

Only one proof by him has survived, for the case $n = 4$, but he never posed the general case. Whether that was a prank to make fun of future generations of mathematicians (my favourite hypothesis) or he really had a proof of it (whether valid or not) is not known.

The first successful proof was released in 1994 by Andrew Wiles, an English mathematician (as usual, the French (or "European") poses the problems, while the English solve them), and formally published in 1995, after 358 years of effort by other mathematicians. Wiles worked on that task for six years in near-total secrecy, covering up his efforts by releasing prior work in small segments as separate papers and confiding only in his wife.

It is called the last theorem because it was the last of Fermat's asserted theorems to remain unproven. It is among the most notable theorems in the history of mathematics and, prior to its proof, it was in the Guinness Book of World Records for "Most difficult mathematical problems".

Further reading
Fermat's Last Theorem: The Story of a Riddle That Confounded the World's Greatest Minds for 358 Years, Simon Singh

GM

Equation 25: The Drake Equation

The Drake equation is a probabilistic argument used to estimate the number of active, communicative extraterrestrial civilizations in the Milky Way galaxy.

$$N = R_* \cdot f_p \cdot n_e \cdot f_l \cdot f_i \cdot f_e \cdot L$$

where: R_* is the average rate of star formation in our galaxy, f_p is the fraction of those stars that have planets n_e is the average number of planets that can potentially support life per star that has planets, f_l is the fraction of planets that could support life that actually develop life at some point, f_i is the fraction of planets with life that actually go on to develop intelligent life (civilizations), f_c is the fraction of civilizations that develop a technology that releases detectable signs of their existence into space, and L is the length of time for which such civilizations release detectable signals into space.

Although written as an equation, Drake's formulation is not particularly useful for computing an explicit value of N. The last four parameters, f_l, f_i and f_c and L, are not known and are very hard to estimate, with values ranging over many orders of magnitude. Criticisms include the fact that it does not take into account colonization, the reappearance factor and the METI factor (messaging to extraterrestrial intelligence).

The major problem is not with the equation but the uncertainties in the inputs that can give values of N from 0 to hundreds of thousands.

The July 2013 issue of Popular Science, as a sidebar to an article about the Daleks of Doctor Who, includes an adaptation of the Drake equation, modified to include an additional factor dubbed the "Dalek Variable". The added variable at the end is defined as the "fraction of those civilizations that can survive an alien attack."

DF

Equation 26: The Principle of Least Action

$$\delta S = 0$$

$$S = \int_{entire\ path} L\ dt$$

Or

$$S = \int_{t_1}^{t_2} L\ dt \quad \text{monogenic, holonomic systems}$$

Not only is this one of the shortest of all equations presented, it is a governing equation for all of the laws of mechanics and is a foundational principle for the laws of quantum mechanics and quantum field theory. In fact, Newton's laws could be developed, in their entirety, from this principle alone.

Originally attributed to Pierre-Louis Moreau de Maupertius (1698 – 1759) who stated that 'Whenever any change takes place in Nature, the amount of action expended in this change is always the smallest possible.' It was left to Leonhard Euler (1707 – 1783) and Joseph-Louis Lagrange (1736 – 1813) to provide the mathematical foundation upon which the laws of dynamics could be extracted. In modern terminology this translates to a statement that the action S of a dynamical system is stationary as it goes from one state to another. In fact, it has been suggested that it be renamed the 'principle of stationary action'.

Values for the action of a system can be computed from the two integrals shown, the first of which formulates Maupertius' statement in its generality (the second will be explained shortly). The parameter L is called the Lagrangian and has already been encountered within the Euler-Lagrange equation. The Lagrangian represents the instantaneous difference in the kinetic and potential energy contained within a closed system. Kinetic energy is associated with rigid bodies of finite mass and potential can be represented by scalar fields capable of doing work on these masses as they move from one position to another (gravity fields, electromagnetic fields, massless springs, etc.). Energy conservation is implicit throughout this discussion. Both integrals represent the total exchange of these two energy types as they transition from one state to another over time.

Simply stated, the principle of least action states that the differences in kinetic and potential energies of a dynamical system when integrated over the time taken to go from one state to another, the value for the true paths that particles travel will not change if a slightly different path is chosen. Mathematically the function describing this path and small variations therefrom is called a stationary function. In practice, for many dynamic systems, it can be shown that any variation will lead to a larger value of the action, hence the principle of least action. In this general form, the variations of the action are more readily analyzed using a system of dynamics formulated by William Rowan Hamilton (1805 – 1865) and this formulation, using 'Hamiltonian mechanics', enables incorporation of relativistic effects even though relativity was not known during his lifetime.

For systems that can be described by a fixed set of coordinates that also fully represent the motion of the system (called holonomic systems) and provided that the potentials can be represented by scalar fields (a condition already stated and called a monogenic system), a simplified formulation emerges and is given by the second integral. For this integral, the action is computed for starting and ending times and positions that correspond to the true motion. Hence deviations from the true path over this time must be zero at the beginning and end of the motion (a condition that is not necessary in the previous, more general, case). It is to be noted that application of the variational operator δ will be different in these two formulations and often a different symbol such as d or Δ is used for the first formulation. This second integral formulation, is now called Hamilton's principle in recognition of his contribution and can be utilized directly to determine equations governing motion of a bewildering array of physical problems, including Hertz's principle of least curvature and Jacobi's form of the principle of least action. The latter of these can be reduced to incorporate Fermats principle for the study of geometric optics and electron optical systems. Finally, it is noted that using a method called the calculus of variations, Hamilton's principle leads directly to the Euler-Lagrange equation.

SS

Equation 27: Lorentz Force

The Lorentz Force equation is used to describe the force of, or acting on a particle due to electrical and magnetic fields. It is the basis of many electromagnetic calculations which have profound importance and much practical application in many everyday items that are taken for granted. When considering the motion of a particle it is considered to follow along a "guiding point" which is a fast circular rotation around an axis. In practical applications, single particle calculations are not representative. A macroscopic calculation is needed containing the complex interactions of the electromagnetic forces of multiple particle interactions.

There were many earlier attempts to define electrical and magnetic forces acting on particles which led up to the discovery of the Lorentz Force such as Johan Tobias Mayer and Henry Cavendish in the mid 18th century who supposed that there was a defined relationship between the force of magnetic poles in the presence of an electric field. This was shown to be true by Charles-Augustine Columb in 1784 ultimately culminating in Hendrik Lorentz's understanding of and derivation of these works into the Lorentz force equation

$$F = qE + qv \times B$$

where F is the force acting on a particle, q is the electric charge, v is the instantaneous velocity, E is the electric field, and B is the magnetic field.

TS

Equation 28: The Light-Scattering Equation

When sunlight travels through the Earth's atmosphere it will collide and interact with atoms of the gases, in a phenomenon known as *light scattering*. The atoms with their electrons act as simple harmonic oscillators each with a natural frequency, and the result of the interaction will depend also upon the frequency of the incident light, ω, and the size of the scattering atoms.

The notion of scattering in this context is not straightforward and needs some explanation. The total amount of energy which would pass through an imaginary target area, usually denoted by σ, is proportional to the incoming intensity and to σ. Now in order to talk about the total amount of intensity which an atom scatters we say that it is equal to the amount which would have fallen on a certain area, so we quantify the amount of intensity by simply stating that area. This will then be independent of the intensity of the incident light, and will have the physical dimensions of an area, because the ratio

$$\frac{\text{total energy scattered per second}}{\text{energy incident per second per sq. meter}}$$

is an area, which is usually called the *scattering cross-section*. We are not thereby implying that the oscillating atom actually has this area, it is just a way of describing the amount of energy coming off the oscillating atom, by referring it to that area which the incident beam would have to hit to account for that much energy coming off. The following equation does this:

$$\sigma_s = \frac{8\pi r_0^2}{3} \cdot \frac{\omega^4}{(\omega^2 - \omega_0^2)^2}$$

where: σ_s is the scattering cross-section, r_0 is the classical electron radius, ω is the frequency of the incident light, and ω_0 is the natural frequency of the oscillating atoms.

Two cases are of interest here. First, for very low or zero ω_0 as with completely unbound particles such as free electrons, the right-hand half of the right-hand side of the equation cancels out to unity, and the cross-section is a constant, called the Thompson scattering cross-section. Secondly, in the case of the particles that make up air, the natural frequencies of the oscillators turn out to be higher than those of the incident light. As an approximation we can then neglect the term ω^2 in the denominator, leaving the scattering cross-section proportional to the *fourth power* of the light frequency. This hugely differentiates between even slightly different frequencies: light which has double the frequency of another colour will be scattered *sixteen times* more intensely. Now blue visible light is roughly double the frequency of red, and is therefore scattered much more than the red. In this way the simple equation above explains not only why the sky is blue whenever we look away from the sun instead of directly at it, but also why the sun itself appears red when it is near the horizon. At these times (sunrise and sunset) the sun's rays have passed through the Earth's atmosphere tangentially rather than radially, and have consequently passed through more air that preferentially scatters their blue component, than they have at midday.

The physicists whose names are most often associated with the equations for light scattering are Lord Rayleigh (1842-1919), Arthur Compton (1892-1962), and J J Thomson (1856-1940).

Reference: *The Feynman Lectures on Physics, vol. II*, R P Feynman, M Sands, R B Leighton

PR

Equation 29: Lamb Waves

Lamb waves, developed by Sir Horace Lamb (English mathematician) in 1917 are elastic waves that propagate in thin plates, where the planar dimensions are much greater than the thickness and where the wavelength is of the order of the thickness. The surfaces provide upper and lower boundaries for the continuous propagation of the waves through the plate.

Lamb waves are formed by the interference of multiple reflections and mode conversion of longitudinal and transverse waves at the surface of the plate. After some travel within the plate, these superpositions cause the formation of wavepackets or Lamb waves. The longitudinal (or compressional) waves are referred to as symmetric waves while the transverse (or flexural) waves are referred to as anti-symmetric waves and each wave type can propagate independently of the other. Each of these waves modes are governed by their own equation. The well-known Lamb dispersion equations are transcendental and are given below.

Symmetric modes
$$\frac{\tan(qh)}{\tan(ph)} = -\frac{4k^2qp}{(k^2-q^2)^2}$$

Anti-symmetric modes
$$\frac{\tan(qh)}{\tan(ph)} = -\frac{(k^2-q^2)^2}{4k^2qp}$$

$$p = \sqrt{\frac{\omega^2}{C_L^2} - k^2}$$

$$q = \sqrt{\frac{\omega^2}{C_T^2} - k^2}$$

where: $h = \frac{d}{2}$, d is the plate thickness, k is the wavenumber, ω is the angular frequency, and C_L and C_T are the longitudinal and transverse wave velocities respectively.

Lamb waves are dispersive and the Lamb wave equations are used to obtain velocity dispersion curves. These dispersion curves can be both phase velocity or group velocity dispersion curves and the corresponding phase or group velocity depends on the material properties. The group velocity, which is also the direction of energy flow, is usually what is measured in experiments.

The dispersion equations and the resulting dispersion curves are used to describe the relationship between frequency, sample thickness and phase or group velocity. Lower order modes exist for all frequencies and the higher order modes appear with increasing frequency. Since Lamb waves and therefore their dispersion curves are dependent on the material properties and the fact that these waves can propagate 10s of meters they are commonly used for non-destructive material characterisation and defect detection through ultrasonic testing and acoustic emission testing.

References
H. Lamb, "On waves in an elastic plate," Proc. R. Soc. Lond. A, vol. 93, no. 648, pp. 114 - 128, 1917.
W. P. Rogers, "Elastic property measurement using Rayleigh-Lamb waves," Res. Nondestr. Eval, vol. 6, pp. 185 - 208, 1995.
J. L. Rose, Ultrasonic Waves in Solid Media, Cambridge University Press, 2004.

CMcK

Equation 30: Interference

$$E = A\cos(\theta_1 + \omega t) + B\cos(\theta_2 + \omega t)$$

$$\langle E^2 \rangle = |A|^2 + |B|^2 + 2|A||B|\cos(\theta_2 + \omega t)$$

Light travels incredibly fast, has a tiny wavelength in the visible (less than a thousandth of a millimeter), and oscillates at 600 THz—far too fast to be detected directly the way that we detect radio waves or sound waves. Miraculously, we can access the wavelength and use it as a unit of measurement, by taking advantage of the principle of superposition: Two light waves can coexist in the same space, and when they do, they interfere with each other. The effect is linear in complex amplitude (well, most of the time), which is lovely. If the two waves are close in frequency, it is easy to observe the interference phase and use the very small wavelength for high-precision metrology. We shine the summed light onto a square-law detector such as the eye or a nice photodiode, and the result is a time-averaged, sinusoidal intensity signal with a phase equal to the difference in phase of the two original waves.

The term "interferometer" appears to have first become popular in the 1880's to describe Michelson's two-beam, division of amplitude interference machine, made famous by two experiments: The Michelson-Benoît experiment to link wavelengths to the standard meter, and the Michelson-Morley measurements that brought down the Ether theory and led the way to special relativity. The equation shown here is for two-beam interference.

For me this equation is especially beautiful not only for the remarkable physics that it describes, but also for its practical value—I have made use of this equation for three decades, to solve problems in optical metrology and instrument design. I will always delight in the appearance of interference fringes.

A short video accompanies this post:
https://www.youtube.com/watch?v=MUy-0NEnfQQ&feature=youtu.be

PdG

Equation 31: Self-Ionisation of Water

Water. It seems rather simple. Take two hydrogens, add one oxygen, connect them in the right way, and you have dihydrogen monoxide. Henry Cavendish and Joseph Priestley first discovered in the latter half of the 18th century that "inflammable air" (now known to be hydrogen) and "common air" could react to form water. Antoine Lavoisier, following up on this work, reacted "inflammable air" with the part of "common air" he had called oxygen, again forming water. He also decomposed water into the previous two substances, confirming that water was not an element, as previously believed, but a compound.

Due to the rather large difference in electronegativity between hydrogen and oxygen, water has some peculiar properties. On one level, there is hydrogen bonding to explain those. But to consider the fact that even pure water conducts electricity, if poorly, you have to go one step further and look at the self-ionisation of said water. As one consequence of that difference in electronegativity, two water molecules can "connect" and pass a hydrogen ion from one to the other, forming a hydronium ion and a hydroxide ion:

$$H_2O + H_2O \rightleftharpoons H_3O^+ + OH^-$$

As a result, we now have some small concentration of hydronium and hydroxide ions in our pure water. The equilibrium constant associated with that is:

$$K_C = \frac{[H_3O^+][OH^-]}{[H_2O]^2}$$

Now, the water concentration in pure water is taken to be 1000 g dm^{-3}, so with a molar mass of 18 that gives us about 55.6 moles of water in one liter, vastly larger than the concentration of either ion. As a result, the equilibrium equation has been reworked somewhat by absorbing the water concentration, taken to be constant, into the value for K_C:

$$K_W = [H_3O^+][OH^-]$$

This, now, is the equation for the self-ionisation of water. At room temperature and pressure, it has a value of $1*10^{-14}$ mol^2 dm^{-6}, resulting in a concentration of $1*10^{-7}$ mol dm^{-3} for each ion. This is where our idea of pH 7 being neutral comes from, because under these conditions that is the case (to convert from concentration to pH you take the negative log of that concentration). As you can see from K_W, though, "neutral" simply means that the ion concentrations are equal – at a different temperature, pH 7 will not be neutral anymore! As the reaction is endothermic, at a higher temperature the equilibrium will be more to the right, resulting in higher concentrations of the ions, and a lower value for a neutral pH.

The equations show two other things of interest. The first thing is that water can act as both an acid and as a base – it is amphoteric. The second is that even if a solution is overall acidic (or basic) there is still a low concentration of base (or acid) present, as K_W is still valid.

OK

Equation 32: Fermat's Little Theorem

$$a^p \equiv a \ (mod \ p)$$

Alternatively expressed as:

$$a^{p-1} \equiv 1 \ (mod \ p)$$

Fermat's Last Theorem is the celebrity of the mathematical world, inspiring documentaries, books and even an article in this prestigious collection of equations (BE 24). Fermat's Last Theorem has the advantage of being mysterious and, like every good X-Factor contestant, the subject of a tragic back-story. "Mysterious", because it remained unproven for 358 years after it was written, and "tragic" as it was discovered by Fermat's son, abandoned in the margin of a book after the mathematician's death. But if the Last Theorem is a musical celebrity, its older sibling, Fermat's Little Theorem, is a composer of elevator music: ubiquitous, but widely ignored.

Contrasted to the Last Theorem, the Little Theorem was proven a paltry 43 years after publication and trivially "found" because Fermat wrote the equation out in a letter to the Parisian mathematician de Bessy. The proof was needed, because Fermat accompanied it in the letter with his now infamous quote: "de quoi je vous envoierois la démonstration, si je n'appréhendois d'être trop long." ("the proof of which I would send to you, if I were not afraid to be too long.")

The Little Theorem describes one of the fundamental properties of prime numbers as applied to number theory. It states that any integer raised to the power of a prime number results in a number which, when subtracted by the original integer, results in a multiple of the original prime. It has the caveat that this doesn't hold if the integer is a multiple of the prime to begin with.

For example, let's use the integer 4 and the prime 5

$$4^5 = 1024$$

Subtracting the original integer gives

$$1024 - 4 = 1020,$$

a multiple of 5.

Fermat's Little Theorem is ubiquitous because of its numerous (in many senses of the word) applications. Rearranging the equation into its popular form, it becomes a useful, if computationally expensive, test if a number is prime or not. This is useful for computer programs seeking new primes. Sadly, not only is Fermat's version computationally expensive (as you need to test the candidate prime with several integers to be confident it wasn't a fluke) but there is a set of "Fermat liars" (Carmichael Numbers) that are not prime, but fool this equation. As a result most modern prime number generators either use a modified version of the equation, or couple it with an additional test to increase the likelihood that the found number is actually prime. One of the reasons it is so important for computers to generate prime numbers is because they are vital to modern cryptography, you're use prime numbers generated with this equation, or one of its variants, to access Facebook and other social media along a secure channel. And the reason primes are so important to modern cryptography? The property of primes defined by Fermat's Little Theorem forms part of the proof for "public key cryptography", the topic of another article.

RFO

Equation 33: The Fermi Function

The Fermi function (*aka* the Fermi-Dirac distribution) applies to fermions (particles with half-integer spin), which must obey Pauli's exclusion principle.

The term 'Fermi energy' usually refers to the energy difference between the highest and lowest occupied single-particle states in a quantum system of non-interacting fermions at a temperature of absolute zero. In a Fermi gas, the lowest occupied state is taken to have zero kinetic energy; in a metal, on the other hand, the lowest occupied state is typically taken to mean the bottom of the conduction band.

The function was named after Italian physicist Enrico Fermi (1901-54). Paul Dirac (1902-84) discovered the statistical distribution independently, though Fermi was the first to define it.

The Fermi function belongs to the class of 'logistical functions'. The latter were named in 1844-45 by Pierre Verhulst, who studied population growth models. The initial phase of growth is roughly exponential. As saturation becomes apparent, growth slows. At maturity, growth stops.

The logistical curve is in turn a class of 'Sigmoid' curve - a tilted S-shaped curve that resembles trends in the life cycle of many living things and growth phenomena.

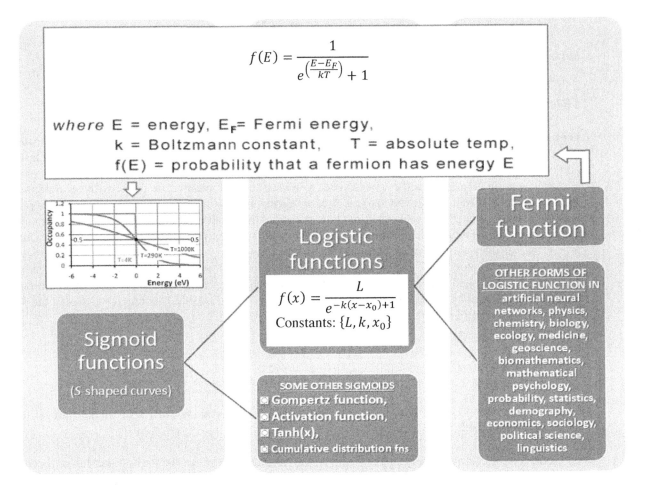

$$f(E) = \frac{1}{e^{\left(\frac{E-E_F}{kT}\right)} + 1}$$

where E = energy, E$_F$= Fermi energy,

k = Boltzmann constant, T = absolute temp,

f(E) = probability that a fermion has energy E

Fermi function

Logistic functions

$$f(x) = \frac{L}{e^{-k(x-x_0)+1}}$$

Constants: $\{L, k, x_0\}$

OTHER FORMS OF LOGISTIC FUNCTION IN artificial neural networks, physics, chemistry, biology, ecology, medicine, geoscience, biomathematics, mathematical psychology, probability, statistics, demography, economics, sociology, political science, linguistics

Sigmoid functions

(S-shaped curves)

SOME OTHER SIGMOIDS
◙ Gompertz function,
◙ Activation function,
◙ Tanh(x),
◙ Cumulative distribution fns

FR

Equation 34: The Fundamental Theorem of Calculus

$$\frac{d}{dx}\int_a^x f(t)\, dt = f(x)$$

In the equation shown, t is an independent variable, x is the value of t at which a continuous function f is to be calculated, and a is an arbitrary value of t that is less than x. The value of a is arbitrary because it cancels out when applying the theory of integral summation of infinitely thin strips of area beneath the function, when illustrated graphically by two orthogonal axes, representing the function (dependent variable) on a vertical axis and the independent variable on a horizontal axis.

The first FTC relates the concept of integration (a summation of infinitesimally small products or areas, often used to calculate areas or volumes) to differentiation (the tangent to a curve or gradient at a specific point, used to calculate rate of change). It shows that the integrated area under a curve (functionally represented as described above) is the reverse operation to differentiation. Until the mid-17th century it was not realised that these two operations were the reverse of each other, and the realisation of this relationship resulted in significant advances in mathematics and the computation of calculus, as it enabled, for the first time, the calculation of complicated integrals as "antiderivatives" without using tedious limit processes, and had a significant influence in the subsequent theory of functions, and ultimately, the development of differential and integral equations (which are the subject of many of the entries in this series). The second FTC follows from the first and basically states that definite integrals can be calculated using derivatives.

34

It is often taken for granted that integration and differentiation are opposite operations, and many of us were taught at school to remember this and to memorise the formulae for differentiation (such as the "elementary power rule") without being derived, and the other rules which also follow from first principles. The elementary power rule for differentiation can be realised through expansion of bracketed terms – for example the function $f(x) = x^2$ results in $\frac{df(x)}{dx} = 2x$ due to the product terms in the expansion of $(x + dx)^2$ and the cancelling of x^2 terms in the formula for differentiation (see below). The general rule for the coefficient being the same number as the power term and the reduction of the power term follows from the same process using progressively higher power expansions.

The integral can be shown to be equivalent to the antiderivative of a function by looking at the problem in a similar way using the concept of a limit of an infinitesimally shrinking quantity. The formula for differentiation is

$$\frac{df(x)}{dx} = \lim_{dx \to 0} \frac{f(x + dx) - f(x)}{dx}$$

which defines the first derivative of a function $f(x)$ as the limiting value of the ratio of the function to its independent variable at a point x. In the case of differentiation we see the convergence towards a tangent to a curve (its ratio or "gradient") when a small change dx in the independent variable tends towards zero. In the case of integration we see, in a similar way, the convergence towards the function itself by considering a small change $f(x)dx$ in the product of the function and independent variable (mathematically represented as an "area strip" below the function). In a mathematical sense the terms converged upon at the limit of vanishing dx, the gradient and the function, are determined as a ratio and a product respectively, and therefore the two are opposite operations. Allowing a thin strip under the function curve to tend towards zero results in

$$f(x) = \lim_{dx \to 0} \frac{A(x + dx) - A(x)}{dx}$$

which is equivalent to the formula for differentiation if $f(x)$ is replaced by $\frac{df(x)}{dx}$ and $A(x)$ is replaced by $f(x)$.

Isaac Newton was well aware of these processes and described them in his calculus in Principia, together with a complete form of the FTC, as did the German mathematician Gottfried Leibniz. Newton, however, used his calculus in a practical sense to solve many real-world problems, including those making use of his important work on infinite series, whereas for Leibniz it was not much more than a mathematical curiosity – although we now use Leibniz' notation as our preferred method of describing integrals and derivatives. Newton did, however, formulate his calculus 8 years before Leibniz, as a recent collection of Newton's mathematical papers now shows, his formulation being derived in 1666 (much of it during his stay in Woolsthorpe while Cambridge was closed due to the outbreak of bubonic plague) and those of Leibniz in 1674 – however, during their lifetimes, Leibniz published his calculus much earlier, resulting in a long-running argument which wasn't resolved by the time both of them had died. Newton and Leibniz both resolved the FTC in its most complete mathematical form, but the first to derive a complete form of the theory was the mathematician Isaac Barrow, following from the work by James Gregory. Barrow was Newton's tutor at Trinity College, Cambridge, and was the first to hold the prestigious Lucasian Chair of Mathematics which he passed over to Newton in 1669, who then held the position for the next 33 years. Barrow's work on tangents to curves also significantly influenced Newton in his development of calculus.

Before the 17th century, the application of integration and differentiation as separate operations has a long history. Evidence of integration as a summation process was first seen in papyrus from Egypt dated around 1800BC. The concept of a tangent (and the other trigonometric functions) originated in India in the 8th century BC. In the 4th to 2nd centuries BC Greek mathematicians such as Eudoxus and Archimedes developed the concepts of the "method of exhaustion" and "infinitesimals". Euclid and Apollonius did work on tangents to curves, which is fundamental to differentiation. In India, infinitesimals were used by Aryabhata (6th century AD) and in the medieval Arab/Eastern world, rates of change by Bhaskara II (12th century AD) and it is now known that Sharaf al-Din al-Tusi worked out cubic derivatives in the 12th century AD. Other advances were made by the Oxford Calculators in the 14th century but "modern" calculus began later with Fermat, Descartes and Pascal in the case of derivatives, and Cavalieri in the case of integrals ("indivisibles"). Another brilliant English mathematician, John Wallis, was instrumental in the development of the infinitesimal calculus in the 17th century that was later generalised and completed by Barrow, Newton and Leibniz[1].

PB

Equation 35: Molar Mass and the Mole

The ancient Greek philosopher Democritus is generally credited with the idea of all matter being made of tiny, indivisible particles. Our word for those particles, atoms, actually comes from the Greek word "atomos", meaning indivisible. Those ideas really did not change much until the 19th century, as you can see when comparing them to the work of Dalton from the early 1800s. The main reason was the lack of equipment that was sufficiently accurate to investigate the problem in a laboratory. However, that changed with the Industrial Revolution. As a consequence, people like Berzelius, Boyle and Lavoisier were able to investigate this much more closely. Atomic weights were determined, as compared to initially hydrogen-1, and now carbon-12.

Up to this point, more or less, for chemical reactions, philosopher-scientists must have mainly relied on trial and error to get the right stoichiometry (relative quantities of reactants and products in a chemical reaction). Once atomic weights were being determined, they could start thinking about a more systematic approach. Gay-Lussac and Avogadro started the work on volumes of gases and numbers of particles in them, and Cannizzaro brought it to conclusion.

Once there was an accepted scale for atomic weights, one could, in principle, start calculating. However, with atoms being so light, a more usable scale needed to be found. Enter the mole. The word is derived from the Latin "moles" or "molis", meaning large mass, or heap, amongst others. The mole was defined in such a way that one mole of an element contained the same number of grams of that element as its atomic weight. This number turned out to be 6.02×10^{23}, and was named Avogadro's number, in honour of his early efforts, although he did not work on this particular part.

To convert a mass of any particle into moles, you simply divide by the molar mass of that particle:

$$n = \frac{m}{MM}$$

where: n is the number of moles, m is the mass in grams, and MM is the molar mass in grams/mole.

---

[1] The mathematical papers of Isaac Newton were published in eight volumes between 1967 and 1981 by Tom Whiteside (a professor of history of mathematics at Cambridge). Whiteside died in 2008 and the latest editions were published that year. He unravelled all of Newton's writings which were in a mess and had sat in the archives at Cambridge for 75 years, virtually untouched (they had been passed on to Cambridge in the 19th century by the Earls of Portsmouth who had been the owners since Newton's family). Each volume is around 600 pages and includes the original manuscripts, footnotes and commentary, translations from Latin and other passages of interest such as reviews on important parts of it by Leibniz and others. Although Newton is known to have formulated his calculus slightly earlier, it is clear that Leibniz did it independently. Fluxions was the terminology Newton used for differential calculus and he had completed a book on it as early as 1671 (but it remained unpublished until 1736, after his death). The term he used to describe integral calculus was "fluents".

Although the mole is an SI unit, Avogadro's number has to be determined indirectly, as there is no equivalent to e.g. the International Prototype Kilogram. This was initially done using coulometry to measure Faraday's constant and then divide by the charge on one electron. More recent measurements involve X-ray crystallography and density measurements of extremely pure samples of silicon.

As a result of this all, we can now compare numbers of moles of particles in chemical reactions, showing exactly how many of each particle need to react to form products. This, in turn, allows one to calculate, rather than find empirically, the masses of each particle that are needed in a particular chemical reaction. The mole, in a sense, is the ultimate accounting tool for chemistry, and underlies any and all practical chemistry done today.

OK

Equation 36: Zernike Polynomials

Zernike Polynomials are defined on a circular unit disc with radius 1, in polar coordinates with (relative) radius ρ and azimuth angle ϕ, the polynomials are given by:

$$\sum_{n}^{m} (\rho, \varphi) = R_n^m(\rho) \cdot \cos(m \cdot \varphi) \quad \text{and} \quad \sum_{n}^{m} (\rho, \varphi) = R_n^m(\rho) \cdot \sin(m \cdot \varphi)$$

here m and n are integers larger than zero.

The typical beauty of the Zernike Polynomials is that the circular periodicity is expressed in the sine-and cosine terms, analogous to the Fourier series of a 1-D function (see BE 14).

The radial part is a typical 1-D polynomial like a Legendre function, that keeps the functions orthogonal, while any radial function can be approximated by it. It is given by:

$$R_n^m(\rho) = \sum_{k=0}^{k=\frac{n-m}{2}} \frac{(-1)^k \cdot (n-k)!}{k! \cdot \left(\frac{n+m}{2}-k\right)! \cdot \left(\frac{n-m}{2}-k\right)!} \rho^{n-2k} \quad \text{for } n-m \text{ even, and zero elseway.}$$

This radial term may appear intimidating, but values are commonly tabulated in (optical) textbooks.

The major use in optics is related to the circular aperture that is naturally present in optical systems consisting of circular lenses and mirrors. If a rectangular area is cut out of a circular aperture, e.g. by a film or CCD (camera), the polynomials no longer work, unfortunately.

Beyond its use in (circular) optics, the Zernike polynomials give an orthogonal set of terms that can approximate any function that is defined on a unit disk. One example is the topography of a circular surface that can conveniently be expressed in Zernike polynomials.

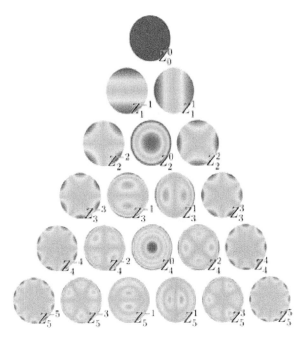

In textbooks it is common practice to depict Zernike polynomials as colourful images in the colour scale ranging from red (+1) to blue (-1), that gives intriguing images for higher order polynomials.

The Zernike polynomials are named after the Dutch physicist Frits Zernike (1888-1966) who developed them to describe optical aberrations. Except for these polynomials, Frits Zernike is known as the inventor of the phase-contrast microscope for which he was rewarded with the Nobel prize in Physics in 1953.

A typical anecdote is that, when he demonstrated his microscope to the Zeiss company the comment was: "if this would have any practical use, we would long have invented it already".

HH

Equation 37: Heat Capacity

The heat capacity equation dictates how the temperature T of an object changes when a given amount of energy Q is inputted (in the form of heat). The relationship between Q and T is linear, and the proportionality is given by mC, where m is the mass of the material and C is the heat capacity per unit mass specific to that material (hence the term specific heat). The value of C corresponds to the amount of energy required to change the temperature of one kilogram of material by one kelvin.

The equation relating the variables is:

$$\Delta Q = mC\Delta T$$

This equation is valid for small changes in temperature. This is due to the fact that the specific heat of a material can actually change as a function of temperature. That is,

$$C(T) = \frac{\delta Q}{\mathrm{d}T}$$

The symbol δ denotes a 'path function,' indicating that the behaviour of the system depends on its current state. Also, the specific heat for any material is zero at a temperature of zero kelvin.

The dependence of the heat capacity to temperature might remind some of the relationship between thermal expansion of a material and its temperature. In fact, this relationship is described in more detail in reference [1]. It is interesting to note that heat was once thought to be an invisible fluid, as opposed to the current interpretation of heat as a transfer of energy. The term *heat capacity* was first coined by Scottish chemist Joseph Black.

[1] Garai J 2006 Correlation between thermal expansion and heat capacity *Computer Coupling of Phase Diagrams and Thermochemistry* 30 354-356

MF

Equation 38: Euler's Equation

$$e^{ix} = \cos x + i \sin x$$

Euler's equation (or formula) is a remarkable bit of mathematics that links circular trigonometric functions to complex exponential functions. In the equation, e has the value of approximately 2.71828..., is the base of natural logarithms and is the limit of $\left(1 + \frac{1}{n}\right)^n$, as n approaches infinity (and is often called Euler's number), i is the square root of -1 (a complex number that only really exists to simplify mathematics – try multiplying a number by itself to get -1 and you will fail) and x is any real number.

Roger Cotes wrote down a similar equation in terms of natural logarithms in 1715 but it was Euler who wrote down today's equation in 1740 (published in 1748). The equation has many uses in pure mathematics, often to simplify calculations, but it is also used in many branches of science and engineering because of its ability to link the exponential function with trigonometry. Re-arranging the equation to give expressions for sine and cosine is especially useful when one is working on differential equations.

There is a special case of Euler's equation when $x = \pi$, but I will leave this for a later equation.

Further reading
Dr. Euler's Fabulous Formula: Cures Many Mathematical Ills (2006), Paul J. Nahin

RL

Equation 39: Bayes' Theorem

Let's say that, at the National Physical Laboratory, 80% of staff are scientists, and 20% of staff are both female and scientists. What is the probability that a random member of staff is female, given that (s)he is a scientist?

If we define the probability of being a scientist as $P(Y)$ and the probability of being female as $P(X)$, we want to know the conditional probability of X, given Y, or $P(X|Y)$.

$$P(X|Y) = \frac{P(X \cap Y)}{P(Y)} = \frac{0.2}{0.8} = 25\%.$$

from this equation, we can derive Bayes' Theorem

$$P(X|Y) = P(Y|X)\frac{P(X)}{P(Y)}.$$

At first glance, the theorem does not seem particularly profound, but, since humans are in fact very poor at estimating conditional probabilities, its application can reveal some surprising facts, especially concerning false positives.

Let's now imagine there is a new test for Ebola. A study reveals that if someone has Ebola, then the test will produce a positive result in 99% of cases, and a negative one in 1% of cases. If the test is used on someone who doesn't have Ebola, then it gives a negative result in 95% cases, and a positive one in 5% of cases. Finally, let's say that Ebola is rare - just 1 in 10,000 have it (0.01%).

You decide to get yourself tested: the result is positive. How worried should you be?

After a lightning-like estimation of conditional probabilities, you might well say "very!" Let's see if you're right: before you had the test, you would have said your chances of having Ebola were 0.0001. Call that $P(X)$. What you want to know is: what are your chances now, given that scary test result Y, or $P(X|Y)$. To calculate that using Bayes' Theorem, we need to know $P(Y|X)$ and $P(Y)$. $P(Y|X)$, the probably of getting a positive result if one has Ebola, is already known; it's 0.99.

So all we need is $P(Y)$, the probability of a positive test result. This is equal to:

$$P(Y|X)P(X) + P(Y|\neg X)P(\neg X).$$

In words, this becomes: "The probability of a positive result" EQUALS "The probability of a true positive multiplied by the probability of having Ebola" PLUS "The probability of a false positive multiplied by the probability of not having Ebola."

So, putting the numbers in, we can calculate $P(Y)$:

$$P(Y) = 0.99 \cdot 0.0001 + 0.05 \cdot 0.9999 = 0.05, \text{ or thereabouts.}$$

Plugging this and the other values into Bayes' Theorem, we find that $P(X|Y)$, the probability of you having Ebola now you've got your test result, is

$$P(X|Y) = \frac{P(Y|X)P(X)}{P(Y)} = 0.99 \cdot \frac{0.0001}{0.05} = 0.002, \text{ pretty much; a mere 0.2\% chance.}$$

So the correct answer to the question "how worried should you be?" is "hardly at all."

The inventor of the bones of this cheering theorem was the Reverend Thomas Bayes; at his death in 1761 his work was left incomplete, but was finished and published by his friend and fellow nonconformist Richard Price in 1763. As a result, Price became a Fellow of the Royal Society two years later.

The theorem finds application in many fields, and was used in the Dreyfus trial, by Turing to crack the Enigma code, by the U.S. Navy to locate a missing H-bomb, to predict the chances of nuclear accidents and to search for a lost submarine (the basis of the book and film *Hunt for Red October*).

further reading

The Theory That Would Not Die (2012), Sharon McGrayne

MJG

Equation 40: Law of Propagation of Uncertainty

$$u_c(y) = \sqrt{\sum_{i=1}^{N} \left(\frac{\partial f}{\partial x_i}\right)^2 u^2(x_i) + 2\sum_{i=1}^{N-1}\sum_{j=i+1}^{N} \frac{\partial f}{\partial x_i}\frac{\partial f}{\partial x_j} u(x_i, x_j)}$$

where,

y is the estimate of a measurand Y, formed by the measurand equation $Y = f(X_1, X_2, \ldots, X_N)$, $u_c(y)$ is the combined standard uncertainty of the measurement result y, x_i is the estimate of an input quantity X_i, $\frac{\partial f}{\partial x_i}$ is the 1st partial derivative of function f (i.e. sensitivity coefficient) evaluated at $X_i = x_i$, $u(x_i)$ is the standard uncertainty associated with the input estimate x_i, and $u(x_i, x_j)$ is the estimate covariance associated with x_i and x_j.

All measurement result should be expressed by an estimate of the measurand (the quantity being measured, e.g. length, time, mass) and the measurement uncertainty - a parameter that characterizes the dispersion of the measured value.

When a lack of international consensus on how to express uncertainty in measurement was first recognized in 1977, the world's highest authority in metrology, the Comité International des Poids et Mesures (CIPM) requested the Bureau International des Poids et Mesures (BIPM) to address the problem with national measurement institutes to make a recommendation. By 1980, a working group made up of metrologists developed a document called "Recommendation INC-1, Expression of Experimental Uncertainties", which later known as the "Guide" or "GUM".

First introduced at 1995, the "Guide to the expression of uncertainty in measurement" establishes general rules for evaluating and expressing uncertainty in measurement that can be followed at various levels of accuracy and in many fields. Rules and routines are documented in a document of 134 pages (2008 version). Standard uncertainty, combined uncertainty, expanded uncertainty etc. are defined and their evaluations are categorized into Type A and Type B methods.

For indirect measurement, where a measurand is modeled by an explicit mathematical function of multiple input quantities, the combined uncertainty is calculated by the equation called the law of propagation of uncertainty.

In this equation, the standard uncertainty of individual input quantity is plugged into a single equation derived from the original measurand model. The contribution to the combined uncertainty from each input quantity is determined by its standard uncertainty as well as the mathematical expression of the corresponding input quantity.

This equation is typically used for uncertainty analysis. It can also be applied to make an uncertainty budget if a quantity of interest can be modeling with a reasonable amount of input quantities. Compared to the error budgeting, a commonly used technique in precision engineering, uncertainty budgeting and analysis should be conducted with all known systematic errors corrected. The "Uncertainty Machine" provided by the United States National Institute of Standards and Technology (NIST) is a software application that can be used to calculate the combined uncertainty based on the propagation equation.

KN

Equation 41: Entropy in Information Theory (Shannon's Entropy)

$$H = E\left[\log\left(\frac{1}{P(X)}\right)\right] = \sum_i P(x_i)\log\left(\frac{1}{P(X)}\right)$$

where X is the random variable describing statistically a source of information, x_i is the ith outcome of the random variable X, $P(x_i)$ is the probability of the i-th outcome, $\log(1/P(x_i))$ is called self-information and E is the average operator.

The entropy of an information source is a measure of the uncertainty related to that source and is defined as the average of the self-information. The self-information formalises a basic concept: the more unexpected the outcome (sometimes called symbol or message) of the random variable X the more information it delivers.

A-priori knowledge of the statistical distribution of the messages coming from an information source and a proper coding scheme allows the optimization of the communication channel or for file compression, by using more resources (e.g. bits) for messages which delivers higher information and less resources for messages that delivers less information.

For example if an information source delivers 4 messages: 'a', 'b', 'c' and 'd' each with probability 1/2, 1/4 ,1/8 , 1/8 . The self-information of each is respectively 1, 2, 4 and 4 bits if the logarithm is in base 2. If no compression is employed by using the coding scheme 00, 01, 10, 11 for the four messages, the average bit per message is 2. However, an example of compression is to store the symbol 'a' as 0, 'b' as 1, and 'c' and 'd' as 10 and 11 respectively. In this case, the average bits per message become 1.25 bits to deliver or store exactly the same amount of information.

The lower the entropy of an information source the better is the results when compressing or encoding the communication.

The entropy and much more was introduced by Claude Elwood Shannon (April 30, 1916 – February 24, 2001) in his 1948 paper "A mathematical theory of communication", the holy bible for all electronic engineers. Shannon's article laid out the basic elements of communication, and it was one of the founding works of the field of information theory.

GM

Equation 42: The Simple Pendulum

$$T = 2\pi \sqrt{\frac{L}{g}}$$

The equation describes the period of a simple pendulum T, where L is the length of the pendulum and g is the acceleration due to gravity. The equation is valid for small pendulum displacements as an assumption is made during its derivation.

It's one of the first measurements many of us make of a physical constant (g) and it's a simple experiment that can be carried out with household items.

All you need is a piece of string, a suitable mass (a nut) and a stop watch. Simply time the swings. To improve accuracy time 10 swings and divide by 10.

Get your kids doing it and see who gets the best result.

The other application of the pendulum is in a grandfather clock where the small angle assumption can result in an error of 15 seconds per day.

Galileo discovered the crucial property that makes pendulums useful as timekeepers, called isochronism; the period of the pendulum is approximately independent of the amplitude of the swing. He also found that the period is independent of the mass of the bob, and proportional to the square root of the length of the pendulum.

DF

Equation 43: Generating Function for Bessel Coefficients

A large number of transport processes can be thought of as arising from a continuous distribution of potential in space. Gradients of these potentials can give rise to a flux that might be considered a flow of whatever phenomena is being considered. Probably the most easily visualized flux would be that of water. This might be flowing due to the gradient of the gravitational potential or because of gradients of pressure (another form of potential). In fact water is a good example because, like many of these flux's, it can often be thought of as a kind of incompressible fluid. In this case, any volume fixed in space and defined by a closed surface boundary placed within this flux (being conceptual only, this boundary doesn't itself have any influence on flow) will have flows going in and out of the surface. Being incompressible, the amount of flux going through the surface into the volume must be matched by an equal amount leaving somewhere else. Hence under these conditions, mathematicians can assume that the divergence of the gradient of the potential is zero (unless there is a source of flux) inside the volume leading to another beautiful equation called the Laplace equation. Examples of potentials that can be continuously distributed in space include temperature, electric and magnetic potentials, pressure in fluids, stress in solids, and gravitational potential. Knowing the distribution of the potentials, the flux can be predicted to determine flows of fluids, electric currents, stress distributions and, even, planets.

Very frequently, it is necessary to determine the above parameters to understand things like flow of fluids in pipes, flow of heat in pipes carrying hot fluids, or, maybe, effects of currents in electrical wires (antennas). Such processes are continuously occurring around us in our everyday lives. For these particular examples the potentials and fluxes are best modelled using three coordinates z, r and θ with each of the numbers represented by these symbols being respectively the distance along the axis of a cylinder, the radial distance from the axis and the angle around the axis. The Laplace equation for these types of models as well as other mathematical equations (in particular the dynamics of circular membranes and plates (the drum)) have solutions that can be expressed as a series of Bessel functions named after Friederich Bessel (1784 – 1846). These Bessel functions (like other naturally occurring functions such as sine and cosine) have very interesting properties (including orthogonality and continuity) that enables them to be combined to form solutions that maintain compatibility with boundary conditions particular to a specific process being modelled. Unlike the more familiar harmonic functions (sine and cosine) elucidating the properties of these interesting functions has provided many challenges to mathematicians and physicists alike for more than two centuries.

A major contribution to the understanding of these functions was introduced by the German mathematician Oskar Schlömilch (1823 – 1901) in 1857 and his equation is called the generating function for Bessel coefficients. In many ways this equation may be thought of as expressing one function (the Bessel function) in terms of another (the exponential function). The importance of this is that the properties of the exponential function are well known and relatively simple to derive. Hence, applying mathematical techniques for transforming the exponential function results in known properties that can immediately be ascribed to the Bessel function. One particular transformation of note is called the Jacobi-Anger transformation that is a foundational mathematical solution for the understanding of communications signals and is itself a beautiful equation.

$$e^{\frac{x}{2}\left(r-\frac{1}{r}\right)} = \sum_{n=-\infty}^{\infty} \sum_{k=0}^{\infty} (-1)^k \frac{\left(\frac{x}{2}\right)^{2k+n}}{(n+k)!\, k!} r^n = \sum_{n=-\infty}^{\infty} r^n J_n(x)$$

where $J_n(x)$ is the Bessel function of the first kind of order n for value of the argument x, e is the exponential coefficient that is an irrational number = 2.718281... and r is the another argument in the exponential function.

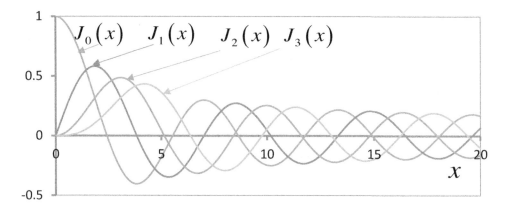

SS

Equation 44: The Nyquist Frequency

$$N = \frac{F}{2}$$

where N is the Nyquist frequency and F is sampling frequency.

Not a very extravagant equation but one that has profound implications for many common applications, especially so in a digital age. An analogue signal technically has an infinite sampling frequency, however any signal (analogue or digital) sampled at a discreet rate or frequency is subject to the Nyquist frequency.

The Nyquist frequency is named after Harry Nyquist, the electronics engineer that discovered the effects of this frequency however the exact date and sequence of events is not clear. Nyquist's work was cited as early as 1928 however Nyquist sampling theorem wasn't directly termed until 1959 and the same effect was quoted as Shannon's sampling theorem in 1954.

What the Nyquist frequency means is that to be able to determine the frequency content of any signal you need to sample the signal at least twice the maximum frequency content to be analysed. If you don't do this you will get an aliased signal where you record a particular frequency but it is not actually recording the real signal. Aliasing manifests itself in many applications and most people will be familiar with some form of aliasing.

The Nyquist frequency or associated aliasing is also why you see TV screens blink or have black bands through them when watching on other TV screens, it is also the reason why some rotary items such as car wheels or plane propellers appear to rotate backwards on TV screens.

It is generally accepted as good practice to operate at a sampling frequency of 10 times the maximum frequency of interest to avoid any clipping or roll off from the measured signal. When processing and analysing such data it is also important to consider what filtering methods are appropriate so that you analyse the data you want but are not subject to signal processing artefacts of information you are not interested in.

TS

Equation 45: The First Fundamental Law of Capitalism

The relentless increase of wealth inequality between rich and poor is the most serious socio-economic problem facing the world today. The French economist Thomas Piketty has devoted himself to analysing data on wealth distribution going back over 200 years from several countries, but chiefly from France and the UK which have the most complete records. He has found that there is a general trend towards greater inequality which has been continuous in that time apart from an interruption that lasted from 1914 thru 1950 due to two world wars.

Piketty focuses attention on two sources of income: income from employment or labour (wages, salaries, sales), and income from capital (rent, interest, dividends from investments), and points out that the richer you are, the greater the fraction of your income and wealth that derive from stored capital as opposed to earnings from labour. The typical rate of return on capital is 6-7%, whereas the rate of growth of income from labour is more or less tied to the growth rate of the economy as a whole – typically 1-2%, and it is the huge disparity between these two figures which is responsible for the diverging fortunes of rich and poor.

The equation which Piketty calls the First Fundamental Law of Capitalism is a simple one:

$$\alpha = r \times \beta$$

where α is the fraction of income due to capital in total national income, r is the rate of return on capital and β is the ratio: total capital / total income from all sources.

The beauty of this equation is that it points us to the parameters which matter when it comes to wealth inequality. It allows us to analyse the importance of capital to an individual person, a corporation, an entire nation, or the whole world. Consider, for example, a company that uses 5 millions' worth of capital (plant, infrastructure, machinery, offices, etc.) to earn 1 millions' worth of product annually, with 600 000 going to pay wages and salaries, and 400 000 remaining as profit. The capital / income ratio for this company is $\beta = 5$ (its capital is equal to 5 years of output), the capital share of income α is 40% (profits are 40% of total earnings), and the rate of return on capital is $r = 8\%$ (400 000 profits from 5 million invested).

Reference: *Capital in the 21st Century*, T Piketty, 2013

PR

Equation 46: Magnetic Permeability

$$\mu = \frac{B}{H}$$

In electromagnetism, the permeability μ of a medium is a measure of the degree to which the medium supports the formation of a magnetic field within itself in response to an applied magnetic field. It reflects how much the organisation of magnetic dipoles in the medium is influenced by the field. The term 'permeability' was coined in 1885 by Oliver Heaviside.

In the relationship $B = \mu H$, B is the magnetic flux density (SI unit: Tesla) and H is the magnetic field strength (SI unit: A•m^{-1}). The SI unit for μ is Henries per metre (H•m^{-1}), or Newtons per ampere squared (N•A^{-2}). For an 'isotropic' medium (one in which the permeability is the same in all directions) μ is a 'scalar' quantity (single number); for an 'anisotropic medium' (one where the permeability is direction-dependent) a second rank tensor (2-dimensional array of several numbers) is needed to describe μ.

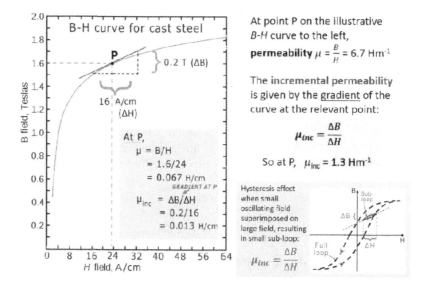

At point P on the illustrative B-H curve to the left, permeability $\mu = \frac{B}{H} = 6.7\ \text{Hm}^{-1}$

The incremental permeability is given by the gradient of the curve at the relevant point:

$$\mu_{inc} = \frac{\Delta B}{\Delta H}$$

So at P, $\mu_{inc} = 1.3\ \text{Hm}^{-1}$

Relative permeability (μ_r) is the ratio of the permeability of a specific medium to the permeability of free space (μ_0), which has the exact value, by definition, of $4\pi \times 10^{-7}\ \text{H} \bullet \text{m}^{-1}$.

A closely related property of a material is *magnetic susceptibility* X_m (where $X_m = 1/\mu_r$), which is a measure of the magnetisation of the material alone, subtracting the magnetisation of the space occupied by the material. The reciprocal of magnetic permeability is magnetic reluctivity.

In general, permeability is not constant. It can vary with location within the medium, humidity, temperature, and other parameters. Permeability as a function of frequency can take on real or complex values. (Complex values are utilised at high frequencies of a varying H-field, when there is a phase lag δ of B behind H, in which case μ can be written as $B_0/H_0\exp(-j\delta)$, where B_0 and H_0 are amplitude values of the B and H fields.)

In ferromagnetic materials, the relationship between B and H exhibits a hysteresis effect. There is not a one-to-one mapping of H onto B: in fact B depends on the material's history. In these situations, it may be useful to consider the material's *incremental permeability* μ_{inc} (or μ_Δ).

One sense in which μ_{inc} is used is as the gradient, dB/dH, of the B-H curve. A different usage is when a small oscillating field is superimposed upon a large non-oscillating field. Small local hysteresis sub-loops occur during each oscillation (see diagram, bottom right), and $\Delta B/\Delta H$ can be quite different from the value of B/H.

FR

Equation 47: The First Law of Thermodynamics

In the equation, and assuming a closed system, delta-U is the change in internal energy, Q is the amount of heat supplied to the system and W is the amount of work the system does on its surroundings. The First Law of Thermodynamics is very simple to state, but has profound consequences. Basically, the Law says that during all the moving and transforming of a system (a simple closed system is some gas in a box) the total amount of energy never changes. And that's it! Another way: energy changes form and moves from place to place but the total amount does not change. This Law is also known as the Conservation of Energy and it is key to many branches of physics and engineering.

You may sometimes see the –ve sign in the equation swapped for a +ve sign – this simply depends on how we set the energy flow conditions for the system – conceptually the two equations are identical.

Development of the Laws of Thermodynamics (there are actually three – we'll get round to the others later) began thousands of years ago. The largest advancements in developing the Laws of Thermodynamics occurred in the mid-1800s. James Prescott Joule proved experimentally that work energy and heat energy are interchangeable and are conserved. His experiment used a falling mass that drove a paddle underwater. The potential energy lost by the mass as it fell matched the heat energy gained by the water. These findings were verified independently by other scientists around the same time. Robert Clausius then took up the reigns and developed the other Laws, but we'll get to those later.

$$\Delta U = Q - W$$

RL

Equation 48: The Wave Equation

The wave equation is a very beautiful one. It has its origins in the 5th-6th century BC when Pythagoras investigated the acoustic effects of different lengths of string. This was developed further by Ptolemy in his theory of harmonics in around 150 AD, but it was not until the 16th-17th century in work by Galileo and others that the relationship between pitch and frequency was realised. The one-dimensional wave equation was derived mainly as a result of trying to understand the vibrations of violin strings. The first physical solution for string movement, considering tension forces on infinitesimally small sections of a string, was worked out in 1708 by the English mathematician Brook Taylor. In an analytical sense, Jean de Rond D'Alembert is credited with the general solution to the one-dimensional wave equation in 1746 and soon afterwards Leonhard Euler further extended the problem to three dimensions to account for spherical waves. The overall development also involved Bernoulli and Lagrange and others. The wave equation has applications in many areas of physics but particularly for describing commonly encountered wave types such as sound, light and water waves – and so the application is particularly important in acoustics, electromagnetism and fluid dynamics. It is also of course essential to an understanding of music, which is fundamentally based on vibrational harmonics.

$$\frac{\partial^2 \psi(x,t)}{\partial t^2} = v^2 \frac{\partial^2 \psi(x,t)}{\partial x^2}$$

The one-dimensional wave equation is the best-known example of a hyperbolic second-order partial differential equation, which describes sinusoidal wave motion. In its most basic form it is linear (this is always an approximation for real physical systems but often a good one), in the sense that the sum of any two solutions by superposition is also a solution, and this property allows for complicated waveforms to be represented by a number of simpler sinusoidal components. This linearity allows for many observed wave phenomena such as interference (BE30). In general, ψ is a scalar function dependent on spatial coordinates and time.

So how can this equation be shown to relate to wave motion? We can start with the simplest solution in the one-dimensional case, which describes a sine wave with amplitude A in a spatial coordinate x. Taking into account the simple harmonic motion of a repeated waveform in the picture shown we can define the function on the left below, representing the waveform on the right

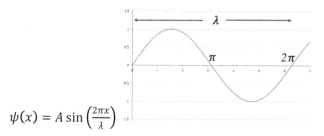

$$\psi(x) = A \sin\left(\frac{2\pi x}{\lambda}\right)$$

where λ is the wavelength. It can be seen that the function ψ is 0 when x is a multiple of $\lambda/2$ (because $\sin(n\pi) = 0$ for integer n) and is A or $-A$ when x is an odd multiple of $n\lambda/4$.

For a wave propagating in a positive direction along the x axis at a velocity v, the function ψ is also time-dependent and therefore it becomes a bivariate function of the form

$$\psi(x,t) = A \sin\frac{2\pi}{\lambda}(x - vt)$$

The velocity of the wave is the distance it travels in unit time, so $v = f\lambda$ where f is the frequency of the wave (number of cycles per second, the reciprocal of the time period). Note that this is sometimes referred to as the basic "wave" equation. It is now useful to define two parameters, the spatial frequency k and the angular frequency ω, where $k = \frac{2\pi}{\lambda}$ and $\omega = 2\pi f$. The wave velocity is now given by $v = \frac{\omega}{k}$ and the function simplifies to:

$$\psi(x,t) = A \sin(kx - \omega t)$$

The derivatives of trigonometric functions are simple to calculate because sines and cosines are derivatives of each other. The second partial derivatives with respect to x and t are:

$$\frac{\partial^2 \psi}{\partial t^2} = -\omega^2 A \sin(kx - \omega t) = -\omega^2 \psi$$

and:

$$\frac{\partial^2 \psi}{\partial x^2} = -k^2 A \sin(kx - \omega t) = -k^2 \psi$$

It is clear that these two equations are parabolic second order differential equations, similar to the Hooke equation for a stretched spring (BE18), or the equation of a simple pendulum (BE42), for example. They can be combined to form the hyperbolic differential equation which includes both the spatial and temporal second derivatives:

$$\frac{\partial^2 \psi}{\partial t^2} = \left(\frac{\omega}{k}\right)^2 \frac{\partial^2 \psi}{\partial x^2}$$

which we can see is the one-dimensional wave equation since $v = \frac{\omega}{k}$. So the equation can be built from a simple sine function, and in general, in the linear case, the solutions will be superpositions of these types of waveforms. In many cases, the Euler formula (BE38) is made use of to simplify calculations. Other methods of solution can be carried out for example by taking Fourier transforms of either side which results in the representation of more complicated waveforms in the form of Fourier series (BE12).

48

By defining the two variables $\xi = x - vt$ and $\eta = x + vt$ d'Alembert showed that a standing wave on a vibrating string can be seen as the sum of right and left travelling wave functions, in which, due to the constraints (boundary conditions), the shape with respect to the spatial coordinate x remains constant. In general there is a unique solution based on set initial conditions and boundary conditions (such as the tethering of a string in two places a distance apart), and this allows for a physical representation of standing waves of musical instruments and their harmonics. The same equation can be used to represent sound waves in a long cylinder, and applied to wind instruments. The wave equation can be extended to two dimensions, for which a basic solution for a circular sheet can be derived to represent, for example, the vibrations of a circular drum. In this case the solution involves Bessel functions in the radial coordinate coupled with trigonometric functions in the angular variable. Extended further to three dimensions, the solutions are spherical harmonics which can be represented using the Bessel functions (BE43) in the radial coordinate, similar to the 2D case, and also Legendre polynomials in the angular components. These are convenient methods of solutions as the Bessel functions and Legendre polynomials have useful properties and are known solutions to particular general forms of second order differential equation. The 3D solutions have applications in electromagnetism and many other areas. Maxwell derived his own version for electromagnetic waves in which the continuous function is expressed as electric and magnetic field strengths – in this case the velocity term v is typically replaced by c the speed of light. The solution to the energy levels of the Hydrogen atom using the Schrödinger equation (BE17, a sort of wave equation but different to the classical one) can be done in a similar way, through separation of variables in spherical coordinates. The description above is the simplest one involving undamped oscillations in a homogeneous medium. A more general form of the wave equation for damped systems with a resistance proportional to the velocity is

$$\frac{\partial^2 \psi}{\partial t^2} + \gamma \frac{\partial \psi}{\partial t} + k\psi = v^2 \frac{\partial^2 \psi}{\partial x^2} + F(x,t)$$

where $F(x,t)$ is an external force function, γ is the damping constant and k is a restoring factor. Most of the familiar examples of waves, such as vibrating strings in violins, guitars and pianos, are linear because the amplitude is small compared with the boundary conditions (e.g. those defining the length constraints on a string). Many wave types can become nonlinear – for example at large enough amplitudes in which a physical system is pushed far enough so the phase becomes dependent on the amplitude - and in these cases there is no general solution and numerical methods may be needed to solve them. In some cases of nonlinear travelling waves there are solutions called "solitons" which exhibit superposition but not in a linear sense – some of these have analytic solutions. A nonlinear variant of the Fourier transform that can be used in such cases is the inverse scattering transform (IST). An example of the successful use of the IST is the solution to the Korteweg-de Vries (KdV) equation for waves in shallow water, and these particular solutions are called "cnoidal" waves (waves with sharper peaks and flatter troughs than sine waves or their combinations) where the sine aspect of the word sinusoidal is replaced by cn, the mathematical representation of the Jacobi elliptic functions that are involved in the solution. The IST method was an important recent advance in mathematics and is now used to solve many types of nonlinear differential equation.

PB

Equation 49: The Nernst Equation

In the late 19th century, Gibbs had developed a relationship to determine whether a chemical reaction was spontaneous, based on its free energy:

$$\Delta G = \Delta G^0 + RT lnQ$$

where ΔG is the change in free energy, ΔG^0 is the same, under standard conditions (room temperature and pressure, aka RTP, 1 mol dm^{-3} concentrations), R is the gas constant, T is temperature in K, and Q is the reaction quotient. ΔG is negative for a spontaneous reaction.

49

In an electrochemical cell a redox (reduction-oxidation) reaction is physically separated, but electrically connected. If the reaction is spontaneous, this would be a battery. The above equation can be further developed, in that Gibbs' free energy is related to the electrochemical cell potential through:

$$\Delta G = -nF\Delta E$$

where n is the number of electrons involved in the reaction, ΔE is the cell potential, and F is Faraday's constant. Combining this with Gibbs' free energy, as Nernst did, leads us to:

$$\Delta E = \Delta E^0 + \frac{RT}{nF} \ln Q$$

Under standard conditions, this can be reduced to:

$$\Delta E = \Delta E^0 + \frac{0.059V}{n} \log Q$$

This equation relates the cell potential under standard conditions to the actual cell potential, depending on the concentrations of the reduced and oxidised species, and the number of electrons involved in the reaction. Q should be expressed in activities (effective concentrations) rather than concentrations, as at higher concentrations not all compounds may fully dissociate, but in dilute solutions (most cases), using concentrations is acceptable.

OK

Equation 50: Fabry-Perot Interferometer

The transmittance T of a Fabry-Perot interferometer, consisting of two flat parallel mirrors at distance L with same reflectance R, is given by:

$$T = \frac{1}{1 + F \sin^2\left(\frac{4\pi L}{\lambda}\right)}$$

where F is defined as:

$$F = \frac{4R}{(1-R)^2}$$

with R the reflectance of both mirrors.

Transmittance T is and reflectance R are the fractions of the light that are transmitted and reflected respectively, i.e. these are numbers between 0 and 1.

The transmittance as a function of the distance L between the plates, for a wavelength

$\lambda = 600$ nm is sketched below:

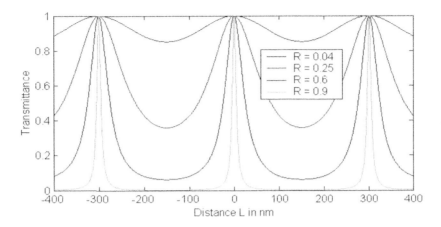

The reciprocal of the full width of a fringe at half of the maximum intensity expressed as a fraction of the distance between two maxima is given by:

$$N_R = \frac{\pi\sqrt{R}}{1-R} = \frac{\pi}{2}\sqrt{F}$$

This term N_R is called the **Finesse** of the interferometer. For example: for $R = 0.9$ one finds $N_R = 30$. This means that 1/30th of a half wavelength can readily be resolved by this interferometer; compare this to 1/2 of a half wavelength using the same criterion for the cosine function in a standard interferometer (see equation 30).

Fabry-Perot interferometers are primarily used in spectroscopy and as laser cavities, they are also used in measuring and generating displacements at the sub-nm accuracy level.

The interferometer was invented and explored around 1900 by the French Physicists Charles Fabry (1867 – 1945) and Alfred Perot (1863 – 1925). They could observe nanometre-level displacements which they generated by adding water droplets to a reservoir which was coupled to one of the mirrors.

HH

Equation 51: Index of Refraction

The speed of light is dependent on the medium, through which the light is traveling. The index of refraction n is a variable that is assigned to a medium describing the behaviour of light as it travels through it. The speed of light through that medium v_m is determined with the following equation,

$$v_m = \frac{c}{n_m}$$

where v_m is the speed of light in medium m, c is the speed of light in vacuum, and n_m is the index of refraction of medium m.

It is clear in the equation above that the index of refraction of vacuum is 1. For media with indices of refraction greater than 1, the speed of light is lower than the speed of light in vacuum. It is interesting to note that the wavelength of light is also dependent on the index of refraction. The relationship between the wavelength of light λ and a refractive index n is given by the following equation

$$\lambda = \frac{\lambda_o}{n}$$

where λ_o is the wavelength of light in vacuum.

The term 'index of refraction' is believed to have been first coined by English polymath Thomas Young in 1807. Prior to his single number convention, the index of refraction was quoted as the ratio of two numbers. The ability to precisely determine the speed of light is limited by the ability to determine the index of refraction of a medium. The speed of light through a meticulously monitored environment (particle composition, temperature, pressure, and humidity) can provide an approximation.

MF

Equation 52: Newton-Raphson Method

Often, in many engineering problems, we are faced with the task of finding the solutions (roots) to an equation of the form $f(x) = 0$ where $f(x)$ is non-linear in x. For example, when solving the diffusion equation in cylindrical coordinates, the roots of various Bessel functions are needed. The most popular method for solving such equations is the Newton-Raphson method which has a fast (quadratic) convergence rate. The origins of the method seem to date back to at least the 12th century, in the writings of the Persian mathematician Sharaf al-Din al-Muzaffar al-Tusi [1]. Around 1600, the French algebraist, Francois Vieta considered solutions to polynomial equations by using a multistep technique that obtained one individual digit at each step. Newton improved upon Vieta's method in 1660s and in 1687, applied his method to a non-polynomial equation [2]. In 1690, Raphson introduced an iterative scheme for finding the roots of polynomials that simplified the root finding of polynomials significantly. The first person to introduce derivatives in the iterative scheme appears to be Simpson in 1740 [1].

The Method:

Given an expression $f(x)$, we wish to find an approximation to the root $\bar{x}$ that satisfies $f(x) = 0$. Suppose that x_o is an approximation to $\bar{x}$. Then, the Taylor series of $f(\bar{x})$ about x leads to:

$$f(\bar{x}) = f(x_0) + f'(x_0)(\bar{x} - x_0) + \frac{1}{2}f''(x_0)(\bar{x} - x_0)^2 + \cdots$$

since $\bar{x}$ is a root of $f(x) = 0$, the above equation leads to:

$$\bar{x} = x_0 - \frac{f(x_0)}{f'(x_0)} + O(h^2)$$

where $O(h^2)$ implies that the truncated terms are second-order accurate with $h = \bar{x} - x_o$. This equation forms the basis for the famous Newton-Raphson method:

$$x_{k+1} = x_k - \frac{f(x_k)}{f'(x_k)}, k = 0,1,2,\dots$$

One starts with an initial guess x_o and use the iterative relation above until convergence is reached. Usually, a convergence criterion is used to terminate the iterations. In general, how fast the solution converges depends on the multiplicity of a given root as well as on the initial guess. When the multiplicity is one, it is easy to show that the error at the kth iteration defined by $\partial_k = \bar{x} - x_k$ is related to the error at the $(k + 1)^{st}$ iteration by:

52

$$\grave{o}_{k+1} = c\grave{o}_k^2$$

which implies that the convergence rate is quadratic. The popularity of the Newton-Raphson method is attributed to this fast convergence rate which essentially assures that the number of new, accurate significant digits double every iteration!

Geometric Interpretation: The Newton-Raphson method has a very simple geometric interpretation: x_{k+1} is the intercept made by the tangent to the curve $f(x)$ at x_k. This is illustrated in the following figure (created using the Computer Algebra System, Maple) for the polynomial $f(x) = 2x^3 + 4x^2 - 5x + 3$. This has a root at -3. The initial guess is 0.4 and the first five iterates are shown in the table next to the figure. As is clear from the figure, the tangent to $f(x)$ at x_0 intersects the x-axis at x_1. The tangent to $f(x)$ at x_1 intersects the x-axis at x_2 and so on with the intercept converging towards the actual root.

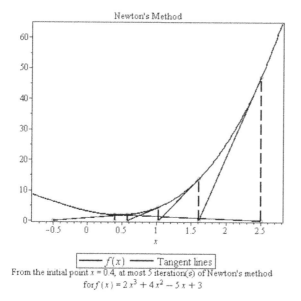

x_0	0.4
x_1	2.50476
x_2	1.61259
x_3	1.02860
x_4	.583192
x_5	-0.495898

From the initial point $x = 0.4$, at most 5 iteration(s) of Newton's method for $f(x) = 2x^3 + 4x^2 - 5x + 3$

Pitfalls: Newton-Raphson method owes its popularity to the quadratic convergence rate and ease of implementation. However, in the presence of multiple roots, the root may converge with less than quadratic rate. Furthermore, the calculation of the term $f'(x)$ in the denominator of can be difficult to find analytically. In such cases, a difference approximation of the derivative term can be used. The method also fails when $f'(x)$ is zero or close to zero. If there are multiple roots, the convergence of the approximate solution to a given root depends on the proximity of the initial guess to the root.

References
Ypma, T. J. (1995). Historical development of the Newton-Raphson method. SIAM review, 37(4), 531-551.
Deuflhard, P. (2012). A Short History of Newton's Method. Documenta Mathematica, Optimization stories, 25-30.

HC

Equation 53: The Explicit Formula for the Prime Counting Function

Why are mathematicians fascinated by prime numbers? Partly because they are the "atoms" of mathematics; any other whole number can be produced multiplying primes, while primes cannot be divided into smaller whole numbers. Partly because large primes are essential to making the codes on which all online financial transactions depend. But mainly because of their sequence. Though there is no obvious pattern to primes, it doesn't take much study to pick out some suggestions of one: primes get further apart the larger they are, and pairs of them often appear together, for instance.

Many mathematicians have spent a great deal of time - in some cases, their whole working lives - trying to tease out the pattern of the primes, and Bernhard Riemann's eponymous Hypothesis, suggested by him in 1859, is by far the most famous suggestion of such a pattern. Some mathematicians believe that the Riemann Hypothesis (or Conjecture) is the most important unsolved problem in mathematics. The person who one day cracks it will win not only everlasting fame, but a cool $1M too, courtesy of the Clay Mathematics Institute which in 2000 selected the proof of the Hypothesis as one of the seven *Millennium Prize Problems of Mathematics*.

But, one might ask, what if there IS no pattern to be found? Could the sequence of primes be as random as the digits of pi? Devoting one's life to a hunt for a well-hidden treasure might be a risk worth taking - but only if the treasure definitely exists. But how can one know a pattern exists without knowing what that pattern is? In the case of prime numbers, one way would be a proven method of calculating the number of primes within any chosen range. And we know just such a method, thanks to the work of Jacques Hadamard and Charles de la Vallée Poussin who, in 1896, independently proved a formula which allows one to calculate the number of primes less than or equal to any given number x. This is called the *Explicit Formula for the Prime Counting Function*, and it is:

$$\pi(x) = \sum_{n=1}^{\infty} \frac{\mu(n)}{n} J(\sqrt[n]{x})$$

where $J(x)$ is

$$J(x) - Li(x) + \sum_{\rho} Li(x^{\rho}) - \log 2 + \int_{x}^{\infty} \frac{dt}{t(t^2 - 1)\log t}$$

$\pi(x)$ is the prime counting function. For instance, $\pi(11) = 5$, because there are 5 prime numbers equal to or less than 11 (2, 3, 5, 7 and 11). It looks like this, for $x \leq 200$.

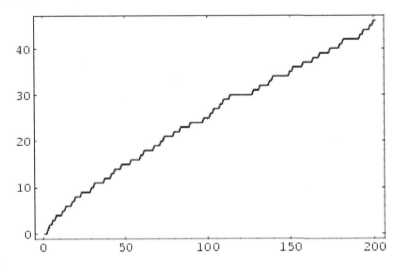

Where n is a positive integer, $\mu(n)$ is the Möbius function, which takes the value -1, 0 or 1 depending on the prime factors n has, $Li(x)$ is the logarithmic integral function, $\int_{0}^{x} \frac{1}{\ln(t)} dt$, and ρ is a non-trivial zero of the Riemann zeta function.

MJG

54

Equation 54: Abbe Number

$$V_D = \frac{n_D - 1}{n_F - n_C}$$

where, V_D is the Abbe number, n_D is the refractive index of material at the wavelength of 589.3 nm (Fraunhofer D spectral line, color: yellow; light source element: sodium) n_F is the refractive index of material at the wavelength of 486.1 nm (Fraunhofer F spectral line, color: blue; light source element: hydrogen) and n_C is the refractive index of material at the wavelength of 656.3 nm (Fraunhofer C spectral line, color: red; light source element: hydrogen).

When a ray of collimated white light enters a positive lens, an ensemble of rays of different colours emerges out and all focus at different points along the optical axis. The focal length of the optical system is wavelength dependent and the spread of colours along the optical axis is called chromatic aberration. Chromatic aberration is a phenomenon of dispersion. First defined by German physicist and optical scientist Ernst Abbe (1840-1905), the Abbe number is a measure of material's dispersion in relation to the refractive index. Abbe number is also known as the V-number or constringence of a transparent material.

Back to the old days with no lasers, bright sources in the visible spectrum were needed to measure the refractive index via refractometry. In day time, the spectrum peak of the eyeball response is 555 nm (green) and the dark-adapted peak shifts down to 513 nm. Consequently, green light is used by lens designers to characterize dispersion. Same reason for the two extreme ends (blue and red) of the visible spectrum appear in the denominator of the equation. In the original definition, the reference refractive indexes are sodium (D line), which is difficult and inconvenient to produce. Therefore, Abbe number can also be defined by Vd, the number with respect to the yellow Fraunhofer d line (Helium) at 587.5618 nm according to ISO 7944 or by the number Ve, the green e line (Mercury) at 546.073 nm for practical convenience.

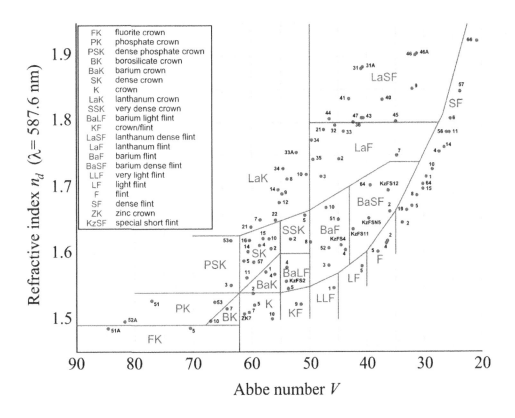

A high Abbe number indicates less colour dispersion and reduces chromatic aberration. For example, a human's eye or the glasses the person wears can be treated as a high Abbe number system when he/she sees colour fringes around a light bulb. An Abbe diagram plots the Abbe number against refractive index for a range of different glasses. In the Abbe diagram, glasses are divided into types like BK, SK, F etc. according to the Schott Glass letter-number codes to reflect their composition and position in the diagram. The Abbe number is widely used as a degree of freedom for optical designers who optimize the optical system characteristics influenced by both geometric and chromatic aberrations. For example, it is used to calculate necessary focal length of achromatic doublet lenses to minimize chromatic aberration.

KN

Equation 55: Current for CMOS Transistor

$$I_{D,Sat} = \frac{W}{L}\mu C_{inv}\frac{(V_G - V_{th})^2}{2}$$

where W is the width of the transistor channel, L is the length of the transistor channel, μ is the charge-carrier mobility, C_{inv} is the gate capacitance, V_G is the voltage difference between the gate (G) and the source (S) terminals.

CMOS transistor is the key block of the whole electronic industry. Every processor, memory, amplifier and other electronic devices have the transistor as elementary block.

CMOS stands for Complementary-Symmetry Metal-Oxide-Semiconductor. The words "complementary-symmetry" refer to the fact that the typical design style with CMOS uses complementary and symmetrical transistors pairs. Metal Oxide Semiconductor refers to the structure of the transistor, namely a layer of metal, than a dielectric (Si oxide usually) and finally the semiconductor (doped Si is the most common). Transistor is essentially a switch: applying voltage on the gate contact (G) creates a channel of electric charges so current can flow between the source (S) and drain (D).

For both type of transistor the current (in the saturation region) can be expressed as a quadratic factor proportional to how much the gate voltage is above the threshold level (namely the voltage to apply on the gate to switch on the transistor). μ is the charge-carrier mobility and is a physical property of the semiconductor. C_{inv} is the gate capacitance and refer to how much charge the gate can attract in the channel. W is the width of the transistor and L is the length of the channel. The channel length is the number used to identify a technology node; in 2014 most semiconductor manufacturer uses the 14 nm technology node. Larger current allows to build faster circuit but to keep the power consumption low the voltage must scale accordingly. Semiconductor manufacturer are trying to keep decreasing the channel length (5 nm announced for 2020), increasing μ and C_{inv} by employing semiconductors other than Si, and high-k material to increase the gate capacitance.

The first patent for the semiconductor transistor principle was filed in Canada by Austrian-Hungarian physicist Julius Edgar Lilienfeld on October 22, 1925, but Lilienfeld published no research articles about his devices, and his work was ignored by industry. However, there is no direct evidence that these devices were built. At the Bell Labs, John Bardeen and Walter Brattain, in collaboration with William Bradford Shockley eventually succeeded in building a triode-like semiconductor device. They made a demonstration to several of their colleagues and managers at Bell Labs on the afternoon of 23 December 1947, often given as the birth date of the transistor. In 1956 Bardeen, Brattain and Shockley were honoured with the Nobel Prize in Physics "for their researches on semiconductors and their discovery of the transistor effect.

GM

Equation 56: The Fifa World Ranking

The FIFA World Ranking is a ranking system for men's national teams in association football and is currently led by Germany. The teams are ranked based on their game results with the most successful teams being ranked highest. Started in 1992 the calculation method has varied. Since 2006 the method below has been used.

The basic equation is:

$$Ranking\ points = 100 \times (Result\ points \times Match\ status \times Opposition\ strength \times Regional\ strength)$$

where:

$$Opposition\ strength\ multiplier = 200 - ranking\ position/100$$

The Result points multiplier is basically 3 points for a win and 1 for a draw with some exceptions for penalty shoot   outs.

Regional strength multiplier is:

$$Regional\ (Team\ 1\ regional\ weighting + Team\ 2\ regional\ weighting)/2$$

Where the regional weightings come from a look up table. For instance Europe is 0.99 and Africa 0.85. Match status is a multiplier where friendlies $= 1.0$ up to matches in World Cup finals $= 4.0$.

Matches played in the last four years are included in the equation but there is also a multiplier based on the assessment period with matches played in the past having less weighting than recent matches. Rankings are published monthly, usually on a Thursday. The deadline for the matches to be considered is usually the Thursday prior to the release date, but after major tournaments, all games up to the final are included

However, the current formula has its problems, particularly that hosts of some major tournaments do not take part in qualifying rounds, and instead participate only in friendlies which offer fewer points. This has been one reason why World Cup 2014 hosts Brazil have fallen to a record low ranking of 22nd in the world.

DF

Equation 57: Jacobi-Anger Expansion

Surprisingly, substitution of Euler's equation (Equation 38 in the beautiful equations series) $r = \pm e^{-i\omega_m t}$ into the generating function for Bessel coefficients (Equation 43 in the beautiful equations series) yields the identities shown in equation (1). This can be expanded to provide the four relationships of equation (2) from which it is apparent that a harmonically varying function within the phase of a harmonic function can be expanded into a series expansion that is in a form of Fourier series (Equation 12 in the beautiful equations series). The surprising aspect of this becomes apparent when considering that a simple harmonic function, such as the cosine, produces its smooth and periodically repeatable curve when the argument of the function linearly increases with time (such harmonics produce the familiar pitch or tone of some perfect musical instrument). What this equation tells us is; if instead of having a linearly increasing argument with time we introduce a harmonically varying argument into the harmonic function, the result is the same as an infinite sum of harmonic functions of integer multiple frequencies with each of these harmonics being scaled by the value of a Bessel function coefficient having its argument equal to the modulation amplitude. Considering a harmonic signal to be a wobble, this may be alternatively stated that a modulating wobble inside a wobble produces an infinite sum of wobbles of integer multiple frequencies.

The last two of these equations were first discovered by Carl Gustav Jacobi (1804 – 1851) in 1836 and expressed in the full form given here by Carl Anger (1803 – 1858) in 1855, both were German Mathematicians.

There is neither time nor space to discuss the full implications of this equation. However, things become especially interesting when the modulation is added to a linearly increasing phase (called a carrier frequency) within the harmonic function. In this case, the effect is to create a signal that contains the origin carrier frequency plus signals with frequencies spaced at the multiples of the modulation frequency about this carrier value (sometimes called sidebands). Often the odd multiple sidebands involve cosine functions while the even multiples are sinusoids. Further to this, the effect of these added frequencies is to create beating effects resulting in low frequency modulations (such as those that are heard when rubbing ones finger around a wine glass rim) that can be readily measured (beating effects will be a future beautiful equation). As a consequence, it is possible to use a lower frequency modulation on top of a carrier frequency that can be used to carry information such as sound in radio or television broadcasting (including digital transmission). In optical systems, the availability of tunable lasers and other electro-optical devices enables similar modulation of optical frequencies. In fact the separation of sine and cosine components of the carrier enables determination of phase shifts in optical interference studies (interference is discussed in Equation 30). In addition to frequency modulation, these are particularly useful in physics for converting between plane and cylindrical waves.

For the reader who wants to pursue the mathematics of these functions further, it is fascinating to explore why the intensity of a modulated signal stays constant independent of modulation depth and also to answer the question of how it is possible for the right hand side equations of the second and third expressions in (2) to contain individual sinusoidal components having an amplitude of greater than 1 when the expressions on the left of all equations cannot exceed this value.

$$e^{ix\cos(\omega_m t)} = \sum_{n=-\infty}^{\infty} i^n J_n(x) e^{in\omega_m t}$$

$$e^{ix\sin(\omega_m t)} = \sum_{n=-\infty}^{\infty} J_n(x) e^{in\omega_m t}$$

Or equivalently

$$\cos(x\sin(\omega_m t)) = J_0(x) + 2\sum_{n=1}^{\infty} J_{2n}(x)\cos(2n\omega_m t)$$

$$\sin(x\sin(\omega_m t)) = 2\sum_{n=1}^{\infty} J_{2n-1}(x)\sin((2n-1)\omega_m t)$$

$$\cos(x\cos(\omega_m t)) = J_0(x) + 2\sum_{n=1}^{\infty} (-1)^n J_{2n}(x)\cos(2n\omega_m t)$$

58

$$\sin(x \cos(\omega_m t)) = -2 \sum_{n=1}^{\infty} (-1)^n J_{2n-1}(x) \cos((2n-1)\omega_m t)$$

where ω_m is the modulation frequency measured in radians per second and assumed to be constant, t is a linear variable (in the discussion presented here it can be thought of as representing time), x is the amplitude of modulation, sometimes called the modulation index, or modulation depth, $J_n(x)$ is the Bessel coefficient of the first kind of order n and argument x (i.e. the modulation amplitude) and sin and cos are the familiar harmonic functions sine and cosine.

Sources

Anger C. T., 1855, Neueste Schriften der Naturf. Ges. Danzig, V, p2.
Jacobi C. G. J., 1836, Journal für Math., xv, p12.
Watson G.N., 1966, A treatise on the theory of Besel functions, CUP.

SS

Equation 58: Reynolds Number

$$Re = \frac{\rho \mathbf{v} D_H}{\mu} = \frac{\mathbf{v} D_H}{v} = \frac{\mathbf{Q} D_H}{vA}$$

The Reynolds number is a very useful dimensionless parameter used in fluid dynamics. It was introduced by George Gabriel Stokes in 1850 but was named after Osborne Reynolds who popularized the parameter in 1883.

The Reynolds number is extremely useful in determining what flow characteristics will be apparent in a fluid dynamics system. With it you can determine the transition phases where flow goes from laminar to turbulent. It can predict turbulent flow and help to design system to either avoid or take advantage of turbulence dependent on system requirements.

A common but conflicting requirement in precision machine design is to have turbulent flow in order to get good temperature mixing of a fluid but then laminar flow into a hydrostatic bearing system. A rule of thumb often used is that the length of a pipe should be at least 42 times its diameter to transition from a turbulent to laminar flow. This assumes that other hydrodynamic conditions have been satisfied. Engineers often refer to the Moody diagram which plots the Reynolds number against friction in a pipe system and can be used to predict flow characteristics in a hydraulic system.

D_H is the hydraulic diameter (ID) of the pipe; its characteristic traveled length, L (m), Q is the volumetric flow rate (m³/s), A is the pipe cross-sectional area (m²), $\mathbf{v}$ is the mean velocity of the fluid (SI units: m/s), μ is the dynamic viscosity of the fluid (Pa·s = N·s/m² = kg/(m·s)), v is the kinematic viscosity (m²/s), ρ is the density of the fluid (kg/m³), A is the cross sectional area and P is the wetted perimeter (perimeter in contact with the fluid in question, not air interfaces)

TS

Equation 59: Centripetal Force

When an object moves in a curved or circular path around some centre of curvature then it experiences an acceleration directed towards that centre. There is therefore the need for a force to provide this acceleration, and this force is called the *centripetal* force. The magnitude of the centripetal force on an object of mass m moving at tangential speed v along a path with radius of curvature r is:

$$F = \frac{mv^2}{r} = ma_c$$

where a_c is the centripetal acceleration.

The direction of the force is toward the centre of the circle in which the object is moving, or the osculating circle (the circle that best fits the local path of the object) in case the path is not circular. This force is also sometimes written in terms of the angular velocity ω of the object about the centre of the circle:

$$F = mr\omega^2$$

 It is sometimes taught or believed that an object travelling in a circular path, such as a stick tied to a string and being swung around someone's head, is being acted on by a *centrifugal force* that seems to want to throw it outwards, away from our head. If we let go of the string then the stick will of course fly outwards (along a tangent to the orbit at the point of letting go) but this is not due to a centrifugal force pulling it outwards, it is because of the *absence* of a centripetal force to hold it in its orbit. The observed effect is simply a consequence of Newton's First Law: that any object not acted on by any forces and travelling with velocity v, will continue travelling with velocity v in a straight line for ever.

The speed in the formula is squared, so twice the speed needs four times the force. The inverse relationship with the radius of curvature shows that half the radial distance requires twice the force. This explains why a car turning a corner of radius r requires a greater force (provided by the friction of the tyres on the road) to keep it on track at a given speed when r is small than when r is big. Likewise, the force required is much larger the faster the speed of the car, for any given radius.

 The term "centrifugal force" is due to Christiaan Huygens who first used it in his 1659 *De Vi Centrifuga* and wrote of it in his 1673 *Horologium Oscillatorium* about pendulums. Isaac Newton was the first to use the term "centripetal force" (*vis centripita*) in his discussions of gravity in his 1684 *De Motu Corporum*.

PR

Equation 60: Public Key Cryptography

Alice has a problem. She wants to send a message to Bob, but it contains extremely private information that she wants to keep secret. Sadly, the local courier service has a reputation for opening letters and reading them in transit! How can Alice be sure that only Bob can read the message?

Alice's first thought is to lock the message in a box with a padlock. But now she has two new problems:
a) How does she get the padlock key to Bob without giving it to the courier?
b) if she only has one type of padlock and Bob decides to give copies of the key away to other people, her messages would be readable to everyone Bob trusts, who might not be the same people Alice trusts.

This was the state of cryptology as recently as the early 1970's, "symmetric key encryption" which required you to give a secret key to everyone you communicated with, was the only way to keep information safe. Users faced the choice of trusting the people they communicated with in secret, or generating a plethora of keys for each group of people they wanted to talk to.

With a bit of lateral thinking, Alice devises a solution. She sends a message to Bob which asks him to send her an open padlock. When Bob's padlock arrives, Alice secures her message with it and sends it back to

Bob. The couriers never see the message nor do they have access to the key. This has the added advantage that Alice and Bob don't need to trust each other either, as no one ever has to give someone else their keys.

Tales of secret messages and untrustworthy couriers might sound like stories from another age, but the modern world is built on the movement of secrets. We identify ourselves to banks and social media using secret passwords. The world economy relies on organisations sharing their intellectual property amongst their staff and suppliers whilst keeping it secret from their competitors. We buy and sell goods and services from all over the world by sending secrets about our credit cards and personal information. We do all of this using one of the wonders of the modern age, a vast, global, information sharing network of computers made up of a patchwork of infrastructure owned by different individuals and organisations, any number of which could potentially try to read our messages and learn our secrets. To do this effectively we need the mathematical equivalent of the open padlock. Some scheme that lets us communicate to people we don't trust, over a network owned by people we don't trust.

The most famous solution to this problem was published in 1977 by Ron Rivest, Adi Shamir and Leonard Adleman. The equations they produced were combined with a series of step by step instructions and published as the "RSA algorithm" a name formed by combining the initial letters of the authors' surnames. Although we now know that they were just beaten to the discovery in 1973 by a GCHQ (Government Communications Headquarters) worker called Clifford Cocks. Sadly Cocks couldn't publicise his discovery until 1997 when the government declassified the information. At the heart of both systems are the two equations for encrypting and decrypting messages.

To continue the padlock analogy, the open padlock is called the receiver's "public key" and the key used to unlock it is the receiver's "private key".

$$M^e (mod\ N) = C$$

$$C^{\frac{1}{e} mod\ (p-1)(q-1)} mod\ N = M$$

where M is the Message (also known as "plaintext"), p and q are the private keys which have to be prime numbers, n is the public key generated by multiplying the private keys together, C is the encrypted message (also known as "ciphertext"), e is a number selected by the receiver and transmitted with the public key.

There are three rules that e must obey: e must be greater than 1, e must be less than $(n - (p + q - 1))$, and e must not be a factor of $(n - (p + q - 1))$.

The relationship that makes this system of equations possible is a generalisation of Fermat's Little Theorem (the subject of an earlier article). With some modification (referred to as the Euler-Fermat Generalisation) Fermat's Little Theorem allows the relationship between the private keys and the public key to be characterised and demonstrates that the two equations are the inverse of each other, an important property if the recipient needs to read the message!

Instead of a proof, here is an example to show how the equations complement each other. Consider Alice wants to send a single character to Bob.
    1) Bob generates his private keys, $p = 11$, $q = 17$
    2) Bob computes his public key $n = 11 \times 17 = 187$
    3) Bob picks an integer e that satisfies the above rules $e = 13$
    4) Bob transmits e and n to Alice
    5) Alice decides to send the letter "X" which, in the standard way computers represent text (ASCII), is the number 88. $M = 88$
    6) Alice encrypts the message C = (88^13) (mod 187) = 165

7) Bob receives the encrypted message and decrypts it using the decryption equation $M = 165^{\wedge}((1/13)(mod\ (11-1)(17-1)))\ (mod\ 187) = 165^{\wedge}37\ (mod\ 187) = 88$

Anyone with an understanding of programming will immediately see a problem – even for a single character, the size of the numbers that we're processing (in particular for C) are huge, much bigger than the standard integer representations that computers use. As a result, specialist libraries are required to allow these huge values to be manipulated.

The reality of public key encryption is that it is much slower than symmetric key encryption. As a compromise, many modern systems generate a separate key for each conversation (a "session key") and transmit that key using public key encryption. This means that they have the speed of symmetric encryption but use public key encryption to solve the problem of how to share the key secretly.

Further Reading
Mathematics of Public Key Cryptography (2012), Steven D. Galbraith

RFO

Equation 61: Root-Mean-Square

$$Y_{rms} = \sqrt{\langle y^2 \rangle}$$

INTRODUCTION.
The root mean square (RMS/rms) value of a set of data, *aka* quadratic mean, is a statistical measure - a type of average - defined as the square root of the mean square value. This in turn is defined as the arithmetic mean of the squares of the individual values. It is a particular case (i.e. with exponent 2) of the 'generalised' mean (see diagram). The RMS value can be calculated for a sequence of discrete values or for a continuous function. The value for a continuous function (e.g. a signal) can be approximated by taking the RMS of a series of equally spaced samples.

One justification for using RMS values as an indicator of average is that we are often interested only in the magnitudes of data values, not their signs. Whereas use of the arithmetic mean would result in a cancelling-out effect between negatives and positives – which is unwanted in many contexts – the squaring involved in calculating RMS values produces only positive outcomes, thus avoiding this problem.

Why do we not use the mean of the absolute values of a data set – a parameter which is more intuitive and easier to calculate than RMS – as an indicator of magnitude-average? This question is at the centre of an ongoing debate about the traditional and almost ubiquitous use – and teaching - of standard deviation (SD) as opposed to mean (absolute) deviation (MD). The principal argument deployed by proponents of SD is that it is much easier to incorporate SD into higher-level calculations than it would be for MD. This is countered by the comment that we are now in an era where the power of computing negates what would otherwise be problematical any advanced mathematical analysis using dispersion data.

An advantage of using RMS values, as far as science is concerned, is that these values may well correspond to actual meaningful and measurable physical parameters. For example, an RMS alternating current is equivalent to a measurable direct current of the same value.

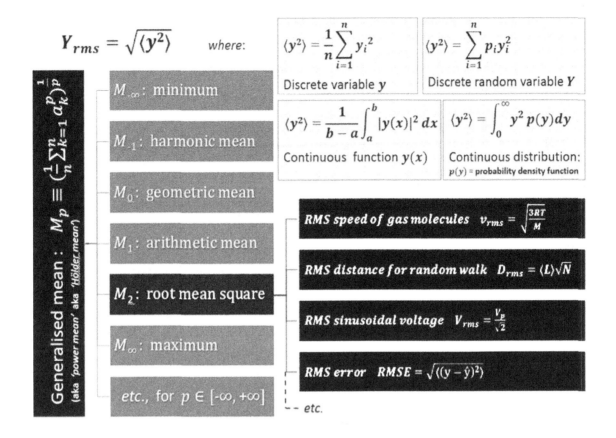

ROOT MEAN SQUARE EXAMPLES

In the formulae (see diagram), the y quantities being rms-averaged can refer to many different physical situations.

(1a) y = displacement (net distance travelled) in a "random walk" of N steps each of mean length L. RMS displacement = $L\sqrt{N}$.

(1b) y = end-to-end distance of a freely jointed linear polymer chain averaged over all conformations and consisting of N segments each of length L. RMS overall length = $L\sqrt{N}$.

(2) y = speed of a gas molecule. RMS speed of all molecules in an ideal gas = $\sqrt{\{3RT/M\}}$: R=molar gas constant, T=absolute temperature, M=molar mass.

(3) y = acceleration G measured by accelerometer. RMS acceleration (GRMS) = $\sqrt{A}$: A = area under acceleration spectral density (ASD) curve. GRMS expresses the overall energy of a particular random vibration event and may be a statistical value used in mechanical engineering.

(4) y = instantaneous voltage or current of cyclically varying signal. RMS value turns out to be its "effective" value, i.e. value of the direct voltage or current that would produce the equivalent power dissipation in a resistive load. For a sinusoidally varying AC, the RMS voltage or current = $y_p/\sqrt{2}$ (subscript 'p' denotes 'peak'). RMS voltage or current when there is a DC offset y_0 and AC amplitude y_p to the sinusoid = $\sqrt{\{y_0^2+y_p^2/2\}}$.

(5) y = stress or strain or sound pressure.

(6) y = the pairwise separation of corresponding atoms along the backbones of two similar superimposed protein molecules. RMSD denotes RMS Deviation.

(7) y = the pairwise differences between two corresponding data items in two sets of equal size and status, e.g. two time series.

ROOT MEAN SQUARE DEVIATION EXAMPLES

Here, the y quantity represents the DEVIATION of a member of a set of values from some single reference quantity, eg the mean. Key: RMSD = RMS Deviation; RMSE = RMS Error; RMSF = RMS Fluctuation.

(8) y = deviation of a selected particle's position from some reference position as a function of time, e.g. backbone atoms during the molecular dynamics simulations. RMSF = measure of average fluctuation.

(9) y = deviation of a measurement from a theoretically predicted value $ÿ$ (this type of deviation is called a residual). The RMSE (aka RMSD) is a measure this 'error'. RMSE = $\sqrt{<|y-ÿ|2>}$. RMSE amplifies large deviations (errors), giving them much higher weight than does the mean absolute error. This means RMSE is very useful when large errors particularly need to be underlined.

(10) y = deviation of a value from population mean. RMS value is the population standard deviation $\sqrt{<|y-\mu|2>}$

(11) y = deviation of displacement or momentum from mean for harmonic oscillator. RMSD = $y_m/\sqrt{2}$: y_m = maximum displacement or momentum (cf RMS voltage or current). Applies to both classical and quantum mechanical harmonic oscillators (though with different interpretations).

(12) y = deviation of the internal energy of a canonical ensemble from the mean. RMSF = $\sqrt{\{kT^2C_v\}}$: k=Boltzmann constant, T=absolute temperature, C_v=specific heat capacity at constant volume.

Other applications of RMSD include geographic information systems, hydrogeology, computational neuroscience, protein nuclear magnetic resonance spectroscopy, economics and experimental psychology.

FR

Equation 62: Trapezoidal Rule

In many applications, integrals of the form $I = \int_a^b f(x)\,dx$ need to be evaluated. For example, given a curve in the functional form $y = f(x)$, the length of the curve can be obtained using $\int_a^b \sqrt{1 + y'^2}\,dx$. Often, an approximation $\tilde{I}$ to I is sought since the form of $f(x)$ makes explicit integration difficult or even impossible. In some cases, the function $f(x)$ itself is unknown and instead, only its values are known at a few points. Many methods exist for obtaining $\tilde{I}$. For almost all of these methods, the general form of the approximation can be written as:

$$I := \int_a^b f(x)\,dx \approx \tilde{I} = \sum_{i=0}^{n} w_i f(x_i)$$

where w_i are called the weights and x_i are called the integration or base points. The Newton-Cotes methods use $n + 1$ equidistant integration points and can exactly integrate polynomials of degree up to n. The trapezoidal rule with $n = 1$, an example of these methods, is one of the oldest and still widely used in practice. The origins of this rule probably date back to the method of exhaustion developed by Antiphon and

Eudoxus of Cnidus in the 5th and 4th centuries BCE and used to calculate areas under curves by inscribing polygons under the curves.

The trapezoidal rule approximates I with the area under the straight-line connecting the points $f(a)$ and $f(b)$ (See Figure 1a):

$$I \approx \tilde{I} = \frac{b-a}{2}(f(a) + f(b))$$

When higher accuracy is desired, the interval $[a, b]$ can be broken up into p subintervals, $[x_0, x_1] \cup [x_0, x_1] \cup ... \cup [x_{p-1}, x_p]$ and the trapezoidal rule is applied on each subinterval $[x_{i-1}, x_i]$ to find $\tilde{I}_t$. The desired result $\tilde{I}$ is then $\sum_{i=1}^{P} I_i$. This is illustrated in Figure 1b for $p = 4$.

The error of approximation is locally $O(h^3)$ where $h = b - a$. When the composite rule is used, the local error of approximation is still $O(h^3)$ with h being the length of a subinterval. However, the global error is

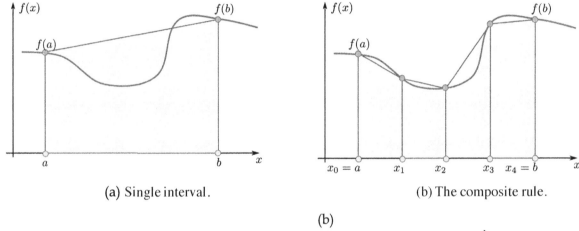

(a) Single interval.

(b) The composite rule.

(b)

Figure 1: The trapezoidal rule for finding an approximation to the integral $I := \int_a^b f(x)\, dx$.

$O(h^2)$ when the subintervals are of equal length. For periodic functions, the convergence properties significantly improve as has been discussed in [1] and [2].

p	I
1	4.0
5	4.56704000000000
10	4.59169000000000
25	4.59866726400000
50	4.59966670400000
100	4.59991666900000
Exact	4.60

p	I
1	1.0
4	0.90
10	0.902780757422973
20	0.902779927767432
40	0.902779927772194
100	0.902779927772194
Exact	0.902779927772194...

EXAMPLES

Consider the integral $I = \int_0^1 (2 + 5x + 6x^3 - 7x^4)dx$. The exact value of this integral is 4.6. The approximate value of the integral is shown in Table 1a for various values of p. The extraordinary convergence properties of the trapezoidal rule for periodic functions is illustrated in Table 1b where the integral (originally due to Poisson; see [2] for more details) calculates the perimeter of an ellipse with axis lengths of $1/\pi$ and $0.6/\pi$.

The trapezoidal rule approximates the area under a given curve in the interval $[a,b]$ with the area under the straight line connecting the two end values of $f(x)$, requires two function evaluations and can integrate a first-order polynomial exactly. On the other hand, Gaussian integration (two-point formula with a total of two function evaluations) can integrate cubic polynomials exactly. Nevertheless, the trapezoidal rule finds extensive use in various applications including the numerical integration of differential equations due to its ease of implementation. For computations where minimizing computational cost is not essential, this method serves as an excellent option since accuracy can be improved by using the composite rule discussed above. When the function is known only at select points, the integration with the trapezoidal rule is easy to implement whether the points are evenly spaced or not. The method converges rapidly when the integrands are periodic. More examples illustrating the fast converging properties of the trapezoidal rule can be found in [1] and [2].

References
Weideman, J. A. C. (2002). Numerical integration of periodic functions: A few examples. American Mathematical Monthly, 21-36.
Trefethen, L. N., & Weideman, J. A. C. (2014). The exponentially convergent trapezoidal rule. SIAM Review, 56(3), 385-458.

Further reading
The History of the Calculus and Its Conceptual Development (1959), Carl B. Boyer

HC

Equation 63: Hess's Law

Hess's Law was published by scientist and physician Germain Hess in 1840, as a result of his work on thermochemistry. It states that the enthalpy change (energy change at constant pressure) between a given set of reactants and products is always the same, regardless of the path taken. Enthalpy change, therefore, is a state function. The best known equations associated with this are the ones involving enthalpy of formation, and combustion. For a spontaneous reaction (considering only enthalpy), ΔH^o_{rxn} has to be negative, meaning that the system is giving off energy.

$$\Delta H^o_{rxn} = \Sigma \Delta H^o_f(products) - \Sigma \Delta H^o_f(reactants)$$

$$\Delta H^o_{rxn} = \Sigma \Delta H^o_c(reactants) - \Sigma \Delta H^o_c(products)$$

Both equations can be used to calculate changes in enthalpy of intermediate steps in processes. One example is the enthalpy of combustion of carbon to carbon monoxide. It is very hard to determine this experimentally, as some of the carbon will generally form carbon dioxide. By considering the combustion of carbon, as well as carbon monoxide, to carbon dioxide, one can then find a value for the process under investigation.

$$C + O_2 \rightarrow CO$$
$$\searrow \qquad \swarrow$$
$$CO_2$$

Alternatively, one can consider the enthalpy of formation of (mono) chloromethane from methane and chlorine. This would be an issue, as repeated chlorination of the methane is a common occurrence under most conditions. By considering the formation of the reactants, as well as the products, from their elements in their natural states at RTP (room temperature and pressure), which can be done independently, one can again isolate the step of interest.

$$CH_4 + Cl_2 \rightarrow CH_3Cl + HCl$$
$$\nwarrow \qquad \nearrow$$
$$C + H_2 + Cl_2$$

One can also use Hess's Law to determine enthalpies of reaction for very slow reactions, where heat loss to the environment may prevent an accurate determination. Lastly, an extension to Hess's Law is a Born-Haber (BH) cycle. These are generally used to determine the lattice enthalpy of ionic compounds. This is the energy given off in the final step in the reaction between two elements. Starting off with the elements in their natural form at RTP, a BH cycle goes through the steps needed for them to end up as their respective ions in the gas phase. This generally involves atomisation and ionisation. The final step is the combination of those gaseous ions to collapse into the crystalline ionic solid. Although the previous steps can normally be measured, this last step in mostly determined through a comparison between the sum of those previous ones and the enthalpy of formation of said ionic compound. Due to the magnitude of the lattice enthalpy, this is generally the determining factor in the stability of the product.

OK

Equation 64: Ellipsometer Equation

Ellipsometry is an optical technique which is used to determine the properties of coating structures in many applications; especially in thin film analysis as its sensitivity is unsurpassed.

The ellipsometer equation is:

$$r(\lambda, \theta) = \frac{r_p(\lambda, \theta)}{r_s(\lambda, \theta)} = \tan\big(\Psi(\lambda, \theta)\big) \exp(i\Delta(\lambda, \theta))$$

The ratio between the incident amplitude E_1 and the reflected amplitude E_2 will be different between light that is polarized parallel to an interface (p-direction) and the direction perpendicular to this (s-direction), as indicated by the arrows in the figure below. In the ellipsometer equation r_p and r_s are the amplitude reflectances for light polarized in the p- and s- directions respectively.

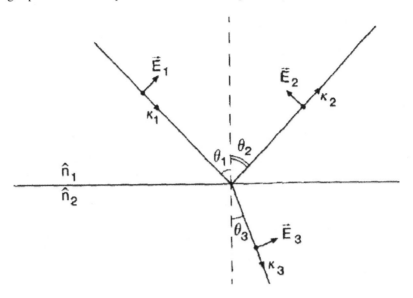

In the figure the beam trajectories are sketched for a wave incident on a plane interface between two media having a refractive index n_1 and n_2 respectively. The incoming beam with amplitude E_1 is reflected by the interface, giving a reflected beam with amplitude E_2 and a transmitted beam with amplitude E_3.

The ellipsometer equation defines the ellipsometric angles Ψ and Δ that can be determined in an ellipsometric measurement. The equation states that the ratio between the reflectances, and the ellipsometric angles Ψ and Δ, will depend on the angle of incidence θ and the wavelength λ for a given interface, whether it consists of a single homogeneous material or is a complicated system consisting of several absorbing and non-absorbing thin films.

Solving the Fresnel equations (see future equation 166) for the interfaces of a given substrate/thin film assembly gives the r of the ellipsometry equation. The inverse problem is less straightforward, so an ellipsometer calculation normally consists of fitting free parameters (such as film thickness, refractive index) to the measured value of r, i.e. the ellipsometric angles Ψ and Δ. In these measurements the angle of incidence θ and the wavelength λ may be varied.

The technique has been known since 1888 by the work of Paul Drude (1863-1906). The term "ellipsometry" was probably first used in 1945.

HH

Equation 65: The Compound Interest Formula

The formula for calculating the total value of a deposit in an interest earning account is dependent on the rate of interest as well as the frequency of compounding. Compounding refers to when interest is earned not just on the original deposit amount but also on the interest earned up to the point of compounding. The formula for calculating the total amount A in your R (R*100 %) interest earning account, which is compounded N times per year, after T years is given by:

$$A = P\left(1 + \frac{R}{N}\right)^{NT}$$

where P is the original (or 'principal') deposit amount, R is the nominal interest rate, N is the number of times the interest is compounded per year, and T is the amount of time (in years) that elapsed since the original deposit.

The chart attached gives an example of how the value of an account, which has an original amount of $1,000 deposited and provides a 20% interest rate, increases over time. The chart also provides the results for various compounding frequencies. It is noticeable that the higher compounding frequency provides more return on the original investment.

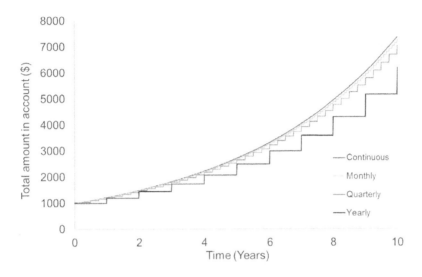

Interestingly, the idea of compounding interest on loans and deposits was considered usury and strictly frowned upon during the Roman times and probably towards the 16th and 17th centuries. The first known in-depth coverage of calculating compound interest was found in a 1613 publication by Richard Witt 'Arithmeticall Questions.'

MF

Equation 66: Definition of Pi

π ("pi") is a fundamental mathematical constant, a geometric relationship, a number defined as the ratio of the circumference C of a perfect circle (shown with centre o) to its diameter d, or half the ratio of C to its radius, r. This ratio is the same number regardless of the size of the circle used to calculate it. The proof of π as a constant arises from its own definition and the definition of a circle as an infinite number of points in a given plane at the same distance from a central point (or the curve traced out by a point that moves at a constant distance): π being a constant is essentially the same thing as saying that all circles are similar, or proportional in their relative dimensions. This can be shown in a similar way to the proof of another fundamental mathematical relationship, that of Pythagoras, by the use of similar triangles – for any same angle between two radii, isosceles triangles can be extended out to ever larger circles and the ratio of the chord (geometric line segment) to the radius remains constant (equivalent to 2× the sine of the half-angle). In a geometrical sense the value of π is an angle representing the number of radii ("radians") in a semicircle, and is therefore inextricably linked with the definition of the radian as a unit of measurement (the value of θ as shown is 1 radian). Since the circumference of a circle can then be calculated as $2\pi r$, it follows from integration over increasingly smaller isosceles triangles that the area is πr^2.

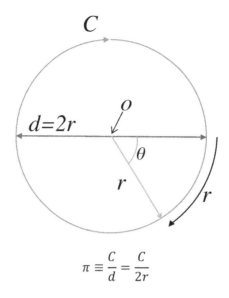

$$\pi \equiv \frac{C}{d} = \frac{C}{2r}$$

Why was the Greek letter π chosen for this number? This was first used by the mathematician John Machin in 1706 as an abbreviation for "perimeter", and then adopted by Euler in 1748 which secured its popularity in that form. Previously it had been named as Archimides' constant and subsequently Ludolph's constant (after the 16th century Dutch mathematician Ludolph van Ceulen, who had π engraved on his tombstone to his achievement of 35 decimal places).

π is an irrational number: it cannot be represented as a fraction, which has many implications. It is also, more specifically, a transcendental number: it cannot be represented as a root of a polynomial equation with rational coefficients. In early times, it made sense to try to represent this number as a rational number or as an unknown number bounded by two rational numbers – the concept of π has been known since ancient Egypt and Babylon, but in around 250 BC Archimedes advanced its calculation by using inscribed and circumscribed polynomials to formulate the relationship $\frac{223}{71} < \pi < 22/7$. This approximated π to about 3.142 in decimal terms. The calculation of π was advanced significantly in the 16th and 17th centuries AD by the use of infinite power series. A formula typically attributed to Gregory and Leibniz is now thought to have originated in India in about 1400 from Madhava, who was the first to develop an inverse tangent series which could be used to calculate π to at least 10 decimal places. Eventually in 1706, 100 decimal places were achieved by John Machin. The advent of computers in the 1940s and '50s greatly advanced the computation of π to tens of thousands of decimal places, often using a formula by Ramanujan from 1910, a much more rapidly converging series which is still used today. Recent calculations of π have now exceeded 13 trillion decimal places. This increasingly large number of digits appears to still be randomly distributed but as yet there is no proof of this. However, the irrationality of π had already been known since 1761 (proved by the Swiss mathematician Lambert), although the transcendence of π was not proved until more than a century later by Lindemann in 1882. It is the transcendence of π that renders the ancient problem of "squaring the circle" impossible without an infinite number of steps. The calculation of π to the current level of accuracy goes well beyond any requirement in the physical world – in fact, just 39 digits is sufficient enough to calculate the volume of the known universe to the precision of one atom. It grew into a kind of record-breaking exercise for mathematicians, and subsequently computer programmers, and in this sense it does have some practical value for testing computers and their algorithms.

Although predominantly associated with geometry, we find that π turns up almost everywhere in mathematics, by the extension of geometrical concepts to the theory of functions, calculus, known solutions to many important definite integrals and other areas including probability distributions. Although its definition is based on two-dimensional geometry it extends naturally to further dimensions and appears intrinsically in the formulae for volumes and surfaces of spheres, cylinders and cones for example. Because it is effectively a representation of half a circular revolution it appears in important limiting values of trigonometric functions, and by extension, exponential functions, and in general the representation of harmonic motions, waves and fields, and naturally has already appeared in many of the equations in this series.

Further reading:
A History of Pi (1976), Petr Beckmann

PB

Equation 67: Newton's Law of Universal Gravitation

$$F = \frac{Gm_1 m_2}{r^2}$$

This compact equation conceals a revolution in scientific thought: F is the force of gravitational attraction between two objects, of masses m_1 and m_2, separated by a distance r. G is a constant. It was developed around 1665 by Isaac Newton, who had been forced to move temporarily from Cambridge to his family home in Woolsthorpe, due to an outbreak of bubonic plague. Specifically - so it is said - in an orchard. According to Newton's first biographer, William Stukeley, "the notion of gravitation came into his mind... occasion'd by the fall of an apple, as he sat in contemplative mood."

What the apple - if there was one - made Newton realise is that it is the same thing that makes objects fall to earth which also keeps the moon in its orbit and the planets in their orbits too (hence, *Universal* Gravitation). This was a radical departure from previous theories which had treated the earthly realm as quite distinct from the celestial one - each subject to its own laws (this was in part an inherited position from the theories of the ancient Greeks, who thought that everything under the Moon was made of the four elements earth, air, fire and water, while everything beyond was made of the fifth element, the quintessence, the nature of which was to move eternally in circles).

Newton was a secretive man, obsessed with hidden messages in the Bible and in the secrets of the alchemists, and he might never have published his revolutionary equation if it were not for the tireless efforts of his loyal friend Edmond Halley, who convinced him to write the *Philosophiæ Naturalis Principia Mathematica*. (*The Mathematical Principles of Natural Philosophy*), perhaps the greatest work of physics of all (as it was, Halley paid for its publication in 1687).

The equation remained the best there was until 1915 when Einstein bettered it with his theory of general relativity (it was clear that Newton's equation could not be right once it was appreciated that the speed of light was unsurpassable- there is no time-factor in the equation, implying that, if two masses came suddenly into existence, each would be affected by the other instantly, no matter what their separation). However it is quite good enough to be used (along with Newton's three laws of motion) to predict the motion of most objects in the Solar System, and to accurately determine orbital trajectories of spacecraft travelling to them as well.

As for the apple - it lives on too, in the form of an ancient apple tree in the grounds of the National Physical Laboratory, a descendent of an apple-pip from Newton's orchard.

Further reading
Gravity (2012), Brian Clegg

MJG

Equation 68: Edlen's Equation

Since its invention, laser has been widely used in science and engineering. One of the major applications of laser is measuring distance or displacement by either interferometric or time-of-flight methods. When traveling in any medium (for example air), the wavelength of the laser beam depends on the refractive index. Therefore, to achieve optimum accuracy of length measurement, the refractive of index of air must be determined at an acceptable level of uncertainty. In 1966, Swedish physicist Bengt Edlen published his empirical model, known as the Edlen's equation to calculate the refractive of index of air in normal laboratory conditions for visible laser radiation.

Edlen's Equation in SI form (in 1966):

$$(n - 1)_{tp} = \frac{p \times (n - 1)_s}{96095.43} \times \frac{[1 + 10^{-8}(0.613 - 0.000998 \times t) \times p]}{1 + 0.0036610 \times t}$$

where t is temperature (unit: °C), p is pressure (unit: Pa).

$(n - 1)_s$ is the refractivity of standard air (i.e. dry and having 78.09% nitrogen, 20.95% oxygen, 0.93% argon and 0.03% carbon dioxide) at 1 atmosphere and 15 °C, given by:

$$(n - 1)_s \times 10^8 = 8342.13 + 2406030(130 - \sigma^2)^{-1} + 15997(38.9 - \sigma^2)^{-1}$$

where σ is the vacuum wavenumber of laser radiation (unit: μm^{-1}).

When the partial pressure f of water vapor is included, n_{tpf} is used to modify n_{tp} by:

$$n_{tpf} - n_{tp} = -f \times (4.2922 - 0.0343\sigma^2) \times 10^{-10}$$

Corrected Edlen's Equation in SI form (in 1993):

$$(n - 1)_{tp} = \frac{p \times (n - 1)_s}{96095.43} \times \frac{[1 + 10^{-8}(0.601 - 0.000972 \times t) \times p]}{1 + 0.0036610 \times t}$$

$$(n - 1)_s \times 10^8 = 8342.05 + 2406294(130 - \sigma^2)^{-1} + 15999(38.9 - \sigma^2)^{-1}$$

$$n_{tpf} - n_{tp} = -f \times (3.7209 - 0.0343\sigma^2) \times 10^{-10}$$

where all symbols are in compliance with the above nomenclature.

In the equation, air temperature and pressure are used to determine the refractive index for a specific radiation. The unit of pressure originally used by Edlen is the torr, which is not in the International System of Units (SI) form. In the picture, the SI form of the original Edlen's equation is given.

Because of the increasing demands of work on standards such as the kilogram and meter, and for precision industries requiring measurement of length at an accuracy level of better than 100 parts in a billion, the uncertainty in the determination of the refractivity of air contributes nearly 10 parts in a billion to the uncertainty of such measurement. In 1993, British metrologists at the National Physical Laboratory (NPL), K.P Birch and M. J. Downs published the updated Edlen's equation to improve the agreement between calculation and experimental results obtaining a smaller experimental uncertainty. In this 1993 version, the increased level of carbon dioxide in air, the adoption of the International Temperature Scale (ITS-90) and a revision of water vapor constants are included so the Edlen's equation has an improved uncertainty approaching +/- 10 part per billion for wavelength range 350 nm to 650 nm.

In 1996, Australian physicist P. E. Ciddor published an equation, which covers the wavelengths from below 350 nm to above 1300 nm and accounts for other extreme conditions of carbon dioxide and humidity. His equation is particularly useful for increasing the precision of geodetic surveys.

At the National Institute of Standards and Technology (NIST) the engineering metrology toolbox website, an interactive calculator of the refractive index of air based on either Edlen's equation or Ciddor's equation is provided. To get a higher accuracy value of the refractive index of air, direct measurement using a dedicated refractometer is recommended.

KN

Equation 69: Operational Amplifier Gain

$$V_{out} = (1 + \frac{R_2}{R_2})V_{in}$$

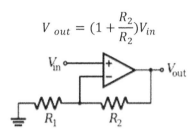

An operational amplifier (usually drawn as a triangle in electronic diagram) is an integrated circuit whose function is to be a comparator, i.e. the output voltage is high or low depending whether the voltage on the positive input is larger or smaller than the voltage on the negative terminal. The picture shows one of the most common configuration: non-inverting amplifier. The output voltage is proportional to the input voltage multiplied by a constant that can be tuned choosing the resistance values of the resistors R1 and R2. Adding capacitance/inductance in series/parallel with R1 and R2 allows having a frequency dependent gain and the circuit become a filters or an oscillators.

The first patent of a device implementing this functionality was filed by Karl D. Swartzel Jr. of Bell Labs in 1941. Throughout World War 2, Swartzel's design proved its value by being used in the M9 artillery director designed at Bell Labs. The artillery director was used to continuously calculate trigonometric firing solutions for use against a moving target. This artillery director worked with the radar system to achieve hit rates close to 90%. Currently, the amplifier is widely used in instrumentation where a small electronic signal has to be detected, and needs to be amplified before any further processing. In a configuration with capacitor/inductance it is a filter, and amplifies only a determined spectral region of the input.

Further reading
Operational Amplifiers with Linear Integrated Circuits (1993), William D. Stanley

GM

Equation 70: Mandelbrot Set

Many of us of a certain age would have played with the Julia and Mandelbrot sets on some of the first home PCs with graphic capability. Playing with these sets enabled some striking images to be produced.

The Julia set can be divided in to two main types either wholly disconnected or wholly connected. The Mandelbrot set refers both to a general class of fractal sets and to a particular instance of such a set. In general, a Mandelbrot set marks the set of points in the complex plane such that the corresponding Julia set is connected and not computable. To tell if the set is connected Julia devised a trick to find out if the set is connected. If the orbit is calculated and goes to infinity the set I disconnected. All points for which the Julia fractal is connected constitute the Mandelbrot set.

The Mandelbrot set is the set obtained from the quadratic recurrence equation:

$$Z_{n+1} = Z_n^2 + C$$

With $Z_0 = C$, where points C in the complex plane for which the orbit of Zn does not tend to infinity are in the set. Setting Z0 equal to any point in the set that is not a periodic point gives the same result. The Mandelbrot set was originally called a molecule by Mandelbrot. J. Hubbard and A. Douady proved that the Mandelbrot set is connected.

A typical plot of the Mandelbrot set in which values of C in the complex plane are coloured according to the number of iterations required to reach $r_{max} = 2$. The kidney bean-shaped portion of the Mandelbrot set turns out to be bordered by a cardioid (heart shaped curve) expressed mathematically as:

$$4x = 2cost - cos(2t)$$
$$4y - 2sint - sin(2t)$$

Mandelbrot was a French and American mathematician who was born in Poland and is noted for developing a "theory of roughness" and "self-similarity" in nature and the field of fractal geometry. With his access to computers at IBM, Mandelbrot was able to use computer graphics to create and display fractal geometric images, leading to his discovering the Mandelbrot set in 1979. By doing so, he was able to show how visual complexity can be created from simple rules.

When Mandelbrot first saw the figure he thought it was a single continent, a cardioid with adjoining circles. What he initially thought to be dust turned out at successive enlargements to be miniature continents.

Concepts such as deterministic chaos and self-similarity typify this wide research area.

Further reading:

Make Your Own Mandelbrot: A gentle journey through the mathematics of the of the Mandelbrot and Julia fractals (Apr 2014), Tariq Rashid

MF

Equation 71: The Linear Equation

$$y = ax$$

The linear equation expresses one of the simplest relationships between two variables (or two sets of variables). Linear relationships between variables appears to have been such an integral part of early mathematical development that no distinct originators can be identified (although the ancient Chinese manuscript The Nine Chapters on the Mathematical Art developed over the 10th to the 2nd century BCE contains solutions to simultaneous linear equations indicating origins preceding this latter date). Considering only the simplest 'single input, single output' form, in the above equation y and x are variables that are linked by the constant H. While mathematically this might be considered a simple linear mapping between two variables (y and x), in many physical models this equation is represented in block diagram form with x being the input, H being the process response and y being the output. This is a foundational concept behind many models spanning a broad array of processes. Examples include Hooke's law for modelling the response of materials to applied forces and Ohm's law for circuit design.

Most remarkably, this representation also applies to any physical model that comprises simultaneous linear differential equations subject to harmonically varying inputs. In this case, all of the parameters of this equation become complex numbers and matrices of numbers if there are multiple inputs and outputs.

One of the most powerful features of this model is that it obeys the principle of superposition. This states that; if the input be comprised from the addition of two independent sources, the output response can, correspondingly, be computed as the sum of responses to each of the independent inputs. Mathematically, this is probably more apparent from the block diagram illustration. Rearranging the linear equation it is apparent that the ratio of the output to input is of course the constant H. In electrical circuit terminology the magnitude of this complex number is often called the gain of the circuit while the ratio of its real and imaginary parts can be used to calculate the delay (or phase) of the response. In mechanical systems H is called the steady-state frequency response, or the transfer function, of the process and often has force as the input and displacement as the output.

Surprisingly, for all smooth and continuous processes, such as those that model electrical, mechanical and thermal systems, the equations governing the input to output relationship will all tend towards linear differential equations for processes that are deviating only a small amount from an equilibrium state.

A lot of people are familiar with the equation for a straight line of the form $y = Hx + C$. While this describes a straight line, it is not linear in terms of the definition based on the principle of superposition. To see this, consider an input $x = A$ and output $y = B$ for which the principle of superposition would imply that for an second input $x = 2A$ then the output should be $y = 2B$. Hence a test for linearity is the affirmative response to the Shakespearean question, $2B$ or not $2B$? Numerically, consider the case where, arbitrarily, H = 3, C = 1and x = 4 = A. In this case, the output y = 13 = B. Doubling A to 8 gives a value 25 ≠ 2B. Setting C to zero and repeating the exercise confirms superposition.

When the constant is finite, the principle of superposition no longer applies and equation analysis becomes more involved. In the world of mathematics, while these equations are still called linear, mapping between these two variables (that can be column matrices of variables) are called affine transformations (after the Latin word 'affinis' meaning connected with) and form an important branch of mathematic for geometric image manipulation. A subset of these affine transformations are homogeneous matrices, both of which are the subjects of future beautiful equations.

SS

Equation 72: Angular Momentum

$$L = r \times mv$$

$$L = I\omega$$

$$L = r$$

Angular momentum is considered to be the reaction force or inertia of an object rotating about a point. It is analogous to the linear components in $F = ma$ in classical mechanics as described by Newton's second law of motion.

The component L angular momentum can be expressed in terms of angular velocity ω and the body's moment of inertia I.

Angular momentum L is often expressed as the linear components of the rotational system, mv is the cross product of the linear momentum and the radius r about which it rotates.

The angular momentum equation is derived from angular velocity which is the distance covered by the orbiting object over a fixed period of time and moment of inertia which is the reaction force around a point, or the force exerted by the object orbiting a point at a defined radius.

The history of angular momentum is a little blurry depending on how it is considered. A French priest named Jean Buridan proposed the impetus principle (circa 1340) in which he stated that a ball that was thrown was given impetus and that defined its motion. The fundamental principles follow the laws of classical mechanics and draw from Isaac Newton's laws of motion (circa 1687). There have been many adaptations on the basic premise of force around a point as defined in electrical systems, particle theory, quantum mechanics as well as the classical mechanics definition stated here.

These equations are for considering rigid body mechanics following a circular path. Angular momentum is a well-defined area of classical physics and is often used in engineering for many purposes.

One of the primary uses of this equation is determining safety factors and speed limits for rotational systems.

TSS

Equation 73: The Quantum Theory of Light

The theory that light consists of electromagnetic waves of a certain frequency and wavelength, travelling at high speed, was able to account for every phenomenon involving light that had been encountered up to the end of the nineteenth century, and was therefore pretty well established in the minds of scientists. The German physicist Max Planck however had been studying the radiation emitted by a hot luminous body, in particular the relative intensities of the different colours of its spectrum, and he found that the only formula he could find that would fit observed data would require a curious and novel assumption: that the light is emitted in the form of discrete bursts or quanta of energy, rather than continuously. Furthermore, he found that the quantum of energy of each burst depends only on the frequency, ν, of the light

$$E = h \times \vartheta$$

where the constant $h = 6.63 \times 10^{-34}$ Js.

Today this constant is named for Planck, and the equation is named for both Planck and Einstein, who in 1905 used Planck's idea to explain the photoelectric effect (that light striking certain conductors can generate an electric current), which was also inconsistent with the wave theory of light. Einstein had to take the additional step of suggesting that light is not merely emitted one quantum at a time, but actually travels thru space as discrete quanta. Despite going on to develop the Special and General Theories of Relativity, Einstein owed his Nobel Prize to this idea and not to them. The notion of quantised energy is so commonplace nowadays that it is hard to appreciate what a huge and original step Planck's simple idea represented. It deserves more than does any other idea, to be truly called the Foundation of Modern Physics.

Further reading
The Bumpy Road: Max Planck from Radiation Theory to the Quantum (2015), Massimiliano Badino

PR

Equation 74: Mass-Energy Equivalence

Perhaps the most famous equation of all time, here E is energy, m is mass and c is the speed of light in a vacuum. The equation basically tells us that mass is just another form of energy or that mass and energy are equivalent. Mass–energy equivalence arose originally from special relativity, as developed by Albert Einstein, who proposed this equivalence in 1905 in one of his *Annus Mirabilis* papers entitled "Does the inertia of an object depend upon its energy content?" Note that one has to be careful about what frames of reference one is using for the equation to hold. In relativity, removing energy is removing mass, and for an observer in the centre of mass frame, today's equation (re-arranged so that $m = E/c^2$) indicates how much mass is lost when energy is removed.

Whenever any type of energy is removed from a system, the mass associated with the energy is also removed, and the system, therefore, loses mass. This mass loss in the system may be simply calculated as $\Delta m = \Delta E/c^2$, and this was the form of the equation historically first presented by Einstein in 1905. However, use of this formula in such circumstances has led to the false idea that mass has been "converted" to energy. If you set light to something combustible and let the heat escape, the escaping heat makes the combustion products weigh less than if they stayed hot. Conversely, if you seal up a nuclear bomb, let it explode, and keep all the products and heat inside the box, the box has the same total mass before and after the explosion. It would have to be some box though!

Where mass is converted to energy may be the case when the energy (and mass) removed from the system is associated with the nuclear binding energy of the system. In such cases, the binding energy is observed as a "mass defect" or deficit in the new system. For example, the nuclear fission bomb that was tested in the Nevada desert during the Second World War had an explosive energy of around 9×10^{13} joules (21,000 tons of TNT). About 1 kg of the approximately 6.15 kg of plutonium in the bomb fissioned into lighter elements totalling almost exactly 1 g less, after cooling. The electromagnetic radiation and kinetic energy (thermal and blast energy) released in this explosion carried the missing 1g of mass.

$$E = mc^2$$

Further reading
E=mc²: A Biography of the World's Most Famous Equation (2001), David Bodanis

RL

Equation 75: Decibels

$$L = log_b \frac{x}{x_{ref}}$$

The neper (Np) and the bel (B) perform similar functions in serving as dimensionless units of signal level or level difference, each of these being an expression of the ratio between two field or power quantities via a logarithmic transformation.

Field quantities are amplitude quantities, such as voltage, current, pressure, electric field strength, velocity, or charge density. *Power* quantities refer to properties such as electrical power, acoustic/luminous intensity and energy density. Power quantities are proportional to the *square* of field quantities.

In the definitional equations $L_{Np} = ln\,(A/A_{ref})$ and $L_B = log\,(P/P_{ref})$, where A and P denote amplitude (field) and power quantities respectively, the L value corresponds to an *absolute level* when the A_{ref} or P_{ref} appearing is a standard reference values and to a *level difference* otherwise. (L_{Np} and L_B denote the neper and bel levels respectively.)

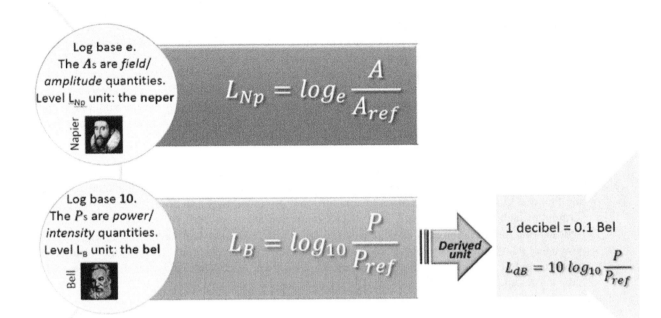

Log base e.
The As are *field/amplitude* quantities.
Level L_{Np} unit: the neper

Napier

$$L_{Np} = log_e \frac{A}{A_{ref}}$$

Log base 10.
The Ps are *power/intensity* quantities.
Level L_B unit: the bel

Bell

$$L_B = log_{10} \frac{P}{P_{ref}}$$

Derived unit

1 decibel = 0.1 Bel

$$L_{dB} = 10 \, log_{10} \frac{P}{P_{ref}}$$

The bel is actually an inconveniently large unit and is hardly ever used as such. Its derivative, the decibel (dB), is by far the most common unit used worldwide for power ratio measurements. Interestingly, the neper ($\approx$ 8.7 dB) is nearly as big as the bel, but its derivative, the decineper, isn't much encountered. The centineper, on the other hand, is closely equal to percentage difference for small differences, and so is useful for calculations involving such differences. Both the neper and the decibel are used in telecommunications and other scientific/engineering/technical disciplines - most prominently in acoustics, electronics, and control theory. In electronics, the gains of amplifiers, attenuation of signals, and signal-to-noise ratios are often expressed in decibels.

The neper refers to an amplitude (field) ratio, is calculated via the *natural* (base e) logarithmic function, and has mainly been used in continental Europe. The unit was so named in honour of John Napier, the mathematician who discovered natural logarithms. The bel, on the other hand, refers to a power ratio, is calculated via the decadic (base 10) logarithmic function, and has mainly been relevant in America and the UK. The bel was so named after Alexander Graham Bell, a scientist, inventor and engineer - and, like Napier a few centuries earlier, Scottish-born.

Neither the bel/decibel nor the neper is classed as an SI unit. However, both are recognised units of the International System of Quantities (ISQ) and are accepted for use alongside SI units.

The decibel symbol is often qualified with a suffix that indicates which reference quantity has been used. For example, dBm indicates a reference level of one milliwatt, dBV refers to a reference level of 1 volt, while dBu is referenced to approximately 0.775 volts RMS.

Although the neper is much less used than the decibel, it has the advantage of being more easily and cleanly dealt with in mathematical analyses, owing to its definitional basis in the *natural* logarithm.

FR

78

Equation 76: Kepler's Second Law of Planetary Motion (Law of Equal Areas)

Johannes Kepler (1571-1630), a strong advocate of the Copernican heliocentric view of the planets, was a German mathematician and astronomer, and an important figure in the scientific revolution which began with Copernicus and concluded with Newton. This article, together with two that follow, makes up a three-part series on Kepler's laws. Kepler's second law is dealt with initially as he derived this before the first. The first law is named thus because it is fundamental in the sense of being a basic statement about planetary orbits being elliptical. The second law can be formulated without this assumption, although in physical terms the variability of speed of an orbit follows in part from the ellipticity. Kepler's laws were derived from astronomical data, specifically the detailed observations of Mars that had been collected by Tycho Brahe. They were invaluable in predicting astronomical events such as transits and eclipses, and were tested as such in the years following Kepler's death. Kepler was also an astrologer so this was important for other reasons. In Kepler's time, the early days of modern astronomy, the subject was still linked strongly to astrology, which resided in the realm of the arts (despite the requirements for mathematics) and attempts to reconcile it with the new "natural philosophy" (physics) was generally not approved of. Kepler's role was critical to the merging of astronomy and physics. His laws turned out to be an invaluable test for Newton's laws of motion and universal gravitation, which were sufficiently accurate to predict most astronomical events (as observable at the time). For this reason, these articles are as much about Newton as they are about Kepler. The whole process was quite a long one: Brahe collected the data over a period of time, Kepler analysed it and finally, Newton explained it and used it to validate his own universal physical laws.

Kepler's second law is called the law of equal areas and states that a planet sweeps out equal areas of its orbit in equal times. One of Kepler's earlier works called Astronomia Nova was published in 1609 and is one of the most important astronomical texts of the whole period. His second law appeared in two forms in chapters 32 and 59, as a distance law (variable speed relating to distance from the sun) and the equal area law which followed. Although Kepler could divide up a planetary orbit into a number of parts and calculate the planet's position from each one, he could not work out the position for any moment in time due to the changing speed (angular acceleration). This was called the "Kepler problem" and was one of the main influences on the development of calculus in its modern form (BE34).

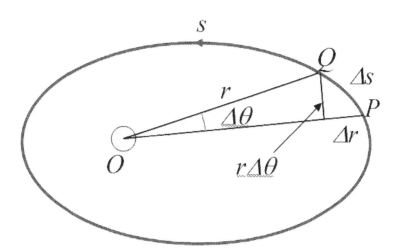

In the picture (in which the ellipticity of the orbit is greatly exaggerated), O is the centre of the sun which is situated at one of the foci of the ellipse. The approximate triangle OPQ is swept out along the orbital direction s as a small area of the ellipse in a certain amount of time by a small angle $\Delta\theta$. In the limit of small Δr, the velocity along Δs is:

$$v_s = \frac{rd\theta}{dt} = r\omega$$

where r is the radial distance and ω is the angular velocity. The rate of area A swept in unit time (from the area of a triangle being $\frac{1}{2} \times base \times height$) is

$$\frac{dA}{dt} = \frac{1}{2} r^2 \frac{d\theta}{dt} = \frac{1}{2} r^2 \omega$$

Kepler found this parameter to be constant (displayed as C in the main equation) from his analysis of the observations. To understand the important implications of this it can be shown that it follows from the conservation of angular momentum and the force of gravity. Angular momentum L (BE72) is defined as:

$$L = mr v_s = mr^2 \omega$$

where m is the mass of the orbiting body, and therefore the rate of area swept becomes:

$$\frac{dA}{dt} = \frac{L}{2m}$$

Extension of Newton's second law (BE1) for angular velocity, assuming negligible change in r in the limiting case:

$$F = ma = mr \frac{d\omega}{dt}$$

leads to the collective torque of all forces on the body as the rate of change of angular momentum:

$$F \times r = mr^2 \frac{d\omega}{dt}$$

Since gravity is the only force present, acting in a radial direction (in the classical approximation) on the orbiting body, the left hand side is zero and so the angular momentum stays constant (and therefore also the areas swept, from the relation above).

The next instalment in this series on Kepler will deal with the equation of an ellipse and the first law, regarding the shape of planetary orbits as eccentric ellipses and their relation to the inverse square law for gravitational attraction.

Because this is such a fundamental rule for rotating astronomical bodies, as a footnote, a bit more background is worth mentioning here. The heliocentric system (sun at the centre) of celestial bodies, including earth, proposed by Copernicus in the modern world of the 16th century, was not the first. It is thought that this was first proposed in a real sense by the ancient Greek astronomer Aristarchus of Samos in the 3rd century BC, but was afterwards replaced again by a geocentric one (probably due to its popularity) by Aristotle and Ptolemy, which held its ground, despite its inadequacy in explaining observations, until Copernicus. In fact, Copernicus referred to Aristarchus in an early statement but left that out in the eventual publication of his work De Revolutionibus Orbitum Coelestium in 1543. Aristarchus had also proposed that the stars were bodies like the sun but appeared small and faint because of the vast distances involved. There is also some evidence, although not clear, that he may have discovered axial precession as well, although this is more readily attributed to Hipparchus of Nicaea in the 2nd century BC. As far as elliptic orbits are concerned, it is likely that Kepler was indeed the first to formulate this but the eccentricity of the sun's location was known about since ancient times, as observed from the irregularity of the sun's motion along the ecliptic. The reason for this was not understood, but after Kepler and Newton's work it was realised that the elliptic orbit with the sun displaced from the centre is the reason why the earth moves faster when closer to the sun (perihelion) and slower when it is further (aphelion). Incidentally, the earth is currently closer to

80

the sun in winter (early January). This is the reason why the sun appears slightly brighter in winter for the same elevation in the sky.

PB

Equation 77: Gibbs Free Energy

Gibbs free energy is the energy given off when a chemical reaction takes place at constant temperature and pressure. In general terms, a chemical reaction will only proceed if the products are in a more stable state than the reactants, hence ΔG must be negative, as the system must have lost free energy. The equation is based on work done by J. Willard Gibbs in the late 19th century, and his "free energy" was eventually renamed in his honour.

$$\Delta G = \Delta H - T\Delta S$$

At standard or room temperature and pressure (STP or RTP) the requisite o or θ superscript is added to the variables to indicate this. The change in Gibbs free energy is given by ΔG, with ΔH representing the change in enthalpy (energy change at constant T and P), T the temperature in kelvin, and ΔS the change in entropy or disorder of the system under observation. In other words, the total energy available from a chemical reaction depends on the change in energy proper as well as the change in disorder.

The main consequence of Gibbs' work is that there are four distinct possibilities for the feasibility of a chemical reaction, given that ΔG must be negative for a reaction to proceed. ΔH can be positive or negative, resulting in an endo- or exothermic reaction respectively. ΔS can also be positive or negative, referring to an increase or decrease of the system's disorder respectively?

ΔH	ΔS	ΔG	Feasible?
-	+	Always negative	Yes
-	-	Negative at low T	Low T only
+	+	Negative at high T	High T only
+	-	Always positive	No

As a result, whether a chemical reaction happens is not only dependent on enthalpy, as is normally taught first, but also on entropy, and can be either enthalpy or entropy driven (2nd and 3rd combinations in table, respectively).

The formation of ammonia from hydrogen and nitrogen in the Haber process, for instance, is exothermic, but decreases in entropy as four moles of reactants only result in two moles of product.

As a result, it is only feasible at (relatively) low temperatures.

OK

Equation 78: Kramers-Kronig Relation

In general, the Kramers-Kronig relations give the imaginary part of the response of a linear passive system if the real part is known at all frequencies, and vice versa. The relations are based on the causality principle and fundamentally link two seemingly independent quantities.

As an example the reflectance from an interface is considered as the complex amplitude reflectance r:

$$r = \sqrt{Re^{-i\phi}}$$

with R the intensity reflectance and ϕ the phase shift.

The Kramers-Kronig relation between the phase shift $\phi(\lambda)$, with λ the wavelength, and the reflectance spectrum $R(\lambda)$ is given by:

$$\phi(\lambda) = -\frac{\lambda}{\pi} P \int_0^\infty \frac{\ln(R(\lambda'))}{\lambda^2 - \lambda'^2} d\lambda' = -\frac{\lambda}{\pi} \int_0^\infty \frac{\ln(R(\lambda')) - \ln(R(\lambda))}{\lambda^2 - \lambda'^2} d\lambda'$$

Here P stands for the Cauchy principal value, which is removed by subtracting the constant $ln(R(\lambda))$ in the right-hand part of the equation.

The equation shows that in principle the reflectance spectrum from zero to infinite wavelength is needed, however the integrand vanishes rapidly for $\lambda' \gg \lambda$ and $\lambda' \ll \lambda$.

The beauty of the equation is that from a single measured spectrum of a quantity $R(\lambda)$, both the real and complex part of the response function can be derived: two quantities for the price of one.

For example: for the perpendicular optical reflectance of a semi-infinite medium, the reflectance follows directly from the complex refractive index $\boldsymbol{n} = n - ik$, using the Fresnel equations (see future equation 166), and this gives for the complex refractive index:

$$n(\lambda) - ik(\lambda) = \frac{1 - R(\lambda) - 2i\sqrt{R(\lambda)}\sin(\phi(\lambda))}{1 + R(\lambda) - 2i\sqrt{R(\lambda)}\cos(\phi(\lambda))}$$

This enables the calculation of both $n(\lambda)$ and $k(\lambda)$ from the reflectance spectrum of a non-transmitting material e.g. glass in the infrared region, or a thick non-transmitting coating such as gold or aluminum.

Applications are in various kinds of spectroscopy. In ultrasonics, Kramers-Kronig relations are used to link attenuation and dispersion.

The equations were derived independently by the Dutch physicist Hendrik Kramers (1894 – 1952) and the German physicist Ralph Kronig (1904-1995), in 1927 and 1926 respectively.

HH

Equation 79: Beer-Lambert Law

The Beer-Lambert Law is an equation that models the attenuation of light in a medium. Despite the name given to the law, the first known mention of it was by French scientist Pierre Bouguer in the early 18th century. The names Beer and Lambert were given to the law due to subsequent work performed by Johann Heinrich Lambert (in 1760) and August Beer in 1852). The equation is very useful in the field of X-ray computed tomography as it can be used to determine the attenuation of X-rays as they propagate through an object.

The simplest form of the Beer-Lambert Law is one in which the X-rays are mono-chromatic, i.e. the wavelength (and energy) of the X-rays is limited to a narrow band in the electromagnetic spectrum, and the object is composed of a single material attenuation. The simplest form of the Beer-Lambert Law is given by:

$$I = I_o e^{-\mu s}$$

where I_o is the initial intensity of the X-rays and I is the intensity of the X-rays after travelling through a material with attenuation coefficient μ and thickness s.

The attenuation coefficient describes how absorption and scattering will attenuate the intensity of the X-rays propagating through the object. The value of the attenuation coefficient depends on the material atomic number, its density, and the energy of the incident X-rays. In the case of a non-homogeneous work piece, that is to say, consisting of multiple materials, equation (1) is modified to account for varying attenuation coefficients:

$$I = I_o e^{-\int_0^s \mu(x)dx}$$

where $\mu(x)$ is the attenuation coefficient that varies with penetration depth x. The exponent is integrated from $x = 0$ to $x = s$ to account for the entire X-ray path through the work piece.

MF

Equation 80: Gauss-Legendre Quadrature

In many computational problems, approximations to integrals of the form $I = \int_{-1}^{1} f(x)dx$ are needed. The numerical approximations need to be calculated efficiently (by minimizing the function evaluations) and accurately. The Trapezoidal rule (Equation 62) requires two function evaluations while being able to integrate only a first-order polynomial exactly. In a method such as the finite element method, integrals of the form $I = \int_{-1}^{1} f(x)dx$ are often evaluated repeatedly hundreds of thousands of times if not more. The functions tend to be very complex and rarely first-order polynomials. Repetitive evaluation of these functions is computationally expensive. Consequently, with some exceptions, the Trapezoidal rule is not the best method for use in these applications.

A numerical technique that keeps function evaluations to as minimum a number as possible while achieving high accuracy is ideal. Gauss-Legendre quadrature is one such method that uses n points (and hence n function evaluations) to accurately integrate a polynomial of the order $2n - 1$. The approximation is written as:

$$I = \int_{-1}^{1} f(x)dx \approx \sum_{i=1}^{n} w_i f(x_i)$$

A clever choice for the weights w_i and the integration points x_i leads to the high accuracy mentioned above. The method, first introduced by Carl Friedrich Gauss in 1814, approximates the given function with an $(n-1)^{th}$-order Lagrange polynomial interpolating the function values at the integration points. The weights and the integration points are listed in the table below for $n = 1, n = 2$ and $n = 3$. In addition, the highest-order p of the polynomial that can be integrated exactly by these quadrature rules is also listed.

	n	Weights (w_i)	Integration Points (x_i)	p
One-Point Rule	1	$w_1 = 2$	$x_1 = 0$	1
Two-Point Rule	2	$w_1 = w_2 = 1$	$x_1 = -\dfrac{1}{\sqrt{3}}, x_2 = \dfrac{1}{\sqrt{3}}$	3
Three-Point Rule	3	$w_1 = w_3 = \dfrac{5}{9}, w_2 = \dfrac{8}{9}$	$x_1 = -\dfrac{3}{\sqrt{5}}, x_2 = 0, x_3 = \dfrac{3}{\sqrt{5}}$	5

Table 1: Gauss points and the corresponding weights for the first three Gauss-Legendre quadrature rules.

The beauty of the method lies in the fact that with just a few function evaluations, it can give very accurate results. It uses a Lagrange polynomial of order $n - 1$ to integrate a polynomial of order $2n - 1$ exactly! Thus, the one-point rule requires one function evaluation and can integrate a first-order polynomial exactly. The two-point rule requires two function evaluations and can integrate a \emph{cubic} polynomial exactly. Clearly, these two rules are superior (in general) to the Trapezoidal rule which requires two function evaluations to integrate a first-order polynomial exactly. A geometric interpretation of the one-point, two-point and three-point Gauss-Legendre quadrature rules is shown in Figure 1.

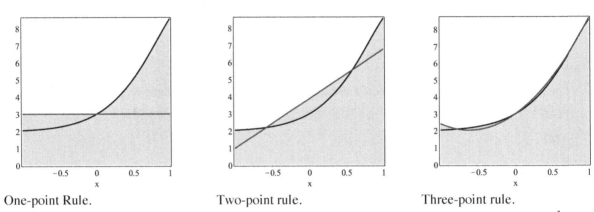

One-point Rule.   Two-point rule.   Three-point rule.

Figure 1: A graphical illustration of the Gauss-Legendre quadrature rule for integrals of the type $I = \int_{-1}^{1} f(x)dx$. The area under the actual curve (black line) is approximated by the areas under the blue curves in Gauss-Legendre Quadrature. The figures are specific to the function: $f(x) = 2 + e^{2.5x}\cos x$

If the integral is of the form $I = \int_{a}^{b} f(x)dx$ where a and b may not be equal to -1 and 1 respectively, a change of variables can be performed as follows:

$$I = \int_{a}^{b} f(x)dx = \frac{b-a}{2}\int_{-1}^{1} f\left(\frac{b-a}{2}\xi + \frac{b+a}{2}\right)d\xi = \int_{-1}^{1} g(\xi)d(\xi)$$

EXAMPLES

Two examples are considered to illustrate the performance of Gauss-Legendre quadrature. These examples are the same as the ones considered in Equation 62. In the first example, we wish to integrate $\int_{0}^{1}(2 + 5x + 6x^3 - 7x^4)dx$. The exact value of this integral is 4.6. The approximate value of the integral is shown in the first table below for the first three Gauss-Legendre rules. The second example considers the integral of a periodic function. Again, the approximate values of the integral for the first three Gauss-Legendre rules are shown in the second table below. Upon comparison with the examples from Beautiful Equation 62, we note that for the first integral, Gauss-Legendre quadrature is significantly more efficient. However, for example 2 where the integrand is a periodic function, Trapezoidal rule is much more efficient with a very fast convergence rate. However, for most practical situations, Gauss-Legendre quadrature is a fast converging method and hence is more commonly used.

Rule	
One-point ($n = 1$)	4.81250000000000
Two-point ($n = 2$)	4.63888888888889
Three-point ($n = 3$)	4.60000000000000
Exact	4.60

Rule	
One-point ($n = 1$)	1.0
Two-point ($n = 2$)	0.812922516006679
Three-point ($n = 3$)	0.955950201948961
Exact	0.902779927772194...

$$I = \int_0^1 (2 + 5x + 6x^3 - 7x^4)\,dx$$

$$I = \frac{1}{2\pi} \int_0^{2\pi} \sqrt{1 - 0.36 \sin^2 \theta}\; d\theta$$

HC

Equation 81: The Frequency of a Stretched String

$$f = \frac{1}{2l} \sqrt{\frac{T}{\mu}}$$

About 2,500 years ago, Pythagoras conducted experiments which showed the relation between the length of a struck or plucked string and the pitch of the note it produced.

Although anyone who makes or plays a stringed instrument knows that tension, as well as length, affects the note a string makes (otherwise, turning pegs to tune stringed instruments wouldn't work), Pythagoras seems not to have attempted to quantify this - and nor did anyone else, for over 1,500 years. Then, in about 1580, Vincenzio Galileo showed that pitch increases with the square root of the tension T. This was the first non-linear physical relationship ever found in nature.

In making this discovery, Vincenzio was guided primarily by experiment, and thus departed from the usual philosophical approach followed by what passed for physicists in his day (thanks to their slavish adoption of what they believed to be the approach of the ancient Greeks). This new way of finding out about the world was applied vigorously, and with great success, by Vincenzio's son Galileo himself to many aspects of physics, including musical acoustics.

We now know that the frequency f also depends on the string's linear density μ (that is, its mass per unit length).

MJG

Equation 82: Biot Number

$$Bi = \frac{hL}{k_{solid}}$$

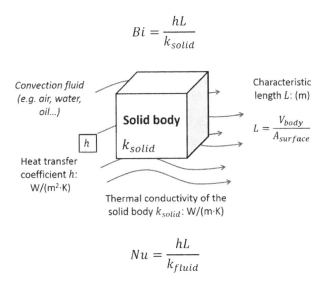

$$Nu = \frac{hL}{k_{fluid}}$$

The Biot number is named after the French physicist Jean B. Biot (1774 – 1862) who is best known for the Biot-Savart law in electromagnetism. Biot once worked on the heat conduction problem around 1802, which was earlier than Fourier's famous 1822 treatise on this subject.

The Biot number is a dimensionless quantity widely used in heat transfer problems. For a solid body, the Biot number is calculated as the ratio of the product of the heat transfer coefficient and the solid body's characteristic length (i.e. volume divided by surface area) to the thermal conductivity of the solid body. The Biot number characterizes a solid body's internal resistance to heat transfer (e.g. conduction) with respect to that body's external surface resistance to heat transfer (e.g. convection).

For example when heat treating metals it is common to heat them up to a high temperature at which the compositional structure will change and then drop the parts into a fluid to rapidly cool them down (called quenching) and thereby 'freeze' in the properties of this different composition. When quenching a workpiece, a small Biot number ($\ll 1$) means that the surface resistance of that workpiece (made from a particular material) in the fluid is high so that the fluid cannot rapidly cool the workpiece. The temperature decreases uniformly within the body as the quenching proceeds, because the internal thermal diffusivity of that workpiece is considerably faster to balance inside the body even though heat is being continuously taken away from the surface. A component having a small Biot number is sometimes referred to as "thermally thick".

Contrarily, a solid body having a high Biot number ($\gg 1$) means that, during a quenching process, the surface heat loss is much faster than the solid body's internal thermal conduction. This leads to a significant temperature gradient from the surface of the body to the centre of the body such as a slab made of aluminium being quenched.

Another dimensionless number, the Nusselt number (Named after the German engineer Wilhelm Nusselt, 1882 – 1957) is occasionally confused with the Biot number. In the equation of the Nusselt number, the thermal conductivity of the fluid is placed in the denominator. The Nusselt number characterizes the heat flux from the surface of a solid body to its surrounding fluid. So it provides insight into the properties of a fluid in a heat transfer scenario.

References:
See website on famous thermodynamics researchers in history at:
http://www.seas.ucla.edu/jht/pioneers/pioneers.html accessed on 2nd March 2015.

KN

Equation 83: Shannon's Juggling Theorem

$$(F + D)H = (V + D)N$$

where F is the time the ball spends in the air (flight), D is the time a ball spends in a hand (dwell), V is the time a hand spends empaty (vacant), N is the number of balls and H in the number of hands.

Shannon's juggling theorem relates the timing quantities involved in a juggling pattern. The equation can be derived by equating the time a ball need to complete a full cycle of a juggling pattern with the time a hand need to wait to see a particular ball again. From the equation can be seen that if F increases (e.g. throwing a ball higher) either N or V has to increase (you can juggle more balls or your hands are empty for longer). On the other hand (pun intended), if the number of ball (N) increases, either the time a ball spends in the air (F) or the number of hands juggling (H) has to increase (you have to make higher throws or need more jugglers).

Juggling is an ancient tradition; the earliest known depiction is in an Egyptian tomb, dating around 1994-

1781 B.C. However, the first mathematics applied to juggling is attributed to Claude Shannon and his effort to build a juggling robot (see https://www.youtube.com/watch?v=sBHGzRxfeJY).

In the last 30 years, the mathematics theory applied to juggling has expanded (see for example Mathematics of juggling: https://www.youtube.com/watch?v=38rf9FLhl-8), starting from the mathematical formalisation of rules to express any possible juggling pattern, aka siteswap.

Latest developments include the extension from mathematical notation of juggling balls to juggling pattern made of balls and anti-balls:
(see https://www.youtube.com/watch?v=_5Q4UhyajC4&feature=youtu.be).

References:
Polster, B. and Behrends, E., 2006. *The mathematics of juggling*. The Mathematical Intelligencer, 28:99-89

GM

Equation 84: Escape Velocity

5...4...3...2...1...we have all seen TV pictures of space flight launches. But why can't you just fly me to the moon?

To escape the earth using a rocket we have to reach the escape velocity. Escape velocity is given by the following equation:

$$ve = \sqrt{\frac{2GM}{r}}$$

where ve is the escape velocity, G is the gravitational constant, M is the mass of the planet (Earth in our case) and r is the distance from the centre of gravity.

The equation does not take in to account atmospheric drag.

This equation is true for ballistic trajectories. Escape velocity is a misnomer as it is actually an escape speed (scalar quantity). What do we mean by ballistic trajectories? A ballistic trajectory is the path of a thrown or launched projectile. A rocket moving out of a gravity well does not have to attain escape velocity to do so. It could achieve the same result at any speed with a suitable mode of propulsion and sufficient fuel. An example would be the proposed space elevator. So in theory, you could fly me to the moon.

On the surface of the Earth, the escape velocity is about 11.2 kilometres per second. Some of this velocity can be gained by using the earth's rotation. That's why it's best to launch from a position close to the equator. It is important to note that the escape velocity is the same for any mass, what would be different for a heavier mass would be the energy required to achieve that speed.

I'll leave it to the reader to calculate the escape speed of other planetary bodies. The moon, Pluto and the Sun are interesting.

At the event horizon of a black hole ve is greater than the speed of light, so light cannot escape (see beautiful equation 4). In the solar system there are many gravitating bodies, a rocket that travels at escape velocity from one body, say Earth, will not travel an infinite distance because it needs an even higher speed to escape the Sun's gravity.

DF

Equation 85: Frequency Response Functions and Cramer's Rule

$$H_{rs}(j\omega) = \frac{1}{\nabla}\frac{\partial \nabla}{\partial e_{rs}}$$

Imagine some arbitrarily complex mechanical mechanism comprising, for example, lots of rigid masses attached to springs in arbitrary orientations like a crazy lumpy mattress with all the cloth material removed. 'Lightly' tapping some of these masses with a hammer would cause the whole structure to wobble incessantly. Surprisingly, this motion can be very accurately modelled using the Euler-Lagrange equation (beautiful equation 6). Further, by immersing this structure in oil or introducing other mechanisms to dissipate energy out of the system (such as electrical eddy current dampers) it becomes necessary to use an expanded form of Lagrange's equation to include a dissipation function as well as accommodating forces being applied to the masses from external sources (i.e. the hammer being wielded by yourself).

Operative in the above sentence is the term 'lightly' tapping. This will cause the system to wobble only small distances from the previous equilibrium state. Under these, or equivalent circumstances, the equations governing the motion of this and any other physical system turn out to be a set of simultaneous linear differential equations. For holonomic systems (see beautiful equation 26), given that there are n independent motions of the system and that there can be a corresponding number of forces that can be applied to drive them, there will be n linear equations. If, instead of tapping the masses, we apply forces varying harmonically with time then this linear system will produce outputs that, after some transient effects have decayed, contain these frequencies (as would be expected from this linear system, see beautiful equation 71). However, as would be expected for the conceptual system imagined above, shaking of one mass will result in many, if not all, of the others to also wobble. Imagining one of the masses (we'll label it r for which this could be any number between 1 and n) being subject to a force Q_r (in the direction of one of the independent motions q_r) and then measuring the motion in another direction q_s. Because this is a linear system, we would expect a motion that will depend on the frequency ω of the excitation force and given by the linear equation $q_s = H_{rs}(j\omega)Q_r$, where j is the imaginary number given by $\sqrt{-1}$. Equations rarely get this simple. All that remains to create a complete model for the steady state frequency response of a physical system of arbitrary complexity subject to small perturbations about its equilibrium state is to compute the constant $H_{rs}(j\omega)$ that can be computed using Cramer's rule (after Swiss Mathematician Gabriel Cramer (1704-1752)).

$$H_{rs}(j\omega) = \frac{1}{\nabla}\frac{\partial \nabla}{\partial e_{rs}}$$

$$e_{rs} = e_{sr} = -a_{rs}\omega^2 + b_{rs}j\omega + c_{rs}$$

$$H_{rs}(j\omega) = H_{sr}(j\omega)$$

$$\nabla = \begin{vmatrix} e_{11} & e_{12} & \cdots & e_{1n} \\ e_{21} & e_{22} & \cdots & e_{2n} \\ \cdots & \cdots & \cdots & \cdots \\ e_{n1} & e_{n2} & \cdots & e_{nn} \end{vmatrix}$$

where ∇ is the determinant of the characteristic equation of the system (not to be confused with a Laplace operator), and the terms a, b, and c represent coefficients of mass, damping (viscous), and stiffness arising from kinetic energy, dissipation forces and potential energy within the system.

It is interesting that all of these matrices are symmetric about the diagonal. This symmetry leads to an interesting phenomena that if a force is applied at some point r resulting in a displacement in another point

s then the same force applied at s will result in the same displacement at r where the original force was applied. This is called Maxwell's (after James Clerk Maxwell) reciprocity theorem and applies to static and dynamically varying forces.

Another remarkable thing with this equation is that all of the responses share a common denominator given by the determinant ∇. Putting in numbers for this determinant it is found that for specific frequencies it can have a small value (even going to zero for systems not having any energy dissipation). At frequencies where the denominator become very small (or zero) the frequency response corresponding to output displacement go very high (or toward infinity!). These frequencies are the resonances (or sympathetic vibrations) and show up as 'poles' in a frequency response. When a system is excited at these frequencies, the output displacements can steadily build up to large values, often resulting in high stresses leading to permanent damage or fatigue failure (both often representing catastrophic failures) and are of particular interest to engineering designers. In contrast, the numerator in the responses will be different and depend on where the forces are applied and where the response is measured. For some numerators there might be frequencies where the value is very low and these are called 'zeros'. At these frequencies application of a dynamically varying force will not be able to move the object no matter how large (others might be wobbling like crazy). Sometimes, surprisingly, this mass might be the one to which the force is being applied and will look as if it is a rigidly fixed in space (Engineers call this an absorber). These and many other interesting phenomena can be extracted from the apparently simple equation.

An example of the amplitudes of response for two masses (i.e. $n = 2$ in this case) connected to one another through a spring while one of them is rigidly connected to a rigid frame through a second spring is shown in the figure. The first figure corresponds to a system without damping for which the poles are infinitely tall and zeros are 0. For real systems have some energy losses the responses are shown in the second graph.

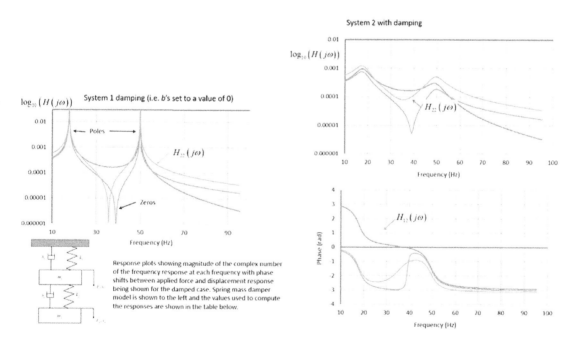

Response plots showing magnitude of the complex number of the frequency response at each frequency with phase shifts between applied force and displacement response being shown for the damped case. Spring mass damper model is shown to the left and the values used to compute the responses are shown in the table below.

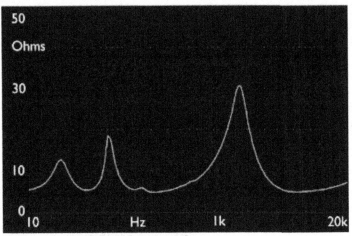

Q Acoustics 2050 impedance graph.

From: http://www.hi-fiworld.co.uk/index.php/loudspeakers/69/93-impedance.html

Table 2: Parameters values used to compute frequency response plots

m_1(kg)	m_2(kg)	b_1(Nsm^{-1})	b_2(Nsm^{-1})	k_1(Nm^{-1})	k_2(Nm^{-1})
0.2	0.1	12	0.8	4000	6000

a_{11}	a_{12}	a_{22}	b_{11}	b_{12}	b_{22}	c_{11}	c_{12}	c_{22}
0.2	0	0.1	12.8	-0.8	0.8	10000	-6000	6000

References: Strutt, J.W., 1873. *Some general theorems relating to vibrations.* Proceedings of the London Mathematical Society 1:357-368; Strutt, J.W., 1990. *On the law of reciprocity in diffuse reflexion.* Philosophical Magazine 49:324-325; Routh, E.J., 1905. Chapter Vii, in *A treatise on the dynamics of a system of rigid bodies.* London: McMillan

SS

Equation 86: Turning Surface Finish: Theoretical Peak to Valley Surface Generated

In diamond turning (and general turning for that matter) a common equation is used to determine what the theoretical peak to valley surface to be expected is. It is very important to know what surface you can expect to generate given particular cutting parameters. This determines what kind of surface finish you can expect to get in the cutting process.

The equation its self is very simple and allows for quick approximation and calculations to be performed.

$$PV = \frac{f^2}{8r}$$

where PV is the peak to valley height, or Rt (the theoretical surface height), f is the feed rate, pitch or increment; and r is the tool nose radius.

The term theoretical is used with this equation because the actual generated surface is always more than the calculated. This is because the generated surface encompasses all machine asynchronous error motions and axis noise as well as any cutting dynamics and tool profile variations.

There are many applications which require a certain level of surface finish to be achieved so calculating what your cutting parameters need to be to achieve this is very important. Furthermore, in general manufacturing processes, these parameters correlate to productivity and ultimately cost, so calculated trade-offs can be made to determine the most efficient cutting process to use.

A worked example:

For example, a tool radius of 6 mm with a feed rate of 0.1 mm per revolution would give the following theoretical *PV* peak to valley surface finish.

$$PV = \frac{f^2}{8r}$$

$$PV = \frac{0.1^2}{8 \times 6}$$

$$PV = \frac{0.01}{48}$$

$$PV = 0.000208$$

In theory, a 200 nm surface height.

TS

Equation 87: Bragg's Law, finding Atoms in Crystals

Bragg's equation is a fundamental law of X-ray and neutron diffraction of crystals. It establishes a simple relationship between the radiation wavelength (λ), the incidence angle of the radiation on the crystal (θ) and the distance between two crystalline planes of the crystal (d) as in the figure:

$$n\lambda = 2d \sin \theta$$

In the equation, n is an integer number and indicates the value at which the diffracted waves from two atoms on two different crystalline planes interfere constructively. The diffracted waves will form diffraction patterns on a detector, either producing stronger peaks when constructively interfering or no peaks when subtracting from each other. The condition for constructive interference is that the waves' phases differ by a multiple of λ.

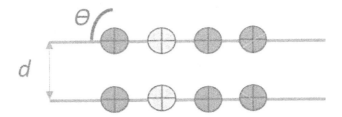

By varying the angle of diffraction – rotating of the crystal under a monochromatic radiation – it is possible to collect information about the distances between the different crystalline planes, determining the positions of the atoms in the crystal.

If there were only two diffracting planes of atoms, then the transition from constructive to destructive interference would be gradual as a function of the angle θ, with broad peaks. However, as crystals have an infinite number of atoms, many planes interfere with each other, producing very sharp peaks surrounded by destructive interference (see in figure).

In highly symmetric crystals – e.g. those with cubic symmetry, such as salt (NaCl) and pyrite (FeS_2) – diffraction patterns have few peaks, because of few constructively interfering atomic planes. On the contrary, low-symmetry crystals – e.g. monoclinic structures – show a large number of peaks.

X-rays diffraction in crystals is possible owing to their short wavelength, of the same order as the interatomic distances in the crystal. Atoms in the crystal scatter the radiation owing to their electron density through a mechanism known as Rayleigh scattering. The lighter the atom, the less strong is the scattering power of the atom, due to a smaller electron density around the nucleus. This is the reason why it is difficult to locate positions of hydrogen atoms in a crystal structure and in proteins, as they are the lightest elements in nature. Neutron diffraction is used for this purpose, together with nuclear magnetic resonance.

Sir William Lawrence Bragg and his father, Sir William Henry Bragg, discovered this law in 1912, 17 years after Rontgen's discovery of X-rays. The law confirmed the existence of real particles at the atomic scale and set the foundation of the diffraction analysis in crystallography. This law paved the way for the discovery of thousands crystal structures of solids in nature, including inorganic materials, metals, polymers, proteins and DNA.

Bragg father and son were awarded the Nobel Prize in physics in 1915, for the discovery of the crystal structure of salt, zinc sulphide and diamond. William Lawrence Bragg was 25 years old at the time, being the youngest Nobel laureate until 2014, when Malala Yousafzai was awarded with the Nobel Peace Prize at the age of 17.

Lawrence Bragg, who was born in Australia, was also the director of the Cavendish Laboratory in Cambridge, supervising Francis Crick's and James D. Watson's studies DNA's structure. Bragg first announced the discovery of DNA's double-helix structure at a Solvay conference on proteins in Belgium on 8 April 1953, but the announcement went unreported by the press. Bragg nominated Watson, Crick and Maurice Wilkins from the King's College for the Nobel Prize in physiology and medicine in 1962, acknowledging Rosalind Franklin's work at King's College London in helping solve the double-helix structure by X-ray diffraction (See picture of Franklin's photograph 51).

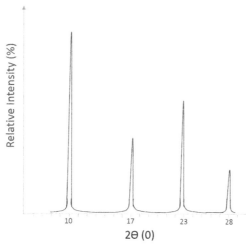
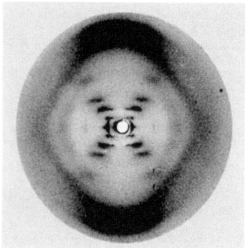

AT

Equation 88: Time-Frequency Analysis and the Spectrogram – Part 1

The Fourier transform is a method of transforming a time domain signal to the frequency domain. If we consider the example below, the time domain signal is a speech signal of the word "Gabor". The Fourier transform is given along with the time domain signal.

While the Fourier transform gives the frequency content of a signal, it does not give any indication of *when* a particular frequency occurred. The short time Fourier transform (STFT) is capable of analysing a signal in both time and frequency domain. Developed by Dennis Gabor (1900 – 1979), a Hungarian/British Electrical Engineer and Physicist, Gabor realised that signals could be presented in two dimensions, with time and frequency acting as coordinates. In other words, they map a one dimensional time domain signal into a two dimensional time and frequency signal which represents the variation of spectral energy over time. This area of signal processing collectively became known as time-frequency analysis with the output of the analysis known as time-frequency representations (TFR). The techniques can be applied to any non-stationary signal, *i.e.* one that changes as a function of time, so there is an endless list of systems that could be analysed in this way.

The STFT contains a windowing function that, when applied to a signal, breaks the signal into segments, where a Fourier transform is performed. This is mathematically defined as

$$STFT(x; \omega, t) = \int_{-\infty}^{\infty} x(\tau)\, h(\tau - t)e^{-i2\pi\omega\tau}\, d\tau$$

where ω is the frequency, t is the time, $x(\tau)$ is the original time domain signal and $h(\tau - t)$ is the window function.

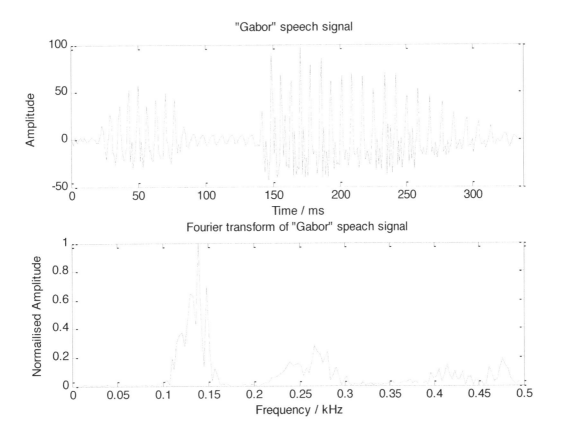

93

The window, $h(\tau - t)$ suppresses the signal around the analysis time point $\tau = t$ and the STFT gives a local spectrum of the signal $x(\tau)$ around t. The output of the STFT is the spectrogram and is the energy density spectrum of the STFT. The window used for the STFT is a Hanning window. The Gabor transform uses a Gausian window.

The resolution in time and frequency of the spectrogram is dictated completely by the window size and type used. A narrow window will give good time resolution and poor frequency resolution while a wide window gives good frequency resolution and poor time resolution. Using the time domain example above, we can see spectral energy at frequencies that match the Fourier transform, but now we can also see when these frequencies occur. We can also see the multimode nature of this particular signal.

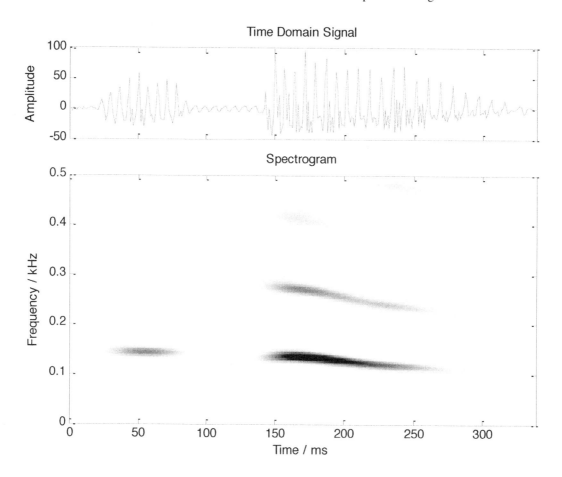

The STFT does however have one major drawback. The output is of limited resolution. This problem will be addressed in beautiful equation 119

References:
Gabor, D., 1946. *Theory of communication. Part 1: The analysis of information.* Journal of the Institution of Electrical Engineers-Part III: Radio and Communication Engineering, 93:429-441
Auger, F., Flandrin, P., Goncalves P. and Lemoine, O., 1996. *Time-Frequency Toolbox: For use with MATLAB*, CNRS, France-Rice University
Hurlebaus, S., Niethammer, M., Jacobs, L. J. and Valle, C., 2001. *Automated methodology to locate notches with Lamb waves.* Acoustics Research Letters Online 2:97-102

CMcK

Equation 89: Group Velocity

The speed of light depends on the index of refraction of the medium in which it is traveling (see beautiful equation 51). For a single wavelength of light λ, we simply divide the speed of light in a vacuum by the index of refraction n. However, the reality is that even nominally "monochromatic" light sources emit over a range or spectrum of wavelengths. The result is a packet of light waves traveling together, as illustrated below:

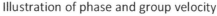

Illustration of phase and group velocity

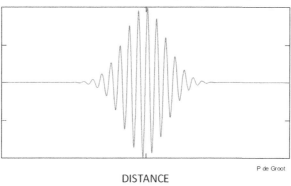

P de Groot

DISTANCE

There are therefore two light speeds of interest—a *phase* velocity v_p, corresponding to the speed of propagation of the wavefront (the sinusoid in the figure), and the *group* velocity v_g, corresponding to the speed of propagation of the wave packet constructed from the sum of all the spectral contributions (the peak of the envelope in the figure). The relationship between the two velocities is given by the following beautiful equation:

$$v_g = v_p - \lambda \frac{dv_p}{d\lambda}$$

We can also define a group index n_g, which for most transparent, linearly-dispersive materials, is larger than the phase index n_p. This means that the packet of propagating waves travels more slowly than the underlying wavefront. The concept of a group velocity dates back at least to Lord Rayleigh and the publication of his "Theory of Sound" in 1877. We also see that the group velocity is the first-order Taylor expansion of a more general description of light propagation in materials.

The idea of a group velocity is important to me because it is relevant to many types of non-contact metrology. Some interferometric methods are sensitive to the light phase, while others are sensitive to the position of the group of waves generated by a spectral distribution (see beautiful equation 30: Interference, 01/29/2015). Examples of the former include phase shifting interferometry, while examples of the latter include swept-wavelength interferometry, coherence scanning interferometry, and multiple wavelength methods. There are instruments that combine these two sensitivities into one metrology principle, leading to interesting problems in accommodating the difference between the group and phase velocities.　　　　PdG

Equation 90: The Second Law of Thermodynamics

$$dS \geq 0$$

In the equation, dS is the change in entropy of a system. The Second Law of Thermodynamics (see equation 47 for the First Law) basically says that "the entropy of a closed system tends to increase". Let's dissect that statement. First of all, by "closed system", we mean a system which does not allow certain types of transfers (such as transfer of mass or heat or other sources of energy) in or out of the system (the easiest closed system to think of, and one that is often used to illuminate physical ideas, is an ideal box of gas). Secondly, "entropy" is a measure of the number of specific ways in which a thermodynamic system may be arranged, commonly understood as a measure of disorder. Lastly, the clause "tends to increase" tells us that the disorder of our closed system will, in all probability, increase. Note that the laws of physics do not preclude a (local) decrease in disorder, but over a suitable long time period, we will inevitably see entropy increase.

Take our ideal box of gas. Let's say all the gas atoms are localised in one corner of our box (it's rectangular) at the start of our time sequence. This is a relatively ordered state. As time increases, the atoms will tend to disperse through the volume of the box, therefore, increasing their disorder (and the entropy of the system). This is the reason we see eggs roll off a table and smash, but we rarely see the opposite, even if the laws of physics allow the opposite case – it is just highly unlikely.

The basis of the Second Law is still a hotly debated philosophical discussion. Why is it so? What about quantum mechanics? Does it imply an "arrow of time"? What is the entropy of the whole universe and how is it different now to what it was at the time of the Big Bang?

Rudolf Clausius was the first to formulate the second law during 1850, in this form: "heat does not flow spontaneously from cold to hot bodies". While common knowledge now, this was contrary to the caloric theory of heat popular at the time, which considered heat as a fluid. From there he was able to infer definition of entropy (1865) (see Wikipedia entry). A little more history of the laws of thermodynamics in general was given in beautiful equation 47.

References:
Atkins, P.W., 1994. *The Second Law: Energy, Chaos and Form*. Lincoln: Potomac

RL

Equation 91: The Kindergarten Constraint

$$\theta + \xi = \gamma$$

For $\theta \in \{1\}, \xi = \theta$ and $\gamma = \sqrt{4}$.

The popularity of mathematics in modern society threatens the progress of this most noble of disciplines. Over-subscribed mathematics courses means that we are now producing 230 % more mathematicians per year than the world can employ, yet we only produce 20 % of the necessary researchers to fact-check unsubstantiated statistical claims. The Kindergarten Constraint is part of a proposed solution to the accessibility of mathematics.

The Kindergarten Constraint brings together several, key techniques. The use of the Greek alphabet makes equations nearly 20 % more likely to be actually correct and makes the probability that a peer reviewer will evaluate or even question the information presented reduce exponentially. By combining "arbitrary set theory" (a branch of mathematics concerned with intentional obfuscation) and an unexplained, yet inspired, refusal to evaluate arbitrary constants within the equation, the Kindergarten Constraint is able to achieve "virtual invisibility". Items with the property of virtual invisibility can still be observed by those outside of the academic niche from which they originate, but the field of view of the observer slips straight over the text without gleaning any information content. Equations that achieve virtual invisibility simultaneously have an information entropy value of 0 and approximately 0.86.

An important property of the Kindergarten Constraint is that it also holds for "true imaginary numbers". Imaginary numbers are those with a no "real" component and are marked with either a letter "i" (everyone else in the world) or a letter "j" (engineers and the morally bankrupt). True imaginary numbers, like the Obscuroff Constant, (ψ = eleventeen) or the Fake ID Ratio, (δ = eighteenty-four), can be comfortably substituted into the Kindergarten Constraint with little impact to its truth cost.

The Kindergarten Constraint, and related equations, are remarkably common in the literature as they make possible the formalisation (which of course, is only ever a good thing) of the common-sense and wholly uninteresting.

The interested reader is reminded that the equations in this text were originally published as part of a daily blog running from the 1st of January, and that the 91st day of the year is April 1st.

RFO

Equation 92: Kepler's First Law of Planetary Motion (Law of Ellipses)

Johannes Kepler presented both the first and second laws in Astronomia Nova in 1609 after several years of work investigating the shape of planetary orbits, from circular through ovoid to elliptical. He initially dismissed the ellipse as a possibility as he thought it was too simple a solution for previous astronomers to have overlooked. The elliptical orbit turned out to be the best fit to the controversial heliocentric system that had been proposed by Copernicus. In the absence of calculus, which would have helped considerably, Kepler split the Mars orbit into 360 parts in his calculations, and worked with ratios of distances. He had to take into account the estimated earth orbit, using Mars and the sun as a reference frame. In 1609 his first law, the law of orbits or ellipses, was stated for Mars only, as it was the astronomical data from Mars observations that were used to derive it. The first law in its most complete form (extension to all planetary orbits) was eventually published in 1622 in book 5 of another important work of Kepler's, the Epitome Astronomiae Copernicanae.

The equation of an ellipse in Cartesian co-ordinates x, y is:

$$\frac{x^2}{a^2} + \frac{y^2}{b^2} = 1$$

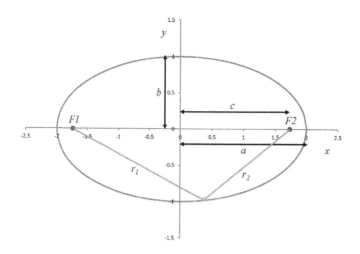

The parameters a and b, the "semimajor" and "semiminor" axes of the ellipse, are shown in the picture above which has been plotted for the arbitrary values $a = 2$ along the x-axis and $b = 1$ along y. Points on the ellipse can be defined by considering circles of radius b and a inside and outside the ellipse respectively and setting $x = a \cos \theta$ and $y = b \sin \theta$, where θ is measured anticlockwise from the positive x-axis. Squaring the terms for x and y and applying Pythagoras results in the equation above. The resulting angles that lines between the points on the ellipse and the centre make with the x-axis are less than θ (except on the axes for $\theta = 0$ or $\frac{n\pi}{2}$) because $a > b$ results in a projection in a positive x and negative y direction. An ellipse can be described as a circle extended in one direction (or contracted in the orthogonal direction, which is equivalent). A circle is a special case of an ellipse where $a = b = r$ (constant radius). An ellipse is also defined as the continuum of points in which the sum of the distances r_1 and r_2 from the two foci (shown as $F1$ and $F2$) to the perimeter is a constant, and equal to $2a$ (in the special case of a circle, the foci coincide and $r_1 = r_2 = r = a = b$). It can be seen that this is the case for the two points where the distances become aligned on the semimajor axis; and by considering one of the two points on the semiminor axis, where $r_1 = r_2 = a$, by Pythagoras it is clear that $c = \sqrt{a^2 - b^2}$, where the parameter c is the distance of one of the two foci from the geometric centre of the ellipse.

For planetary orbits, and explaining them using Newton's Laws, it is convenient to describe the ellipse in polar form (r,θ), in which r varies with θ, with the coordinate origin at one of the foci (equivalent to the approximate position of the sun or other large body being orbited). An important parameter, the "eccentricity" of the ellipse, is defined as $\varepsilon = c/a$: the ratio of the focus-to-centre distance with the semimajor axis, which for the example above (with ellipticity greatly exaggerated) is $\sqrt{3}/2 \approx 0.866$. The eccentricity is zero for the special case of a circle and approaches 1 as the ellipse flattens out towards a line. The coordinates $x = c + r \cos \theta$ and $y = r \sin \theta$ can be substituted in the Cartesian equation which, after factoring out b and c with some algebra that is a bit lengthy to show here, results in an equation that relates the elliptical constants a and ε to the variables r and θ:

$$\frac{a(1 - \varepsilon^2)}{r} = 1 + \varepsilon \cos \theta$$

This is a beautiful equation because in this form it can be used to show that the elliptical orbits result from Newton's inverse square law of gravitation, also making use of the conservation of angular momentum, previously shown to be the physical reason for Kepler's second law (BE76). This can be done by considering the total energy of the system (kinetic and potential) or equating the force (gravity) with the outward and inward accelerations. We can equate Newton's second law (BE1) with the law of gravitation (BE67)

$$m\frac{d^2r}{dt^2} - mr\omega^2 = -\frac{GMm}{r^2}$$

Where the angular momentum $L = mr^2\omega$, a change of variable $u = 1/r$ results, after some manipulation, in the differential equation in (u, θ)

$$\frac{d^2u}{d\theta^2} + u = \frac{GMm^2}{L^2}$$

which can be solved to get

$$\frac{1}{r} = \frac{GMm^2}{L^2} + A\cos\theta$$

where A is a constant of integration that can be evaluated from the initial conditions. With a bit of rearrangement and substitution this is seen to be the same form as the polar ellipse equation, and the elliptical parameters a and ε are related to three physical constants of the system: the mass, the gravitational constant and the angular momentum. This explains why most celestial bodies have elliptical orbits. The reason they are typically not circular is because a circle is a very special case of an ellipse and all the conditions would have to be right – such as the initial speed and distance of capture by the larger mass without any other interference. Exceptions are artificial satellites, for example, in which the orbit can be controlled accurately enough to be approximately circular. Planetary and natural satellite orbits usually turn out elliptical because of this and collisions, influence of other bodies in the vicinity, etc. However, most of the planetary orbits are near-circular: their orbital eccentricity is small, nothing near the example given above. The exceptions are Mercury and Pluto, whose eccentricities are above 0.2. The eccentricity of the earth's orbit is 0.017, and for Mars it is 0.093, large enough for Kepler to do accurate calculations. The sun's centre is a good approximation for the focus of elliptic planetary orbits because the planets are relatively small in mass compared with the sun. The actual centre of rotation (the barycentre) is the centre of mass of both (and other) bodies), and there is a very slight wobble in the sun's rotation due to the planets rotating around it. Newton proposed that orbits are always conic sections, and in some cases such as comets they can be approximate parabolas or hyperbolas. In the hyperbolic case the orbiting object will swing around the larger mass only once and not return, as is also the case with spacecraft using planets to slingshot around the solar system. Conic sections will be the subject of a later beautiful equation. The third law, relating to periods of orbit, will be the final article in this series on Kepler.

PB

Equation 93: Stellar Magnitude

OVERVIEW

It may seem counter-intuitive that a so-called magnitude 1 star is brighter than a magnitude 6 star. Hipparchus, who bequeathed Ptolemy and then us his classification system, looked upon his magnitude numbers as rankings according to star size (he assumed that the *size* accounted for brightness). Thus the brightest stars he referred to as being 'of the first magnitude' and so on down to the just-about-visible stars, which he described as being 'of the sixth magnitude'.

Additional magnitudes were later added, starting with Galileo who was able to make out otherwise invisible stars with the telescope he had constructed. In 1856 the Oxford astronomer Norman Pogson devised a mathematical system (quickly adopted) that fitted and extended Hipparchus's scheme. He proposed that a difference of five magnitudes be defined as a brightness ratio of exactly 100 to 1. Thus a difference of one magnitude between two sources corresponded to a brightness ratio between them of exactly the fifth root of 100 (≈ 2.512), now known as Pogson's ratio. With this definition, it can easily be shown that the difference in the apparent magnitudes m and M of two stars can be expressed by the formula:

$$m - M = -2.5 \log \frac{q}{Q}$$

where q and Q denote the 'quantities' of radiation measured from the m and M magnitude stars respectively.

If the magnitude M star referred to above is a standard reference star, m for the other star can be thought of as an absolute magnitude value. In this case, the magnitude formula can be rewritten as

$$m = k - 2.5 \log q$$

where the calibrating constant k is determined by the measured quantity of light from the reference star (typically though not always Vega) and whichever magnitude value (typically though not always zero or near zero) one decides to allocate to that star. In practice, the value of k is modified by correctional factors to do with atmosphere, colour etc.

QUANTITY OF RADIATION

The rather vague term 'quantity of radiation', used above and represented by the letters q and Q in the formulae, alludes to one of several precisely defined quantities, typically energy flux (units W/m^2), photon flux (photons/s/m^2), or flux density ($W/m^2/Hz$ or Jy).

The terms 'flux' (symbol F), 'intensity' (I), and 'apparent brightness' (b) are commonly used interchangeably in place of 'energy flux' as in the preceding paragraph. Some authors have pointed out, though, that using the term 'brightness' in this context is misleading. When dealing with an extended luminous object such as the Moon, the Sun, a nebula or a galaxy, the geometry of the situation means that the observed surface brightness is unlike the flux in that it is independent of the source-observer separation. Thus the red giant Betelgeuse only appears as bright as it does because it is a relatively extended object as stars go, subtending a much larger solid angle than most stars.

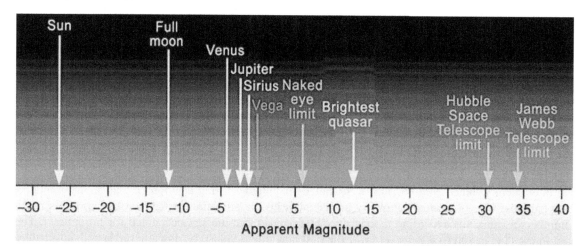

PASSBANDS

It is often desirable to allocate an individual magnitude to each passband for a given star. A passband is the wavelength range that can pass almost unattenuated through a specified filter. Photometry is the measurement and study of light. A 'standard photometric system' is the conceptual and physical setup needed to make these measurements in a way that is repeatable and comparable across different locations. Such systems comprise: [1] a set of well-defined discrete passbands; and [2] a set of carefully measured primary standard stars. In the commonly used Johnson-Morgan, or UBV, system, for example, each of a set of main sequence stars, which includes Vega, has been assigned a magnitude of 0.03 for each of its three passbands (corresponding to UV, Blue and Visible). The system has been extended to incorporate two additional passbands (Red and IR). The corresponding passband magnitudes M are identified using subscripts (U, B, V, R, I).

Note that a star which is 'bright' in the visible spectrum (Sirius is the brightest) may be relatively-speaking 'dim' in another part of the spectrum, and vice versa. In the near-infrared J-band, it is Betelgeuse that turns out to be the 'brightest' star.

The magnitude formulae can now be written so as to take account of passband-specific measurements:

$$m_p - M_p = -2.5 \log \frac{q_p}{Q_p}$$

and

$$m_p = k_p - -2.5 \log q_p$$

Sometimes the bands themselves are given subscripts to designate to a particular photometric system, e.g. R_i and R_c for the Red passband in the Johnson and Cousins systems respectively.

Differences between magnitudes in two passbands can often be measured more accurately for a star than individual absolute magnitudes can. Astrophysicists often refer to this difference for the star relative to the corresponding magnitude difference for, say, Vega.

SUBJECTIVE BRIGHTNESS
Suited as the logarithmic magnitude scale is to the quantification of objectively measurable quantities, it seems not so suitable for representing subjective 'brightness'. Recent experiments indicate that visual, auditory, etc. perceptions are characterised by power-law relationships in response to stimuli rather than the logarithmic ones. Thus a star that looks to the eye halfway in brightness between 2.0 and 4.0 will have a magnitude of about 2.8 rather than 3.0.

FR

Equation 94: Energy of an Electron in the nth State in a Hydrogen Atom

In the late 19th and early 20th century, various avenues of research came together to provide a much better picture of the hydrogen atom. By 1885, Balmer had identified the visible light emissions, and developed a formula to describe these. In the same period, Rydberg was making similar efforts for alkali metals, and realised there were similarities. As a consequence, in 1888 he published the now well-known Rydberg formula:

$$\frac{1}{\lambda} = R \left(\frac{1}{n_f{}^2} - \frac{1}{n_f{}^2} \right)$$

Here, $1/\lambda$ is the wave number, R is the Rydberg constant, and n indicates the energy level (*final* and *initial* respectively). By 1908, Paschen identified similar infrared emissions, and by 1914, Lyman added ultraviolet ones to the picture. All fitted Rydberg's formula. Around the time of Lyman's work, Bohr worked out a theoretical model for the energy levels in the hydrogen atom, and consequentially identified the basis for the Rydberg constant. The equation for the energy of the individual levels in a hydrogen atom, according to Bohr, is as follows:

$$E_n = \frac{m_e c^2 \alpha^2}{2n^2}$$

101

where m_e is the mass of the electron, c the speed of light, α the fine structure constant as defined below, and n the energy level.

$$\alpha = \frac{1}{4\pi\varepsilon_o}\frac{e^2}{\hbar c}$$

Here, e is the charge on an electron, ε_o the permittivity of a vacuum, and $\hbar$ Planck's constant divided by 2π.

Pulling all this together, one can derive a formula for the Rydberg constant that is:

$$R = \frac{m_e e^4}{8h^2\varepsilon_o^2}$$

where h is Planck's constant, and the other symbols are as given before. Both Bohr's formula and the Rydberg constant have been developed and refined further since then.

In everyday life, one can see effects similar to the ones that led to this work in anything from neon signs and plasma TVs, to fireworks and flame tests in the chemistry laboratory.

OK

Equation 95: Van der Pauw Method

Van der Pauw has shown that the resistivity of a sheet or a thin conductive coating of any shape can be measured by putting four contacts A, B, C and D clockwise at the circumference and make current flow between two contacts while measuring the voltage over two others, and switch the voltage/current in a defined way.

A resistance is defined as $R_{AB,CD} = V_{CD}/I_{AB}$, where V_{CD} is the voltage over contacts C-D caused by the current flowing between A and B. For the resistivity ρ of a coating or a disk of thickness d the following beautiful equation holds:

$$e^{\frac{-\pi d R_{AB,CD}}{\rho}} + e^{\frac{-\pi d R_{AD,CB}}{\rho}} = 1$$

In practice this equation is solved by defining

$$\rho = \frac{\pi d}{\log 2}(R_{AB,CD} + R_{BC,DA})f(R_{AB,CD}/R_{BC,DA})$$

where the function $f(x)$ is calculated, e.g. by successive approximation starting with $f(x) = 1$, from the transcendental equation:

$$f(x) = \frac{\log 2}{\log\left(2\cosh\left[\left\{\frac{x-1}{x+1}\right\}\frac{1}{f(x)}\right]\right)}$$

The beauty of the method is its independence of the coating shape and the positioning of the contacts. Today it is in general used for determining the electrical properties of thin films.

The beauty of the derivation of the method is the sequence of small, almost trivial, steps, where the application of complex function theory leads to the equation. Before that, symmetry considerations already lead to $f(x) = 1$ for a symmetric configuration.

Conditions are that the thickness and the contact areas are small in comparison to the surface. The original paper gives equations that estimate the deviations this may give. Also the surface should not contain any holes.

Leo Johan van der Pauw (1927-2014) worked at the Philips Research Laboratories in Eindhoven, The Netherlands. In 1968 he obtained his PhD at the Delft University of Technology.

References:
van der Pauw, L.J., 1958. *A method of measuring the resistivity and Hall coefficient on lamellae of arbitrary shape*. Philips Technical Review 20:220–224

HH

Equation 96: Radiative Equilibrium of the Earth

Radiative equilibrium is a concept in which the net amount of heat transfer between an object and its surroundings is zero. That is, the amount of radiative heat flux absorbed is equal to the amount of radiative heat flux released. In the case of our planet (Earth), the incoming heat flux is in the form of short wavelength radiation from the Sun, whereas the outgoing heat flux is characterized by long wave radiation. An indicator of the Earth being in radiative equilibrium is the overall stability of global temperatures. The topic of global warming is directly related to the concept of radiative equilibrium: global temperatures are observed to increase over time, meaning that there is more incoming heat flux than outgoing heat flux and the planet is, therefore, not in radiative equilibrium. Climatologists suggest that the increase in greenhouse gases being released into the atmosphere by human (and bovine, see http://www.ibtimes.com/cow-farts-have-larger-greenhouse-gas-impact-previously-thought-methane-pushes-climate-change-1487502) activity has resulted in the trapping of heat that would have otherwise been emitted.

The equation for radiative equilibrium is given by

$$Q_i + Q_o = 0$$

where Q_i is the incoming heat flux and Q_o is the outgoing heat flux.

The general concept of heat flux being absorbed and radiated by objects was first studied by Pierre Prevost in 1791. The definition of radiative equilibrium was introduced in the field of astrophysics by Karl Schwarzschild in 1906. Interest in studying the Earth's radiative flux was strengthened in the 1980s with the launch of several Earth-orbiting satellites by the National Aeronautics and Space Administration (NASA), courtesy of the U.S. taxpayer.

Below, an informational poster on the Earth's energy budget (provided by NASA) is presented. (Source: http://science-edu.larc.nasa.gov/energy_budget/)

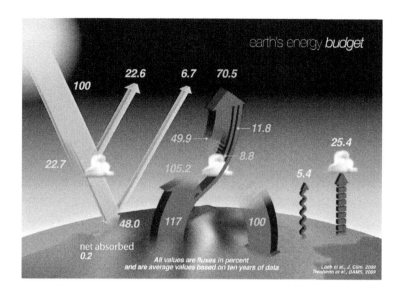

MF

Equation 97: Stress-Strain Relations for Isotropic, Linear Elastic Solids

Mechanics of solids is the study of deformations, motions and internal force distributions in a given structure (body) due to the application of external forces and temperature changes. Applications of mechanics of solids can be found in almost every engineering discipline where there are some type of load-carrying members. Some examples are the design and analysis of aircraft structures, automobile components, bridges, and hip implants.

The deformations involve changes in lengths and changes in shapes. These are characterized by normal strains (changes in lengths per unit length) and shear strains (changes in angles between any two curves emanating from a point in a body). The internal force distributions are characterized by stresses which consist of normal stresses (force per unit area perpendicular to a surface) and shear stresses (force per unit area parallel to a surface). Usually, the stresses and strains are referenced with respect to a coordinate frame. For example, if the coordinate frame is rectangular, the strains and stresses are written in matrix form as follows:

$$\varepsilon = \begin{bmatrix} \varepsilon_{11} & \varepsilon_{12} & \varepsilon_{13} \\ \varepsilon_{21} & \varepsilon_{22} & \varepsilon_{23} \\ \varepsilon_{31} & \varepsilon_{32} & \varepsilon_{33} \end{bmatrix}$$

and

$$S = \begin{bmatrix} S_{11} & S_{12} & S_{13} \\ S_{21} & S_{22} & S_{23} \\ S_{31} & S_{32} & S_{33} \end{bmatrix}$$

Here 1 represents the x direction, 2, the y direction and 3, the z direction. The diagonal components in the strain matrix represent changes in lengths per unit length (normal strains) in each of the coordinate directions. The off-diagonal components $\varepsilon_{ij}, i \neq j$ represent changes in angles (shear strains) between two initially perpendicular lines parallel to ith and jth axes. The diagonal components of the stress matrix represent the normal stresses perpendicular to each of the three coordinate planes. The components $S_{ij}, i \neq j$ represent shear stress on the ith coordinate plane in jth direction. Both the matrices are symmetric, i.e., $\varepsilon_{ij} = \varepsilon_{ji}$ and $S_{ij} = S_{ji}$.

104

When a given body is subjected to external forces, it can deform elastically or plastically. Elastic deformation is the deformation that can be recovered completely when the applied forces are removed. On the other hand, plastic deformation implies that the body does not return to its original shape or size when the applied forces are removed. Typically, a given material deforms elastically first and under increased loads, can deform plastically. Many practical applications require the applied loads to be such that the deformations are elastic only.

For elastic deformations of many isotropic (i.e., material response is the same in all directions) materials, there is a beautiful equation called the Generalized Hooke's law that relates stresses to strains. It is given by

$$\varepsilon_{ij} = \frac{1}{E}\left[(1+v)S_{ij} - vS_{kk}\delta_{ij}\right], \text{ with } S_{kk} = S_{11} + S_{22} + S_{33} \text{ and } i,j = 1,2,3$$

In the above, E is the Young's modulus and v is the Poisson's ratio. δ_{ij} is equal to unity when $i = j$ and zero when $i \neq j$. Both are material properties and are determined through simple experiments such as a uniaxial tension test. The law was first proposed by Augustin-Louis Cauchy in a memoir communicated to the Paris Academy in 1822. However, Hooke's name is associated with this law probably because a one-dimensional version of it was originally due to Robert Hooke (see beautiful equation 18) in 1660.

This equation is beautiful for the reason that it is simple and has several physically significant implications. We list a few in the following.

The shear strains are given by

$$\varepsilon_{12} = \frac{1+v}{E}S_{12}, \varepsilon_{23} = \frac{1+v}{E}S_{23} \text{ and } \varepsilon_{31} = \frac{1+v}{E}S_{31}$$

This shows that normal stresses do not affect shear strains. The quantity $E/2(1+v)$ is called the shear modulus and represented by μ. Then, the above becomes

$$\varepsilon_{12} = \frac{1}{2\mu}S_{12}, \varepsilon_{23} = \frac{1}{2\mu}S_{23} \text{ and } \varepsilon_{31} = \frac{1}{2\mu}S_{31}$$

Similarly, one can show that shear-stresses do not contribute to normal strains. In uniaxial tension, the only non-zero stress present is S_{11}. Let us set this to equal to σ. Then, the shear-strains are zero and the normal strains are given by

$$\varepsilon_{11} = \frac{\sigma}{E}, \varepsilon_{22} = \varepsilon_{22} = -v\frac{\sigma}{E} = -v\varepsilon_{11}$$

From the above, we see that E is the slope of the $\sigma - \varepsilon$ curve and the Poisson's ratio v has the physical meaning of the negative of the ratio of lateral normal strains to the normal strain in the direction of uniaxial loading. Suppose we take a bar of length L and a cross-sectional area A. If we pull this bar with a force P, then $\sigma = P/A$. If we denote the change in length by δ, then, $\varepsilon_{11} = \delta/L$. From the above expression for ε_{11}, we find that $P = k\delta$ with $k = EA/L$. This is the same as the Hooke's law discussed in beautiful equation 18. The bar can be thought of as a spring with a spring constant k. If the normal strains are summed, we find that

$$\varepsilon_{11} + \varepsilon_{22} + \varepsilon_{33} = \frac{1-2v}{E}(S_{11} + S_{22} + S_{33})$$

The left-hand side has the physical meaning of change in volume per unit volume and is often denoted by e. If the hydrostatic pressure is defined as $p = -(S_{11} + S_{22} + S_{33})/3$, then the above relation becomes

$$e = -\frac{3(1-2v)}{E}p, \text{ or } p = -\frac{E}{3(1-2v)}e$$

The quantity $E/(3(1-2v))$ is the well-known bulk modulus and often denoted by κ. An interesting outcome of this is that when $v = 0.5$, e is zero which implies that the material is incompressible.

HC

Equation 98: Binet's Equation for the Fibonacci Sequence

In 1936, Albert Einstein wrote that "the most incomprehensible thing about the universe is that it is comprehensible," alluding to the mysterious truth that the universe operates on mathematical principles, despite the fact that mathematics is apparently in large part the product of human reason.

Perhaps the most celebrated simple example of a mathematical principle found in nature is the Fibonacci sequence, discovered in 1202 by Italian mathematician Leonardo Pisano (later known as Fibonacci because centuries after his death, part of the title of his handwritten masterpiece, Liber Abaci, was thought to be his name. The phrase actually reads "filius Bonacci"; "son of Bonaccio").

The sequence could hardly be simpler; each term is the sum of the two previous ones:

$$1, 1, 2, 3, 5, 8, 13, 21 \ldots$$

Remarkably, most tree branches, flower petals, leaf patterns, and pine cones scales come in Fibonacci numbers (which is "why" four-leaf clovers are proverbially rare).

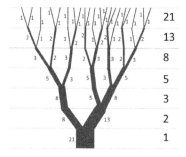

Even idealised honeybee family trees exhibit them: because every male bee is the progeny of an unmated female and every female the progeny of a fertilized female, it follows that each male bee has 1 parent, 2 grandparents, 3 great-grandparents, 5 great-great-grandparents, and so on.

The Golden Ratio can be derived from the Fibonacci sequence, by setting numerators to Fibonacci terms and denominators to the next highest terms:

$$\frac{1}{1} \, \frac{2}{1} \, \frac{3}{2} \, \frac{5}{3} \, \frac{8}{5} \cdots$$

The limit of this sequence is the Golden Ratio ϕ, about 1.61804.

The sequence is defined as $F_{n+2} = F_n + F_{n+1}$. In 1843, Jacques Binet solved this difference equation to give Binet's formula (though he was not the first to do so), which gives F_n, the value of the nth Fibonacci number.

$$F_n = \frac{\phi^n - (1 - \phi)^n}{\sqrt{5}}$$

106

References:
Posamentier, A.S. and Lehmann, I., 2007 *The Fabulous Fibonacci Numbers,* Amherst: Prometheus

MJG

Equation 99: Airy and Bessel Points

When lying in a horizontal plane, any macro scale object with finite stiffness and mass will sag under its own gravitational load unless supported at infinite points along longitudinal direction. In dimensional metrology, this phenomenon can cause considerable errors when a length bar or rod is used as a physical length standard or as a straightedge. The sagging of a bar leads to tilt of its two end faces with respect to one another. So when the bar is measured at both ends by a mechanical probing instrument like coordinate measuring machine, different measurement results will be obtained dependent on the location of the contacting probe on the end faces. One solution in history is to float a length bar (made of metal) in mercury so a zero bending is technically achieved by this infinite number of supports. However this solution it is too hazardous and difficult to implement.

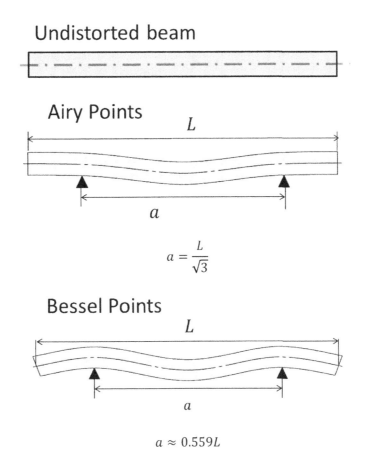

In practice, this problem is addressed by supporting the length bar at two points. Suitable selection of the positions of the two supports along the beam will result in the end of the bar being parallel with each other. These two points are called "Airy Points" named after George Airy, who solved this problem for a more general case of multiple supports in 1846 (on the Mem. Roy. Astron. Soc. 15, 157 – 163). As shown in the figure, the Airy Points are symmetrically arranged around the centre of the length bar and are separated by a distance given as a fraction of the bar's nominal length. Another useful two-point supporting configuration is called "Bessel Points" in which the difference between the measured length of the bar with bending distortions is matched to the same length that would be measured if the bar were ideally supported.

References:
Phelps, F.M., 1966. *Airy points of a meter bar*. American Journal of Physics 34:419-422.
Mitutoyo, 2015. *Quick guide to precision measuring instruments*, Sakado: Mitutoyo

KN

Equation 100: XOR Operator

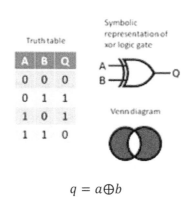

$$q = a \oplus b$$

where a and b are Boolean variables, and $\oplus$ is the XOR Boolean operator.

Exclusive-or or XOR, is a logical operation that outputs true if inputs differ.

The simplest use in computer science is to tell when two bits are equal. Other uses are for example bit flipper (the deciding input decides whether to invert the data input), or it tells if there are an odd number of '1' bits ($a \oplus b \oplus c \oplus d$ is true if an odd number of bits are true).

The case of bit flipper is of use in cryptography. A message is XORed with the 'key' to flip only certain bits. To decrypt the message the encrypted message is XORed again with the 'key' to flip back the bits to the original value.

Checking whether the bits in a bit train are even or odd is useful in transmission error detection. Some communication protocols add a parity bit to the transmitted data, to make the number of bits in a message always even, and therefore the parity bit is the XOR of all the bits in the message. Two examples of transmission are: 101**0** and 100**1** where the parity bit is in bold. If a bit is wrongly transmitted the error is detected and the message usually sent again. Of course, the error detection fails if the wrong transmitted bits are two (or an even number), which is less likely to happen than a single error.

More complex error detection schemes allow for error correction as well. For example the Hamming code, achieves error detection for whatever numbers of errors occurring, including error in the parity bit transmission, it tells you which bits are wrong and therefore they can be corrected, and it is optimised to achieve maximum code rate (minimise the amount of redundancy in the bits transmitted, see beautiful equation 41).Integrated circuits which implement this logic function were commercialised quickly after the integrated circuit manufacturing process was developed (1958) along with other Boolean logic functions. In the 60s a very common integrated circuit family was the 7400 series which implemented all the basic Boolean operators.

GM

Equation 101: Doppler Shift

$$f' = \left(\frac{v \pm v_r}{v \mp v_s}\right) f_0$$

The Doppler shift or Doppler effect was first proposed in 1842 by the Austrian physicist Christian Doppler and is the change in frequency of a wave for an observer moving relative to its source. It is most commonly observed when listening to the sound of the siren of a passing emergency vehicle. Compared to the emitted frequency, the received frequency is higher when the vehicle is approaching, identical at the instant of passing by, and lower as it gets further away.

The Doppler equation expresses this mathematically where v is the velocity of the waves in the medium, v_r is the velocity of the receiver relative to the medium and v_s is the velocity of the source relative to the medium and f_0 is the emitted frequency. Note that these are velocities and not speeds so sign is important. Positive means that the receiver is moving towards the source (and vice versa). A source moving away from the observer has a positive velocity (and vice versa).

In practice the values of v_s and v_r are usually small relative to the speed of the wave thus allowing some simplification of the equation.

Applications of this effect apart from sirens include astronomy, radar, medical imaging, flow measurement, satellite communication, audio and vibration measurement. Those more advanced readers may wish to investigate the inverse Doppler effect.

DF

Equation 102: de Broglie Relation

$$p = \frac{kh}{2\pi} = \frac{h}{\lambda} = \frac{mv}{\sqrt{1 - \frac{v^2}{c^2}}}$$

where v is velocity (ms^{-1}), h is Planck's constant ($= 6.62607 \times 10^{-34}$ Js), m is mass (kg), p is momentum (kgms^{-1}), λ is wavelength (m), c is the velocity of light ($= 299792458$ m·s^{-1}).

For centuries, scientists (or natural philosophers) have discussed the nature of light. Translating these ideas into what might be considered a modern scientific, or mathematical, model was primarily initiated by Isaac Newton and Christiaan Huygens in the 17th century. Newton put forth the idea that light was made from fundamental corpuscles while Huygens favored an interpretation in terms of waves. With advances in the science of optics and optical interactions with matter, it appeared for a while that Huygens wave theory could be used to explain a broader range of scientific observations thereby leading scientists to discount the corpuscular theory. However, the discovery of Max Planck (at the turn of the 20th century and for which he received a Nobel Prize in 1918) that light emits or absorbs packets of energy having discrete values made it apparent that a simple characterization of light as either particle or wave was unacceptable. For some reason, it was not particularly difficult to see why, after centuries of discussion, it finally had to be concluded that, on a fundamental level, light behaves as both a wave and a particle. After all, how can light be absorbed into molecules if its wavelengths (visible light has a vacuum wavelength of about 500 nm) were so much bigger than the molecules absorbing the energy packets (molecules are typically fractions of a nanometre or about 1000 times smaller than the wavelength)?

Having connected the energy of photons (beautiful equation 73) to their wavelength and knowing that energy can be equated with mass (beautiful equation 74), it was in some ways natural to think of light as being made up of 'particles' that were capable of 'wavelike' interactions. In particular, the particle-like nature of photons is particularly pronounced when looking at their interaction with molecules in solids. A major component of this realization was established with measurements of the photo electric effect originally explained by Einstein (a few years after Planck's theory of the quantization of energy transfer), an achievement that played a major part in the development of wave mechanics and also the reason he was awarded his Nobel Prize in 1921. In fact, Max Planck had not connected, and even rejected, his discovery of the quantization of light as involving photons as actual particles. It was not until Einstein met Planck at the 1911 Solvay Conference and discussed the growing body of evidence supporting the light quanta hypothesis that he finally acceded to this new physics.

Surprisingly, while the composition of atoms as involving the aggregation of 'fundamental' particles such as electrons and protons was firmly established at this time, it wasn't until nearly a quarter of a century later that Louis de Broglie (1892 - 1987) set out to determine the fundamental nature of these particles. In fact, because these particles had a measureable mass, it was simply assumed that they could be modelled as a solid only smaller than the rocks and other apparently indivisible objects that we are familiar with at human scales. Hence it was quite a revolutionary idea when, in 1924, de Broglie suggested that all other particles (electrons, protons, neutrons, and, even, a rock) will also have an associated wave structure in which the wavelength of a particles wavelike nature is given by the above equation. The reason for this suggestion was based on the development of a relativistically invariant wave function that would include all energy of a moving particle quantized to the discrete energy levels determined by Planck's constant.

Acceptance of the dual wave-particle nature of electromagnetic interactions might not be considered conceptually difficult in view of the experimental evidence and, of course, the fact that the wavelike nature had been established with an extensive mathematical and experimental framework extending over 150 years. On the other hand, there simply was no wavelike mathematical framework for the dynamics of particles with finite rest mass. Not only this, there appeared to be no way in which to include this behaviour within the long established equations of dynamics developed by Newton with the recent modifications to include relativistic effects. It took a further two years for a brilliant German Physicist named Erwin Schrodinger who, in 1926, presented a suitable wave equation (beautiful equation 17) that incorporated both the connection between frequency and energy determined by Max Planck (beautiful equation 73) and also between wavelength and momentum suggested by de Broglie. This new equation provided the foundation of wave mechanics that has since been developed into the modern mathematical structure of quantum theory that is, in turn, the foundation of current endeavours to produce a grand unified theory of everything.

The example in the figure shows the wavelength of an electron that is accelerated in a vacuum using an accelerating voltage, V. The velocity is derived by equating the potential energy change of the electron (having mass $m_e = 9.109 \times 10^{-31}$ kg and charge $e = 1.602 \times 10^{-19}$ C) to the kinetic energy gain. At high voltages it is necessary to correct for relativistic increase of the mass. This changes the wavelength by around 2.5 % at 50000 volts and 8.5 % at 200000 volts. At voltages of over 10 million volts the wavelength becomes comparable to that of the nuclei of atoms. Particles, such as the proverbial rock, that we are used to at human scales are so much larger than the constituents of atoms that their wavelike nature is too small to be perceptible, even to the most sensitive of experiments. This, in part, is why the behaviour at atomic scales is so alien to our everyday interactions with the physical world.

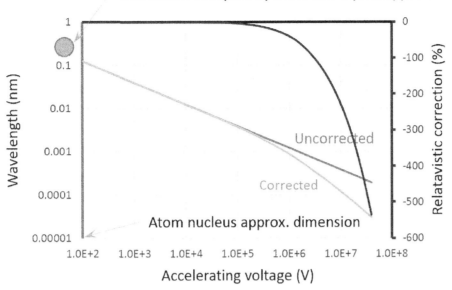

Wavelength of an electron when accelerated with a voltage V

SS

Equation 103: Young's Modulus

Young's modulus is a measure of the stiffness of an elastic material defined by the ratio of stress and strain. Young's modulus, commonly specified as the E value or E number of a material can be used to determine the performance of a material used in a structural system.

Young's modulus is named after British scientist Thomas Young who popularized its use in the 19th century however Leonhard Euler developed the concept in 1727.

$$E = \frac{F L_0}{A_0 \Delta L}$$

Where E is Young's modulus, F is the force applied to an object or structure, A_0 is the original cross-sectional area through which the force is applied, ΔL is the amount by which the length of the object changes and L_0 is the original length of the object.

The two main components 1) tensile stress and 2) extensional strain can be considered as 1) force per unit area and 2) the deformation over the area due to the applied force, thus Young's modulus is an expression of a ratio of these two components. The ratio only holds true while the components follow a linear function, in practice there are limits and material yield points where this ratio no longer holds true, therefore Young's modulus must only be considered valid while below the elastic limit of a material. This is visualized in the graph below showing a typical stress v strain curve for a steel material. The line in the first part of the graph shows the area over which Young's Modulus can be considered valid.

One of the most common uses of Young's modulus is in a beam bending calculation where structural engineers need to know how much a structure will sag or deform due to the loads the structure is supporting.

As this value is a ratio it is interesting to note that stress and strain can values can change by a factor of two or more and still have the same Young's Modulus. As such, the Young's Modulus value gives no indication as to some of the other materials properties such as hardness - as pointed out by Stuart Smith.

References:
Beautiful equations 97 and 18 (for expansion on this topic)

TS

Equation 104: Combinations and Permutations

There are many situations in life in which we have to deal with a number of things or objects according to a certain rule, and it becomes necessary to count how many ways of doing this are possible. For instance, we might wish to know how many different arrangements are possible for seating five children on five chairs, one child per chair. For the first chair we have 5 children to choose from, but after we have seated that first child, we have only 4 children left from which to choose the occupant of the second chair. Each of the five choices for the first chair's occupant can be combined with any of the 4 choices for the second chair's occupant, giving us a total of 5×4 = 20 possible arrangements for just the first two chairs. For each of these 20 arrangements, there are now 3 children from which to select the occupant of the 3rd chair, after which there are 2 children from which to choose one for the 4th chair. The one remaining child then has to go on the 5th chair. We thus have a total of 5×4×3×2×1 (which is written as 5! and called "5 factorial", and has the value 120) possible arrangements. In general, the number of different ordered arrangements, or *permutations*, of n distinguishable things is $n!$.

If we label the children A, B, C, D and E, and forget the chairs, and consider how many different ways there are to select two children from the five, then the ways to select are permutations in case the order of the two children is important (as was the case when we were seating them one to a chair, so that the arrangement AD would be distinguished from the arrangement DA) . The arrangements are called combinations in case the order of the children is unimportant, in which case AD would be identical to DA. The number of permutations of r things chosen from n, and the number of combinations of r things chosen from n, are denoted by the symbols $_nP_r$ and $_nC_r$ respectively, and the expressions for them are given below.

$$_nP_r = \frac{n!}{(n-r)!}$$

$$_nC_r = \frac{n!}{r!(n-r)!}$$

Nowadays there is scarcely a branch of science or technology, from physics and chemistry to engineering, biological science, economics, psychology, and sociology, which remains untouched by probability theory and statistical mathematics. The original impetus for much early work done in probability theory, however, came from the urgent practical problems experienced by the aristocratic gambling and betting community in seventeenth-century Europe. The mathematical theory of probability can in fact be precisely dated to 1654, when the French gentleman Chevalier de Mere approached the mathematician Blaise Pascal for help with a question about betting odds in a certain game of dice.

References:
Weaver, W., 1963. *Lady Luck*, Mineola: Dover

PRA

Equation 105: Taylor Series

The Taylor Series, sometimes called the Taylor Expansion, is a way of representing a function as a power series. Of course, not all functions can be represented as a power series, but many can. Assume the function, $f(x)$, has a power series representation about $x = x_0$ and that all the derivatives of $f(x)$ can be computed, then the equation in the figure is the Taylor Series, where $f^{(n)}$ represents the nth derivative of $f(x)$ with respect to x and $n!$ is n factorial ($2! = 1 \times 2$, $3! = 1 \times 2 \times 3$, etc.).

The Taylor Series was first discovered by James Gregory (a Scottish mathematician) but Brook Taylor (an English mathematician) was the first to publish it in 1715. The Taylor Series has a number of applications in the physical sciences including calculations for elasticity, wave phenomena and thermal behaviour. However, it is mostly used to form an approximation and there are far more applications of the linear approximation, where only the term in $n = 1$ is used. Application examples include the approximation of awkward integrals, perturbation theory and finding error functions.

As an example of the use of the Taylor expansion, the first two terms in the expansion for the cosine function are $\cos x = 1 - (x^2/2)$ – this often gives enough numerical precision for many engineering applications.

$$f(x) = \sum_{n=0}^{\infty} \frac{f^{(n)}(x_0)}{n!} (x - x_0)^n$$

RL

Equation 106: Stoichiometric Combustion of Hydrocarbons

$$C_m H_n + \left(m + \frac{n}{4}\right) O_2 \to mCO_2 + \left(\frac{n}{2}\right) H_2 O$$

INTRODUCTION

You could classify chemical reactions according to 6 basic types: combination; decomposition; single displacement; double displacement; acid-base; and combustion. Combustion occurs between a fuel and an oxidising agent, producing energy, usually in the form of heat, light and smoke. The heat produced can make combustion self-sustaining.

Many types of combustion reaction are possible, depending on the fuel and the oxidising agent (aka oxidant or oxidiser). Although the oxidant is usually oxygen, there are other elements that can oxidise: fluorine is actually a more powerful oxidant than oxygen.

Reactions of this type are essential to life, and are exploited to generate power, to provide heat, to run motorised vehicles, and so on.

The most familiar combustion processes involve the burning of hydrocarbons, which combine with oxygen in the air to form carbon dioxide and water. A hydrocarbon is an organic compound consisting entirely of hydrogen and carbon, having the general formula $C_m H_n$. Fossil fuels are largely made up of hydrocarbons, the majority of which occur naturally in crude oil. Extracted hydrocarbons in a liquid form are referred to as 'petroleum', whereas gaseous hydrocarbons are referred to as 'natural gas'.

1 mole of $C_m H_n$ fuel needs exactly $m + \frac{n}{4}$ moles of O_2 for complete combustion.

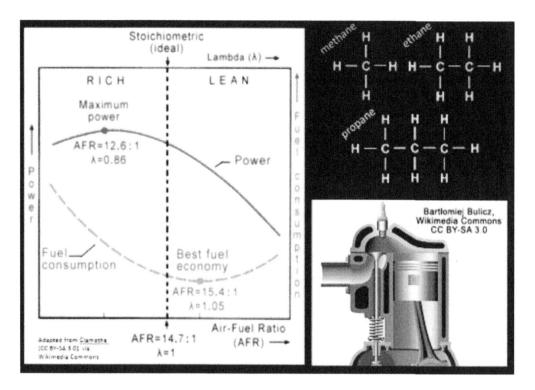

EXAMPLES OF HYDROCARBON COMBUSTION IN *OXYGEN*
Methane, the simplest hydrocarbon burns in oxygen as follows:

$$CH_4 + 2O_2 \rightarrow CO_2 + 2H_2O$$

For pure octane, we have:

$$C_4H_{18} + 12.5O_2 \rightarrow 8CO_2 + 9H_2O$$

STOICHIOMETRY
'Stoichiometry' is the calculation of relative quantities of reactants and products in chemical reactions. The term was first used by Jeremias Benjamin Richter in 1792 and is derived from the Greek 'stoicheion' (element) and 'metron' (measure).

'Complete combustion' is a chemical reaction in which all of the carbon atoms in the fuel are consumed. When the oxidiser is available in exactly the right quantity to burn all the fuel, we say that the mixture is 'stoichiometric'. 'Stoichiometric air' means the minimum air needed for complete combustion.

The term 'air-fuel ratio' (AFR) is commonly used for mixtures in internal combustion engines and is also utilised for industrial furnaces. It commonly alludes to masses, but can allude to volumes. The masses refer to ALL constituents that compose the air and fuel, whether combustible or not. In certain applications, the term fuel-air ratio (FAR) is used instead of AFR.

The 'air-fuel equivalence ratio' λ is the ratio of actual AFR to stoichiometric AFR. For rich mixtures $\lambda < 1.0$; for lean mixtures $\lambda > 1.0$. The fuel–air equivalence ratio ϕ is the reciprocal of λ.

INTERNAL COMBUSTION ENGINE
Although catalytic converters are designed to work best when the exhaust gases are the result of nearly perfect combustion, in practice stoichiometric combustion in an internal combustion engine is never quite achieved, due primarily to the very short time available for each combustion cycle. Furthermore, for

114

acceleration and high load conditions, a richer mixture (lower AFR) is used to prevent possible engine damage.

For petrol (gasoline fuel), the stoichiometric air-fuel ratio is roughly 15:1 (15 g air to 1 g fuel). For pure octane, the stoichiometric AFR is close to 14.7:1.

EXAMPLES OF HYDROCARBON COMBUSTION IN *AIR*
For methane in air, we could write:

$$CH_4 + 2O_2 + 7.52N_2 \rightarrow CO_2 + 2H_2O + 7.52N_2$$

For pure octane in air:

$$C_4H_{18} + 12.5O_2 + 47.0N_2 \rightarrow 8CO_2 + 9H_2O + 47.0N_2$$

The general equation for burning hydrocarbons in air, bearing in mind that air is 79% nitrogen and 21% oxygen (with trace amounts of other gases), is:

$$C_mH_n + \left(m + \frac{n}{4}\right)\left(O_2 + \frac{79}{21}N_2\right) \rightarrow mCO_2 + \left(\frac{n}{2}\right)H_2O + \left(m + \frac{n}{4}\right)\left(\frac{79}{21}\right)N_2$$

since air is 79 % nitrogen and 21 % oxygen (with trace amounts of other gases)

For the latter situation, the stoichiometric AFR would be given by

$$\left(m + \frac{n}{4}\right)\left(M_0 + \frac{79}{21}M_n\right) \div (mM_c + nM_h)$$

where M_0, M_n, M_c & M_h are the molar masses of O_2, N_2, c and H respectively.

References:
Bhatt, B.I., 2010. *Stoichiometry, 5th Edition*. New Delhi: McGraw-Hill

FR

Equation 107: Kepler's Third Law of Planetary Motion (Law of Harmonies)

Kepler's third law, the law of harmonies or periods, was presented in its final form in 1619 in book 5, chapter 3 of a work called Harmonices Mundi (translated as the Harmony of the World). An earlier version had appeared in the first part of Epitome Astronomiae Copernicanae in 1615. The first four chapters of Harmonices Mundi dealt with polygons, polyhedra and the congruence of these figures, and harmonic proportions in music and configurations in astrology, before finally discussing the harmony related to the motions of the planets. This was all part of Kepler's attempts to understand the motions of heavenly bodies in the sense that he perceived them. The concepts arose from medieval philosophical ideas about musical harmonies relating to the relative distances of the planets ("music of the spheres"). The ideas in fact stretch back to Ptolemy and his work Harmonica, and Kepler was aware of this work but his book took into account the heliocentric system and elliptical orbits and therefore was a considerable advancement. Kepler saw these harmonies as something more than simple musical notes, a natural phenomenon that interacts with human existence on earth and was therefore inextricably linked with astrology. Despite the inconsistencies of astrology with modern astronomy which are obvious to us now, Kepler was a brilliant mathematician with a deep knowledge of geometry and spent years undertaking rigorous calculations on astronomical data and other pursuits. Among his side-line ventures into mathematics, he formulated what became the Kepler Conjecture, on the packing of spheres, which was not proved until 1998.

The third law states that the square of the orbital period is proportional to the cube of the semi-major axis of the ellipse

$$T^2 \propto a^3$$

The beautiful equation displayed as the last equation in this article is the later version in which the constant of proportionality had been evaluated from application of Newton's inverse square law. This can be shown readily for the special case of a circular orbit (most of the planetary orbits are near-circular with slight elliptical eccentricity). The period is

$$T = \frac{2\pi}{\omega}$$

where ω is the angular velocity $d\theta/dt$. Squaring gives

$$T^2 = \frac{4\pi^2}{\omega^2}$$

The radial (centripetal) acceleration is the second derivative of a harmonic function of $\theta = \omega t$:

$$\ddot{r} = (-)r\omega^2$$

(where the "$-$" sign conventionally relates to the inward direction and $\ddot{r}$ is used to represent radial acceleration here to distinguish it from the semi-major axis a of the ellipse above)

This can equated to the gravitational acceleration GM/r^2 (assuming the orbiting body is small compared with the primary mass, e.g. the sun) and therefore

$$\omega^2 = \frac{GM}{r^3}$$

and

$$T^2 = \frac{4\pi^2}{GM}r^3$$

In the elliptical case we make use of the relationship between the gravitational constant, mass and angular momentum and the elliptic parameters shown in beautiful equation 93, where we find

$$\frac{L^2}{GMm^2} = a(1 - \varepsilon^2)$$

Using the area of an ellipse, which can be shown to be πab, and the angular momentum in the expression above we find that b and ε cancel in the resulting expression for the period resulting in

$$T^2 = \frac{4\pi^2}{GM}a^3$$

The important result of this is that the orbital period depends only on the semi-major axis of the orbit (i.e. the size of orbit) and not its ellipticity. For larger orbiting bodies M is replaced by $M + m$. Because of the mass of the orbiting body the constant of proportionality is not the same for each planet but the above expression is a good approximation for the planets in the solar system.

Approximate examples for the nine planets in the solar system are displayed in the table below, showing that the constant of proportionality is very similar in all cases. Pluto is now classified as a dwarf planet along with similar objects such as Eris and Haumea; none of the planets beyond Saturn were known about in Kepler's time.

Planet	$a/10^{10}$ m	T/years	T^2/a^3
Mercury	5.8	0.241	2.99
Venus	10.8	0.615	3
Earth	15	1	2.96
Mars	22.8	1.88	2.98
Jupiter	77.8	11.9	3.01
Saturn	143	29.5	2.98
Uranus	287	84	2.98
Neptune	450	165	2.99
Pluto	590	248	2.99

PB

Equation 108: Van der Waals Equation

As described in beautiful equation 20, the ideal gas law is named as such because certain assumptions have been made about the behaviour of the gas. The particles are assumed to be point masses, and undergo only completely elastic collisions. In other words, they do not take up space, and there are no forces between them. While under the right circumstances (low pressure, high volume), the ideal gas law is a perfectly reasonable approximation for practical use, when deviating from these conditions, it has to be adapted. In 1873, Van der Waals proposed the following equation (in one of its guises) to compensate:

$$\left(P + \frac{n^2a}{V^2}\right)(V - nb) = nRT$$

In this equation, the main variables have the same meaning as in the ideal gas law. In the part dealing with pressure, $(P + n^2a/V^2)$, the ratio between n and V is a measure for the density of the gas, with a compensating for the forces between the gas particles. In other words, the pressure would have been higher without those interactions. In the part dealing with volume, $(V - nb)$, b compensates for the volume taken up by n gas particles, thereby reducing the actual volume available.

OK

Equation 109: Kirchhoff Radiation Law

The Kirchhof law of radiation states that in thermal equilibrium at temperature T for any object, medium or material:

$$\alpha(\lambda, T) = \varepsilon(\lambda, T)$$

where $\alpha(\lambda, T)$ is the absorptance: the fraction of radiation with wavelength λ that is absorbed and $\varepsilon(\lambda, T)$ is the emittance: the fraction of radiation that is emitted relative to an ideal black body.

The equation follows from energy conservation and the concept of thermal equilibrium.

117

Inside a room without any heats ources and where all objects have a temperature T, the objects should not be heated up by absorbing heat radiation from other objects, or cooled down by emitting more radiation to the environment than it received. For example a silver ball will reflect almost all radiation. That means that in order to keep the same temperature it should not radiate more than the small fraction it does not reflect.

A similar argument holds for e.g. a sheet of more or less absorbing glass, for which can be stated:

$$r(\lambda, T) + t(\lambda, T) + \alpha(\lambda, T) = 1$$

Where r is the reflectance and t is the transmittance. This is logic: radiation is either reflected, transmitted or absorbed. However, for this sheet to maintain thermal equilibrium with its surroundings it should not emit more or less than it absorbs. For an opaque material $t(\lambda, T) = 0$, and it can be stated that:

$$r(\lambda, T) = 1 - \alpha(\lambda, T) = 1 - \varepsilon(\lambda, T)$$

This means that the emittance of an object at temperature T can be calculated from the (infrared) reflectance spectrum weighed by the Planck radiation function it temperature T, assuming that this reflectance is not temperature-dependent. The measurement of the infrared reflectance spectrum of a material is a common method to determine its emittance.

Gustav Robert Kirchhoff (1824 – 1887) was a German physicist who contributed to a variety of fundamental physical concepts, e.g. electrical circuits, spectroscopy, and the emission of radiation by objects, for which he developed the 'black body' concept.

HH

Equation 110: Attack Damage Equation in Role-Playing Games (RPGs)

There is a series of video games known as role-playing games, often referred to by their acronym 'RPGs.' In RPGs, the player assumes the role of one or more characters through a narrative, or story. The player 'controls' the actions of the character(s) throughout the game and can make critical decisions on the advancement of their character(s) as the story unravels. One of these advancements is the 'Level', or physical characteristics, of each character, such as their 'Hit Points (HP)', 'Magic Points (MP)', 'Attack Strength', and 'Defense Strength', as well as what armor the character wears or what weapons they have equipped. Some RPGs can have violent elements, such as the occurrence of battle scenes with vagabond creatures in search of mischief, or with evil villains convinced they will either rule the world or cause its destruction. In these battle scenes, the physical characteristics of the characters and of the enemies are used to determine whether good or evil will reign supreme.

For example, HP is a quantitative measure of a character's health – when a character's HP reaches zero, that character is considered dead. Luckily for the player, the games are often designed such that a special potion or magic spell can be used to revive a dead character. No one said the video game was realistic. MP, on the other hand, is a quantitative measure of the amount of magic spells that a character can cast. Similarly to HP, when MP reaches zero, the character can no longer cast magic spells – unless a special potion is used to regenerate the MP. Many RPGs consist of turn-based battles, whereby actions are alternated between the character and the enemies. During a battle, when a character attacks, the amount of damage inflicted, i.e. the amount of HP lost by the enemy, can be calculated by the following equation:

$$Total\ Damage = Damage_{weapon} \times \left(\frac{Attack\ Strength_{character}}{Defense\ Strength_{enemy}} \right)$$

where *Damage* (also known as base damage) is a value attributed to the weapon equipped on the character, *Attack Strength* is one of the physical characteristics of the attacking character, and *Defence Strength* is a physical characteristic of the enemy being attacked.

The same general equation is used when an enemy attacks one of the characters. It should be noted that the equation above is a simplistic example of attack damage equations. With the advancement of processors in video game consoles, the equations can include many more parameters. One such parameter is a random number generator, which provides for some variance in the damage inflicted when attacks are repeated. In the case of a damage equation with a random number generator, the occurrence of a 1 is often considered as a 'Critical Hit' as it corresponds to the highest possible damage given the other parameters are kept fixed. Some of the earliest electronic RPGs were developed in the 1970s.

MF

Equation 111: The Bakshali Square Root Formula

The Bakshali manuscript was discovered in the village of Bakshali near Peshawar in 1881. It is generally agreed that this manuscript was written sometime between the 2nd Century BCE and 3rd Century CE. The manuscript contains several rules and techniques for various mathematical problems [1]. One method described in this manuscript is the rational approximation to square roots of integers. This method is particularly captivating and intriguing due to the speed and accuracy with which square roots of positive numbers can be calculated.

THE METHOD
Consider a real, positive number a whose square root we wish to find. Let us denote the square root by x. The Bakshali manuscript (see [1]) gives the approximation $\tilde{x}$ to x as

$$\tilde{x} = x_0 \left[1 + \frac{\grave{o}}{2} - \frac{\left(\frac{\grave{o}}{2}\right)^2}{2\left(1 + \frac{\grave{o}}{2}\right)} \right]$$

with

$$\grave{o} = \frac{a - x_0^2}{x_0^2}$$

The equation was originally proposed for rational approximations to the square roots of integers and x_0 is the square root of the nearest perfect square. If the third term in the above equation is omitted, then the square root approximation is simply the Heron's method [1]. Thus, it appears that the Bakshali formula is an improvement on the Heron's method, also known as the Babylonian method. In fact, it turns out that the method is simply the Heron's method with two iterations.

The Bakshali square root formula is beautiful for many reasons.

The method involves only the elementary algebraic operations of addition, multi- plication and division.

The method in its original form provides a very good approximation to the square root of an integer. For example, suppose $a = 175$. Then, $x_0 = 13$ and the Bakshali approximation to the square root of 175 is 12.2287567084... which is accurate to the 7th decimal place. And this accuracy is achieved in one step! Equally astonishing is the fact that if we take $x_0 = 10$ instead of 13, the square root approximation is 13.2386363636... which is still accurate up to the third decimal place.

119

An iterative method based on the formula has been developed recently [2]. The method has the following form:

$$\delta_n = \frac{a - x_n{}^2}{2x_n}$$

and

$$x_{n+1} = x_n + \delta_n - \frac{\delta_n{}^2}{2(x_n + \delta_n)}$$

With the above form, the method can be shown to be equivalent to two consecutive applications of the Newton-Raphson method (beautiful equation 52) to the equation $(x) = x^2 - a$. Thus, the convergence rate of the method is quartic!! Note that the method was in use some 1300 years before Newton and Raphson.

Furthermore, the method can be applied to non-integers and the initial guess x_0 does not have to be the square root of the nearest perfect square.

Though a trivial point, both the positive and negative square roots can be obtained with the method. If the initial guess is negative, the method provides the negative square root. On the other hand, if the initial guess is positive, the positive square root is obtained.

AN EXAMPLE TO ILLUSTRATE THE FAST CONVERGENCE OF THE METHOD

An example calculation using the iterative version of the Bakshali square-root formula is shown in the following table for finding the square root of 2000.011. For comparison purposes, results from the quadratically convergent Newton-Raphson method are also shown. The table lists only the error between the calculated values and the exact value at each iteration. The table clearly shows the incredible quartic convergence rate of the scheme and confirms that it is simply the Newton-Raphson method applied twice in succession. The latter is clear from the fact that the error at every iteration in column 2 is exactly the same as the error in every other iteration in column 3.

Iteration	Error (Bakshali)	Error (Newton-Raphson)
0	-3.4721482533565454e+01	-3.4721482533565454e+01
1	1.7302604484562070e+01	6.0279067466434546e+01
2	6.1786355305506965e-02	1.7302604484562070e+01
3	2.0311038756486958e-11	2.4134182084912339e+00
4	2.3784278197403190e-49	6.1786355305506965e-02
5	4.4722012713624055e-201	4.2622544051122679e-05
6	5.5904504114390440e-808	2.0311038756486958e-11
7	1.0000000000000000e-998	4.6123056749929930e-24
8	—	2.3784278197403190e-49
9	—	6.3246102021203336e-100
10	—	4.4722012713624055e-201
11	—	2.2361271449984011e-403
12	—	5.5904504114390440e-808
13	—	1.0000000000000000e-998
14	—	1.0000000000000000e-998

Table 1: Comparison of the Bakshali formula and the Newton-Raphson method in calculating the square root of 2000.011. The exact value of the square root is 4.4721482533565454e+01. Initial guess was 10.0. The error in each case is calculated from $x_n - x$.

References:
http://en.wikipedia.org/wiki/Methods_of_computing_square_roots accessed on 2nd March 2015.
Bailey, D. H. and Borwein, J. M., 2012. *Ancient Indian square roots: an exercise in forensic paleo-mathematics*. The American Mathematical Monthly 119:646-657

HC

Equation 112: The Euler Characteristic

While the science of the Ancient Greeks has long been superseded, their geometrical discoveries remain as the foundations of our own - it is only a few decades since maths students were taught directly from Euclid's *Elements*, written over two millennia ago.

One geometrical legacy of the Greeks is the identification of the five Platonic solids (regular convex polyhedra whose faces are regular polygons), together with the proof that there can be no more than five of them. (Fig. 1)

The solids fascinated the ancient Greeks, who believed that the five elements were composed of atoms in the shapes of the solids, which conferred on them their characteristics (fire was made of spiky tetrahedral atoms, water of unstackable, rollable icosahedrons, and so on). In 1596, Kepler was still sufficiently in awe of the solids as to theorise that their nested shapes provided the dimensions of the Solar System. (Fig. 2)

Surprisingly, given such interest in the solids, it was not until 1750 that the simple mathematical relationship that describes them was discovered, by Leonhard Euler (arguably the greatest mathematician of all). He found that

$$faces + vertices - edges = 2$$

where 2 is known as the Euler Characteristic. In fact, 2 is the Euler Characteristic of any three-dimensional geometrical shape whose edges do not cross and whose faces do not intersect. For shapes in which edges do cross and/or faces do intersect (nonconvex polyhedra), the Euler Characteristic can take different values. The octohemioctahedron, for instance, has 12 vertices, 24 edges and 12 faces, and its Characteristic is therefore 0. (Fig. 3)

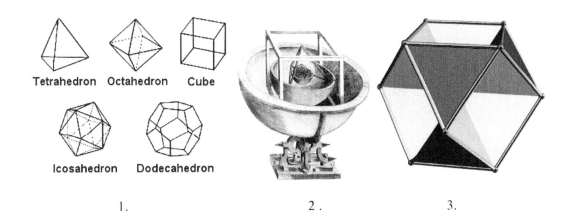

Tetrahedron Octahedron Cube

Icosahedron Dodecahedron

1.   2.   3.

MJG

Equation 113: The Grating Equation

$$n(\sin\theta_d - \sin\theta_i) = m\frac{\lambda}{\Lambda}, (m = 0, \pm1, \pm2, \dots)$$

where n is the refractive index of medium, θ_i is the angle between the incident beam and the grating surface normal, θ_d is the angle between the diffracted beam and grating surface normal, m is the diffraction order, λ is the wavelength of incident light and Λ is the spatial period of the grating.

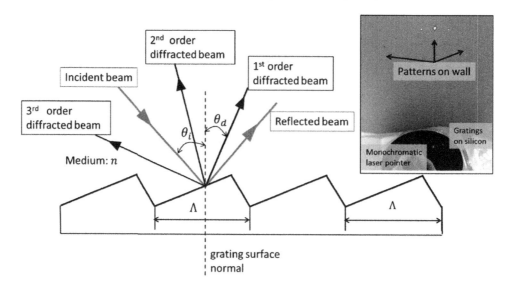

In optics, a grating is a diffraction component with a periodic structure, which splits incident light into beams causing different wavelengths to travel in different directions. In history, the German scientist Joseph von Fraunhofer (1787 - 1826) is one of the pioneers who studied the grating and established the diffraction theory. Fraunhofer established a company and built the first ruling engine (an engine that would scribe closely spaced parallel lines onto a flat surface) to make the gratings that he used to determine the wavelengths of light. His effort marked the start of spectral analysis, which has a significant impact in science and engineering. As pointed out by G. R. Harrison, a founder of the Spectroscopy Laboratory at MIT, in his paper J. Opt. Soc. Am. 39, 413-426 (1949), "No single tool has contributed more to the progress of modern physics than the diffraction grating, especially in its reflecting form."

The grating equation shows when a monochromatic light incidents on a grating surface at a given angle, there are a series of diffracted lights be generated. The diffraction orders (denoted by the integer m) are determined by the wavelength of the light and the grating spatial period. The up right corner of the figure displays a diffraction pattern on a wall by shooting a monochromatic laser to a grating.

Gratings are divided according to several criteria: their geometry, their material, their efficiency and the method of manufacturing them. In the figure, a reflection grating with a saw tooth profile is shown. Reflection gratings are often covered with a highly reflective material that can be sensitive to specific frequencies in the spectral regions of the incident light. Gratings that are reflective in one spectral region can be transparent in another.

The making of gratings with a spatial period at micro meter range, known as "ruling" has long history going back to Fraunhofer. The ruling method is to burnish a large number of fine grooves, normally parallel and equally spaced onto an optical surface. Building a ruling engine has itself been a driving force in the development of precision engineering design. Among those scientists that worked on design of these

122

engines, Henry Rowland (1848-1901) showed exceptional skill and made the largest and most accurate gratings during the early developments in the late 1800's.

References:

Lowen, E.G. and Popvo, E., 1997. *Diffraction gratings and applications* New York: Marcel Dekker

KN

Equation 114: Boolean Derivative

$$\frac{\partial f}{\partial x} = f(0, y) \oplus f(1, y)$$

where $f(x, y)$ is a Boolean function of the Boolean variables x and y, and $\oplus$ is the XOR Boolean operator (1 if the operands are different, 0 if equal).

Boolean algebra is a sub-field of algebra in which the variables can assume two values: true or false, often referred as 1 or 0. Boolean algebra has been fundamental in the development of digital electronics, and all modern programming languages have dedicated operators to work with it. Boolean algebra was born as a mathematical formalisation of logical reasoning and was first introduced by George Boole in 1847 in his book "The mathematical analysis of logic". A Boolean function is a relationship between an input and an output set of Boolean variables. The input variables are combined through Boolean operator such as AND, OR, NOT, XOR in order to provide a Boolean output. In a similar way as for function defined between set of real numbers, Boolean functions have derivative.

The derivative of a Boolean function with respect to a variable x is defined as the XOR (exclusive-or see beautiful equation 100) of the function values when $x = 1$ and $x = 0$. The term $f(0, y)$ and $f(1, y)$ are called the cofactors of f with respect to x.

An application of the Boolean derivative is in logic network fault detection. For example given a logic function $f(x, y, z) = x(y + z)$ and its implementation on chip we want to decide what input send to the network to test for a fault on the variable x path. Applying the Boolean derivative definition we found that $df/dx = (y + z)$. The derivative is 1 if either y or z is equal 1. Therefore by setting an input which has y or z equal to 1 the output is expected to change when x switch from 0 to 1, if not a fault is present on the variable x logic path. Another practical application is edge detection in binary or grey level images.

The Boolean derivative is part of a whole field called Boolean differential calculus (BDC), which originated from the treatment of electrical engineering problems in the areas of error-correcting codes and of design and testing of switching circuits. The development into a mathematical theory was achieved in 1959.

[Agaian, S.S.; Panetta, K.A.; Nercessian, S.C.; Danahy, E.E., "Boolean Derivatives With Application to Edge Detection for Imaging Systems," Systems, Man, and Cybernetics, Part B: Cybernetics, IEEE Transactions on , vol.40, no.2, pp.371,382, April 2010

S.B. Akers, "On a theory of Boolean functions" SIAM J., 7 (1959) pp. 487–498

A.D. Talantsev, "On the analysis and synthesis of certain electrical circuits by means of special logical operators" Avtomat. i Telemeh., 20 (1959) pp. 898–907 (In Russian)

Petzold, C., 2000. *Code: The Hidden Language of Computer Hardware and Software*. Redmond: Microscoft

GM

Equation 115: Thin Lenses

A thin lens is a lens with a thickness that is negligible compared to the radii of curvature of the lens surfaces. If we know the focal length of a thin lens and the position of the object there are three methods of determining the position of the image graphical construction; experiment; and use of the thin lens formula. The thin lens formula is

$$\frac{1}{s} + \frac{1}{s'} = \frac{1}{f}$$

where s is the object distance, s' is the image distance and f is the focal length measured from the centre of the lens.

The above is the lens formula in the Gaussian form. Karl Friedrich Gauss (1777-1855) published the first general treatment of the first order theory of lenses in 1841 in the now famous Dioptrische Untersuchungen.

The thin lens approximation ignores optical effects due to the thickness of lenses and simplifies ray tracing calculations.

The derivation is based on the fact that certain rays follow simple rules when passing through a thin lens, in the paraxial ray approximation:

- Any ray that enters parallel to the axis on one side of the lens proceeds towards the focal point F on the other side
- Any ray that arrives at the lens after passing through the focal point on the front side, comes out parallel to the axis on the other side; and any ray that passes through the centre of the lens will not change its direction.

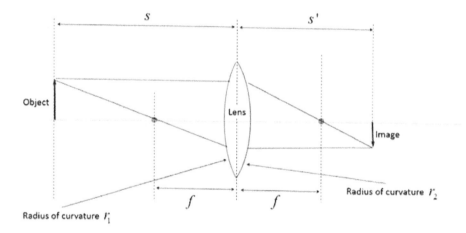

DF

Equation 116: The Cross Correlation Function

For two signals represented by x and y that vary with time (or any other linear variable), the cross correlation function $R_{xy}(\tau)$ at an arbitrary delay τ is given by:

$$R_{xy}(\tau) = E[x(t)y(t - \tau)]$$

124

$$R_{xy}(\tau) \approx \frac{1}{T - \tau} \int_0^{T-\tau} x(t)y(T - \tau)dt$$

Where the first equation uses the function $E[x(t)y(t - \tau)]$ to represent the expected value of the product of x and y with y being shifted and arbitrary time τ. The second equation represents an approximate method for computing the first equation when the signal is known only for a length of time, T. This equation is attributed to the English Polymath Francis Galton (1822 – 1911) who provided a vivid account of its inception in 1888 while sheltering from rain in Naworth Castle (near Carlisle in Northern England) in his 1908 memoir.

In general, if the two signals are random with zero average value then each might be represented by a list of random numbers with equally distributed positive and negative numbers. If both lists are multiplied together, then it would be expected that the result would also be a random number whose average, or expected, value would be zero. Not very interesting. However, if there is a common shape (even if it is random in nature or any other conceivable signal) in these two signals, when the two signals are lined up, then the numbers of the common signal will be either both positive or both negative. Hence the product of the matching signals will always be positive and the expected value becomes the mean square. To illustrate this, let's play a mathematical game of 'Where's Waldo'. Firstly, Waldo is here a burst of activity acting for a duration of about 4 seconds. However, he has been buried in a signal containing random noise and is almost impossible to see in the second illustration. However, he can be exposed by using a copy of Waldo as the signal y and multiplying it with the noisy signal x. By moving Waldo along the noise and computing the above integral, the cross correlation can be determined and is shown in the third figure. From this it is apparent that there is a distinct correlation at 12 second along the noise which is where he was originally buried. Clearly, this is an excellent tool for extracting patterns from signals and finds applications as diverse as determining optical signals sent to the moon and back to the optical mouse on your mouse pad. This latter device uses a digital camera to photograph the surface under the mouse. As the mouse is moved, cross correlation between the adjacent photographs provides a measure of the magnitude and direction of its movement.

It might also be intuitive that a mathematical transformation that is sensitive to the content of a signal might contain information relating its inherent characteristics. In fact, if a random (or other) signal is compared with itself, then for zero shift, the resultant expected value is simply the mean square of the signal. Shifting the signal and again multiplying it with an unshifted copy of itself, the chances of positive and negative signals coinciding will increase. Eventually, depending on the frequency content of the noise, the products of the two signals will look like the product of two random signals having once again an average of zero. Consequently, the Auto-Correlation Function will contain information about the power in the signal and how the power is distributed with frequency. In fact, it can be shown (and will be in a future beautiful equation) that the correlation functions are a precursor for extracting the spectral content of a signal. This is illustrated in the final pair of figures that show the autocorrelation, $R_{uu}(\tau)$ of a frequency band limited signal having a variance $\sigma_u{}^2$. Also shown in the power spectral density, $S_{uu}(\omega)$ of the signal for frequencies ω and the relation between bandwidth and the time constant, τ_0 in the autocorrelation function. The interested reader will find more detailed descriptions in literature on signals and random process analysis. The relation between correlation functions and the spectral content of signals will be established using the Weiner-Kinchine theorem, a future beautiful equation.

The cross correlation function plays an enormous part in the mathematical characterization of signals, in particular signals that do not contain distinct frequency components. Such signals never repeat and pose particular problems when it is desired to understand the power in the signal spanning particular bands of frequencies. The reader will notice the paradox in the above statement that is talking about frequencies that are periodic when considering signals that are not. This is a complex and fascinating topic of itself that leads directly to some of the most important mathematics behind quantum theory and is exemplified by Heisenberg's confounding principle presented as beautiful equation 15.

125

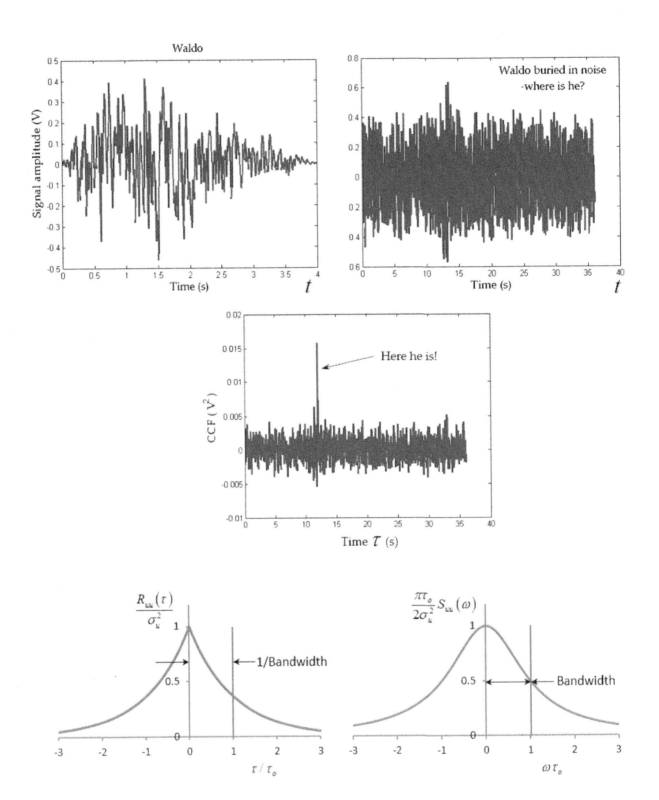

References:

Galton, F., 1888. *Co-relations and their measurement chiefly from anthropocentric data*. Proceedings of the Royal Society 45:135-145

Stigler, S.M., 1989. *Francis Galton's account of the invention of correlation*. Statistical Science 4:73-86

SS

Equation 117: Simple Harmonic Motion

$$F_{net} = m\frac{d^2x}{dt^2} = -kx$$

where m is the inertial mass of the displaced oscillating system, x is the displacement of the object (from its equilibrium state or mean position) and k is the spring constant of the system.

Simple harmonic motion is a type of regular motion of an object that is defined in classical physics and obeys Newton's second law of motion $F = ma$ (see beautiful equation 1). A common definition of simple harmonic motion is the mass spring system, however many rigid bodies and other mechanical systems also exhibit simple harmonic motion.

Simple harmonic motion defines a resonant frequency which an object will vibrate at if displaced from its equilibrium or non-excited, non-displaced state. It can be used to define motion in any mass spring system, an example of this is pendulum motion (see beautiful equation 42). The resonant frequency of an object or mechanical system is critical for many applications. In precision machine design it is vital to design systems that will not be excited by their respective resonant modes. Finite Element Analysis (FEA) software is typically used in precision machine design to engineer systems that have preferable resonant frequencies so that process dynamic conditions do not excite the mechanical structure.

Simple Harmonic motion is largely derived from Hooke's law (see beautiful equation 97). Robert Hooke was born in 1635 and published many significant papers on elasticity and vibration motion of springs. Simple harmonic motion also draws heavily on Newton's second law of motion published circa 1687. Both key laws from Hooke and Newton frame how simple harmonic motion

The one dimensional simple harmonic motion can be given by the equation below; this is a second-order differential equation with linear solutions.

Further differential solving of this equation can give the speed, force and phase relationships of the oscillating system.

TS

Equation 118: Dalton's and Henry's Laws of Partial Pressures

The English chemist, physicist and meteorologist, John Dalton, formulated this law in 1801, stating that:

"In a mixture of non-reacting gases at constant temperature and volume, the total pressure exerted by the mixture is equal to the sum of the pressures of the individual components as if they occupied that volume alone".

$$P(mixture) = p(1) + p(2) + \ldots p(n)$$

where, $p(i)$ is the pressure exercised by the individual gaseous component, also called partial pressure.

This law was followed by the law formulated by Dalton's colleague William Henry in 1803, which says that

"At a constant temperature, the amount of a gas that dissolves in a liquid of a specific volume, is directly proportional to the partial pressure of the gas in equilibrium with the liquid".

$$p(i) = kc(i)$$

where c is the concentration of the gas component in the liquid and k is the Henry's constant, which varies with temperature.

Henry's law implies that the solubility of the gas in the liquid is directly proportional to the partial pressure of the gas above the liquid. It always applies to the opening of a can of fizzy drink, such as Coke. In the sealed can, the concentration of carbon dioxide (CO_2) into the liquid is achieved by a balancing partial pressure of CO_2 in the "head space" of the can. When the can is open, CO_2 partial pressure drops, with the consequent release of an amount of CO_2 from the liquid solution, corresponding to the hiss produced by the can when opened.

SCUBA DIVERS

Henry's and Dalton's laws play an important role in scuba diving, where the effect of gas mixtures at high pressures could be fatal to divers. If we consider air being composed by 80 % nitrogen and 20 % oxygen in first approximation, the partial pressure exercised by nitrogen in the lungs is four times that of oxygen, according to Dalton's and Henry's laws. According to the ideal gas law (see beautiful equation 20), an increase in pressure corresponds to a decrease in volume of a gas, at constant temperature. Divers experience an increase of about 1 atmosphere in pressure for every 10 metre descent into the sea. At this depth, as the pressure of air doubles in their lung, the volume will shrink by half the volume at ambient pressure. Hence, in order to keep a constant volume of air in their lungs, divers need to breathe double amount of air from the tank. Analogously, at a depth of about 40 metres, the amount of air breathed will be about five times that at ambient pressure, corresponding to a fivefold increase in nitrogen concentration in the blood compared with that at ambient pressure. This will kill the divers. As a remedy, divers' tanks contain a helium-oxygen gas mixture, as helium solubility in blood is much lower than nitrogen.

The phenomenon of "the bends", or decompression sickness, is related to the formation of bubbles in divers' body as gas dissolved in their blood is quickly released when they re-surface too rapidly. This phenomenon can be fatal to divers. It is a consequence of Henry's law, in virtue of which a great variation between the partial pressure of a gas and its pressure within the solvating liquid (also called pressure gradient) causes a rapid release of gas from the liquid, leading to the formation of bubbles. This is experienced when a can of fizzy drink is shaken and then opened, and when water is boiled at high temperature: this is accompanied by bubbles forming at the bottom of the container as the temperature speeds up water molecules which in turn push dissolved gas out of the water.

AT

Equation 119: Time-Frequency Analysis and the Spectrogram – Part 2

Beautiful equation 88 showed how the spectrogram can be applied to a non-stationary multimode signal, but it was also evident that the resulting spectrogram suffered from limited resolution. This is due to the trade-off between time and frequency resolution and while a window size can be chosen that gives the best compromise, what is produced is still a somewhat blurry image.

The reassignment method goes a step further and focuses the energy in the time-frequency data to a central point. To use a mechanical analogy, it can be considered as focusing the mass of an object to its geometric centre.

The reassignment method was first developed by Auger (French Engineer) and Flandrin (French Physicist) and provides a method for "cleaning up" the spectrogram. In the reassignment method, energy of the signal is moved from its original location (t, ω) to a new location $(\hat{t}, \widehat{\omega})$ reducing the spread of the spectrogram and improving its resolution by concentrating its energy at a "centre of gravity". The reassigned coordinates $\hat{t}$ and $\widehat{\omega}$ for a spectrogram are

$$\hat{t} = t - \Re\left(\frac{S_{Th}(x,t,\omega).\overline{S_h(x,t,\omega)}}{|S_h(x,t,\omega)|^2}\right)$$

and

$$\widehat{\omega} = \omega - \Im\left(\frac{S_{Dh}(x,t,\omega).\overline{S_h(x,t,\omega)}}{|S_h(x,t,\omega)|^2}\right)$$

where $S_h(x,t,\omega)$ is the standard STFT of the signal x using window function $h(t)$, $S_{Th}(x,t,\omega)$ is the STFT using a time ramped version of the window $t \cdot h(t)$, $S_{Dh}(x,t,\omega)$ is the STFT using the first derivative of the window function $dh(t)/dt$, ω is the frequency, and t is the time.

The reassignment method can be considered a two-step process:

129

- Smoothing – the purpose is to smooth oscillatory interference but has the disadvantage of smearing localised components.
- Squeezing – the purpose is to refocus the contributions that survived the smoothing.

The reassigned version of the spectrogram given in beautiful equation 88 is shown in figure 1 and it is immediately clear that the detail is much sharper.

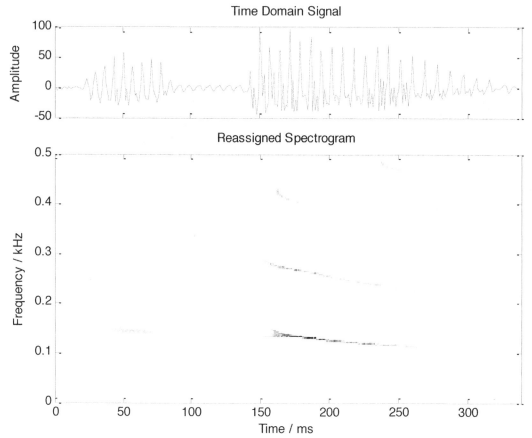

The drawback with the reassigned time-frequency method is that interference is caused by two closely spaced components. With time-frequency representations, there is a trade-off between resolution and localisation. If more than one component is seen with a time-frequency smoothing window, a beating effect occurs, causing interference fringes. This limits the reassigned time-frequency method from being a "super resolution" process.

References:
Gabor, D., 1946. *Theory of communication. Part 1: The analysis of information.* Journal of the Institution of Electrical Engineers-Part III: Radio and Communication Engineering, 93:429-441
Auger, F., Flandrin, P., Goncalves P. and Lemoine, O., 1996. *Time-Frequency Toolbox: For use with MATLAB*, CNRS, France-Rice University
Auger, F., Flandrin, P., 1995. *Improving the readability of time-frequency and time-scale representations by the reassignment method,"* IEEE Transactions on Signal Processing, 43:1068-1089
Fitz, K.R. and Fulop, S.A., 2009. *A unified theory of time-frequency reassignment.* arXiv preprint arXiv:0903.3080

CMcK

Equation 120: The Babylonian Method for Calculating Square Roots

In beautiful equation 111, the Bakshali square root formula, dating back to about 1800 years ago, was presented as a remarkably fast converging method for calculating the square roots of positive numbers. An equally remarkable formula, much older than the Bakshali formula, and still the most heavily favoured method of computing the square roots of positive numbers is the Babylonian Method (also known as the Heron's method). This method appears to have been known to the Babylonians as early as 1800 BCE. Heron of Alexandria, a Greek mathematician and engineer, was the first person to formalize the method in the first century CE. Pictures of a clay tablet from the Babylonian times showing the calculation of the square root of 2 are shown in the Figure below.

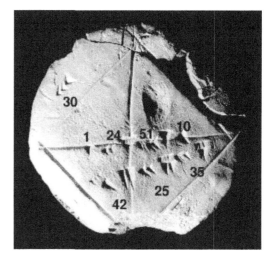

Figure 1: A clay tablet from the Babylonian times showing the Babylonian method of calculating the square root of 2. The tablet is known as the ybc7289 and is part of the Yale Babylonian collection. The left shows the approximation to the square root of 2 using the Babylonian symbols for numbers. The annotated right figure shows the same in modern notation. Note that the numbers are in the sexagesimal system that the Babylonians used for arithmetic. Photo credit: Bill Casselman (http://www.math.ubc.ca/~cass/Euclid/ybc/ybc.html) and the original holder of the tablet is the Yale Babylonian Collection.

Suppose that a is a positive real number, the square root of which we wish to find. The Babylonian method for computing the square root uses the following equation (in modern notation):

$$x_{n+1} = \frac{1}{2}\left(x_n + \frac{a}{x_n}\right)$$

$$n = 0,1,2,3 \ldots$$

In the above, x_{n+1} is the current value and x_n is the previous value. The iteration process starts with an initial guess x_0 and continues until the desired accuracy is reached. It is worth pointing out that the right hand side above is the same as the first two terms of the Bakshali approximation. The Babylonian method appears to have been known to the Indian mathematicians in the 8th century BCE and the Bakshali formula came much later. Thus, it is likely that the Bakshali formula was devised as an improvement over the Babylonian method.

As with Bakshali approximation, the Babylonian method also involves only elementary arithmetic operations. However, less number of operations are needed per iteration. It provides a rational approximation to the square root of a number. The method is equivalent to the Newton-Raphson method

131

for finding a root of the equation $f(x) = x^2 - a$ and therefore, is quadratically convergent. Convergence is guaranteed for any $x_0 > 0$. (A negative initial guess will converge to the negative square root of a). It is this quadratic convergence coupled with the fact that the calculations involve only elementary operations is what makes the Babylonian method very efficient and popular [2].

As an example, suppose we wish to find the square root of 28000. The following table shows the estimates of the root at each iteration for three different initial values. Clearly, the speed with which the square root is calculated depends on the proximity of the initial guess to the actual root. Thus, for fast calculation of the square root, a good initial guess is essential. Special techniques exist for choosing the initial guess (see [1]).

n	$x_0 = 1$	$x_0 = 100$	$x_0 = 150$
1	14000.5000000000	190.000000000000	168.333333333334
2	7001.24996428699	168.684210526316	167.334983498350
3	3502.62462507461	167.337425075950	167.332005333318
4	1755.30931521346	167.332005394584	167.332005306815
5	885.630459864862	167.332005306815	
6	458.623177642513		
7	259.837739008355		
8	183.798648682636		
9	168.069634081591		
10	167.333623969854		
11	167.332005314644		
12	167.332005306815		

Table 1: The Babylonian method for finding the square root of 28000 for three different initial guesses. The exact root is 167.332005306815 to twelve decimal places.

References
https://en.wikipedia.org/wiki/Methods_of_computing_square_roots accessed on 2nd March 2015.
Kosheleva, O., 2009. *Babylonian method of computing the square root: Justifications based on fuzzy techniques and on computational complexity.* Annual Meeting of the North American Fuzzy Information Processing Society, Cincinnati, USA, 1-6

HC

Equation 121: The Backpropagation Equations

Someone gives you the height and mass of a person and tells you that the person in question is either a jockey or a basketball player. Could you determine their profession from that information? The answer is "probably". We can generalise that most jockeys are short and light, whilst most basketball players and tall and heavy. This generalisation comes from experience that you have learned over time by seeing people with the label "jockey" or "basketball player". This ability to learn how to generalise if an item belongs in one set or another, just from examples of what those sets are made up of, is called the "learning classifier problem" by computer scientists and it is a popular use of a technique from artificial intelligence called neural networks.

In this example a given person would be represented by the collection of elements $\{h, m, c\}$ the height h (a number of centimetres), mass m (a number of kilograms) and "class label" c ("jockey" or "basketball player", assigned to some numerical value, 0 for jockeys or 1 for basketball players).

132

Neural networks have long been a staple of artificial intelligence. Neural networks are collections of "neurons". For a single neuron, the output is a weighted sum of its inputs, passed through some function (referred to as the "transfer function") that is selected by the designer. An example of a neuron is given in Figure 1.

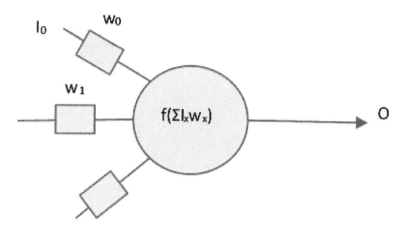

Figure 1: A single, three-input neuron. The output is some function of a weighted sum of the inputs.

A common transfer function is a "sigmoid function", an S-shaped curve that acts in a similar way to thresholding the input, so any value beneath the threshold is 0 and any value greater than or equal to the threshold is 1. Indeed, the equations presented assume that the transfer function is the sigmoid function and need to be updated if a different one is used. Networks of neurons "learn" by adjusting the weights between neurons until the output of the network classifies the examples presented to it correctly.

Classifiers that are only made up of a single neuron are referred to as "single-layer perceptrons". They enjoyed a brief period of popularity as learning classifiers as there were a number of well-established techniques for selecting the appropriate weights for a given problem. Sadly, there was a significant problem with single-layer perceptrons, as there were a large number of problems that they simply could not solve. These problems became known as "non-linearly separable problems".

To understand why, it's necessary to understand that the effect of the function that a neural network produces. The weighted sigmoid is equivalent to separating space into two regions and deciding that everything of one class belongs in one region and everything else belongs in the other. If your input space is only one dimensional (i.e. we're only looking at a number line representing the height of our jockeys and basketball players) then the "space" can be divided by a single point on the line, where everything above is one class and everything below is another. For the mass and height of our sportspeople our input space is two dimensional and looks like a flat surface (like a piece of paper). To separate that space we can draw a straight line to divide it into two. The learning process of adjusting the weights can be viewed as moving and rotating a line until space is divided into regions that correctly classify the examples that you have. This concept can be generalised into higher dimensional space as a "decision hyperplane", (moving from using lines to divide 2D space to using planes to divide 3D space etc.) but difficult for a human to visualise beyond 3 dimensions.

The trouble with "decision hyperplanes" is that some problems are not "linearly separable", i.e. it is not possible to draw a straight line that correctly classifies all of the inputs. Some problems require curved lines, or even bounded shapes to separate the input space into the correct partitions, something that a single neuron cannot do.

For example, instead of jockeys and basketball players, let's look at a dataset that we have all of the answers for - logical operations. Consider the two logical functions *AND* and *XOR* (the topic of a previous article).

Table 1: Truth Table for AND

i_0	i_1	O
0	0	0
0	1	0
1	0	0
1	1	1

Table 2: Truth Table for XOR

i_0	i_1	O
0	0	0
0	1	1
1	0	1
1	1	0

To allow a neural network to "learn" these operations we can say that our input tuple is the set $\{i0, i1\}$ where i_n is either 0 or 1. The class labels we want our classifier to learn are "things that make my logic gate say '0'" or "things that make my logic gate say '1'" which we'll simplify to the values 0 and 1.

The *AND* function is linearly separable - i.e. you can draw a line where everything on one side is a 1 and everything on the other is a 0. *XOR* is not linearly separable as there is no such line. See Figure 2.

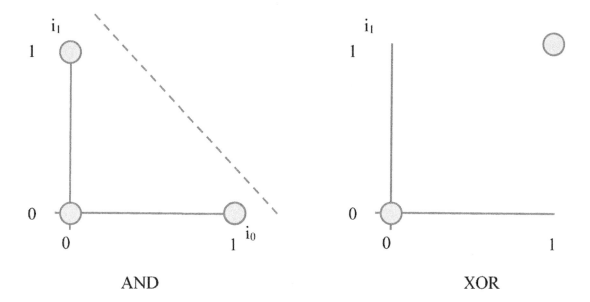

Figure 2: A graphical representation of the AND function and the XOR function input space. Circles are points where an output of 0 is required and crosses are points where an output of 1 is required. For the AND function a possible decision hyperplane is displayed as a dashed line. Note that no such line can be drawn for XOR.

For many years solving non-linearly separable problems (such as the *XOR* problem) was an impossible task for neural networks. Some people believed that by building up more complex networks of neurons the resulting decision boundary wouldn't be a hyperplane, but more complex shapes like curves and bounded shapes, thus being able to solve the linear separation problem. However, there was no way of reliably calculating the weights for multiple layers of neurons connected together, and so neural networks were considered by many to be a research dead-end, unable to solve anything other than toy problems in a lab.

After a nearly twenty year stall in neural network research caused by this seemingly intractable problem, Rumelhart, McClelland and Williams published their backpropagation equations that could determine the appropriate weights to apply to more complex networks of neurons. As the name "backpropagation" suggests, the equations start by altering the weights that feed into the output neuron(s), then systematically applying weight changes to progressively earlier network layers.

For the weight linking neuron k to neuron i, the change in value of the weight is given by:

$$\Delta w_{ki} = \alpha \delta^k x_k^p$$

For output neurons the term δ^k is given by:

$$\delta^k = x_k^p (1 - x_k^p)(t_k^p - x_k^p)$$

For non-output neurons the term δ^k is given by:

$$\delta^k = x_k^p (1 - x_k^p) \sum_{j \in I} \delta^j w_{kj}$$

135

where α is a learning constant, t_k^p is the target value for neuron k in the presence of pattern p, x_k^p is the current output of node k in the presence of pattern p, I is the set of neurons that the output of neuron k connects to.

This radically changed the face of artificial intelligence and made it possible for neural networks to address non-linearly separable, real-world problems. Neural networks have since been applied to handwriting recognition to automate postal systems, automated cancer diagnosis for mass medical screening programs and are even used by banks to decide if you are eligible for a loan or not!

RFO

Equation 122: Seismic Moment of an Earthquake

$$M_0 = \mu As$$

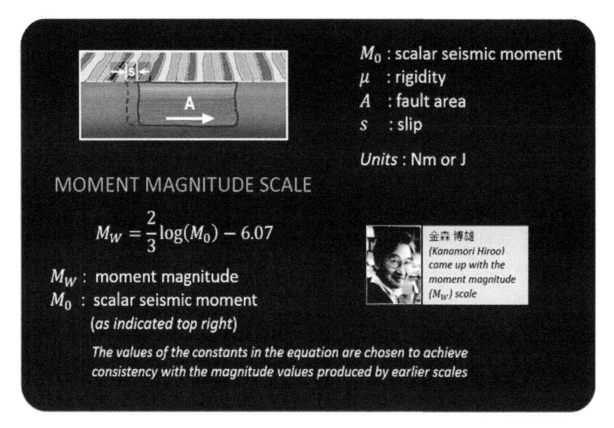

SEISMIC MOMENT & THE MOMENT MAGNITUDE SCALE
The concept of 'seismic moment' M_0 – a quantitative means of measuring the amount of energy released by an earthquake – was introduced in 1966 by Keiiti Aki.

M_0 takes into consideration such factors as the length and depth (area A) of the rupture along the causative fault, the rigidity μ of the displaced rocks, and the distance the rocks slip (displacement s).

M_0 can be calculated from either (1) recorded seismic waves or (2) field measurements of the size of the fault rupture.

In the 1970s, not long after M_0 was introduced, Hiroo Kanamori and Tom Hanks used M_0 as the basis for defining the moment magnitude scale MMS (symbol M_w or M) to succeed both the local magnitude (M_L) –

136

commonly known as Richter – scale and the surface wave magnitude (M_S) scale. M_w is the base 10 logarithm of M_0, the formula being tweaked by means of constants so that the familiar Richter magnitude values are retained:

$$M_w = \left(\frac{2}{3}\right) \log(M_0) - 6.07$$

Scientists consider M_0, and hence M_w, a more reliable, uniform and soundly based (on classical mechanics) indicator of earthquake energy than botzh the local magnitude (M_L) scale and the surface wave magnitude (M_S) scale.

HISTORICAL MEASURES OF EARTHQUAKE SIZE

The M_L scale is defined as $M_L = \log\left[\frac{A}{A_0}(\delta)\right]$, where A is the maximum excursion of the seismograph, the empirical function A_0 depends only on the epicentral distance of the station, δ.

The latter was developed in 1935 by Charles Richter and Beno Gutenberg with the goal of quantifying medium-sized earthquakes in Southern California. The scale was based on the ground motion measured by a particular type of seismometer 100 km from the epicentre. The M_L scale corresponded well with the observed local damage, but was unreliable for large earthquakes and for measurements taken at a distance of more than about 600 km from the epicentre.

In 1950s Gutenberg and Richter subsequently developed the surface wave magnitude (M_S) scale, based on measurements in Rayleigh surface waves. This scale is currently used in People's Republic of China as a national standard.

Moment magnitude values are often reported as 'Richter' values, even though a different formula is used to the older ones.

FR

Equation 123: Stellar Fusion Part 1: P-P Chain Reactions

Net process

$$6\,^1H + 2e^- \longrightarrow\, ^4He + 2\,^1H + 2v_e + energy$$

Stars are the sources of all of the elements in the periodic table except for the very lightest. Before the first stars formed only hydrogen, primordial helium and their isotopes were in existence. (Some lighter elements such as lithium and beryllium can also be produced by the interaction of matter with cosmic rays.) The relative abundances of elements and ageing of stars is explained by nucleosynthesis in the cores of stars. The idea of stellar fusion of hydrogen to form helium was first proposed by Arthur Eddington in 1920, and Eddington also speculated that stars are the source of heavier elements. Much of this was theory, treating a star as an ideal gas, but turned out to be quite relevant. The rate of the nuclear reactions was derived by George Gamow in 1928. By 1939, the German nuclear physicist Hans Bethe had defined the main processes of the source of energy in stars, the proton-proton (p-p) chain reaction which was initially proposed by Gamow, and the carbon-nitrogen-oxygen (CNO) cycle. Bethe proposed an intermediate deuterium production stage in the p-p chain reaction which allowed for the highly unstable diproton state to be accounted for via beta decay. The theory of production of heavier nuclei was addressed by Fred Hoyle in 1946; later in 1954, in a major piece of work, Hoyle described the nucleosynthesis of the elements from carbon to iron. The observed relative abundance of the elements was accounted for by Hoyle and others in 1957. Further advances were made by Alistair Cameron and Donald Clayton in the following years. Of all these contributors to this important theoretical area of astrophysics, only Hans Bethe eventually received the

Nobel Prize for physics in 1967 – mainly for his discovery of the *CNO* cycle at Cornell University in the 1930s. Empirical evidence of these theorised processes is largely borne out by measurements of the sun's core temperature and luminosity, and solar neutrino detection experiments which began in the 1960s and are ongoing. Raymond Davis shared the Nobel Prize in 2002 for neutrino detection as a contribution to astrophysics. Discrepancies between the results of neutrino experiments and solar theoretical models over the last few decades were recently resolved by neutrino oscillation (change in neutrino type).

The two principal processes of hydrogen fusion in stars are the p-p chain reaction, the dominant fusion reaction in stars up to approximately the mass of the sun, and the *CNO* cycle, more important in larger stars but still responsible for about 7% of the sun's energy output. A part of the p-p chain reaction is deuterium burning, which occurs at a much lower temperature, and can also proceed from primordial deuterium as an independent process in substellar objects such as brown dwarfs, and explains why these objects shine for about 100 million years until their deuterium supply is exhausted.

The p-p chain reaction is summarised in this article, and the beautiful equation shown here is the net process. The energies given below in MeV are approximate to the nearest 0.1 MeV. The first reaction is the fusion of two protons under compression (note that at the high temperature and pressure of the stellar core, all atomic nuclei are stripped of their electrons)

$$^1H + {}^1H \longrightarrow {}^2He$$

The diproton ^2He is very unstable and in most cases decays immediately back to two hydrogen nuclei (proton emission). However, the rare weak nuclear $\beta+$ decay of the diproton to deuterium is responsible for the next step towards helium

$$^2He \longrightarrow {}^2D + e^+ + \nu_e$$

0.2% of these reactions involve an electron on the left side of the equation (the "pep" reaction) instead of a positron on the right; otherwise the positron is quickly annihilated with a neighbouring electron releasing two gamma rays of total energy 1 MeV. The neutrino carries off an energy of 0.4 MeV. The rare occurrence of this reaction compared with the decay back to hydrogen makes the first stage of nucleosynthesis very slow, in fact the half-life for this fusion stage in the sun is about a billion years, which explains why a star such as the sun keeps shining for so long and will continue to do so for several billion years. Only one reaction in 10^{19} results in a fusion chain reaction.

Once the deuterium is available it does not last long and the next step towards helium is deuterium burning involving the strong nuclear interaction, and occurs within seconds

$$^2D + {}^1H \longrightarrow {}^3He + \gamma$$

The energy of the gamma ray released here is 5.5 MeV. The light isotope of helium lasts for about a million years, after which the final step occurs mostly as follows, in the temperature range 10-14 million Kelvin

$$^3He + {}^3He \longrightarrow {}^4He + 2\,{}^1H$$

A further 12.9 MeV of energy is released in this reaction. It can be seen that the production of a helium nucleus requires two of the chains described above, and the total energy released (thermal and radiative) is about 26.7 MeV which is equivalent to the difference in mass between four protons and a helium nucleus (alpha particle), by Einstein's mass-energy equivalence (BE74). The time interval between production of radiative energy and the emergence of visible light photons in the star's photosphere (after much scattering, absorption and re-emission) is around 30,000 years. The last process shown above is the dominant one and in most of the remaining 14% of cases, at temperatures up to 23 million K, the reaction involves beryllium and lithium nuclei as intermediates, and in a small smaller number of cases at even higher temperatures boron is produced before decaying back to beryllium and helium – this last reaction is not significant in terms of overall energy production but is of interest in observations of solar radiation as it generates high energy (14 MeV) neutrinos.

In larger stars the CNO cycle becomes more significant and will be covered in the next article by this author.

PB

Equation 124: Banzhaf Power Series

The Banzhaff Power Series, sometimes known as the Penrose-Banzhaff Power Series was invented by Lionel Penrose, FRS (11 June 1898 – 12 May 1972) (Father of Sir Roger Penrose), a British psychiatrist, medical geneticist, mathematician and chess theorist, and was later used by John F. Banzhaf III (b. 1940), an America Law professor to define by the probability of changing an outcome of a vote where voting rights are not necessarily equally divided among the voters.

During the 2015 General Election, the media speculated there may be a second Hung Parliament, i.e. there is the possibility no one party will win enough seats to form a government, so a coalition would have to be formed to allow a minority government to pass legislation.

The tool for calculating how power is distributed in these circumstances is the Banzhaf Power Index given below:

$$B_i = \frac{t_i}{\sum_{k=1}^{N} t_k}$$

where B_i is the Banzhaf Power Index (BPI), t_i is the number of times a party is a critical party, t_k is the total number of times any party is a critical party.

The analysis below was conducted by Dr Hannah Fry, a lecturer in Mathematics at University College London. The equation calculates the proportion of cases where a party could provide the swing vote to make a losing side a winning side.

Scenario 1			Scenario 2		
Party	**Seats**	**B$_i$**	**Party**	**Seats**	**B$_i$**
Con or Lab	280	32.6%	Con or Lab	285	39.7%
Lab or Con	275	24.2%	Lab or Con	270	20.1%
SNP	39	21.3%	SNP	39	20.1%
LibDem	30	7.1%	LibDem	30	9.8%
DUP	9	7.1%	DUP	9	5.1%
UKIP	4	2.9%	UKIP	4	2.1%
PC	3	2.4%	PC	3	1.4%
SDLP	2	1.1%	SDLP	2	0.7%
Greens	1	0.7%	Greens	1	0.5%
Ind (UU)	1	0.7%	Ind (UU)	1	0.5%
Sinn Fein	5	-	Sinn Fein	5	-
Speaker	1	-	Speaker	1	-

It makes no difference if the Labour Party or Conservative Party is the larger of the two main parties. It is also assumed that the five predicted seats to be won by Sinn Fein will not be taken up by their MPs. This gives a threshold for a Commons majority of 323 seats.

The two big parties would need to win 273 seats if the Lib Dems won 30 seats and the smaller parties won 20 seats, to not require the backing of the SNP. Considering the two scenarios in the table above, the SNP power score is much higher than its proportion of seats in scenario 1 and it is the same as the lower of the top two parties in scenario 2. If the Lib Dems win 30 seats, they are in a much weaker position than the SNP who could win 39 seats.

In these scenarios, the Lib Dems could not provide the swing vote for a majority without the assistance from smaller parties, but the SNP could just provide this swing vote. If one of the two big parties is still short of a majority, but has a large enough lead, the Lib Dems would move into the same position as the SNP. Likewise, if one of the main two parties is very close to a majority, any of the other small parties would start to catch up.

M. Rosenbaum, "Election 2015: Doing the maths on a hung Parliament," 13th April 2015. [Online]. Available: http://www.bbc.co.uk/news/32264701

CMcK

Equation 125: The Drude-Lorentz Model of Electrical Conduction

The Drude-Lorentz model gives the relation between electrical and optical properties of a conductive material. It explains - for example - why metal surfaces are shiny.

The model can be expressed in different ways, we give one of them:

$$\varepsilon(\omega) = \big(n(\omega) - ik(\omega)\big)^2 = \varepsilon_\infty - \omega_p^2 \cdot \frac{\varepsilon_0}{(\omega^2 - i\omega\gamma)}$$

Here ω is the circular frequency $\omega = 2\pi c/\lambda$, with λ the wavelength of light, n and k are the real and complex part of the referactive index, ε_∞ is the high-frequency limit of the permittivity, ε_0 is the vacuum permittivity and c is the speed of light. ω_p is the plasma frequency given by:

$$\omega_p^2 = \frac{n_- \cdot e^2}{\varepsilon_0 \varepsilon_\infty m_{eff}}$$

with n_- the electron density, e the elementary charge and m_{eff} the effective electron mass. The relaxation frequency γ is related to the electron mobility μ by:

$$\gamma = \frac{e}{m_{eff}\,\mu}$$

For frequencies well below the plasma frequency, the normal reflectance R can be approximated by:

$$R(\omega) = \frac{(n(\omega) - 1)^2 + k(\omega)^2}{(n(\omega) + 1)^2 + k(\omega)^2} = 1 - 2 \cdot \sqrt{2\rho\omega\varepsilon_0}$$

140

This is known as the Hagen-Rubens relation (also beautiful). The reflectance will approach unity with decreasing circular frequency and with decreasing resistivity ρ of the material. The high reflectance of e.g. silver in the visible spectrum is well explained by this. Light is absorbed near the plasma frequency, and a material becomes transparent for frequencies above the plasma frequency.

The theory was first developed by Drude in 1900 and extended by Lorentz in 1905.
Hendrik Antoon Lorentz (1853 – 1928) was a Dutch physicist who was a nestor in Physics in the times when the relativity theory and later the quantum physics was developed. He made important contributions to these fields (Lorentz force, Lorentz transformations, etc.).
Paul Karl Ludwig Drude (1863 – 1906) was a German physicist specializing in optics.

HH

Equation 126: The Principle of Buoyancy

The principle of buoyancy describes the phenomenon that some objects float when immersed in a liquid, while other objects sink. When an object is placed in a liquid, the weight of the object (due to gravity) pulls it downward; however, at the same time, the liquid asserts a repulsive force caused by its displacement by the object. When the weight of the object is greater than the force of the displaced liquid, the object will sink. Alternatively, if the weight of the object is less than the force of the displaced liquid, the object will be 'buoyed', or pushed upwards towards the surface of the liquid. There can exist a state of equilibrium, whereby the weight of the object equals the repulsive force of the surrounding liquid. In this case, the object will be immersed in the liquid at the depth corresponding to this equilibrium state.

The concept of buoyancy was first recorded by Greek know-it-all Archimedes at some point in his life, which is believed to be between 287 – 212 BC. He claimed,

"Any object, wholly or partially immersed in a fluid, is buoyed up by a force equal to the weight of the fluid displaced by the object."

The equation of buoyancy for an immersed object can be written as follows.

$$F_{net} = F_{buoyancy} - F_{weight}$$

where F_{weight} is the downward (due to gravity) force caused by the weight of the object, $F_{buoyancy}$ is the repulsive or 'buoyant' force of the surrounding liquid, and F_{net} is the net force imposed on the object.

A positive F_{net} results in the object being buoyed upwards towards the surface of the liquid, a negative F_{net} results in the object sinking, while a zero F_{net} indicates a state of equilibrium of the object whereby it does not change its depth in the liquid.

The principle of buoyancy can be described in terms of the relative densities of the object and liquid. When the average density of an object is greater than the average density of a liquid, the object will sink. When the average density of an object is less than the average density of a liquid, the object will float. When the average densities are equal, the object is in buoyant equilibrium.

MF

Equation 127: Viviani's Curve

Consider the parametric equations:

$$x(t) = a\cos^2 t, y(t) = a\sin t\cos t, and\ z(t) = a\sin t$$

where t is a parameter taking on values from 0 to 2π. These equations describe the intersection curve of a sphere of radius $2a$ and a cylinder of radius a. The cylinder is internally tangent to the sphere and passes through the centre of the sphere. The intersection curve (figure-of-eight on a sphere) is known as the Viviani's curve named after the Italian mathematician Vincenzo Viviani who found this curve in 1692 as a solution to an architectural problem [1]. Incidentally, Viviani was a student of Torricelli and a disciple of Galileo. The Viviani's curve has recently found applications in motion planning algorithms for implementation in controls systems for robotic mechanisms [2]. The curves were also considered as possible candidates in defining spatial kinematics for insect-like flapping wing micro air vehicles [3].

Suppose we have a sphere of radius $2a$ with centre at (0,0,0) and a cylinder of radius a centred at (a,0,0) (see Figure 1). The resulting Viviani curve is shown in blue in this figure. Various views of the curve are shown in Figure 2.

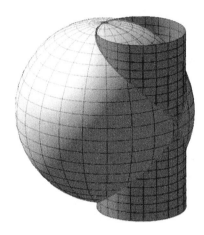

Figure 1: Viviani's curve is the curve traced by the intersection of a cylinder with a sphere

Figure 2: Front, top and right views of the Vivian Curve.

References
Caddeo, R., Montaldo, S., & Piu, P. (2001). The möbius strip and Viviani's windows. The Mathematical Intelligencer, 23(3), 36-40.
Svinin, M., & Hosoe, S. (2008). Motion planning algorithms for a rolling sphere with limited contact area. Robotics, IEEE Transactions on, 24(3), 612-625.
Galinski, C., & Zbikowski, R. (2005). Insect-like flapping wing mechanism based on a double spherical Scotch yoke. Journal of the Royal Society Interface, 2(3), 223-235.

HC

Equation 128: Bernoulli's Equation

Take a strip of paper about 3 × 10 cm, hold it by one end just under your lower lip, and blow...

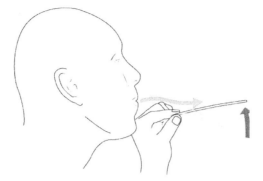

... and that's how planes fly. Well, partly, at least. This, and a multitude of other party tricks, are understandable thanks to the work of Daniel Bernoulli (1700 - 1782), a member of a celebrated family of Dutch mathematicians. One might imagine that blowing along a paper or other light flexible object would force that object out of the air-jet, but the reverse is in fact the case. The reason is that energy is conserved: increasing the velocity of air increases its kinetic energy ($= \frac{1}{2}mv^2$) and hence its pressure must fall ($pressure = \frac{energy}{volume}$).

Bernoulli published the mathematical form of his Principle in 1738:

$$\frac{v^2}{2} + gz + \frac{p}{\rho} = Constant$$

where: v is the velocity at a point, g is the acceleration due to gravity, z is the height, p is the pressure at a point and ρ is the density.

Bernoulli's mathematical interests were wide-ranging: among many other highly original pieces of work he developed a mathematical theory economics, based on his belief that the moral value of the increase in a person's wealth is inversely proportional to the amount of that wealth.

Bernoulli's life was somewhat blighted by his father Johann, also a mathematician, who was so jealous at his son's greater brilliance that he banished him from his house when they both entered a mathematical competition in 1734 and were awarded equal honours.

MJG

Equation 129: Hertz Contact

In 1882, German physicist Heinrich Hertz (best known for his pioneering work of proving the existence of electromagnetic waves) published his analysis of the distribution of stress between two contacting solid bodies. His work, which was apparently developed during a Christmas vacation in 1880 when attempting to determine elastic deformation of lenses in contact with flat surfaces, marked the start of classical contact mechanics theory.

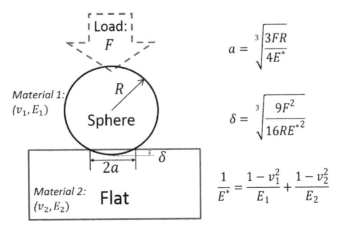

$$a = \sqrt[3]{\frac{3FR}{4E^*}}$$

$$\delta = \sqrt[3]{\frac{9F^2}{16RE^{*2}}}$$

$$\frac{1}{E^*} = \frac{1-v_1^2}{E_1} + \frac{1-v_2^2}{E_2}$$

where a is the contact radius, δ is the depth of deformation, E^* is the effective elastic modulus, F is the vertical force load, E_1 and v_1 are the elastic modulus and the Poisson's ratio of material 1 respectively and E_2 and v_2 are the elastic modulus and the Poisson's ratio of material 2 respectively.

As illustrated in the figure, a stationary vertical force is applied to a sphere that is in contact with a horizontal flat surface. In this simplified case no adhesive mechanism between the two solids is being considered and the contacting surfaces are assumed to be smooth and frictionless. The equations express the contact radius zone (a), the depth of deformation (δ) as function of the loading condition, geometric feature and material properties of the two solid bodies.

Using the Hertzian analysis, the deformation between contacting elements with finite stiffness can be calculated, for example the contact between two spheres, the contact between two crossed cylinders of equal radius and so on. For instance, in precision mechanism design, a kinematic mount is commonly used to ensure that a specimen can be placed back into a machine or instrument after it has been removed for some experimental purpose. This mounting device is typically constructed by attaching spheres or cylinders to the object holding the specimen. Upon relocation, these spheres locate in flats and/or V grooves that are machined into the base of the instrument. The Hertzian equations can be used to determine stresses and deformations and, therefore, performance of the kinematic mount. A handy toolbox for calculating elastic compression between spheres and cylinders can be found at NIST's engineering metrology online toolbox.

References:
S.T. Smith, D. G. Chetwynd. Foundations of ultraprecision mechanism design. Taylor & Francis Books Ltd. 1992, ISBN: 2-88449-001-9.
NIST   Engineering   Metrology   Toolbox, http://emtoolbox.nist.gov/Elastic/Documentation.asp
K.L. Johnson, Contact mechanics, Cambridge University Press, 1987, ISBN 0-521 34796 3.

KN

Equation 130: Small Signal Model for CMOS Transistor

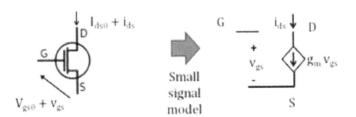

$$i_{ds,sat} = \left| \frac{\partial I_{ds,sat}}{\partial V_{gs}} \right|_{V_{gs0}} * V_{gs} = \frac{W}{L} \mu C_{inv}(V_{gs0} - V_{th}) * V_{gs} = g_m V_{gs}$$

where I_{ds0} and V_{ds0} are the large signal current and gate voltage, i_{ds} and v_{gs} are the small signal current and gate voltage, W is the transistor width, L is the transistor length, μ is the electron mobility, C_{inv} is the gate capacitance, V_{th} is the threshold voltage and g_m is the transistor transconductance.

The current flowing through a CMOS transistor is a function of the applied voltage between the Gate terminal (G) and the Source (S), and is described more in details in Equation 55. The CMOS current equation can be linearised (if the applied voltages do not vary much) in order to approximate the non-linear transistor by a linear electric network. When linearised, the CMOS transistor in the saturation region is a current generator where the parameter g_m (transconductance) express the transistor current gain, i.e. the infinitesimal current change associated with the gate voltage change v_{gs}. The small signal model of the CMOS transistor is widely employed in the design of analogue integrated circuits, where each transistor is set in the operational region by the bias voltage and current v_{gs0} and I_{ds0}, and the variation around those operational points are the electronic signals to process. More advanced models includes the non-ideal effects of the CMOS transistor such as the parasitic capacitances in order to characterise the small signal behaviour in the frequency domain.

GM

Equation 131: 1+1=2

The most trivial equation. But how do you prove it? You could argue that it's not an equation merely an equality.

It's the first basic mathematics we learn. I give you one thing and then I give you another thing and we define that state as having two things. However, this is just us using language and doesn't help us in mathematics. In mathematics, we need a proof.

If we look more detail, we have to think what do we mean by oneness and what exactly is this operator addition.

The proof is too long to put here but can be found at http://tachyos.org/godel/1+1=2.html. As stated on this website this certainly isn't the longest proof that $1 + 1 = 2$: that honour probably goes to Alfred North Whitehead and Bertrand Russell in Principia Mathematica, where they develop mathematics from an abstract version of set theory, and get around to proving $1 + 1 = 2$ on page 362.

Actually, they weren't there yet. After 378 pages, they were able to describe how you could prove that $1 + 1 = 2$. But they couldn't actually do it yet, because they hadn't yet managed to define addition
It is interesting to read what Mark Dominus has to say on his blog about Whitehead and Russell's proof in particular this statement

"Principia Mathematica is an odd book, worth looking into from a historical point of view as well as a mathematical one. It was written around 1910, and mathematical logic was still then in its infancy, fresh from the transformation worked on it by Peano and Frege. The notation is somewhat obscure, because mathematical notation has evolved substantially since then. And many of the simple techniques that we now take for granted are absent. Like a poorly-written computer program, a lot of Principia Mathematica's bulk is repeated code, separate sections that say essentially the same things, because the authors haven't yet learned the techniques that would allow the sections to be combined into one."

DF

Equation 132: The Wiener-Kintchine Theorem

$$R_{xy}(\tau) = \int\limits_{-\infty}^{\infty} S_{xy}(j\omega)e^{j\omega\tau}d\omega$$

$$S_{xy}(j\omega) = \frac{1}{2\pi} \int\limits_{-\infty}^{\infty} R_{xy}(\tau)e^{-j\omega\tau}d\tau$$

Where $R_{xy}(\tau)$ = the cross correlation function between signals x and y at a time delay τ, $S_{xy}(j\omega)$ = the cross spectral density of signals x and y at frequency ω, e is the exponential factor (2.7182818...) and j is equal to $\sqrt{-1}$.

When listening to music, it appears possible to determine the tones that are present and the loudness of each. However, in the development of a mathematical procedure, this presents a number of dichotomies. When considering a tone, it is implied that the signal contains both a specific frequency and amplitude and that both persist forever. However, for a musical piece the tones are transient and, even for the most repetitive music, the sound will not be truly periodic.

Truly periodic signals can be perfectly broken down into discrete frequency components that are exact multiples of the fundamental period (see beautiful equation 12). However, when the signal does not repeat, it becomes necessary to develop a mathematical technique to extract the frequency components assuming a periodicity of infinite duration, a procedure that results in the Fourier Transform (a future beautiful equation). Signals analysed using this transform indicate a continuous distribution of frequencies. However, the Fourier transform breaks down for a single frequency and returns an amplitude of infinity over an infinitely thin band of frequencies. Not only this, the Fourier transform also breaks down if random signals, for which there is no periodicity, are used. So how is it possible to determine frequency components of a signal?

A solution to this dichotomy is to consider the energy content, or power, in the signals. For this consideration, the total energy will be bounded and the mathematical formulation of the Fourier transform will produce results that, at least, predict power in the signal. Based on this, we define a power spectral density, $S(\omega)$, that measures the amount of power in a signal over a frequency band by computing the area under the power spectral density plotted against frequency. Because this does not provide single frequency information, it does appear to remove most of the previous dichotomies.

In practice, it should be possible to determine the continuous spectra of the power in a signal using a Fourier transform. The question becomes, what is the signal that is used in the Fourier transform to determine the power spectral density? Amazingly, using various theorems related to the Fourier Transform, it is possible to demonstrate that the original signal must be first transformed into its auto-correlation function (see beautiful equation 116) after which the Fourier transform produces the power spectral density. In fact, it is possible to use the Fourier transform pair (shown above) to predict the cross correlation function from the power spectral density and vice versa. This also validates the previous statement in beautiful equation 116 that the cross, and self (or 'auto'), correlation functions contain all of the frequency content of non-periodic or random signals. The proof of this theorem was first presented by the American Mathematician Norbert Wiener (1894 – 1964) in 1930 with a more general solution being developed four years later by the Russian mathematician Aleksandr Yakovlevich Khinchin (1894 – 1959) (whose first name is often spelled Alexander and his surname also appears in literature as Kintchine).

146

Khintchine, Alexander (1934). "Korrelationstheorie der stationären stochastischen Prozesse". *Mathematische Annalen* **109** (1): 604–615. doi:10.1007/BF01449156.
Wiener, Norbert (1930). "Generalized Harmonic Analysis". *Acta Mathematica* 55: 117–258. doi:10.1007/bf02546511

SS

Equation 133: Continuous Wavelet Transform

Wavelet transforms and wavelet analysis are useful tools for processing data when looking for specific signatures buried within a noisy signal. Wavelet transforms are typically compared to a normalized sinc function. Wavelet functions equate to a value of 1 where the compared signal matches the wavelet. Therefore signals with values close to 0 do not exhibit the signature being looked for and signals with values closer to 1 do exhibit the signature or characteristics being looked for. The biggest distinction between more traditional signal processing analysis techniques and wavelet techniques is that wavelets look for a specific signature within a signal, whereas most other analysis techniques reduce a signal down to specific components that make up the signal.

A continuous wavelet transform is defined by the following equation

$$\psi(t) = 2\operatorname{sinc}(2t) - \operatorname{sinc}(t) = \frac{\sin(2\pi t) - \sin(\pi t)}{\pi t}$$

The normalised sinc function is as follows as used in digital signal processing, an unnormalised sinc function is also used as shown below. The normalised sinc function is used to make the equating parts of the signal (the definite integral) that match the appropriate wavelet equal 1.

$$\operatorname{sinc}(x) = \frac{\sin(\pi x)}{\pi x}$$

$$\operatorname{sinc}(x) = \frac{\sin(\pi x)}{\pi x}$$

Many wavelet techniques exist to break a signal into different parts or spectrums for appropriate analysis to be conducted.

Wavelet analysis and wavelet theory started with Alfréd Haar's work on the first wavelet in 1909. There were no significant breakthroughs or advancements in this field until the mid-20th century. It was not until that later part of the 20th century when many adaptations and analytical methods around wavelets were proven. The continuous wavelet was discovered by George Zweig in 1975 and improved upon by Jean Morlet in 1982.

STS

Equation 134: Haplodiploidy in Social Insects

Social insect species of the order *Hymenoptera* (membrane-winged insects: ants, bees and wasps) are in many respects pinnacles of social organisation in the animal kingdom. They lack man's intelligence and the culture of human society, but with regard to social cohesion, caste specialisation, and individual altruism, they are beyond comparison. Some of the social characteristics of honeybees, such as bravely defending their colonies against attack, and the tendency to store almost unlimited supplies of food, may be well-known, but they are not easy to understand. Charles Darwin, for instance, realised that the evolution of worker castes, which being infertile leave no offspring, presented a fatal objection to his theory of natural selection. In fact it is the peculiar mode of sex-determination in social insects, known as *haplodiploidy*, which is largely responsible for these features and which ultimately also explains them.

Honeybees are divided into three castes: first, proper females (queens), next, proper males (drones), and lastly sterile females (workers), who cannot mate, and cannot lay eggs except in unusual circumstances (if the colony is queenless, eg.) Egglaying is the job of the queen, who by virtue of her special sperm sac which stores for up to two years the sperm received during mating, is able to lay either fertilised or unfertilised eggs. The former will hatch into females (either queens or workers) and the latter into males. A male drone bee, coming from an egg unfertilised by any sperm and therefore bearing only the genes of his mother, has no father. Only worker and queen bees carry genes of both a mother and a father.

The average fraction of genes shared between two individuals by virtue of common descent is known as the coefficient of relationship, denoted by r. In mammalian species, for example, any individual receives half his genes from each of his parents, and so does his sibling – although not exactly the *same* half: in some cases the individual may receive nearly the same set of genes from a parent as his sibling does, in other cases nearly a completely different set. Averaging over many individuals we can see that the mean number of genes which 2 siblings have in common from one parent is one-half of the total number each got from that parent (which in turn is one-half their total genes). They similarly have the same number in common with each other from the other parent. So the total fraction they have in common from both parents is

$$r = \left(\frac{1}{2} \times \frac{1}{2}\right) + \left(\frac{1}{2} \times \frac{1}{2}\right) = \frac{1}{2}$$

Thus in the case of normal sex determination, two full siblings share ½ their genes in common with each other, the same fraction as each shares in common with either parent. With the haplodiploid mode of sex determination however, each female shares in common with her sisters *all* of the genes she has from her father, since her father, having only a mother but no father himself, is homozygous (having only one set of chromosomes) rather than dizygous (two sets of chromosomes). Each female in addition shares with her sisters one-half of the genes from her mother. The average fraction of genes shared through common descent between two sisters is now

$$r = \left(1 \times \frac{1}{2}\right) + \left(\frac{1}{2} \times \frac{1}{2}\right) = \frac{3}{4}$$

This strange calculus makes it beautifully clear how it is that worker honeybees of the same colony are more closely related to each other than they are to either of their parents. In this light we also see how the worker-bees' altruistic stinging behaviour, which is usually fatal to them and painful to the victim but protective of the colony, can be explained by the fact that the sacrifice of an individual worker confers fitness on the rest of the colony, made up largely of her sisters.

Reference: *The Insect Societies*, Edward O. Wilson, 1971.

PR

Equation 135: The Fresnel Equations

These four equations describe what happens when light is incident on a dielectric surface, i.e. one that cannot conduct electricity, for example, glass. They are critical to many optical systems and an integral part of the design of lenses, prisms and many other optical components.

When light is incident on a dielectric surface at an angle (θ_i), a proportion, R, will be reflected and the remaining, T, is the proportion of the light that is transmitted into the medium (at an angle of θ_t). A is the amplitude of the incident light. Note that because we are only considering dielectrics, there is no absorption of the light in the medium so $R + T = 1$. The reason for two values of R and T (one in the x-direction and

148

one in the y-direction), is that we are considering linearly polarised light. If we consider light as a sinusoidal oscillation moving in the z-direction of a Cartesian coordinate system, then the direction of linear polarisation is either in the x- or y-direction. The equations represent both directions. The n values in the equations are the refractive indices of medium 1 (in which the light enters and is reflected) and medium 2 (in which it is transmitted).

Note that in practical cases, light is rarely linearly polarised and one has to take this into account. One case to note is when the angle of incidence is zero, i.e. the light is incident perpendicular to the surface of medium 2. In this case, all the cosine terms become 1 and the two reflection equations become equal, as do those for transmission. The Fresnel equations need to be modified when one of the mediums is non-dielectric, i.e. a conductor or semi-conductor. In this case, the refractive index becomes complex, but the equations still hold.

These equations were derived by Augustin-Jean Fresnel in 1823 as a part of his comprehensive wave theory of light. However, the Fresnel equations are fully consistent with the rigorous treatment of light in the framework of the Maxwell equations.

$$T_x = \frac{2n_1 \cos\theta_i}{n_2 \cos\theta_i - n_1 \cos\theta_t} A_x$$

$$T_y = \frac{2n_1 \cos\theta_i}{n_1 \cos\theta_i - n_2 \cos\theta_t} A_y$$

$$R_x = \frac{n_2 \cos\theta_i - n_1 \cos\theta_t}{n_2 \cos\theta_i + n_1 \cos\theta_t} A_x$$

$$R_y = \frac{n_2 \cos\theta_i - n_1 \cos\theta_t}{n_1 \cos\theta_i + n_2 \cos\theta_t} A_y$$

RL

Equation 136: The Osmotic Pressure Equation

$$\Pi = icRT$$

Where Π is the pressure on one side of a semipermeable membrane, i is the van 't Hoff factor, c is the molarity, R is the universal gas constant and T is the thermodynamic temperature.

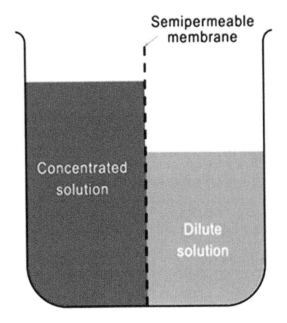

Jacobus Henricus van 't Hoff
(1852-1911)
first winner of the
Nobel Prize in Chemistry

Harmon Northrop Morse
(1848-1920)
Avogadro Medal
first person to
synthesize paracetamol

This equation - sometimes called the *Morse equation*, but more properly called the *van 't Hoff equation* - gives an estimated value for the osmotic pressure Π (or P) on one side of a semipermeable membrane of an ideal solution (cf ideal gas) at low concentration. c represents molarity, R the gas constant and T the absolute temperature. The *van 't Hoff factor i* (aka *dissociation factor*) is the ratio of the number of moles of solute actually in solution to the number of moles of solute before it got dissolved.

The osmotic pressure at a particular temperature may be defined in a number of ways: (a) the minimum pressure which needs to be applied to a solution to prevent the inward flow of water across a semipermeable membrane; (b) the measure of the tendency of a solution to take in water by osmosis; and (c) as the excess hydraulic pressure that builds up when the concentrated solution is separated from the dilute solution by a semipermeable membrane. For a given solvent, the osmotic pressure depends only upon the molar concentration of the solute, not on its nature.

I have put forward the Morse equation because of its correspondence to the gas equations - and I like to see the patterns which often crop up in science.

HISTORY
In 1877 pioneering plant physiologist Wilhelm Pfeffer made the first direct measurements of osmotic pressures in plants using a semi-porous membrane he had developed.

Dutch chemist Jacobus van 't Hoff utilised Pfeffer's experimental data to propose that a substance in solution behaves just like a gas: that the osmotic pressure of a dilute solution is equal to the pressure which the solute would exert if it were a gas at the same temperature and occupying the same volume. There are a series of Van 't Hoff laws of the same form as Boyle's, Charles', Amonton's (aka Gay-Lussac's) and Avogadro's gas laws. These laws culminate in his equation $V = nRT$, or $\Pi = cRT$, where the molarity $c = n/V$. In order to cater for the effect of ionisation in electrolytes, he introduced a factor (the van 't Hoff factor i), giving $\Pi = icRT$.

The American chemist Harmon Morse took these insights further. Morse's contribution was to: (a) indicate that Pfeffer's cells had been leaky at high pressure, giving rise to unreliable data; (b) design an improved set-up for measuring osmotic pressure; and (c) show experimentally that a better approximation to osmotic pressure is given by the equation

$$\Pi \ = \ \rho bRT/1000$$

in which the molality b replaces the molarity c and ρ is the solvent density.

APPLICATIONS
Osmotic pressure is necessary for many plant functions, and is the basis of filtering ("reverse osmosis"), a process commonly used to purify water.

FR

Equation 137: Stellar Fusion Part 2: CNO Cycle

The carbon-nitrogen-oxygen (CNO) cycle is a catalytic process for fusion of hydrogen into helium at stellar core temperatures above 15 million degrees Kelvin (K). At 17 million K it becomes the dominant source of energy. The sun, with a core temperature of 15.7 million K, makes a small amount of its helium from the CNO cycle. The luminosity of stars of several solar masses (eg stars such as Sirius and larger) rely mainly on the CNO cycle. The CNO cycle was theorised independently by Weizsacker and Bethe in the late 1930s.

$$4\,^1H + 2e^- \ \longrightarrow \ ^4He + 2v_e + energy$$

$$(\text{catalysed by } {}^{12}C \ \to \ {}^{13}N \ \to \ {}^{13}C \ \to \ {}^{14}N \ \to \ {}^{15}O \ \to \ {}^{15}N \ \to \ {}^{12}C)$$

The cycle consists of 6 reactions and needs carbon-12 to start off, which is regenerated in the last reaction enabling the cycle to repeat itself. Positrons generated in the 2nd and 5th reactions are, as in the p-p chain process (BE 123), annihilated by interaction with electrons. Neutrinos are also generated in these two reactions. The 5th reaction is the fastest (seconds) as it involves an unstable isotope of oxygen produced in the 4th reaction. The 1st, 3rd and 4th reactions result in gamma ray emission, which eventually contributes to luminosity. In a small percentage of these reactions nitrogen-15 generates oxygen-16 rather than carbon-12 in the last reaction, and a further cycle of 5 reactions ensues including fluorine as an intermediate which eventually produces more nitrogen-15 to repeat the process. In massive stars, part of this second cycle involves larger isotopes of fluorine as additional processes. At higher temperatures and pressures, such as in novae, additional proton capture becomes dominant over decay processes resulting in alternative catalytic sequences ("hot" CNO cycles).

So where does the initial carbon-12 come from to allow these catalytic processes to begin? The carbon-12 is already available in second generation stars, and would have been produced in the final stages of first generation stars that reached core temperatures of around 100 million K. The production of carbon by the triple alpha process, and subsequent production of heavier nuclei, will be described in the final article in this series.

PB

Equation 138: The Law of Mass Action

A chemical reaction is all about different chemical compounds – or one unstable compound – reacting to give new chemical compounds. The compounds, which react in specific proportions, are called reactants. The new chemical compounds, which form in specific proportions, are called products.

A simple equation governs chemical reactions: the "law of mass action". It states that, when a chemical reaction reaches equilibrium, the ratio between the product of the molar concentrations of the products and the molar concentrations of reagents is constant.

$$K_c = \frac{[Products]}{[Reagents]}$$

Where [ ] is the symbol of the molar concentrations.

For a simple bimolecular reaction

$$aA + bB \rightleftarrows cC + dD$$

The law of mass action becomes

$$K_c = \frac{[C]^c[D]^d}{[A]^a[B]^b}$$

In principle, the products of a chemical reaction could also play the role of reagents of the inverse reaction. In practice thermodynamics will determine in which direction a reaction moves, whether towards the products or back to reagents.

Norwegian chemists and mathematicians, Cato Maximilian Guldberg and Peter Waage came up with this law between 1864 and 1879, as result of their studies on the kinetics of chemical reactions. They discovered that chemical equilibrium is dynamic, with forward and backward reactions proceeding at equal rates. A few years later, Jacobus Van't Hoff, Nobel Prize for chemistry in 1901, discovered that the rate of a chemical reaction is proportional to the product of the molar concentrations of the reactants.

IMPORTANCE

K_c is a fundamental parameter to predict the extent of a chemical reaction. A high value of K_c indicates that the reaction will proceed to completion, with almost all reactants transforming in products. This is the case of strong acids, like sulfuric acid, which has a K_c of about 10^{10} for its almost complete dissociation in water. Another example of a reaction proceeding to completion is the ozone dissociation into oxygen, which has a K_c of about 10^{55}!

Vice versa, a very small value of K_c indicates that the reactants will not tend to react. This is the case of the stable compound hydrofluoric acid, which does not decompose to hydrogen and fluorine gases, as the K_c for the decomposition reaction is about 10^{-13}.

Many reactions in chemistry, however, occur at equilibrium, with only partial formation of products. One example is the formation of ammonia from nitrogen and hydrogen, a reaction that has a K_c of about 10. This reaction is also used in the Haber process for the industrial production of ammonia. An iron/potassium hydroxide catalyst is used in order to speed the reaction, but this does not affect the equilibrium of the reaction, as the catalyst speeds both the forward and backward reactions at the same time.

LE CHÂTERLIER PRINCIPLE AND BUFFER SOLUTIONS

One of the most important consequences of the law of mass action is how the equilibrium changes by varying the concentration of reactants and products in the reaction. French chemist, Henry Louis Le Châtelier, discovered that when a system at equilibrium is subjected to changes in concentration, temperature, volume or pressure, it will re-adjusts itself to counteract the effect of the applied change, so that a new equilibrium is established.

This means that:

- An increase in concentration of reactants will shift the equilibrium towards the products (i.e. it will lead them to react to form more products)
- A decrease in concentration of reactants will shift the equilibrium towards the reactants (i.e. it will promote the backwards reaction)

152

- An increase in concentration of products will also promote the backwards reaction to occur.
- A decrease in concentration of products will promote the forward reaction, making more reactants form products. This is often the case in industrial processes, in which product synthesis is maximised by continuously removing the product from the batch.

Le Châtelier's principle is a consequence of the law of mass action, which requires the K_c of a reaction to be approximately constant.

Le Châtelier's principle controls the mechanism of buffer solutions, which consist of an appropriate solution of a weak acid and the salt of its conjugate basis. A typical example of a buffer is the solution of acetic acid (CH_3COOH) and sodium acetate (CH_3COONa).

Buffer solutions are important because their ability to keep the pH – the concentration of hydrogen ions in the solution - almost unchanged when a moderate amount of acids or alkalis is added to the solution. Buffer solutions are particularly important for life organisms, including us, as they thrive only in relatively small pH range. A big variation in pH would slow down enzymes catalytic activity, eventually completely stopping them from working. A buffer of carbonic acid (H_2CO_3) and bicarbonate (HCO_3^-) is present in blood plasma to maintain its pH between 7.35 and 7.45. Buffer solutions are also used in industry, in fermentation processes, colouring fabrics and chemical analysis.

AT

Equation 139: The Integrating Sphere

An integrating sphere is a hollow sphere with highly diffusely reflective material at the inside. Small openings are made for light detectors and for enabling light to enter and/or exit. Often baffles are required to ensure that the detector only detects diffusely radiated light. A cross section is depicted in the figure.

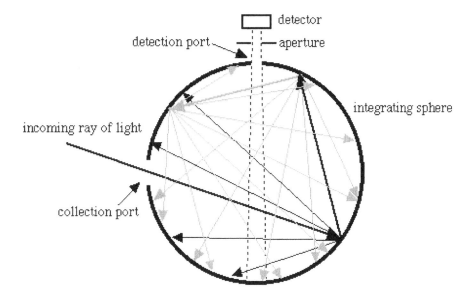

Figure: Example of a cross section of an integrating sphere

The irradiance L on the inside of the sphere, due to a flux ϕ entering the sphere is given by:

$$L = \frac{\phi}{\pi A_s} M = \frac{\phi}{\pi A_s} \frac{\rho}{1 - \rho \cdot (1 - f)}$$

153

Where A_s is the sphere area, M is the so-called sphere multiplier, ρ is the reflection coefficient of the material inside the sphere $(0 < \rho < 1)$ and f is the fraction of the sphere area that is not covered by the reflective material, i.e. ports for detectors, specimen etc.

The beauty of the sphere principle is that very high irradiances inside the sphere can be obtained by reducing the area covered by openings, detectors, etc., and by a highly diffusely reflective material inside the sphere. A contradictory aspect is that in principle the factor M could approach infinity, but in that case there cannot be an opening to let any light in, and there cannot be any detector measuring that light by absorbing it.

Applications are e.g. the measurement of light source intensities, making a diffuse light source, measuring total transmittance or reflectance of a surface, i.e. both the specular and diffusely reflecting or transmitting component of a surface.

The integrating sphere is also called "Ulbricht Sphere", after the German engineer and professor Friedrich Richard Ulbricht (1849–1923), who published on this subject in 1900.

HH

Equation 140: Logistic Growth Equation

The logistic growth equation can be used to describe population dynamics. In particular, the differential form of the equation shown below allows one to determine the change in population $\frac{dP}{dt}$ given a current population P, rate of increase r, and carrying capacity K.

$$\frac{dP}{dt} = rP\left(1 - \frac{P}{K}\right)$$

When applied to a real scenario, for example the human population, the rate of increase r could be a value corresponding to the average fertility rate; higher fertility rates in a population result in larger r value. The carrying capacity K is the maximum population achievable; this parameter could be dependent, for example, on the available resources to sustain human life.

According to the equation, growth slows down as the current population approaches the carrying capacity, i.e. the population is reaching its maximum achievable value. The logistic growth equation was first published by Pierre Verhulst in the 1840s.

MF

Equation 141: The Courant-Friedrichs-Lewy Condition

Dynamic behaviour of materials is governed by a class of partial differential equations (PDEs) known as the hyperbolic partial differential equations. A prototypical equation for this class of PDEs is the classical one-dimensional wave equation (Beautiful Equation 48) given by

$$\frac{\partial^2 u}{\partial x^2} = \frac{1}{c^2}\frac{\partial^2 u}{\partial t^2}$$

with x denoting the spatial coordinate and t denoting the time. This equation governs the vibrations of a string fixed at both the ends, or the tension/compression waves traveling in a bar due to external disturbances applied on them. For these cases, $u(x, t)$ is the displacement of a particle in the string or the bar and c is the speed with which a disturbance propagates.

A commonly used method for obtaining a solution to the hyperbolic PDEs is an explicit finite difference method. In this method, the spatial domain (ex: the length of the bar or the string) is replaced by a set of uniformly spaced points $x_i, i = 0,1,2, \ldots, I$ with the spacing between any two successive points denoted by Δx. The time variable is also replaced by a set of uniformly spaced points $t_n, n = 0,1,2, \ldots, N$ with Δt being the interval length between any two points t_n and t_{n+1}. Let u_i^n denote the displacement at (x_i, t_n). Then, given a set of initial data and boundary conditions, the solution u_i^n is obtained from the PDE by first replacing the spatial and time derivatives in the PDE by appropriate finite difference approximations. For example, for the one-dimensional wave equation mentioned in the above, a centered difference scheme in both space and time leads to the following approximation to the PDE:

$$u_i^{n+1} = 2u_i^n - u_i^{n-1} + p^2(u_{i+1}^n - 2u_i^n + u_{i-1}^n)$$

where $p = \frac{c\Delta t}{\Delta x}$ is called the Courant number. The right-hand side of the above equation involves terms at the time intervals n and $n - 1$ whereas the left hand side involves only the solution at the $(n + 1)^{st}$ time interval. Thus, using the initial conditions on displacement and velocity (which give u_i^0 and u_0^{-1}), one can generate the solution u_i^n for all $n \geq 1$ incrementally.

The choice for Δx or equivalently I is usually at the user's disposal. The physics of the problem usually governs the choice. However, Δt cannot be prescribed arbitrarily. The explicit finite difference method requires that Δt satisfy a stability criterion given by

$$p = \frac{c\Delta t}{\Delta x} \leq 1$$

which is the famous Courant-Friedrichs-Lewy (CFL) condition proposed by Courant, Friedrichs and Lewy in 1928 [1]. This equation is beautiful for its simplicity and since it provides an upper bound for the allowable time-step and has the physical implication that a stable solution is obtained when the distance travelled by a disturbance in time Δt is less than or equal to the spacing Δx of the discrete points.

So, what happens if the CFL condition is not met? Let us assume $c = 1$ and investigate this. Consider a bar of unit length fixed at the right end. Suppose the initial displacement and velocity are zero. The boundary condition at $x = 0$ is taken to be a square displacement wave of finite duration and unit amplitude. The numerical results are shown for 4 different times in the next page for the cases $p = 1$ and $p = 1.01$ respectively. Note that the figures in the left column and the right column do not correspond to the same times. Clearly, when $p = 1$, the square wave imposed at the left end travels to the right undistorted and is reflected as a wave with opposite sign and the same amplitude. Thus, with $p = 1$, the shape and amplitude of the pulse are preserved. On the other hand, when $p > 1$, the solution is unstable. It becomes oscillatory with the amplitudes growing rapidly. What is amazing is that p in the second case is only 1% different from the maximum stable value and yet, the solution is so incorrect and radically different.

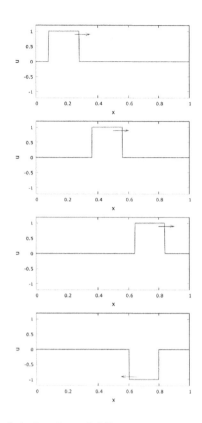 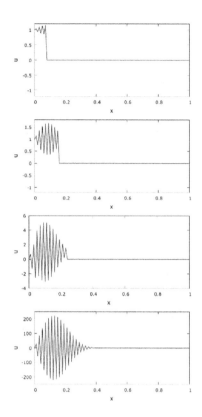

Stable Solution ($p = 1.00$).   Unstable Solution ($p = 1.01$).

References

Courant, R., Friedrichs, K., & Lewy, H. (1967). On the partial difference equations of mathematical physics. IBM journal of Research and Development, 11(2), 215-234.

HC

Equation 142: Stefan's Law

The temperatures of the vast majority of stars range from about 3,600 K to 50,000 K: just over an order of magnitude. One might perhaps assume - as 19th-century astronomers did - that their powers would cover a proportionate range (if corrected for size variation). In fact, the powers of the hottest stars are more than 30,000 times greater than the coolest stars of the same size.

This remarkable fact is described by Stefan's Law:

$$j^* = \sigma T^4$$

in which j^* is power radiated per unit area, σ is Stefan's constant and T is absolute temperature. It applies strictly only to black bodies (objects which radiate or absorb all forms of electromagnetic radiation). However, the Sun and most other stars closely approximate such bodies.

Joseph Stefan derived the equation in 1879 from experimental data. Five years later Ludwig Boltzmann derived it from consideration of a heat engine. It is therefore often known as the Stefan-Boltzmann Law.

With the advent of quantum theory, Stefan's Law was derived Planck's Law, and it also became possible to relate Stefan's constant to universal constants:

$$\sigma = \frac{2\pi^5 k^4}{15c^2 h^3}$$

(where k: Boltzmann's constant; c: speed of light ; h: Planck's constant)

In the 19th century, many nations were seeking a hypothetical "Northwest Passage": a navigable route around the northern coastline of North America, which would be a valuable trading route. In fact, people had been seeking such a route since the 15th century. Numerous ships had come to grief in the process of exploration, by being trapped, and in some cases crushed, by ice. (At the time, all such searches were vain: though occasional expeditions have made the journey since 1905, it was never a practical trade route. Today however global warming means that there is such now such a route, albeit a seasonal one).

Since, by Stefan's time, there was a considerable amount of data on temperature and ice thickness along the route, he was able to use a modified version of his law to allow ice thickness to be predicted, which made such voyages far safer. In its modern version, the formula states that the expected ice accretion is proportional to the square root of the number of degree days below freezing.

MJG

Equation 143: Divergence Theorem

$$\iiint_V (\nabla \cdot \boldsymbol{F}) \, dV = \oiint_S \boldsymbol{F} \cdot d\boldsymbol{S}$$

Where ∇ is the differential operator acting on the components of a vector field in Cartesian coordinate system. ∇ can be written in a vector form as $\nabla = (\partial/\partial x, \partial/\partial y, \partial/\partial z)$. $\boldsymbol{F}$ is the vector field of a quantity in three dimensional space, V represents a volume in three dimensional space, $\boldsymbol{S}$ is the closed boundary surface of the volume. The normal of $\boldsymbol{S}$ is outward pointing orientated.

The theorem says that the integral of the divergence of a vector field ($\boldsymbol{F}$), which is given by the sum of partial derivatives of the vector field in each coordinate ($\nabla \boldsymbol{F} = \frac{\partial \boldsymbol{F}}{\partial x} + \frac{\partial \boldsymbol{F}}{\partial y} + \frac{\partial \boldsymbol{F}}{\partial z}$) over a volume ($V$) equals to the surface integral of that field over the boundary of that volume ($\boldsymbol{S}$).

The historical origin of the equation can be traced back to French scientist Pierre Laplace in 1762 and later German scientist Carl Gauss in 1813. In 1828, George Green, a British mathematician who was almost entirely self-taught published his *"An Essay on the Application of Mathematical Analysis to the Theories of Electricity and Magnetism"*, where the equation was expressed. The divergence theorem is later commonly referred as Green's theorem.

Application of the divergence theorem is widely spread in physics and engineering such as electromagnetism and fluid mechanics. For example, the theorem can be vividly interpreted geometrically as the sum of all faucets (assuming no sinks) within a volume gives the sum of the liquid flows out of the surface of the volume. Another application of the divergence theorem is the Gauss's Law of electrostatic filed. The law states that the electric flux through a closed surface is a measure of the total electric charge within the volume.

Reference:
David J. Griffiths Introduction to Electrodynamics 3rd edition, 1999 Prentice Hall

KN

Equation 144: Comparator-Based Relaxation Oscillator Frequency

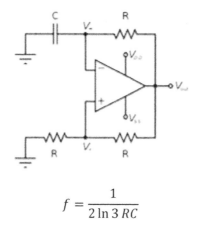

$$f = \frac{1}{2 \ln 3 \, RC}$$

where f is the oscillatory frequency, and R and C are the resistance and capacitance shown in the figure.

The circuit shown is a comparator-based oscillator belonging to the class of astable multivibrator. Multivibrator means that the circuit has two states, and astable that none of them is stable, i.e. the circuit continuously oscillates between those two states. The active non-linear component is a comparator, i.e. the output voltage is high or low depending on whether the voltage on the positive input is larger or smaller than the voltage on the negative terminal. The passive component (capacitance and resistances) causes the circuit to relax after each switching. The system is in unstable equilibrium if both the inputs and outputs of the comparator are at zero volts. The moment any sort of noise brings the output of the comparator above zero the comparator output saturate at the positive or the negative supply voltage (V_{dd} or V_{ss}). This induces a capacitor charge or discharge transient, which brings the voltage on the negative input asymptotically equal to the voltage on the comparator output. On the other hand the positive input voltage is a fraction of the output voltage due to the two resistors. When the charged/discharged capacitor has an applied voltage that is larger/smaller than the voltage on the positive input the comparator switches its state changing the voltage on its output and the process is repeated. The result is a square wave signal on the V_{out} terminal with a frequency given by the charge plus the discharge transients and given in the equation.

The astable multivibrator was invented in 1917 by French engineers Henri Abraham and Eugene Bloch. They called their device a *multivibrateur* because the square-wave signal it produced was rich in harmonics compared to the sinusoidal signal of other oscillators.

GM

Equation 145: The Screw Thread Equation

The screw thread, after fire and the wheel, is arguably one of the most important discoveries of mankind. It is ubiquitous. It enables us to fix items together, convert rotary motion in to linear motion, pump water and to provide a mechanical advantage – think cider press.

A screw thread is basically a helix and the equation for a circular helix of radius a and pitch $2\pi b$ is described by the following parametrisation:

$$x(t) = a \cos (t)$$
$$y(t) = a \sin(t)$$
$$z(t) = b(t)$$

158

As the parameter t increases, the point $(x(t), y(t), z(t))$ traces a right-handed helix about the z-axis, in a right-handed coordinate system.

Screw-threaded bolts are believed to have been used by the Romans, however this assertion is based on one threaded nut displayed in the Provincial Museum of Bonn and dated to AD180-260 (Rybczynski,W. 2000). Stronger evidence comes from Josephus of the first century AD, who described iron tie rods reinforcing supporting columns, "The head of each rod passed into the next by means of a cleverly made socket crafted in the form of a screw....... These were held by these sockets, the male fitting into the female". (Humphrey *et al*).

Screw threads then disappeared until medieval times. The Mediaeval Housebook of Wolfegg Castle contains an early illustration of a screw-cutting lathe dating from between 1470 and 1490.

Real advances were made by Henry Maudslay who 'In his system of screw-cutting machinery, his taps and dies, and screw-tackle generally, he laid the foundations of all that has since been done in this essential branch of machine-construction.' In 1800 he produced a large screw-cutting lathe for industrial applications. (tilthammer.com, 2001).

Standardization of screw threads has evolved since the early nineteenth century work of Whitworth and Clements. The standardization process is still ongoing; in particular, there are still competing metric and inch-sized thread standards in common use.

There are many equations used in the design and measurement of screw threads and the interested reader can refer to the screw thread booklet available for free download from the NPL website (website: www.npl.co.uk).

Do screw threads occur in nature? Well the obvious answer is to point to our own DNA which is based on a double helix structure. A bit tenuous I know but better still a musculoskeletal system so far unknown in the animal world was discovered in 2011 in weevils. The hip of Trigonopterus oblongus does not consist hinges, but of joints based on a screw-and-nut system. This first biological screw thread is about half a millimetre in size and was studied in detail using synchrotron radiation. The discovery is reported in Science magazine. (DOI:10.1126/science.1204245)

DF

Equation 146: The Mott Transition

$$N_C^{1/3}\, a_H = K$$

Where N_c is the critical number of impurities at the metal to non-metal transition, a_H is the first Bohr radius of the impurity. Today, a more realistic isotropic Bohr radius a_H^* is used. K is a constant ($= 1/4$ for the original Mott model).

Human societal development is often defined by our ability to utilize specific materials to create technology. Starting with the 'Stone' age, subsequent periods are identified as Bronze, Iron (Steel really), and, today, the Silicon age. Based on this, most of our civilized existence has been dominated by our ability to produce metal parts. It makes sense that historically a significant human effort must have been expended trying to understand what makes a material a metal. We know that it has a shiny lustre, can be malleable, and conducts both heat and electricity rather well. In fact, only metals will conduct at zero absolute temperature. Beyond these simple considerations, a comprehensive definition remains elusive. However, with the discovery of atomic structure at the turn of the 19th century and development of wave mechanics in the early twentieth century, it looked promising that a scientifically based definition for the metallic state might be forthcoming. Further, because metals form regular and simple crystal structures, it is possible to break the model down to a periodic array of atoms containing nuclei concentrated in a small region near to the centre of each atom that produces a positive electric field. For the most part, the volume of the atom is empty space filled by the orbiting electrons. These electrons orbit with quantised energies and each allowed orbit can only be occupied by two electrons with their magnetic axes reversed (called the spin of the electron). Typically these electrons will fill up the allowed energy levels until the number of electrons matches the number of protons in the nucleus. The problem can be solved exactly using methods of quantum mechanics. In particular, provided the potential energy in the region of the atom is known, Schrodinger's equation can be integrated to provide solutions from which the properties can be determined. Unfortunately, Schrodinger's equation is too complicated to solve completely for all but the simplest of atoms in isolation from any neighbours, even of the same kind. However, early on in the development of quantum theory at the age of 23 Felix Bloch demonstrated a wave solution (now called a Bloch wave) in 1928 that would enable an electron to move freely throughout a solid. It has been mentioned that the success of this seminal work changed the question from 'why are materials metals?' to 'why are materials insulators?'

In 1949 the British physicist Neville Mott (1905 – 1996) presented his idea about the conditions for the metallic state of matter. For this and subsequent developments he was awarded a Nobel Prize in 1977 alongside two American physicists Phillip W. Anderson and John J. van Vleck. Mott considered a material that has some impurities introduced that can bond with the native material but leave a single outer electron in a large weak binding orbit (strictly these wave function shapes are called orbitals). However, as the number of impurities increases it is possible for adjacent orbits to interact (a correlation exists between their respective wave functions) and this can, in turn, act to screen the effect of the positive charge of the nucleus. Meanwhile, the other electrons bound to the core of the atom can also distort (measured as the polarizability, κ) their orbits resulting in a further shielding of the nucleus potential. Mott's major contribution was to realize that, as the number of the impurities per unit volume, N, increased up to a critical value, N_c, there could arise a condition such that when the average orbital radius ($1/\gamma$) of the screening sphere would be smaller than orbital radius of the outer electron the material could transition from an insulator to a metal. Equating these two radii results in the Mott condition.

Subsequent analysis revealed a number of flaws with the relatively simple reasoning presented by Mott. However, having demonstrated that a higher level reasoning consistent with possible wave mechanics solutions to this problem could produce valuable physical insights, this work inspired future generations of physicists to develop more complete theories. Two of note were the development of British physicist John Hubbard (1931 – 1908) almost 20 years after Mott who used energy based arguments and came up with a value for K of approximately $1/5$. In a surprising twist, in a second development it was realized in the mid 1970's that an older, pre-quantum mechanics, theory by the physicists D. A. Goldhammer (?) and Karl Hertzfeld (Austrian physicist, 1892 – 1978) each publishing in 1911 and 1927 respectively, that a sudden rapid change in the polarization of the material could be incorporated into quantum theoretical models to provide a more complete model for the transition. Based on this, a new value of $K = 0.376$ results. What is most surprising is that experimental measurements of the two parameters reveal that the Mott condition appears to be a reliable predictor over values ranging by many orders of magnitude (see plot). No doubt research will continue on this topic and greater insights will emerge. However, all future developments will have to explain the success of Motts original idea.

One example of particular importance is the doping of the semiconductor silicon by phosphorus 'impurities' that will result in a relatively free electron (called an n-type material because the 'extra' electron provides a semi-mobile negative charge). It transition can be seen in the plot taken from the paper of Edwards and Sienko. Another conclusion from Motts work was that materials will become metallic if they are compressed sufficiently. In fact, silicon itself will transition to a metal at room temperature under a pressure of around 12,000,000,000 newtons per meter squared. This is about one thirtieth of the pressure at the centre of the earth which is, of course, much hotter.

160

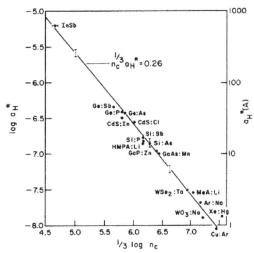

FIG. 1. Metal-nonmetal transition in condensed media. A plot of $\log a_H^*$ vs $\frac{1}{3}\log n_c$ (symbols defined in the text). The points represent experimental systems in which both a_H^* and n_c are known (Table I). Typical error estimates in n_c and a_H^* are shown (solid error bars) for the representative systems InSb and Si:As, respectively. A complete breakdown of error estimates (where available) in both n_c and a_H^* for all the experimental systems is contained in Fig. 2 and Table I. The vertical, broken-line, error bars represent the limits $n_c^{1/3}a_H^* = 0.23$ and 0.29.

Figure and caption taken from the 1978 paper of Edwards and Sienko listed in the references.

References
Edwards P.P. and Sienko M.J., 1978, Universality of the metal-nonmetal transition in condensed media, Phys. Rev. B, **17**(6), 2575 – 2581.
Edwards P.P., Johnston R.L., Rao C.N.R., Tunstall D.P., and Hensel F., 1998, The metal-insulator transition, *Phil. Trans. Roy. Soc. Lond. A*, **356**, 5 – 22.
Edwards P.P., Lodge M.T.J., Rao C.N.R., Hensel F. and Redmer R., 2010, '… a metal conducts and a non-metal doesn't', *Phil. Trans. Roy. Soc. Lond. A*, **368**, 941 – 965.

SS

Equation 147: Preston's Removal Rate for Force Controlled Polishing of Glass

Preston's removal rate (r) is an empirically developed equation used to take the guess work and "black magic" out of fine abrasive lapping and polishing processes. It provides a function for the amount of material removed for a given contact force or pressure (P) between a polishing or lapping apparatus applied over an area (A) on a substrate with a surface velocity (V) between the substrate and polishing head. As the equation is empirically developed a constant (K) is used that is valid for specific setup conditions such as the nature of the polish or lap and substrate and system mechanics.

$$r = K \cdot P \cdot V \cdot A$$

Where K is a constant, P is the contact pressure, V is the velocity and A is the area.

Preston's removal rate is used in modern automated computer controlled polishing (ccp) processes where science and automation is used in place of art and experience. Modern automation has achieved great things in this field however the skilled touch of an experienced tradesperson can often achieve better results.

Preston, F,W. Developed this empirical equation in 1927 when experimenting with the process parameters and here say in polishing processes to remove the uncertainty from the process.

TS

Equation 148: Arrhenius Equation

The rate of a chemical reaction – how fast a chemical reaction proceeds – is proportional to the consumption of reactants or the formation of products. The proportionality constant k, in turn, depends on the temperature, according to the *Arrhenius Equation*:

$$k = Ae^{\frac{-E_a}{RT}}$$

where R is the universal gas constant, T is the temperature, A is a pre-exponential factor and E_a is the activation energy of the reaction.

Arrhenius equation describes the dependence of the chemical reactions on the temperature. This equation applies to evaluate the temperature influence on a variety of processes, including diffusion, population of crystal vacancies, creep rates.

Swedish physicist and chemist, Svante Arrhenius, proposed this equation in 1889, based on the work of Dutch chemist Jacob van 't Hoff, who found the temperature dependence of the chemical equilibrium constants in 1884. Arrhenius also won the Nobel Prize for Chemistry in 1903, but not for his equation, but for developing his theory of electrolytic dissociation.

MICROSCOPIC INTERPRETATION
The exponential factor in the equation contains the ration of two terms, which are both energies (active energy and thermal energy), and it is, consequently, a numerical factor. This means that A and k have the same units.

The interpretation of the equation from a microscopic point of view is related to the collision theory. In this theory, k is the number of collisions between molecules of reactants that leads into a reaction per second. The pre-exponential factor, A is the total number of possible collisions, while the exponential factor represents the probability that a collision has a sufficient energy to lead to the formation of products – the activation energy E_a (Figure 1).

Consequently, the rate of the reaction will increase either by increasing the temperature or by decreasing the activation energy.

CATALYSIS
The second strategy is what is used in catalysis in order to speed up a reaction. This consists of introducing a catalyst (often a solid material, but it can also be a liquid species), which changes the pathways of the reaction by participating in it, but without

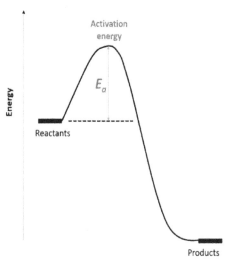

Figure 1: Activation energy for an exothermic chemical reaction.

transforming itself in a different product once the reaction has reached completion.

The catalyst has the effect of lowering the activation energy of the reaction, therefore speeding up the chemical reaction. Catalysts can be either heterogeneous, such as platinum solid metal in for the conversion of exhaustion gases in carbon dioxide and nitrogen (Figure 2), or homogeneous, such as

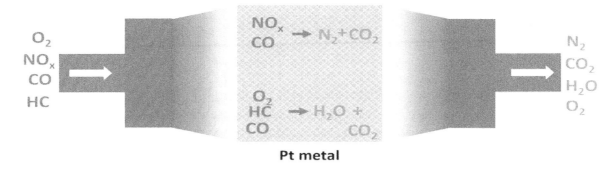

Pt metal

Figure 2. Schematic illustration of a 3-way, Pt-based catalytic converter.

enzymes which control a series of chemical reactions in our body.
Catalysts are widely used in very important chemical industrial processes, such as the synthesis of ammonia, the production of hydrogen gas from methane, the production of polyvinyl chloride polymers and the cracking of gas oil.

AT

Equation 149: The Wheatstone Bridge

Regardless of its name, The Wheatstone Bridge was invented by Samuel Hunter Christie (1784 – 1865) (British scientist and mathematician) in 1833. The circuit was improved and popularized by Sir Charles Wheatstone (1802 – 1875), (British scientist and inventor) in 1843.

The Wheatstone bridge can be used in numerous ways to measure electrical resistance, capacitance and inductance. If we consider only electrical resistance, the Wheatstone bridge can be used for:
1. the determination of an unknown resistance by comparison with known resistances;
2. the determination of relative changes in electrical resistance.

It is this second point which makes the Wheatstone bridge suitable for measurements involving strain gauges or resistance temperature detectors.

The Wheatstone bridge consists of four arms, made up with resistors or resistive elements R_1 to R_4. An excitation voltage is applied to the upper and lower points, which may be an AC voltage or more commonly a DC voltage and the bridge output is obtained from between the left and right hand side points as shown in the figure.

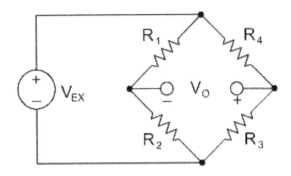

If the bridge is out of balance, different voltages from each of the electrical resistances are caused and the bridge output can be calculated using the equation given in the second image.

$$\frac{V_{EX}}{V_O} = \frac{R_1}{R_1 + R_2} - \frac{R_4}{R_3 + R_4}$$

V_{EX} is the excitation voltage, V_O is the output or detected voltage and R_1 to R_4 are the resistive elements making up the bridge. If $\frac{R_1}{R_2} = \frac{R_3}{R_4}$, the bridge is balanced and the output voltage is zero.

It is assumed that the source resistance of the supply voltage is negligible and the internal resistance of the measurement instrument is very high and is not responsible for loading the bridge.

Finally, for strain measurements, The Wheatstone Bridge can be used in three main configuration categories:
3. quarter bridge (one strain gauge and three known resistor values),
4. half bridge (two strain gauges and two known resistor values),
5. full bridge (four strain gauges).

There are a large number of options when deciding which configuration to choose and these are based on what exactly is to be measured.

References: K. Hoffmann, "Applying the Wheatstone bridge circuit," HBM.
"The Three-Wire Quarter-Bridge Circuit," Vishay Precision Group, 2010

CMK

Equation 150: The Fourier Transform

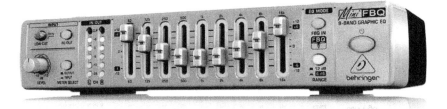

In popular culture we have become entirely comfortable with the concept of a decomposition of a signal according to frequency. We have a foundational understanding, from a very early age, of linear instrument transforms—wherein a signal is transformed into the frequency domain, multiplied by a transfer function, and then transformed back into a time-domain signal. This popular understanding reached its peak in the 1970's, when it was impossible to purchase even the simplest bit of audio equipment without a graphic equalizer.

Although the concept is evidently simple, it is not quite so clear how we can perform the miracle of frequency composition and decomposition mathematically. Joseph Fourier published the solution in 1822, in the *Théorie analytique de la chaleur*, showing how at least some functions can be expanded in a series of sine and cosine functions with appropriate weights. What we call a Fourier transform today is often written

$$F(v) = \int_{-\infty}^{\infty} f(t)e^{-2\pi ivt}\, dt$$

where $F(v)$ is the strength of a sinusoidal component of the signal $f(t)$ at the frequency v. The way this works is quite clever: Because the complex function

$$e^{-2\pi ivt} = \cos(2\pi vt) - i\sin(2\pi vt)$$

oscillates with a period $1/v$, the integral from $-\infty$ to ∞ will eventually average to zero unless the function $f(t)$ or one of its components has exactly a sinusoidal behavior at the frequency v. Cool! We can just as easily reconstruct the signal using

$$f(t) = \int_{-\infty}^{\infty} F(v)e^{-2\pi ivt}\, dv$$

which is just a way of saying that the signal $f(t)$ is the sum of a continuous series of sinusoids of frequency v. Fourier transforms enable calculation of the coefficients of a Fourier Series, as described by BE 12.

For many functions, there are closed-form analytical solutions to Fourier integrals. There is a fairly complete catalogue already in the Wikipedia article on Fourier Transforms. Many solutions have already been worked out for integrals that extend in 2 or more dimensions, or that include a window or weighting function to modify or limit the extent of the signal $f(t)$. One of the more famous results is that the Fourier Transform of a Gaussian is another Gaussian:

$$\int_{-\infty}^{\infty} e^{-\alpha t^2} e^{-2\pi ivt}\, dv = \left(\sqrt{\pi/\alpha}\right) e^{-\pi^2 v^2/\alpha}$$

Fourier Transforms are immensely useful and probably the most powerful mathematical tool that I ever learned, after add, subtract, multiply and divide. I use it to analyse optical systems using Abbe theory, calculate far-field diffraction, and to perform complicated digital signal processing to extract useful information from interferometric signals, which are naturally oscillatory. I use it for environmental vibration characterization, electronic circuit design, optical spectrum measurements, and surface structure analysis. Thanks, Joseph!

PdG

Equation 151: The *Ra* Parameter

$$Ra = \frac{1}{lm} \int_{x=0}^{x=lm} |y|\, dx$$

Surface texture is defined in the CIRPedia (a published collection of engineering terms) as "geometrical irregularities present at a surface. Surface texture does not include those geometrical irregularities contributing to the form or shape of the surface." Surface texture is produced by a combination of the materials characteristics of the surface (e.g. grain boundaries) and the processes used to manufacture the surface; both the process itself (e.g. grinding or additive manufacturing) and any errors in the process (e.g. chatter or over-melting). Surface texture is clearly important in nearly all engineering disciplines as it is often the surface characteristic that most strongly affects the function of the components, e.g. friction in a cylinder liner or scattering off an optic such as a lens or mirror.

The Ra parameter (for roughness average) was one of the first attempts to give a quantitative description of surface texture. In the early part of the 20th century, a contact stylus was run over a surface and the heights of the surface measured. Such stylus instruments outputted a profile graph and the Ra parameter was simple to read off using a planimeter - a measuring instrument used to determine the area of an arbitrary two-dimensional shape. F. H. Rolt at the National Physical Laboratory in the UK was the first to advocate the use of parameters but it is not entirely clear who was the first to use Ra, however, it was likely first used by engineers in Taylor Hobson in Leicester, UK. Such a parameter allows one to compare surfaces and to specify a desired surface on an engineering drawing. In the equation, lm is the length in the x-direction over which the surface is evaluated, and y is the surface height measured from a mean line as a function of x. The S symbol with the dx represents an integral; basically it is a continuous summation – indeed, in practice the equation would be calculated over a finite sum due to the finite sampling of the instrument. The sum runs from $x = 0$ to $x = lm$. The vertical lines around y mean that only positive height values are considered – all values below the mean line are effectively multiplied by -1 to give a positive value (this is called the absolute value of y). The combination of the summation and the division by lm gives a mean of all the absolute y values. Finally, to calculate Ra, it is necessary to filter the surface to remove unwanted high and low spatial frequency features.

The absolute value of y means that the Ra parameter does not give any information about how the heights are distributed about the mean line. For this reason, it can only be used to show a change from one surface to the next, and gives limited functional information. Since the advent of the Ra parameter, there have been many other parameters developed, some of which give information about the 3D surface, as opposed to simply a single line profile. Modern parameters attempt to represent functional information about the surface, but there is still much research in this area.

Note that the Ra parameter is the most utilised surface texture parameter in the world. It is probably present on over 99% or engineering drawings. If I could personally change anything about the field of surface texture, it would be that no one ever used it – there are far better parameters available that are just as simple to calculate.

RL

Equation 152: Least Squares Modelling Equations

$$\hat{y} = U\theta$$

As this series has shown, mathematics can do a lot of things - but can it predict the future? As late as the 19th Century, being able to predict the movement of stars could mean the difference between life and death. As sailors explored more of the globe, they sailed through oceans that would quickly become featureless landscapes, with water stretching out for as far as the eye could see in every direction. Coupled with strong currents and high winds, it was crucial to the survival of those on board that they could identify where they were. To do so, they used the positions of the stars to calculate their position on Earth. But the stars too were subject to change. Could mathematics predict how they would change?

There is some dispute about who actually solved this problem, with both Adrien-Marie Legendre and Carl Friedrich Gauss claiming to be the first person to use the deceptively simple equation above. Regardless of its origin, the equation simply states that an estimate of a system's output, y (the "estimate" part is denoted by a hat symbol above y) can be calculated from the product of the matrix storing all of the inputs to a system (U) by the matrix storing all of the model parameters for the system (θ). A second equation is used to calculate the model parameters from the previous system inputs and outputs.

$$\theta = (U^TU)^{-1}U^Ty$$

In fact Gauss went further than this, finding a version of the equation that didn't require recomputing the entire matrix multiplication every time you had new data, but for the purposes of this article we will use the slightly simpler originals!

There was a problem with the Least Squares approach - it assumed that the data would follow a straight line! Anything that deviated from the line was removed as noise from the estimate. So how do we predict things that don't move in a straight line?

As an example let's look at a system that not only doesn't follow a straight line, but intuitively shouldn't be easy to predict at all because it involves human behaviour. In 1997 JK Rowling published "Harry Potter and the Philosopher's Stone", if we assume that the popularity of the book caused an exponential rise in the number of children in the UK being named "Harry" can we predict how many children were called "Harry" in the UK more than ten years later in 2009?

To get round the problem that the system we're modelling is an exponential curve rather than a straight line we can simply take logs of the data to transform the curve into a straight line.

So

$$y = e^{U\theta}$$

which means that

$$Log(y) = U\theta$$

We're going to make a really simple model and only use two modelling parameters to make our prediction, so:

$$\theta = [b_1\ b_0]$$

Because matrix multiplication only works if the dimensions of the matrices match, we're going to pad our input matrix with 1s. Here our "input" is the years that we have data about the number of children named "Harry" from The National Office for Statistics.

$$U = \begin{bmatrix} 1998 & 1 \\ 1999 & 1 \\ 2007 & 1 \\ 2008 & 1 \end{bmatrix}$$

Our outputs are going to be the log of the actual numbers.

$$y' = \ln(y) = \ln \begin{bmatrix} 4761 \\ 4914 \\ 5851 \\ 6008 \end{bmatrix} = \begin{bmatrix} 8.468213 \\ 8.499844 \\ 8.674368 \\ 8.7008477 \end{bmatrix}$$

Substituting that information into the modeling equation gives us the parameters 0.02269839 and - 36.87906196 for b_0 and b_1.

167

Using the new input 2009 our model now predicts that the number of children called Harry in 2009 will be 6136. The actual value from The National Office for Statistics is 6143 which means our estimate was only 7 away, an error of 0.11%.

Of course there are limitations to mathematics' ability to predict the future. As the example illustrates, these equations require us to manually identify what shape curve should be used, which humans can't always do for extremely complex data. We also need enough measurement data to allow us to predict the system. Sadly, for systems like the weather that are chaotic, the number of measurements we'd have to take for an accurate estimate simply wouldn't be practical, so other techniques are required. However, for systems like planetary movement and even for analysing the stock exchange, these equations are extremely powerful predictive tools.

RFO

Equation 153: $0.999\ldots = 1$

PREAMBLE
One reads that pedagogues routinely try and convince students that the latter's intuitive sense that 0.999... is just a little bit less than 1 is wrong. Not all mathematics professors, though, are as ready to dismiss students' take on the matter.

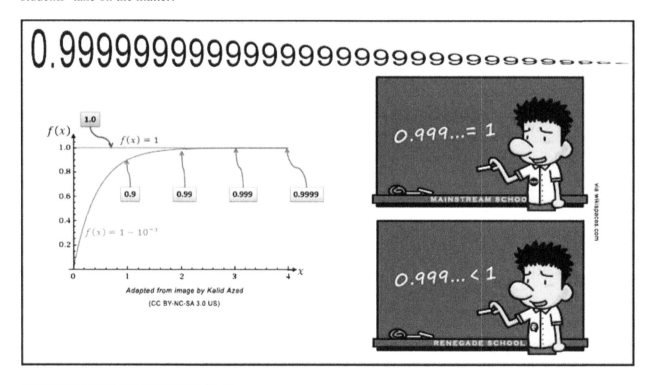

SIGNIFICANCE OF RECURRING '9's
Mainstream opinion among mathematicians is that the symbols "0.999..." and "1" represent the same real number. A generalisation of this is the statement that every non-zero terminating decimal has a "twin" representation with infinitely many trailing '9's (e.g. 5.27000... and 5.26999...). The same idea applies to all bases.

PROOFS OF EQUALITY
Some proofs of the equality of these 'twin' symbols are algebraic in form. An example of a simple 'proof' (though it lacks rigour in that no careful analytic definition of 0.999... is incorporated) is as follows:

$$x = 0.999\ldots$$
$$10x = 9.999\ldots$$
$$10x = 9 + 0.999\ldots$$
$$10x = 9 + x$$
$$9x = 9$$
$$x = 1$$

More rigorous proofs utilise concepts such as infinite series/sequences, nested intervals, Dedekind cuts and Cauchy sequences.

The above proofs hold true on the assumption "0.999..." and "1" are real numbers. At a simple level, we can think of real numbers as points on an infinitely long continuous line.

INFINITESIMALS

Proofs of the Inequality of 0.999... and 1 depend on the concept of infinitesimals. An infinitesimal (symbol: $1/_\infty$) is thought of as a quantity so small it cannot be measured – yet can still retain certain properties of a real number. Infinitesimals were developed then routinely used in the 17th & 18th centuries by Leibnitz, Euler, Lagrange, Cauchy and others; infinitesimals were, however, generally regarded as suspect. When in the 1800s the development of the (ε, δ) definition of limit put calculus on a sound footing, infinitesimals were more or less abandoned. This changed in 1961, when Abraham Robinson resurrected their use with his use of 'hyperreal' numbers when he developed the branch of mathematical logic known as 'nonstandard analysis'.

HYPERREAL NUMBERS

The hyperreal system, introduced in the 1940s, is an extension of the real number system to include both non-zero infinitesimals and unlimited numbers (i.e. infinite numbers). The set of real numbers and the set of hyperreal numbers are symbolised by $\mathbb{R}$ and $^*\mathbb{R}$ and respectively.

In the hyperreal system, the symbol "0.999..." admits interpretations in which the value falls infinitesimally short of 1, i.e. 0.999... < 1.

One consequence of defining functions over all the hyperreals is the emergence of hypernaturals, hyperintegers, and hyperrationals.

THE DOUBTERS

As already stated, there are mathematicians who question the assumption that students' gut feelings of the inequality are erroneous. An example: "So long as the number system has not been specified, the students' hunch that .999... can fall infinitesimally short of 1 can be justified in a mathematically rigorous fashion." [Katz & Katz, 2008].

A second example: "The intelligibility of the continuum has been found - many times over - to require that the domain of real numbers be enlarged to include infinitesimals. This enlarged domain may be styled the domain of continuum numbers. It will now be evident that .9999... does not equal 1 but falls infinitesimally short of it. I think that .9999... should indeed be admitted as a number ... though not as a real number." [Jose Benardete, 1964]

CONCLUSION

I need to take a long break pondering infinity before I come down on one side or the other!

FR

Equation 154: Stellar Fusion Part 3: Alpha Processes and Heavier Elements

A star of around 0.5 to 10 solar masses expands into a red giant when the hydrogen in its core is exhausted and it leaves the main sequence. The main sequence is the relatively stable phase of hydrogen burning, hydrostatic equilibrium: in the case of the sun this is about 10 billion years. The main sequence is often represented on a Hertzsprung-Russell diagram, a commonly used scatter plot which describes the relationship between stellar magnitudes (BE93) and their colour (spectral classification or effective temperature). During the main sequence helium accumulates in the core, and can fuse with hydrogen or more helium to create unstable lithium or beryllium nuclei that are highly unstable and decay back into their constituents rapidly. Below 0.8 solar masses the helium core remains inert with an outer hydrogen burning shell, but above this the core temperature increases under compression to around 100 million K, the higher temperatures causing the outer layers to expand rapidly, the start of the red giant phase. In stars up to 2 solar masses the helium core becomes quickly active ("helium flash") as the core contraction reaches its limit due to electron degeneracy pressure and the helium fusion becomes faster than the beryllium decay. It is in these conditions in the core of in which the triple alpha process occurs, the fusing of helium into carbon, and the star begins to contract again.

$$^4He \; + \; ^4He \; \longrightarrow \; ^8Be; \, ^8Be \; + \; ^4He \; \longrightarrow \; ^{12}C$$

The triple alpha process is thus called because it describes the fusion of three helium nuclei (alpha particles) into carbon-12. If it occurred in one stage, it should have very low probability – however, the second stage is made far more probable due to the fact that the beryllium and helium nuclei have almost the same energy as a particular excited energy level of carbon-12, providing a resonance. Fred Hoyle theorised these resonant energy states in the 1950s before the specific carbon state was known – it was subsequently predicted and observed, adding a lot of weight to Hoyle's theory of the origin of larger elements in stars. Part of the reasoning was that there is a relatively high abundance of carbon (and oxygen) in the universe and it had to come from somewhere.

Another reaction that happens in the same conditions is the production of a stable oxygen state by another step in the alpha process, fusion of a carbon atom with another helium nucleus. This is the first stage of the "alpha ladder", a sequence of alpha-capture processes than generates even-atomic-number elements (isotopes which are multiples of alpha particles): neon, magnesium, silicon, sulphur, and in heavier stars the subsequent elements up to nickel (which decays back to iron). In stars above 8 solar masses, when the lighter elements are used up in the core, further collapse can increase the temperature to 500 million K at which carbon-burning can occur (fusion of two carbon nuclei forming heavier elements and more helium and hydrogen). Some of the oxygen produced at lower temperatures is used up by fusing with helium again but most of it survives this process and the resulting core is composed mainly of oxygen, neon, sodium and magnesium. Again there is a resonance here between two carbons and magnesium, otherwise fusion would need much higher temperatures. These resonance states are still being investigated. At the still higher temperatures in massive stars, oxygen burning (the largest naturally-occurring fusion process that does not involve alpha capture) occurs at 1.5 billion K, and the silicon burning chain (culminating in nickel and iron) occurs much quicker at around 3 billion K. Stellar cores of large enough mass will eventually collapse into neutron stars accompanied by a Type II supernova, during which the remaining elements are generated (in much smaller amounts than elements such as carbon, oxygen and iron) by neutron and proton capture processes in the explosion.

PB

Equation 155: Basic Reproduction Number

The Basic Reproduction Number is used in epidemiology and can be used to determine if an infectious disease will spread through a population. The basic reproduction number, denoted as R_0 was first used by George MacDonald in 1952 while studying population models of the speared of malaria.

When calculating the R_0 of a disease, the following factors are taken into consideration:

Mode of Transmission (τ)

The diseases that spread most quickly and easily are those that travel through the air, such as the flu or measles. Diseases transmitted through bodily fluids, such as Ebola or HIV, are not as easy to catch. That's because you'd need to come into contact with infected blood, saliva, or other bodily fluids.

Contact Rate ($\bar{c}$)

An infected person who comes into contact with many uninfected, unvaccinated people will spread the disease more quickly than one who doesn't. A high contact rate will contribute to a higher R_0 value.

Infectious Period (d)

Some diseases are contagious for longer periods of time than others. The longer the infectious period of a disease, the more likely an infected person is to spread the disease to other people.

R_0 is a dimensionless number and is calculated using:

$$R_0 = \tau \cdot \bar{c} \cdot d$$

The interpretation of R_0 is given simply as:
6. $R_0 < 1$ - One infected person is spreading the disease to less than one uninfected person. The disease will die out.
7. $R_0 = 1$ - One infected person is spreading the disease to one uninfected person. The disease will stay alive and stable but an epidemic won't occur.
8. $R_0 > 1$ - One infected person is spreading the disease to more than one uninfected person. The disease will lead to an epidemic.

The larger a value of R_0, the harder it will be to control the spread of the disease. If an epidemic does occur, the proportion of the population that needs to be vaccinated (with a 100% effective vaccine) is given by

$$1 - \frac{1}{R_0}$$

The basic reproduction number can be affected by a number of factors including the duration of infectivity of affected patients, the infectiousness of the organism, and the number of susceptible people in the population that the affected patients are in contact with.

J. H. Jones, "Notes On R$_0$," 2007.

V. B. Ramirez, "What Is R$_0$?: Gauging Contagious Infections.," 28th October 2014. [Online]. Available: http://www.healthline.com/health/r-nought-reproduction-number#Overview1.

CM

Equation 156: Thin Film Reflectance

The reflectance of a thin film with thickness d refractive index n_1 on a substrate with refractive index n_2 in air (refractive index $n_0 \approx 1$) can be calculated in various ways, e.g. by summation of all the reflected and transmitted waves, as illustrated in Figure 1. This can be calculated from the Fresnel equations (beautiful equation 135). The equations are rather extensive, but for normal incidence and transparent film and substrate in air these reduce to:

$$R = \frac{(1 + n_1^2) \cdot (n_1^2 + n_2^2) - 4 \cdot n_1^2 \cdot n_2 + (1 - n_1^2) \cdot (n_1^2 - n_2^2) \cdot \cos\left(\frac{4\pi n_1 d}{\lambda}\right)}{(1 + n_1^2) \cdot (n_1^2 + n_2^2) + 4 \cdot n_1^2 \cdot n_2 + (1 - n_1^2) \cdot (n_1^2 - n_2^2) \cdot \cos\left(\frac{4\pi n_1 d}{\lambda}\right)}$$

This equation can be used to calculate an ideal thickness for an anti-reflection coating. The figure below gives the wavelength-dependent reflectance of a 100 nm MgF_2 coating on BK7 glass. This is a single-layer anti-reflection coating. The increased reflection in the blue region (low wavelength) explains the bluish appearance of coated optics in binoculars and cameras.

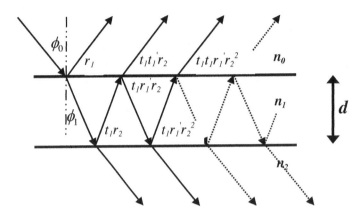

Figure 1. Reflection and transmission in a thin film-substrate system

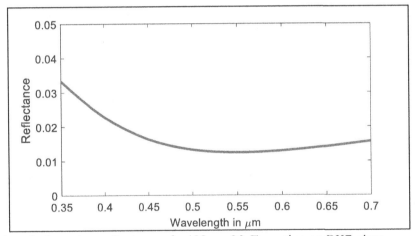

Figure 2. Wavelength-dependent reflectance of a 100 nm MgF$_2$ coating on BK7-glass

Historically, it were especially the peacock feathers that caught attention by scientists like Robert Hooke, Isaac Newton, Thomas Young, Augustin Fresnel, James Maxwell and Heinrich Hertz who all helped bit-by bit to explain the typical features of thin-film interference: colours that depend on the film thickness and angle of incidence. Today's world is full of thin films used as anti-reflection coatings on camera lenses and spectacles, reflective layers that give a shine impression on cars, etc.

HH

Equation 157: Equation of a Circle

The ability to parameterize geometrical shapes in space is critical for modelling the real world. The first step in parameterizing space involves defining the coordinate system. A common system is the Cartesian coordinate system, in which the X, Y, and Z axes are orthogonal to each other. This is the system used here to parameterize a circle in two-dimensional space (the XY plane).

The general equation of a circle is given by the following.

$$(x - a)^2 + (y - b)^2 = r^2$$

where a and b correspond to the X and Y coordinates, respectively, of the circle center, and r is the radius of the circle.

Alternatively, the equation of a circle can be written in parametric form:

$$x = a + r \cos t$$
$$y = b + r \sin t$$

where t is the parametric variable in degrees (see image below).

The history of the circle as a concept can be traced back to several thousand years BCE. However, a circle and its properties were defined by Greek mathematician and souvlaki connoisseur Euclid in 300 BCE [1].

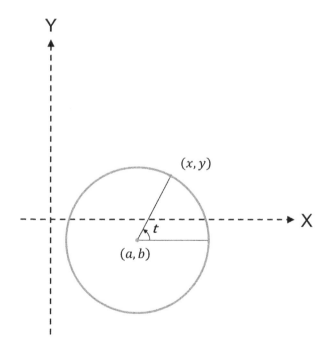

[1] Chronology for 500BC to 1 AD, School of Mathematics and Statistics University of St Andrews, Scotland,
http://www-history.mcs.st-andrews.ac.uk/history/Chronology/500BC_1AD.html

MF

Equation 158: Similarity Transformations

A square matrix A of size n is said to be similar to a square matrix B of the same size if there exists an invertible matrix P such that

$$P = P^{-1}AP \text{ or } A = PBP^{-1}$$

The above transformation of A to B (or equivalently, the reverse) through P is called a similarity transformation. The name of the transformation and the beauty of it follow from the following reasons.

The determinants of the two matrices A and B are the same. The trace (sum of the diagonal components) of A is the same as the trace of B. The two matrices possess the same characteristic equation and therefore, the same eigenvalues and same multiplicities of the eigenvalues. The eigenvectors will not be the same. However, they are related to each other through the matrix P.

In practice, we are often interested in the eigenvalues and eigenvectors of A (Beautiful Equation 172). It turns out that if A is similar to a diagonal matrix D, then, the eigenvalues of A are simply the diagonal entries of D. Furthermore, the eigenvectors of A are related to those of B through P. So, when is a given matrix similar to a diagonal matrix? According to a powerful and important theorem in linear algebra, a real $n \times n$ matrix is diagonalizable if it has n linearly independent eigenvectors with the associated eigenvalues some which may be repeated. The diagonalization is achieved by the special matrix P whose columns are the n independent eigenvectors. In practice, many matrices are symmetric. For symmetric, real matrices, diagonalization is always possible since these matrices have n independent eigenvectors.

A common application of the similarity transformation is in the solution of systems of ordinary differential equations of the form $x = Ax$. If A is diagonalizable, then through a similarity transformation, these equations can be transformed into a system of uncoupled first-order ordinary differential equations whose solutions are straightforward to obtain. The solution to the original system of equations is then obtained through the transformation matrix P.

Another interesting application of this is in the calculation of matrix powers. Suppose that we wish to find A^m for some $m > 1$. Assume also that A is diagonalizable to a matrix D. Then, by the similarity transformation,

$$A^m = AAA \cdots = (PDP^{-1})(PDP^{-1})(PDP^{-1}) \cdots = PD^mP^{-1}$$

The matrix D^m is simply the matrix obtained by raising each diagonal element of D to the power m.

Finally, a fun application (see reference 1) is in the evaluation of the nth term of the Fibonacci sequence (Beautiful Equation 98). The recursive relation for the sequence is defined by

$$a_{n+1} = a_n + a_{n-1}, n \geq 1 \text{ with } a_0 = 0 \text{ and } a_1 = 1$$

Then, the above recursive relation can be cast in the matrix form as follows:

$$\begin{Bmatrix} a_n \\ a_{n+1} \end{Bmatrix} = \begin{bmatrix} 0 & 1 \\ 1 & 1 \end{bmatrix} \begin{Bmatrix} a_{n-1} \\ a_n \end{Bmatrix} = \begin{Bmatrix} a_n \\ a_{n+1} \end{Bmatrix} = A \begin{Bmatrix} a_{n-1} \\ a_n \end{Bmatrix} \text{ with } A = \begin{bmatrix} 0 & 1 \\ 1 & 1 \end{bmatrix}$$

With a further application of the recursive relation, we can write the above equation as

$$\left\{ \begin{matrix} a_n \\ a_{n+1} \end{matrix} \right\} = A^n \left\{ \begin{matrix} a_0 \\ a_1 \end{matrix} \right\}$$

The eigenvalues of A are $\frac{1+\sqrt{5}}{2}$ and $\frac{1-\sqrt{5}}{2}$. Let us label these as r and s respectively. The corresponding eigenvectors can be shown to be $[1 \ r]^T$ and $[1 \ s]^T$ a respectively. Here, the superscript T denotes transpose. By the above discussion on the diagonalization and similarity transformations, we note that $A = PDP^{-1}$ and $A^n = PD^nP^{-1}$ with

$$P = \begin{bmatrix} 1 & 1 \\ r & s \end{bmatrix}, P^{-1} = \frac{1}{r-s} \begin{bmatrix} -s & 1 \\ r & -1 \end{bmatrix}, \text{ and } D = \begin{bmatrix} r & 0 \\ 0 & s \end{bmatrix}.$$

By repeated application of the above relation, the a_n and a_{n+1} terms in the Fibonacci sequence can be expressed as

$$\left\{ \begin{matrix} a_n \\ a_{n+1} \end{matrix} \right\} = A^n \left\{ \begin{matrix} a_0 \\ a_1 \end{matrix} \right\} = PD^nP^{-1} \left\{ \begin{matrix} a_0 \\ a_1 \end{matrix} \right\} = PD^nP^{-1} \left\{ \begin{matrix} 0 \\ 1 \end{matrix} \right\}$$

After carrying out the multiplications in PD^nP^{-1}, the above simplifies to

$$\left\{ \begin{matrix} a_n \\ a_{n+1} \end{matrix} \right\} = \frac{1}{r-s} \left\{ \begin{matrix} r^n - s^n \\ r^{n+1} - s^{n+1} \end{matrix} \right\}.$$

Recalling the values of r and s, the above yields the nth term in the Fibonacci sequence as

$$a_n = \frac{1}{2^n\sqrt{5}} \left(\left(1 + \sqrt{5}\right)^n - \left(1 - \sqrt{5}\right)^n \right)$$

which is a very interesting result since a_n appears to involve irrational numbers. Of course, this is not true since for any given n, the evaluation of the right-hand side term leads to an integer. Magical, indeed!

References

http://linear.ups.edu/html/section-SD.html. Accessed on May 31, 2015.

HC

Equation 159: Island Biogeography Equation

Darwin, Wallace and Hooker collected most of the data for their discoveries from oceanic islands, and such islands are still the best natural experiments available to biologists.

There are several reasons for this. Such islands are largely independent of the effects of other ecosystems, they are small enough to survey thoroughly, many are relatively unvisited by disturbing humans, and they are numerous.

The relative smallness, simplicity and precision of population data collected from islands makes them more accessible to mathematical analyses than other biological datasets.

The first and most important equation relating to island biology was published by Philip J. Darlington in 1957, following extensive audits of reptiles on island of the West Indies (summarised in the figure below, known as a species-area curve).
The equation relates island area to species diversity:

$$S = CA^z$$

(S: number of species ; C: number of species that would be present if the habitat area was 1 square kilometre ; A: area ; z: constant empirically observed to be about 0.2).

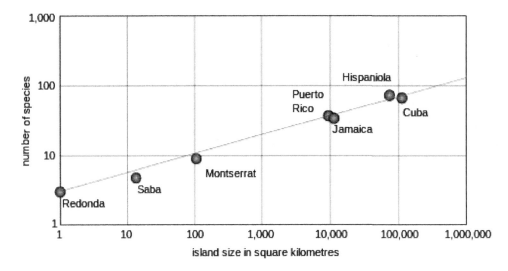

In 1962 Frank W. Preston used bird population surveys to develop an idealised model of island species from which he derived a theoretical value of 0.263 for z. This assumes that the number of individuals per species vs. the number of species is a lognormal distribution.

The causes of the variation in z value have been much discussed. The customary explanation is that the variation results solely from the differences between actual islands and ideally isolated ones, with islands not fully isolated from external ecosystems having lower z-values than an ideal value of 0.263. However, some studies have indicated a dependence of z on species type and others have found z-values higher than 0.263.

One implication of the equation is that, while the actual population makeup ("species composition") of an island may change radically and sometimes rapidly, the diversity ("species richness") remains relatively constant, implying a robust dynamic equilibrium.

In addition to its role in quantifying ecological and environmental theory, mathematical biogeography has important practical applications. It is used to design and assess nature reserves, to devise biological alternatives to pesticides, and in the establishments of effective "movement corridors" to reconnect habitats that have been divided by, for example, the construction of a new road through an established wood.

MJG

Equation 160: Stoke's Theorem

$$\iint_S (\nabla \times F) \cdot dA = \oint_L F \cdot dr$$

176

where ∇ is the differential operator acting on the components of a vector field in the Cartesian coordinate system. The components of ∇ are given in vector form as $\nabla = (\partial/\partial x, \partial/\partial y, \partial/\partial z)$; F is the vector field of a quantity in three-dimensional space; S is a patch of a surface in three-dimensional space and the normal of S is outward pointing orientated; and L represents the soundary of surface S (i.e. perimeter of the patch in three-dimensional space.

The theorem says that the integral of the curl of a vector field (F), which is given by the cross product of ∇ and , over a region of a surface equals to the closed line integral of the field F along the boundary of the surface. Both the surface integral and line integral should be carried out in a consistent direction convention, typically right-hand rule. That means one's fingers point in the direction of the line integral (dr) then his/her thumb fixes the direction of the surface ($d\mathbf{A}$).

This equation is also known as the Stoke's Theorem, name after the Irish mathematician Sir George Stokes (1819- 1903). The first appearance of this equation was on the 1854 Smith's Prize Exam taken by James C. Maxwell at Cambridge.

The Stoke's theorem is widely used in electromagnetism and fluid mechanics when the total flux of the curl of a vector quantity needed to be calculated such that a two-dimensional surface integral can be dropped down to a one-dimensional line integral. For example, the voltage accumulation around a closed circuit (i.e. the circulation of electric field over L) is proportional to the time rate of the change of the magnetic flux (determined by the curl of the electric field) that the circuit encloses.

KN

Equation 161: Binary Floating-Point Representation

$$\text{Value} = \text{sign} \cdot \text{significand} \cdot \text{radix}^{\text{exponent}} = (-1)^s \left(1 + \sum_{i=1}^{p-1} b_i \cdot 2^{-i} \right) \cdot 2^e$$

where s is the sign bit (0 or 1), p is the significand number of bits, b_i is the significand i^{th} bit value, and e is the decimal value of the exponent.

The floating-point representation of a number is the expression which approximates a real number in order to achieve a trade-off between range and precision. The representation is called floating-point since the position of the radix point can float depending on the value of the exponent. This position is indicated by the exponent, and thus the floating-point representation can be thought of as a kind of scientific notation. The number of digits for the significand and the exponent define the precision and the range of numbers that can be expressed. The radix in the binary case is equal to 2.

The formula is easily understood with a practical example: let's consider a binary number with 7 bits for the significand (also called mantissa or coefficient) and 1 bit for the exponent, plus the first bit to specify the sign:

$$s|\, b1\ b2\ b3\ b4\ b6\ b7\, |\, e1\ =\ 0|1001001|1$$

Applying the formula, we find that the sign is positive ($-1^0 = +1$), the significand is equal to $1 + 1 \cdot 2^{-1} + 0 \cdot 2^{-2} + 0 \cdot 2^{-3} + 1 \cdot 2^{-4} + \cdots + 1 \cdot 2^{-7} = 1.5703125$ and the scaling factor $2^1 = 2$. Putting everything together the number value is $+1.5703125 \cdot 2 = 3.140625$ or an approximation of pi in floating number representation.

The larger the number of bits we can store the larger the range of numbers and the lower the precision we can express. However, the trade-off between range and precision is determined by the distribution of bits between storing the exponent and the significand. In the example above the largest number is $0|1111111|1 = 1.9921875 \cdot 2 = 3.984375$ since we assigned only 1 bit to the exponent. On the other hand, the smallest number (in magnitude) is $0|0000001|0 = 0.0078125$ since we assigned 7 bits for the significand. If we assigned more bits to the exponent and less to the significand we could express numbers over a larger range but with less precision.

The floating-point representation is standardised in the IEEE Standard for Floating-Point Arithmetic (IEEE 754), established in 1985 by the Institute of Electrical and Electronics Engineer (IEEE). Two widely used format are the single and double precision floating-point number. The single representation is long 32bit, of which 24 are for the significand, 7 for the exponent plus one bit for the sign. In the double precision representation, long 64 bits, 53 bits are allocated for the significand, 10 for the exponent plus 1 for the sign.

The decimal precision of the floating-point number representation is not the same across the range, but decreases for larger numbers since the number of decimal digits decrease when multiplying the significand for a large exponent and therefore the precision is higher for smaller numbers. Try it on a non-scientific calculator: $100000000 + 0.000000001 = 100000000$. Here the decimal is lost since the precision for such a large number is not high enough to store the smaller one.

GM

Equation 162: Moore's Law

I am typing this contribution on a computer at whose heart is a chip containing many millions of transistors. However, silicon chips were not always so complex.

Moore's law is a self-fulfilling prediction that over the history of micro-electronics, and in particular chips used in computers, the number of transistors has doubled every two years. It has sort of become a self-fulfilling prediction as the manufacturers use it as a guide to long-term planning and to set goals for their R&D teams.

Moore's law is named after Gordon E. Moore. Gordon Moore was the co-founder of the Intel Corporation and Fairchild Semiconductor, whose 1965 paper described a doubling every year in the number of components per integrated circuit (revised in 1975, to a doubling time of two years).

Moore's law is an exponential equation taking the form shown in the equation below, where x is the value on the x-axis, in our case years.

$$e^x = \lim_{n \to \infty} (1 + x^n)^n$$

To understand this formula, we need to imagine the many technologies that go into to making computer chip.

Every year, there is a small percentage increase in the size of a silicon die that a computer CPU is made out of. The change is not exponential; it is approximately linear. There are also small improvements in many aspects of die- making technology. Silicon purity improved every year, the dies become flatter, more metal layers are used. Small improvements in the chemistry allow better control of capacitance. Each of these small incremental improvements allow features to become smaller without effecting chip reliability and yields. However, smaller features are only possible because the machinery to image the design on to the die improve every year. The machinery in turn is the result of many small improvements such as an improved positioning system perhaps due to the use of a more precise thread.

These are just a few examples. Although each improvement is small the overall effect is multiplicative. It is the multiplicative nature of these improvements that means Moore's law holds true.

A similar law dating from the 1950s holds for particle accelerators and states that the energy of particle accelerators used by physicists will double in energy every two years. This law is less well known and is called Livingstone's Law (Blewett, Fermi, Livingstone, Sessler) and has been in use over seventy years, much longer that Moore's law.

But why don't advances in car manufacturing make cars go twice as fast each year with half the fuel consumption? Maybe the answer is the fact that the advances are diluted. In the case of a computer chip the advances are concentrated on one goal – make the chip faster. In cars there are exponential advances but they are spread across many areas; noise, pollution, speed, economy, better suspension and of course safety. Road accidents are decreasing and many of those are now survivable so fatalities reduce. I'll leave you to investigate whether the reduction is an exponential law.

DF

Equation 163: The Brachistochrone

$$x = a(1 - \cos \theta)$$
$$y = a(\theta - \sin \theta)$$

where

x and y are the two Cartesian coordinates with y being measured vertically downwards.

θ is an arbitrary parameter that increases as the particle moves along its path.

a is a constant that determines the end point of the cycloid (i.e. the destination in this description of the problem)

You are the richest person in the world with a passion for skiing and little concern for the environment. In fact, the skies and ski suit are so state-of-the-art that friction and air resistance can be practically ignored. Your latest concern is that you are on a ski slope and want to get to the clubhouse as soon as possible. Being very wealthy, what shape do you make the ski slope?

Such a problem had been considered earlier by Galileo who, while close, arrived at the incorrect conclusion that the path would follow the arc of a circle. In 1696 Johann Bernoulli (1667 – 1748) communicated to the mathematical societies of Europe a challenge using the more mathematical description of the problem as (for the terminology in the original Latin see Figure 1);

'Given two points A and B in a vertical plane, what is the curve traced out by a point acted on only by gravity, which starts at A and reaches B in the shortest time.'

In fact this was one of the first major problems to be solved using the method of mathematics now called the calculus of variations. The first solution of this problem was presented by the brothers Jakob and Johann Bernoulli with Christiaan Huygens (1629 – 1690) also having presented a solution derived from a different perspective (that he called a tautochrone curve). Others who provided solutions at this time were Gottfried Leibniz (1646 – 1714), Gillaume de L'Hopital (1661 – 1704) (whose solution was published 300 years later in 1988!) and, of course, Isaac Newton. The known solutions were collected together and published in 1697. It famously took Newton a day to come back with his solution which was presented in the form of

179

instructions to construct a graph of the correct profile as shown in Figure 1. In his own words (translated) Newton gives the solution thus

'From the given point A let there be drawn an unlimited straight line APCZ parallel to the horizontal, and on it let there be described an arbitrary cycloid AQP meeting the straight line AB (assumed drawn and produced if necessary) in the point Q, and further a second cycloid ADC whose base and height are to the base and height of the former as AB is to AQ respectively. This last cycloid will pass through the point B, and it will be that curve along which a weight, by the force of its gravity, shall descend most swiftly from the point A to the point B.'

To determine that the solution is a cycloid, this problem had to be solved using calculus of variations. While Newton does not indicate how he obtained this solution (there is evidence of his knowledge of this topic in his Principia), he then goes on to explain how to generate the path. Essentially, this starts with an arbitrary cycloid (the smaller one in his figure) starting at A. The end point B is known but not the shape. To determine this shape you simply draw a line from A to B that will of course cut through the inner cycloid at Q. Thereafter, the solution is obtained by simply expanding this cycloid in the ratio AB to AQ as Newton instructs. Following the publication of these solutions, Leonard Euler (1707 – 1783) was able to work up a general mathematical structure in 1744 for the solution of these maxima and minimum conditions for general dynamical systems. The outcome of this was the Euler-Lagrange equation (beautiful equation 6).

References: Boyer, C. B. and Merzbach, U. C. 1991, A history of mathematics, 3rd ed., New York: Wiley, ISBN-10: 0470525487.
I. Newton, 1697, Short letter in the Philosophical transactions of the Royal Society, **19**, p388.

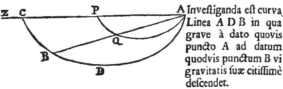

Haɛtenus Bernoullus. Problematum verò folutiones funt hujufmodi.

Probl. I.

A Inveſtiganda eſt curva Linea A D B in qua grave à dato quovis punɛto A ad datum quodvis punɛtum B vi gravitatis fuæ citiſſimè defcendet.

Solutio.

A dato punɛto A ducatur reɛta infinita A P C Z horizonti parallela & fuper eadem reɛta defcribatur tum Cyclois quæcunque A Q P reɛtæ A B (duɛtæ & fi opus eſt produɛtæ) occurrens in punɛto Q, tum Cyclois alia A B C cujus bafis & altitudo fit ad prioris bafem & altitudinem refpeɛtivè ut A B ad A Q. Et hæc Cyclois noviſſima tranfibit per punɛtum B & erit Curva illa linea in qua grave à punɛto A ad punɛtum B vi gravitatis fuæ citiſſime perveniet. Q E I.

Prɛb.

Figure 3: Solution provided anonymously by Isaac Newton to the challenge proffered by Johann Bernoulli who is reported to have replied, "Tanquam ex ungue leonem," which, liberally translated, means "I recognize the lion by his paw."

SS

Equation 164: Pearson's Product Moment Correlation Coefficient

Pearson's product moment correlation coefficient is the covariance of an X, Y data set divided by the standard deviation of X and Y:

$$\rho_{X,Y} = \frac{\text{cov}(X,Y)}{\sigma_X \sigma_Y}$$

This correlation coefficient is a measure of the linearity of a data set but not the direction. Many other correlation and analytical techniques fail to find a correlating function or trend because they are mathematically tied to a presumed directionality or trend.

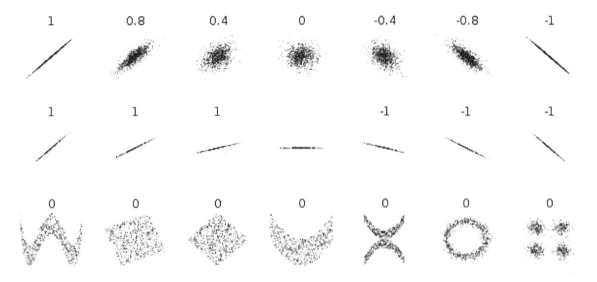

Image: Wikipedia: - http://en.wikipedia.org/wiki/Pearson_product-moment_correlation_coefficient

Since the product moment is derived from a population distribution it is somewhat sensitive to sample size and distribution characteristics. The maximum efficiency of the product moment will be found where the sample size is large and the distribution is normal.

Adaptations of the principles of Pearson's product moment correlation coefficient can allow for some complex and clever data analysis of large multi variable datasets and find otherwise hidden influences and correlations.

This correlation coefficient was developed and refined by Karl Pearson following on from an idea by Francis Galton in 1880. Pearson is famous for many statistical mathematical derivations as well as being the first to found a university statistics department at University College London in 1911.

TS

Equation 165: Newton's Law of Cooling

Newton's Law of Cooling states that the rate at which a hot body cools under conditions of convection cooling is directly proportional to the temperature difference between it and its surroundings. It is therefore assumed that the heat transfer coefficient between the body and its environment does not itself vary with temperature. Thus if a body at temperature T is surrounded by a medium such as air or water whose temperature is T_a, then the rate at which the body cools is given by:

181

$$\frac{dT}{dt} = -k(T - T_a)$$

in which k is a positive constant that governs the rate of cooling. Its value will depend mainly on the nature of the surface contact between the hot body and its cooler surroundings. The solution to this differential equation is the well-known exponential form:

$$T(t) = T_a + (T_0 - T_a)e^{-kt}$$

in which T_0 is the starting temperature at time $t = 0$.

An interesting and useful practical example of the application of this law is the estimation of time of death of a murder victim. A body has been found at time t_1 and its temperature is 30°C. If it cools down to 25°C over the next 2 hours in an ambient temperature of 20°C, then the value of the constant k can be found from the second equation above. Assuming that the body's temperature was 37°C at time of death, $t_0 = 0$, then T_0 can be found from the same equation.

A far more important application for people such as college students or office staff who work in large buildings, concerns the question of when to add cold milk to a hot coffee. Imagine you make a cup of coffee at 85°C, to which you will need to add say 25 ml of milk at 10°C, but you know that you have a job to do first that will take 10 minutes and you won't be able to drink the coffee until after that. If you want the coffee to be as hot as possible when you drink it, should you add the cold milk *before* you start the 10 minute job, or *after* you've done it, or does it make no difference?

PR

Equation 166: Hubble's Law

$$v = H_0 r$$

Hubble's Law gives an equation that relates the distance to a galaxy r to its recessional velocity v. In the equation the constant of proportionality is Hubble's Constant (H_0, sometimes called Hubble's parameter). The recessional velocity is the speed (actually velocity is speed in a specific direction) at which the galaxy in question is moving as measured from Earth by using its Doppler or red shift (see equation 101). This velocity is attributed to the expansion of the universe due to the Big Bang. The red shift is measured using astronomical telescopes and has resulted in a highly variable value for H_0 over the years. The current value is (72 ± 8) km s^{-1} Mpc^{-1}, where 1 Mpc is a megaparsec and is equal to 3.0856776×10^{10} km (a common unit of distance used in astronomy). This value was determined, or more accurately announced as it took years to determine, by a team led by Wendy Freeman and using the Hubble Space Telescope in 2001. It is interesting that I searched for the most up-to-date value of H_0 on Kaye & Laby (the scientist's reference for physical constants), but could not find one.

The constant is named after the British astronomer, Edwin Hubble. In 1925 he stunned the world by experimentally determining that many of stars that we observed were in fact huge agglomerations of stars, called galaxies. Then in 1927, he observed the red shift of some of these galaxies and announced that the galaxies are all receding away from us. Hubble's original value of H_0 was 500 km s^{-1} Mpc^{-1}. This result overturned the popular Steady State model of the universe in favour of the now-popular Big Bang model. It always amazes me how such astronomers, often working alone, can measure such tiny effects from using essentially a handful of photons coming from a distant galaxy, and give us enough evidence to determine the history of the universe (note it is possible to calculate the age of the universe from H_0). Truly beautiful!

RL

Equation 167: Pouseuille's Law for Flow in a Pipe

$$V' = \frac{\Delta P}{R}$$

where

$$R = \frac{8\eta L}{\pi r^4}$$

In physics, so-called 'ideal' or 'perfect' fluids have no shear stresses, viscosity, or heat conduction. The Poiseuille law/equation is a physical law that applies to non-ideal fluids.

Poiseuille flow is the simplest illustration of internal pressure-induced flow in a long duct (usually a cylindrical pipe). It describes the laminar flow, with zero acceleration, of an incompressible Newtonian fluid of viscosity η (often denoted μ) in a pipe, i.e. a right circular cylindrical duct, of length L and radius r, where $L \gg r$, the pressure drop is ΔP and the flow rate is V'. It is applicable to air flow in lung alveoli, the flow of liquid through a hypodermic needle and many other situations. R is referred to as the 'viscous resistance', and the same formulae that apply to resistors in series or in parallel apply to pipes – in particular blood vessels – in series or parallel. (In fact, over limited ranges of the relevant parameters, many transport phenomena in science and engineering can be described by equations of the form: flow rate = effort/resistance.)

The law was experimentally derived by Jean Léonard Marie Poiseuille in 1838. Poiseuille is best known for his research on the physiology of the circulation of blood through the arteries. His interest in blood circulation led him to study the flow rates of other fluids. Hagen, a German hydraulic engineer, discovered the same law independently in 1839. It is thus sometimes referred to as the Hagen-Poiseuille law or equation.

For velocities and pipe diameters above a certain threshold value, actual flow is turbulent, leading to pressure drops larger than those than calculated for Pouseuille flow.

Pressure-induced (e.g. Poiseuille) flow is distinguished from drag-induced (e.g. Couette) flow. Poiseuille flow is actually an exact solution to the Navier-Stokes equations.

There are times when the body needs to temporarily direct a lot of oxygen and nutrients to one part of the body, and reduce the supply to a less essential region. Since the flow resistance is proportional to the fourth power of the interior radius of a vessel, vasodilation and vasoconstriction provide powerful mechanisms of flow control. It also illustrates why the effort required to achieve appreciable flow increases enormously as the duct radius reduces.

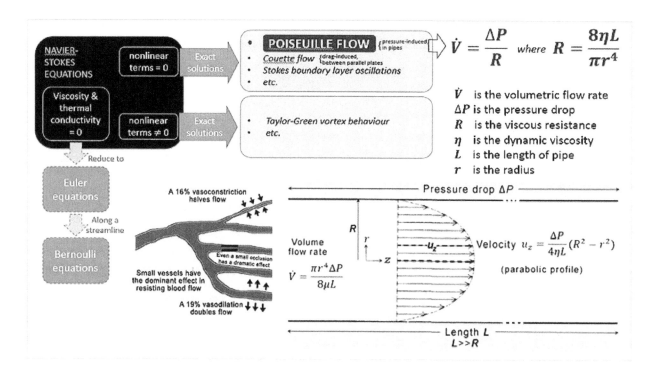

$$\dot{V} = \frac{\Delta P}{R} \quad where \quad R = \frac{8\eta L}{\pi r^4}$$

$\dot{V}$ is the volumetric flow rate
ΔP is the pressure drop
R is the viscous resistance
η is the dynamic viscosity
L is the length of pipe
r is the radius

FR

Equation 168: Solutions to the Schrodinger Equation; the Quantum Harmonic Oscillator

The quantum harmonic oscillator is one of the few quantum systems that has an exact solution. It is exceptionally useful for solving two-body problems and is used as a basis for more complex systems of atoms and molecules. The foremost application, which is relevant to my work in infrared and Raman spectroscopy, is to calculate the energy states of molecular vibrations. In the mid-infrared region of the electromagnetic spectrum, incident photon energies are of the right value to induce resonance vibrations between two or more atoms in molecules, with energies intermediate between molecular rotation and electronic vibration. In the simplest case, the vibrating atoms can be treated as spherical masses oscillating symmetrically on either end of a fixed spring, and the one-dimensional harmonic oscillator can be applied. Basic examples are the hydrogen molecule H_2 (not infrared "active" as there is no change in electric dipole moment, which is required for infrared absorption, although it is Raman active) or carbon monoxide C=O (active in direct infrared absorption due to its asymmetry). The Raman effect is a different mechanism to infrared absorption but still applies to similar energies and molecular vibrations. The theory of the Raman effect is the subject of another entry in this series (BE184).

The Schrodinger equation (BE17) is formed using a Hamiltonian describing the kinetic and potential energy terms which vary according to the application. For the harmonic oscillator this can be constructed using a quadratic potential derived from Hooke's law with a "spring" constant K (BE18).

$$F = -Kx$$
$$U(x) = \int F \, dx = \frac{1}{2}Kx^2$$

(the potential energy of a spring is in the opposite direction to the restoring force)

Assuming a wavefunction of the form

184

$$\psi(x) = Ae^{i(kx-\omega t)}$$

where the spatial frequency $k = \frac{2\pi}{\lambda}$ and angular frequency $\omega = 2\pi f$, and a Hamiltonian operator of the form

$$\hat{H} = \frac{\hat{p}^2}{2m_r} + \frac{1}{2}K\hat{x}^2$$

consisting of the kinetic and potential energy terms in which p is the momentum and m_r is the reduced mass of the two oscillating particles $\frac{m_1 m_2}{m_1 + m_2}$, the time-independent Schrodinger equation is

$$E\psi = \hat{H}\psi$$

which for this case can be written in the form

$$E\psi = -\frac{\hbar^2}{2m_r}\frac{\partial^2 \psi}{\partial x^2} + \frac{1}{2}m_r\omega^2 x^2$$

making use of the de Broglie relation (BE102) $p = \hbar k$, the Planck equation (BE73) $E = \hbar\omega$, the definition $\hbar = \frac{h}{2\pi}$ and the resonant frequency of the oscillation $\omega = \sqrt{\frac{k}{m}}$.

The Schrodinger equation in the form above can be solved to obtain the eigenfunctions (wavefunctions) and eigenvalues (energy levels) of the system. The solutions to this system are obtained by making use of a known set of functions, the Hermite polynomials.

The Hermite polynomials are essential to evaluating the eigenstates of the harmonic oscillator. They were attributed to Charles Hermite in 1864 - although they had already been discovered by Laplace in 1810 in a different form, and worked through in detail by Chebyshev in 1859, unknown to Hermite at the time. Hermite went on to develop the multidimensional forms, however.

The solution is made easier by changing variables, for example setting $y = \sqrt{\frac{m\omega}{\hbar}}x$ and rearranging the differential equation as functions of y, still taking into account that the wavefunction has to remain finite as y tends to ∞. This results in solutions of the form

$$\psi_n(x) = \frac{1}{\sqrt{2^n n!}}\left(\frac{m\omega}{\pi\hbar}\right)^{1/4} H_n\left(\sqrt{\frac{m\omega}{\hbar}}x\right)e^{-\frac{m\omega x^2}{2\hbar}}$$

where the Hermite polynomials are

$$H_n(z) = (-1)^n e^{z^2}\frac{d^n}{dx^n}\left(e^{-z^2}\right)$$

It should be noted that the "physicists' Hermite polynomials" as used here are different to the probabilists' by a scaling factor. The physicists' Hermite polynomials are solutions to a scaled version of the Hermite equation arranged as an eigenvalue problem. The corresponding energy levels are equally spaced (unlike the electronic states of the hydrogen atom, for example). The ground state is not zero (as would be the case in a classical oscillator) as it has to satisfy the Heisenberg uncertainty principle (BE15). This state is called the zero-point energy, and has many implications in physics, particularly quantum field theory.

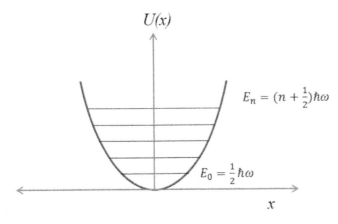

The solution is approximately symmetric for the lowest energy levels. Because the Morse potential for atomic interactions is asymmetric, higher energy levels require additional terms as they become progressively "anharmonic". Depending on molecular symmetry groups and bond strengths, an anharmonicity constant, characteristic of a pair of atoms or molecular group, can be determined to account for the overtones of the fundamental vibrations typically being less than an integral multiple of the fundamental frequency. The overtones (and combinations) of molecular vibrations extend into the near infrared region of the electromagnetic spectrum, and some of the higher overtones (which are progressively weaker due to their lower probability) overlap with the visible region (which is the part of the spectrum mainly accounting for the electronic vibrations of individual atoms at higher energies). The fundamental vibrational frequencies depend on the atomic masses of the components. Covalent chemical bonds such as C-C, C-O and C-N have their fundamental stretch vibrations in the 900-1200 cm^{-1} frequency (11-8 μm wavelength) region. For bonds involving hydrogen, a much lighter atom, such as the organic groups C-H, O-H and N-H, stretching occurs in the significantly higher frequency range 2800-3100 cm-1 (3.5 – 3.2 μm), and their overtones occur in the near infrared range.

PB

Equation 169: Raoult's Law

In 1882, French physicist, Francois-Marie Raoult, established an empirical law stating that, in an ideal mixture of liquids, the partial vapour pressure of each component is equal to the vapour pressure of the pure component multiplied for its mole fraction in the liquid mixture. The law is mathematically expressed as:

$$p_i = p_i^* \cdot \chi_i$$

where, p_i^* is the vapour pressure of the ith component and χ_i is the molar fraction in the liquid mixture. The mole fraction is equal to the mole concentration in the case of an ideal mixture.

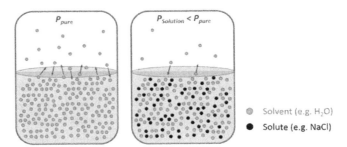

Figure 1. Schematic representation of vapour pressure

186

Vapour pressure is the pressure exerted by a vapour in *thermal equilibrium* with its condensed phase (either a liquid or a solid), at a given temperature, in a closed system. For example, in a bottle of still water at room temperature, some water molecules on the liquid surface will escape in the gas phase and some in the gas phases will re-enter the liquid surface, establishing an equilibrium process (Figure 1). The pressure exercised by the water molecules in the gas phase on the liquid surface is the vapour pressure of water. This process also occurs in a liquid mixture.

Once the mixture is at equilibrium, the total vapour pressure of a liquid mixture can be calculated by combining Roult's law and Dalton's law of partial pressures. For a mixture of two components, A and B, the total vapour pressure is:

$$p_{TOT} = p_A + p_B = p_A{}^* \cdot \chi_A + p_B{}^* \cdot \chi_B$$

COLLIGATIVE PROPERTIES: SPREADING SALT ON STREETS IN WINTER

If the mixture corresponds to a solution with a non-volatile solute (with a zero partial pressure), the total vapour pressure of the mixture will be only the result of the partial pressure of the pure solvent multiplied by its mole fraction. Hence, it will be lower than that of the pure solvent. Consequently, when we add salt (NaCl – non-volatile solute) in water, forming a saturated solution, the saturated vapour pressure curve will be lower than the curve for the pure water at all temperatures. This corresponds to a shift of the liquid-vapour curve in the phase diagram of water (Figure 2). As illustrated in the new diagram, at a given pressure (e.g. 1 atm), the temperature required for the liquid-vapour phase transition of the solution is higher than that required for the same transition in pure water.

The shift in the liquid-vapour curve has also the effect of lowering also the position of the triple point – the point where the three phases, solid, liquid and gas co-exist. This effect implies that the liquid-solid transition of the solution will occur at a lower temperature than that of the pure solvent. By spreading salt on the streets during winter, we can form a saturate solution water (snow) and salt that will freeze at a lower temperature than 0 °C, temperature at which pure water turns into ice.

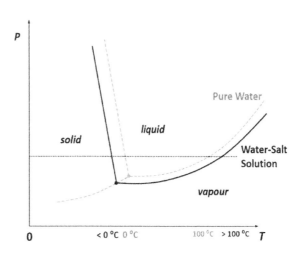

Figure 2. Phase Diagram for water and water+salt solution.

IDEAL VS *NON*-IDEAL MIXTURES: WHY VODKA IS DIFFERENT

Raoult's law is valid across the whole range of mole fractions for ideal mixtures.

From a microscopic point of view, an ideal mixture – or ideal solution – is one in which the two components have identical physical properties. For example, a mixture of two isotopes – e.g. pure water and deuterated water – represents an ideal solution. Raoult's law applies almost exactly to ideal solutions. Another example of ideal solution is a mixture of hexane and heptane, two hydrocarbons which have very similar molecular structure. In the case of ideal solutions, the intermolecular forces between two different molecular species are almost the same as those between two identical molecules – i.e. the interaction between a hexane and a heptane molecule is almost the same as that between two molecules of hexane or two molecules of heptane.

In *non*-ideal solutions such as vodka, which mainly consists of ethanol and water in almost equal proportion, the interaction between two molecules of water substantially differs from that between one molecule of water and one of ethanol. As pure ethanol and pure water have substantial vapour pressures, we might expect that the total pressure of the gas above the liquid is the sum of the vapour pressure of the ethanol and that of water, according to the equation that combines Dalton's and Raoult's laws. However, it is higher, because the molecules are breaking away more easily from the liquid mixture than they do in their respective pure liquids.

For a two-component liquid mixture A-B, the sum of the mole fractions is equal to 1. Therefore, the mole concentration of one component (say A) can be expressed as $1 - \chi_B$, where χ_B is the mole fraction of B. The total vapour pressure of the mixture becomes then:

$$p_{TOT} = p_A{}^* \cdot (1 - \chi_B) + p_B{}^* \cdot \chi_B = (p_B{}^* - p_A{}^*) \cdot \chi_B + p_A{}^*$$

This is the equation of a line in the diagram p_{TOT} vs χ_B. The straight line indicates a linear behaviour over the whole range of composition of the liquid mixture and this is valid only for ideal mixtures. In the case of *non*-ideal mixtures, there can be positive or negative deviations from linearity in the intermediate range of mole fractions (Figure 3).

Positive deviation is observed for a water/ethanol mixture, while negative deviation is observed in the case of a nitric acid/water mixture. In this case, the dissociation of nitric acid in water, generates anionic species and protonic species, which interact electrostatically with each other, making the species more difficult to leave the liquid phase of the mixture compared to their respective pure phases. However, in the case of *non*-ideal liquid mixtures, Raoult's law is still valid for very low mole fractions of one component.

By measuring the deviations from linearity in the diagram vapour pressure *vs* mole fraction a two-component liquid mixture, we can determine the strength of the intermolecular forces between the two components in the mixture. Raoult's law and its deviation from linearity underpin the process of distillation.

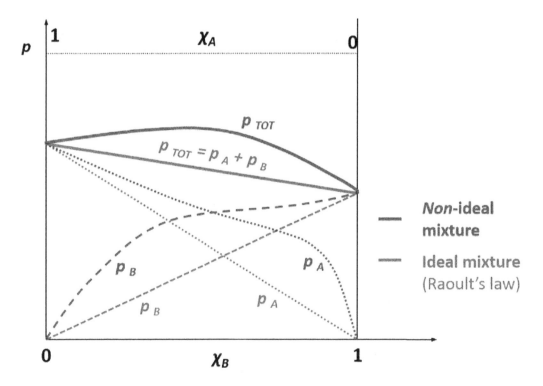

Figure 3. Raoult's Law for ideal liquid mixture and deviations for non-ideal liquid mixture.

AT

188

Equation 170: Equal Arm Balance

The figure below gives the schematic of the equal-arm balance. The balance can rotate around the knife-edge at point O. One mass m_1 is positioned at length a_1, mass m_2 is positioned at length a_2. The indicator is represented by mass m_j with arm length c.

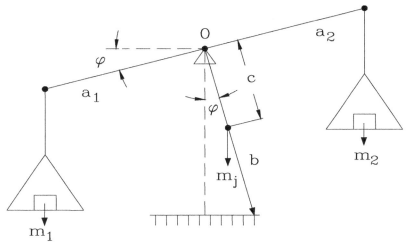

Figure: schematic of an equal arm balance

When the balance is in equilibrium, the sum of moments must be zero. This gives:

$$m_1 g a_1 \cos \varphi = m_2 g a_2 \cos \varphi + m_j g c \sin \varphi$$

With approximation for small angles: $\cos \varphi \approx 1$, $\sin \varphi \approx \varphi$ and equal arm length $a_1 = a_2$ this gives:

$$m_1 = m_2 + \frac{m_j c \varphi}{a}$$

This means that the indication is proportional to the mass difference. The constant c can be determined by calibration, the sensitivity can become very high by taking small m_j and c, and large a, and especially a sharp knife-edge. The difference in arm length can be calibrated and corrected by exchanging masses m_1 and m_2.

All these aspects make this principle nearly unsurpassed in sensitivity and applicability. It is a major reason that the kilogram is defined as an artefact until late 2018: better means of realizing and comparing a mass standard are much (much !) more complicated (Watt balance).

The history goes back to the beginning of civilization; the equal-armed balance can already be found on ancient Egyptian pictures. In the illustrated case it is used to depict the weighing of the soul of a deceased person, in order to determine its future.

HH

Equation 171: Equation of an Ellipse

In equation 157 a circle was parameterized. Here, the equation of an ellipse is presented. In the interest of conciseness, ellipse orientation will not be parameterized; that is, the axes of the ellipse are parallel to the X and Y axes of the coordinate system.

The general equation of an ellipse is given by the following.

$$\frac{(x-a)^2}{c^2} + \frac{(y-b)^2}{d^2} = 1$$

where a and b correspond to the X and Y coordinates, respectively, of the ellipse centre, and c and d are the half-lengths of the two axes of the ellipse. The term major axis is assigned to the larger of the two axes and term minor axis is assigned to the other. For example, if $c > d$ (as shown in the figure below), then the major axis is parallel to the X axis and the minor axis is parallel to the Y axis. When $c = d$ the equation for an ellipse becomes the equation of a circle.

Alternatively, the equation of an ellipse can be written in parametric form:

$$x(t) = a + c \cos t \cos \alpha - d \sin t \sin \alpha$$

$$y(t) = b + c \cos t \sin \alpha + d \sin t \cos \alpha$$

where t is the parametric variable in degrees (see figure below) and α is the angle from the X-axis to the major axis. That is, $\alpha = 0$ when the major axis is parallel to the X-axis and $\alpha = \frac{\pi}{2}$ when the major axis is parallel to the Y-axis.

A brief history of ellipses is given in reference [1]. Ellipses are first known to be studied in the 4th century BCE by Greek mathematician Menaechmus. The name was given to the shape by Greek geometer Apollonius of Perga either in the end of the 3rd century BCE or at the beginning of the 2nd century BCE. In ancient Greek, the word ἔλλειψις (pronounced élleipsis) means "omission" [2].

[1] Ellipse, School of Mathematics and Statistics University of St Andrews, Scotland, http://www-history.mcs.st-and.ac.uk/Curves/Ellipse.html
[2] Ellipse – From Wikipedia, the free encyclopedia https://en.wikipedia.org/wiki/Ellipse

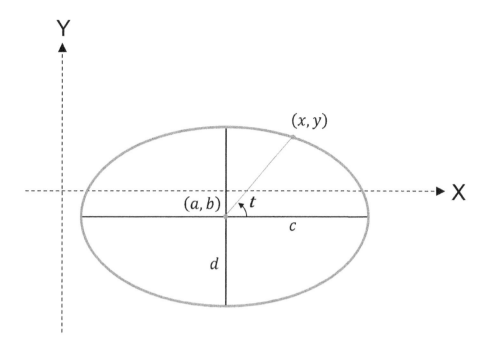

MF

Equation 172: The Generalized Eigenvalue Problem

The generalized eigenvalue problem is stated as follows. Given two $n \times n$ matrices **A** and **B**, determine a non-trivial vector **x** such that

$$\mathbf{A}\mathbf{x} = \lambda\mathbf{B}\mathbf{x}$$

or equivalently

$$(\mathbf{A} - \lambda\mathbf{B})\mathbf{x} = \mathbf{0}$$

for some real or complex number λ. The vector **x** is called the eigenvector of **A** relative to **B** and λ is called the (corresponding) eigenvalue of **A** relative to **B**. When **B** equals **I**, the identity matrix, then, the generalized eigenvalue problem is referred to as the standard eigenvalue problem. The pair $(\lambda, \mathbf{x})$ is called an eigenpair of **A** relative to **B**.

The eigenvalues are obtained by requiring that $\det(\mathbf{A} - \lambda\mathbf{B}) = 0$ where det denotes the determinant. This leads to an nth-order characteristic polynomial of **A** relative to **B** and therefore, n eigenvalues some of which may be repeated. The set of eigenvalues is called the spectrum of **A** relative to **B**. The eigenvector corresponding to each eigenvalue is obtained by solving the above matrix equation.

The generalized eigenvalue problem appears in many physical problems. For example, most structures can resonate (i.e., oscillations at very high amplitudes) when subjected to forces at frequencies that coincide with the natural frequencies of the structures. Consequently, structures are often equipped with dampening mechanisms to damp out the high amplitude vibrations.

The natural frequencies and mode shapes of structures are governed by the balance of linear momentum equations along with Hooke's law (BE97). The frequencies and mode shapes depend on the stiffness of the

191

structure, geometry and the density. Numerical methods such as the finite element method convert these equations to a system of equations of the form

$$\mathbf{M\ddot{d}} + \mathbf{Kd} = 0$$

Here, $\mathbf{M}$ is the mass matrix, $\mathbf{K}$ is the stiffness matrix and $\mathbf{d}$ is the degree-of-freedom vector that contains the displacement components at a set of points (nodes) of the structure. Setting $\mathbf{d} = \mathbf{v}e^{i\omega t}$ and substituting in the above leads to

$$(\mathbf{K} - \omega^2 \boldsymbol{M})\mathbf{v} = 0$$

which is the same as the generalized eigenvalue problem defined as above. The vector $\mathbf{v}$ is the eigenvector or the mode shape of the structure corresponding to the eigenvalue ω^2. ω is the natural frequency corresponding to the mode shape $\mathbf{v}$. The accuracy of the eigenpair $(\omega^2, \mathbf{v})$ depends on the size of the mass and stiffness matrices which in turn depends on the number of points (nodes) considered.

As an application of the generalized eigenvalue problem, consider a tuning fork which is used in many applications. For example, it is used to tune musical instruments and in medicine, to test hearing of patients and detect possible bone fractures. The natural frequencies and mode shapes are obtained by solving equation presented above. The fundamental mode and the clang mode of vibration are shown in the figure below. The clang mode is the sound one hears initially when the tuning fork is struck. The fundamental mode is heard clearly after a few initial seconds when the clang mode has died away.

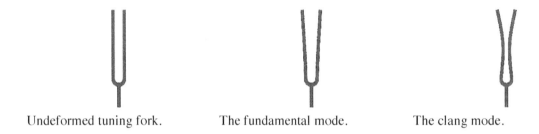

Undeformed tuning fork.   The fundamental mode.   The clang mode.

Figure 1: The tuning fork and its two most commonly observed modes of vibration.

HC

Equation 173: Parametric Array Equation

When a sound wave is produced by a loudspeaker (or by any other source), it will tend to radiate in all directions as long as its wavelength is much longer than the loudspeaker (source) size. Hence, if you speak in the open air, your lower tones can be heard by someone standing behind you, because they are a few hundred Hz, corresponding to wavelengths of a few decimetres.

(Frequencies are related to wavelengths through the well-known equation $v = f\lambda$ (v: velocity ; f: frequency; λ: wavelength)).

The highest-pitched components will not be very audible, as they are several kHz in frequency, and so around a centimetre in length. Since they are therefore smaller than the mouth opening, they are quite directional.

Underwater, since sound speeds are much higher than in air (typically about five times higher), the wavelength of a wave of a given frequency is proportionately large. This means that, unless sources are many metres across, all audible-frequency underwater sound waves tend to radiate in all directions.

In many applications, it is desirable to generate directional sound beams, such as for communication or for pulse-echo detection techniques. Narrow beams reduce off-axis noise levels, travel great distances with little energy loss, produce few confusing side-echoes and are more suitable for transmitting sensitive information.

However, in all media, higher-frequency sound waves are absorbed over shorter distances than lower ones, and for this and other reasons one often wishes to transmit relatively low-frequency sounds in directional beams.

An elegant way to make a low-frequency directional sound source is the parametric array. If two sound sources generate waves which differ just a little in frequency, then waves with that difference frequency will be produced, along with others whose frequency is the sum of those of the sources. The wavelength of the difference wave can be a long as required, but it maintains the directionality of its parent waves.

Parametric arrays exploit the fact that sound velocity depends on density. At high sound powers, the pressure in the compressions becomes very large, increasing density significantly and therefore speeding up the sound wave briefly; the reverse happens in the rarefactions. The effect of these velocity changes is to distort the wave from its usual sinusoidal form, as shown in the figure below

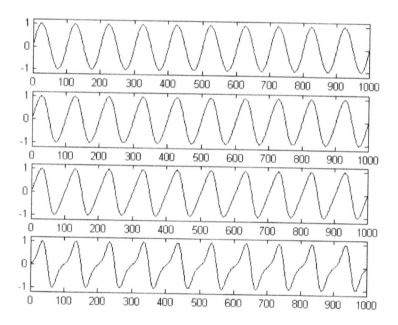

As Auguste Fourier showed, a non-sinusoidal wave is equivalent to a sum of component sinusoids (see equation 12). In the case of the parametric array, these components include the original waves, together with the sum and difference waves; the difference wave being the one of interest:

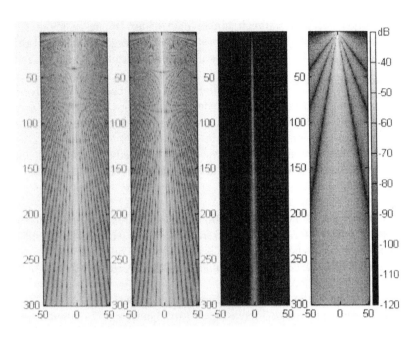

These four plots are cross-sections of the sound field of a parametric array (axes are labelled in metres). The sound source is at 0. The first plot is of a 100 kHz primary beam, the second of an accompanying 105 kHz primary beam and the third shows the secondary beam, which has the difference frequency of 5 kHz. The narrowness of this secondary beam is the point of the array: the fourth plot shows how widely spread a conventionally-generated 5 kHz beam is.

Mathematically, the directionality of a sound wave is defined through a directivity function, which describes the intensity (or other measure) of the sound as a function of angle from the source axis. For a parametric array, the directivity function is

$$D(\theta) = \left[1 + \frac{k_d^2}{\delta_m^2}\sin^4\left(\frac{\theta}{2}\right)\right]^{-1}$$

where $D(\theta)$ is the intensity as a function of angle θ from the source axis in radians, k_d is the wave number of the difference frequency $\frac{2\pi}{v}|f_2 - f_1|$, and δ_m is the mean absorption frequency of the primaries ($= \frac{\delta_1 + \delta_2}{2}$; δs are amplitude decrements)

The parametric array effect was discovered by chance when, in 1951, acoustician Peter J. Westervelt was working at the Office of Naval Research in London. He noticed that an experimental superheterodyne (frequency-mixing) radio receiver was generating directional audible sound at low frequencies. It was not until 1960 that Westervelt was able properly to explain the effect, which he did at a meeting of the Acoustical Society of America, and in a published paper a few years later.

Today, parametric arrays have numerous underwater applications, and they are also used in museums and galleries for the "acoustic spotlight" system in which recorded descriptions are beamed at appropriate exhibits and can only be heard by people close to the exhibits. A parametric array is also the basis of the LRAD (Long Range Acoustic Device), used to direct painfully load sounds at, for example, Somalian pirates. Underground, sound beams from parametric arrays are used for prospecting.

MJG

Equation 174: Van Cittert-Zernike Theorem

$$A_c = \frac{L^2 \lambda^2}{\pi d^2}$$

Where d is the diameter of the light source, L is the distance away from the light source, λ is the average wavelength of the light source and A_c is the spatial coherence area.

The concept of coherence plays an essential role in modern physics. It describes the condition of forming a stationary interference from two wave sources in terms of the sources' temporal and spatial properties, which is usually mathematically expressed by a correlation function.

The van Cittert-Zernike Theorem is about a measure of spatial coherence in light fluctuations at two points in an optical field. The conceptual work was first presented by the Dutch physicist Pieter Hendrik van Cittert (1889 -1959) in 1934 and later formulated by Frits Zernike (1888-1966) in his 1938 paper, where the degree of coherence and mutual intensity were introduced.

The equation states that for a realistic light source (i.e. with finite size rather than geometrically ideal "point"), the father away from the source the bigger an area of interference pattern will be formed. In another words, the larger the area of a clearly visible interference pattern the more spatially coherent is the light. The phenomena predicted by the van Cittert Zernike theorem can be visualized by observing a group of ducks (finite size of wave source) enter the pond. In the following picture, a circular pattern (spatially coherent wave) is formed as the waves propagate away from the center of duck group. That means you cannot tell if the ripple is generated by a duck or a group of ducks when you observe only the coherence area!

One application of van Cittert-Zernike theorem is in determining the aperture size of a telescope, which is used for imaging distant stars. Because only by collecting waves or light from an area significantly larger than the coherence area, one can distinguish between a point source and an extended object. Therefore, the aperture of telescope is usually tens of meter in diameter since the coherence area for nearby stars is in the range of a few to tens of meters.

References
[1] P.H. van Cittert (1934). "Die Wahrscheinliche Schwingungsverteilung in Einer von Einer Lichtquelle Direkt Oder Mittels Einer Linse Beleuchteten Ebene". *Physica* **1**: 201–210.
[2] F. Zernike (1938). "The concept of degree of coherence and its application to optical problems". *Physica* **5**: 785–795.
[3] "Early Days of Coherence Theory"
http://www.optics.rochester.edu/~stroud/BookHTML/ChapIV_pdf/IV_29.pdf
[4]"You could learn a lot from a ducky: the van Cittert-Zernike theorem"

http://skullsinthestars.com/2010/06/12/you-could-learn-a-lot-from-a-ducky-the-van-cittert-zernike-theorem/

[5] Nicholas Fang, "Review for Optics", MIT OpenCourseWare 2.71/2.710 Introduction to Optics

KN

Equation 175: Carriers Concentration In Intrinsic Semiconductor

$$n_e = N_C(T) \cdot \exp\left(\frac{E_F - E_C}{kT}\right)$$

and

$$n_h = N_V(T) \cdot \exp\left(\frac{E_V - E_F}{kT}\right)$$

where N_C, N_V are the effective density of states for electrons and holes respectively, E_C, E_V are the conduction and valence band energy levels, E_F is the Fermi energy level, k is the Boltzmann constant, and T is the temperature.

Semiconductors are materials whose electric properties are in between a perfect conductor and a perfect insulator. Elements of the 4th column of the periodic table (such as C, Si, Ge) are semiconductors. The materials' properties depends on how the atom's electrons behave when the atoms are arranged in a lattice structure; in particular for a semiconductor very few electrons are available to move therefore leading to very poor conductivity.

The available charge carries can be electrons, or holes, i.e. an electron vacancy. Holes are an abstraction since physically only the electrons are moving. However, the utility of the hole concept is easily understood by comparing electrons, and holes to a moving crowd. In an empty room a few people moving around would be the equivalent of electron moving in a semiconductor. However, in a crowded room would be much easier to follow an empty spot rather than each persons, and the empty spot would be the equivalent of an hole. The main difference between electrons and holes is how easily they can move; electrons usually can move more easily since there is no repulsion from other electrons. On the other hand, holes have lower mobility since an hole moving implies an electron moving in the opposite direction toward other (repulsive) electrons.

The two equations express the concentration of electrons and holes in an intrinsic (non-doped) semiconductor. N_C and N_V are the effective density of states for electrons and holes, parameters which depend on the semiconductor and the temperature T; the exponential is the Maxwell-Boltzmann probability distribution, where E_C and E_V are the conduction and valence band energies, which are the the energy levels where electrons and holes, respectively, are free to move; E_F is the Fermi energy level and is a characteristic of a semiconductor.

The solid state physics of semiconductor is a theory that has been developed by solving Schrödinger equation in a periodic lattice and it has been developed in the first half of the 19th century.

GM

Equation 176: Lotka–Volterra Equation

The Lotka–Volterra equation describes the dynamics of a predator/prey situation.; for instance, if we have a large population of foxes and a larger population of rabbits the fox population will grow because of the abundance of food. How do we describe this mathematically?
Well first we have to make some assumptions namely.

1. The prey population finds ample food at all times.
2. The food supply of the predator population depends entirely on the size of the prey population.
3. The rate of change of population is proportional to its size.
4. During the process, the environment does not change in favour of one species and genetic adaptation is inconsequential.
5. Predators have limitless appetite.

The listed assumptions are taken from the excellent Wikipedia article on the equations.

The Lotka–Volterra system of equations is an example of a Kolmogorov model, a more general framework that can model the dynamics of predator-prey interactions taking into account competition, disease, and mutualism. The equation system is:

$$\frac{dx}{dt} = \alpha x - \beta xy$$

$$\frac{dy}{dt} = \delta xy - \gamma y$$

where x is the number of prey and y is the number of predators. The term t represents time and the differentials the rate of change of population with time. The terms α, β, γ and δ are positive real numbers that represent the interactions between the species.

If we look at the prey equation, $dx/dt = \alpha x - \beta xy$ we have an exponential growth rate αx minus a term that relates the probability of the two species meeting. So the change in prey numbers relates to its own growth rate minus that rate at which it is preyed upon.

If we look at the predator equation, $dy/dt = \delta xy - \gamma y$ it has some similarities to the prey equation with different values for the constants. The term γy relates to the loss of predator population due to death or emigration. The term is exponential and expresses the decay in the predator population as growth fuelled by the food supply, minus natural death.

The solutions are periodic as one might expect. As predator population grows, prey population dwindles until a point where lack of prey causes predator populations to dwindle and prey populations to grow.

The Lotka–Volterra predator–prey model was initially proposed by Alfred J. Lotka in the theory of autocatalytic chemical reactions in 1910. [Wikipedia] This was effectively the logistic equation (see equation 140).

Lotka was a bio-mathematician and a bio-statistician, who sought to apply the principles of the physical sciences to biological sciences as well. His main interest was demography, which possibly influenced his profession as a statistician at Metropolitan Life Insurance.

DF

Equation 177: The Bell curve, Gaussian, or normal distribution

$$y = \frac{1}{\sigma\sqrt{2\pi}} e^{-\frac{(x-m)^2}{2\sigma^2}}$$

197

where σ is the signal standard deviation, x is the value of the outcome, m is the shift of the origin of the normal distribution plot. If m is equal to the mean of x then the above curve is centred about the origin. y is the value of the normal distribution at location x.

Flipping a coin to make a decision will provide a binary outcome that we consider to be random. For a coin flip, the nature of each event is considered to be influenced by numerous, non-repeatable interactions so that the outcome is not predictable. The result is a 50% probability of an outcome being either heads or tails, mathematically speaking, a probability of ½ for each event. When throwing an unbiased gaming dice it is reasonable to expect probabilities of 1/6 for each face to come to rest at the top. Examples of other random events with discrete outcomes abound.

Extending the coin flip experiment to look at the outcomes of multiple coin flips, things start to become interesting. If we perform two coin flips, there will be three outcomes; 2 heads, 1 head and one tail, one tail and 1 head, and 2 tails (or no heads). The associated probabilities now become ¼, ½ (or ¼ + ¼), and ¼ respectively.

Further extending this to 10 coin-flip sets and looking at the number of heads, the probabilities of the 11 outcomes can be determined using the Binomial distribution (BE 294). However, for our purposes, we will ignore this equation and do this experimentally. To do so, after each 10 flips the number of heads will be recorded and this will be repeated for 10000 sets of flips. Following this we can count how many sets had a specific number of heads (labeled 'Frequency of occurrence' in Figure 1). Fortunately, it is possible to do this using a computer for which the outcome is shown in Figure 1.

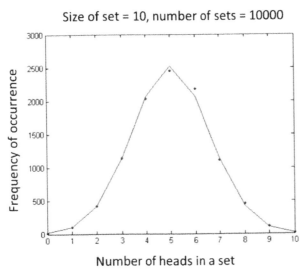

Size of set = 10, number of sets = 10000

Figure 1: Number of times that a specific count of heads was observed during sets of 10 coin tosses for 10000 sets of experiments. Solid line corresponds to the theoretical prediction based (incorrectly) on the equation of the normal distribution multiplied by the number of sets (i.e. 10000). The standard deviation is the square root of the number of coin tosses in a set divided by 4 (i.e. $\sigma = \sqrt{10/4} = 1.581, m = 0$). Note that the sum of the frequency of occurrence values = 10000.

Mathematically, it is possible to extend the binomial distribution for large numbers of large sets resulting in the above equation for the continuous normal distribution with the standard deviation being related to the number of coin flips in each set (see caption of Figure 1). Figure 2 shows a similar plot to Figure 1 for a set length of 1000 coin flips repeated 100000 times with the occurrence of heads divided by the total number of sets (100000) and the horizontal axis normalized to the standard deviation of the sets (= 15.81 in this case). For this large number of experiments, it is apparent that the theoretical model of the distribution is converging to a precise and accurate statistical model of the measured data.

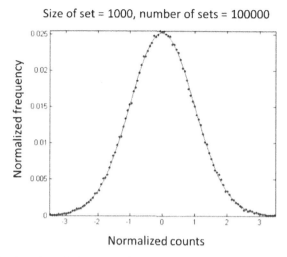

Size of set = 1000, number of sets = 100000

Normalized frequency

Normalized counts

Figure 2: Similar plot to that shown in Figure 1 with large numbers of both coin flips in each set and number of sets measured, exhausting if computers were not available for such a task ($m = \bar{x} = 500$, $\sigma = 15.81$). Note that the sum of normalized frequency values = 1.

As the size of the set and number of sets tend to infinity, the series of points merge to a continuous graph, see Figure 3. This is often called the 'unit normal distribution' and the random variable, correspondingly scaled, is called a 'standard normal deviate', z. As with the normalized histogram for which the sum of all points must be equal to 1 (i.e. the probability that an outcome will be within the range of all possible values), for the continuous distribution, this curve is often referred to as the probability density function. Instead of adding up normalized counts of events, it now becomes necessary to determine the probability of events occurring between two values by integrating this probability density function as indicated in Figure 3.

How often do random processes resemble this distribution? The answer to this is probably more than any other distribution but less than might be inferred by its ubiquitous application across almost all disciplines involving quantitative analysis of data. Most random events tend to have a histogram of data that features a hump. Also, if the event outcome is influenced by a large number of significant random inputs, it can be shown that the resultant scatter will tend to a normal distribution almost independently of the shape of the distributions of disturbances that sum to produce the deviation (a loose description of the phenomena referred to as the 'central limit theorem'). Hence, whenever the distribution looks reasonably symmetric and the underlying causes of random fluctuations are unknown, the 'go to' default is almost always the normal distribution.

Additionally, many physical phenomena (thermal models, diffusion, ground state of a quantum harmonic oscillator, BE 168 to name a few), natural features of organisms (such as height), and a broad spectrum of data obtained by sampling from large populations have characteristics accurately modelled by the normal distribution. Many social or cultural phenomena (frequency that trains depart relative to their designed time, likelihood of being married at a certain age) tend to be skewed.

The first presentation of this distribution is attributed to the French Mathematician Abraham de Moivre (1667 – 1754) in an unpublished work dating to 1733. This early presentation was taken up by Pierre-Simon Laplace (1749 – 1827) and Carl Friederich Gauss (1777 – 1855) over a period spanning the late eighteenth and early nineteenth centuries and the curve is now usually associated with the latter Mathematician. In fact it is common to hear people calling this a Gaussian distribution, normal distribution, or Bell curve.

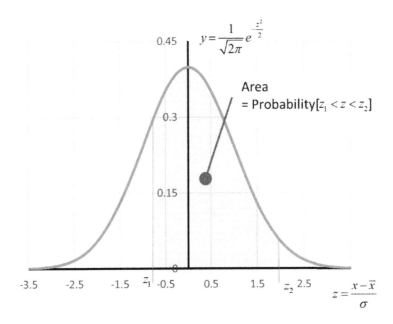

$$y = \frac{1}{\sqrt{2\pi}} e^{\frac{-z^2}{2}}$$

Area
$= \text{Probability}[z_1 < z < z_2]$

$$z = \frac{x - \bar{x}}{\sigma}$$

Figure 3: The standard normal curve. The fact that this has a mean of 0 and standard deviation of 1 is often referred to as a normal variate N[0,1]. Note that the total area under the curve = 1.

SS

Equation 178: Brewster Angle

The brewster angle (often referred to as the polarization angle) is the angle at which light travels through a transparent dielectric medium with no reflection. Light the hits a transparent medium at any angle is usually reflected to some degree as described by the Fresnel equations and is due to the polarization of the light. The Brewster angle calculates the angle at which the light is said to be perfectly polarized. Brewster's angle varies with the polarization and wavelength of light.

$$\theta_B = \arctan\left(\frac{n_2}{n_1}\right)$$

where n_1 is the refractive index of the initial or origin medium and n_2 is the refractive index of the transmission medium.

This angle of incidence effect is named after Scottish physicist David Brewster (1781-1868) who is famous for inventing the kaleidoscope among many other significant contributions to the field of optics.

Brewster's angle is used in many practical applications such as polarized sunglasses and windows to reduce glare from certain commonly reflected surfaces.

TS

Equation 179: The Fenske Equation

Fractional distillation is an important chemical engineering process for the production of commercial products such as gasoline, liquid oxygen, liquid nitrogen, vodka and many others. Industrial distillation takes place in large, vertical cylindrical columns known as 'distillation or fraction towers', which can reach heights over 60 metres, with diameters up to 6 metres. Feed is continuously added to the distillation column, while products being continuously removed through a continuous, steady-state separation process.

A useful way to calculate the minimum number of theoretical plates of a distillation tower – operating at total reflux – for the separation of a binary feed (a liquid solution), operated is the *Fenske equation*:

$$N = \frac{\log\left[\left(\frac{\chi_d}{1-\chi_d}\right)\left(\frac{1-\chi_b}{\chi_b}\right)\right]}{\log[\alpha_{avg}]}$$

Where N is the minimum number of theoretical plates required at total reflux; χ_d is the mole fraction of the more volatile component in the distillate at the top of the column; χ_b is the mole fraction of the more volatile component at the bottom the column and α_{avg} is the average relative volatility of the two components.

$$\alpha_{avg} = \sqrt{\alpha_t \alpha_b}$$

and

$$\alpha_{avg} = (Z_B/\chi_B) / (Z_A/\chi_A)$$

where B is the more volatile component and A is the less volatile component, commonly called light key (LK) and heavy key (HK). Z_i is the vapour-liquid concentration of the component i in the vapour phase, while χ_i is the vapour-liquid equilibrium concentration of the component i in the liquid phase.

In the expression of the average relative volatility, α_t is the relative volatility at the top of the column and α_b is the relative volatility at the bottom of the column. The average relative volatility is used if the relative volatility of the light key to the heavy key is not constant from the top to the bottom of the column.

Merrell Fenske, professor at the Pennsylvania State University, formulated this equation in 1932, 7 years after Warren McCabe and Ernest W. Thiele at the Massachusetts Institute of Technology (MIT) had derived a method to analyse the distillation of binary mixtures. McCabe and Thiele method is based on the fact that the composition of each theoretical tray in the fractional distillation tower is completely determined by the mole fraction of one of the two components of the mixture.

DISTILLATION TOWERS

The general principle of operation of fractional distillation column is the following. A heat source at the bottom of the column raises the temperature of the initial liquid mixture up to its boiling point. At this temperature, part of the mixture vaporises, while changing composition and becoming richer in the most volatile component, and rises along the column. At the same time, and a temperature gradient is established within the column, low temperature at the top and high temperature at the bottom. As the vapour mixture rises along the temperature gradient, it partially condenses falling towards the bottom of the column, until it re-vaporises due to the hotter conditions. Every time the condensate vaporises again its composition becomes richer in the more volatile component of the binary mixture.

The condensation process takes place on glass plates – also known as trays – and runs back down into the liquid at the bottom, refluxing the distillate. The mixture enriched with the more volatile component exits as gas at the top of the column and then passes into the condenser and cools down to liquefaction (Figure 1).

The higher the number of trays used in the distillation column, the greater is the separation process.

The reflux of distillate is used to achieve a more complete separation of the product, increasing the efficiency of the process. By flowing downwards along the column, the reflux liquid provides additional cooling, which is needed to condense the vapour rising upwards. The more reflux is provided for a given number of theoretical plates, the better is the separation of the lower boiling component. Alternatively, the more reflux provided, the lower is the number of theoretical plates required for a given desired separation.

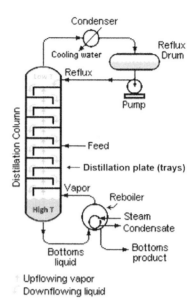

Figure 1. Schematic illustration of the functioning principle of distillation towers.

THERMODYNAMIC PRINCIPLES OF DISTILLATION
The thermodynamic principles of the distillation process lie in the Raoult's law for the vapour pressure of liquid mixtures and the deviations from the law.

Ideal mixtures

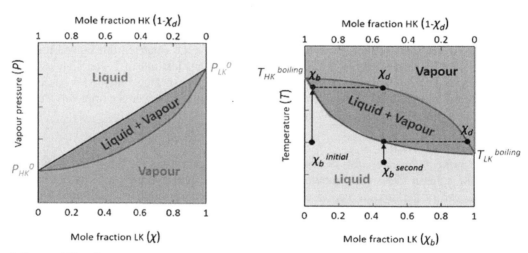

Figure 2. P-χ and T-χ diagrams and distillation for ideal mixtures (e.g. hexane-heptane)
In order to understand the underlying principle of fractional distillation, let us examine the overlap of the diagram pressure *vs* mole fraction of the liquid solution (represented by a straight line) and that for the vapour solution.

The resulting diagram has 3 areas corresponding to 3 different states of the binary mixture (Figure 2):

1) liquid state; 2) vapour state; 3) liquid+vapour state, in which the two phases co-exist.

The diagram temperature *vs* mole fraction associated to the binary mixture is similar, but with liquid and vapour states swapped (Figure 3).

202

Non-ideal mixtures: positive deviations

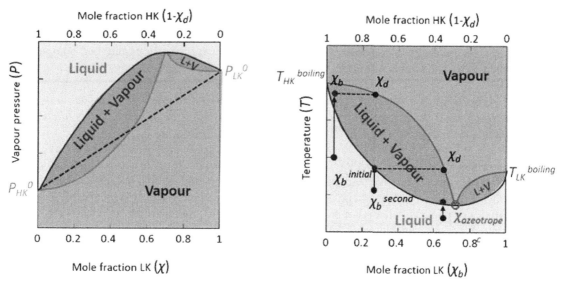

Figure 3. P-χ and T-χ diagrams and distillation for non-ideal mixtures (e.g. ethanol-water)

When heating up a liquid mixture with an initial composition χ_1, the temperature will the value corresponding to the curve liquid+vapour. At this temperature, the liquid mixture (still with composition χ_1 co-exist with a vapour phase with composition χ_2, which is richer in the most volatile component. If this vapour mixture is condensed in the liquid phase and then re-heated, the temperature will reach a new value, lower than the previous one, at which the liquid mixture with composition χ_2 will co-exist with a vapour mixture with another composition χ_3, which is further enriched in the most volatile component. Eventually, if this process of condensing and re-heating continues along the boiling point curve for the liquid composition, it will produce a mixture in the vapour phase that is very rich in the most volatile component of the mixture.

The curve vapour pressure *vs* composition of an ideal binary mixture is represented by a straight line according to Raoult's law, while deviations for non-ideal mixtures are represented either by a positive or a negative curvature of the line (Figure 4). This has important implications for distillation. For a non-ideal mixture with a strong (positive) deviation – for example, water-ethanol mixture – the curvature indicates that the maximum value of the vapour pressure for the mixture does not correspond to the vapour pressure of the most volatile pure component (like for the ideal mixtures), but to a given intermediate composition of the mixture. This maximum value for the vapour pressure will corresponds to a minimum value for the boiling temperature of the mixture in the diagram temperature *vs* mole fraction. The resulting diagram changes compared with that of ideal mixtures, with the liquid and vapour curves overlapping at the lowest boiling point.

At this point, the distillation process is no longer possible, because the liquid mixture boils as if it was a pure solution, retaining the same composition of the mixture in the gas phase, at the same temperature. This composition is known as *azeotropic* and the corresponding liquid mixture is known as *azeotrope*. Therefore, it will never be possible to obtain pure volatile component (for example, ethanol) for a liquid mixture with a composition lower than the azeotropic composition. This composition corresponds to the maximum amount of more volatile component we can separate through the distillation process.

AT

Equation 180: Bell's Theorem And Bell's Inequalities

Bell's Theorem was developed by John Stewart Bell (28 June 1928 – 1 October 1990), a Northern Irish physicist who published his most famous work "*On the Einstein Podolsky Rosen paradox*" in 1964 while at Stanford University. He is known as the man who proved Einstein wrong and at the time of his death, and unknown to him, he was nominated for a Nobel Prize. Today, Bell's theorem is said to have laid the foundation for quantum information technology and this area of cryptography and computing is used in cyber security and the financial services industries.

Bells theorem draws a distinction between quantum physics and classical physics, and in its simplest form, states "*No physical theory of local hidden variables can ever reproduce all of the predictions of quantum mechanics*"

Einstein, Podolsky and Rosen (EPR), in 1935 published a paper which concluded that quantum mechanics was incomplete due to so called "hidden variables". These are fundamental properties of particles or photons (spin, polarisation, etc) that cannot be observed as dictated by the Heisenberg Uncertainty Principle (BE 15). The particles or photons, released in pairs, are linked through entanglement, and even over vast distances, a measurement on one particle or photon immediately affects the other. This violated Einstein's theory of special relativity which he called spooky action at a distance. Bell called this phenomenon "nonlocal". Bell's theorem showed that hidden variables disagreed with quantum mechanics.

Bell inequalities are concerned with measurements made on pairs of particles or photons that have interacted and separated. Consider the polarisation of individual photons and assume all the photons are linearly polarised. If we have an experiment, where a light source gives off two photons, both with the same polarisation. Two observers have three polarisers each, A, B & C set at 0°, 120° & 240° respectively. According to EPR, the photon must contain information, or hidden variables, that tell it whether it will go through the polarisers.

If two polarisers are selected at random, there are eight combinations of these hidden variables (*a*, *b* & *c*) as to whether the photon goes through the polarisers. If the observers pick a polariser each which allows the photon to pass through both or neither, they get the same result (σ). Likewise, if the observers pick a polariser each which allows the photon to pass through one, but not the other, they get a different result (δ)

Combination	a	b	C	AB	BC	AC	Probability of photon going through or not going through both polarisers
1	✓	✓	✓	σ	σ	σ	1
2	✓	✓	✗	σ	δ	δ	$1/3$
3	✓	✗	✓	δ	δ	σ	$1/3$
4	✓	✗	✗	δ	σ	δ	$1/3$
5	✗	✓	✓	δ	σ	δ	$1/3$
6	✗	✓	✗	δ	δ	σ	$1/3$
7	✗	✗	✓	σ	δ	δ	$1/3$
8	✗	✗	✗	σ	σ	σ	1

From the table, we can see the probability of the photons going through both polarisers is at least $1/3$ even if we don't know which of the eight combinations are in place. The fact that matches occur greater than or equal to $1/3$ of the time is called Bell's Inequality and is represented mathematically by:

$$\rho(A,C) - \rho(B,A) - \rho(B,C) \leq 1$$

where ρ is the correlation between measurements of the polarisation of the pairs of photons and $A, B, \& C$ are three arbitrary settings of the two analysers.

Quantum theory actually tells us that matches occur greater than or equal to $1/4$ of the time. This clearly is not greater than or equal to $1/3$ and the above analysis of hidden variables can't be true. If there were absolute determined variables that tell us whether or not the photon goes through the polariser, this rules out the possibility of hidden variables, leaving instantaneous communication i.e. the particles are entangled. Whether this is true or not has yet to be explained and still remains one of physics unsolved mysteries.

A. Einstein, B. Podolsky and N. Rosen, "Can quantum-mechanical description of physical reality be considered complete?," *Phys. Rev.*, vol. 41, p. 777, 1935.
J. S. Bell, "On the Einstein Podolsky Rosen paradox," *Physics*, vol. 1, no. 3, p. 195, 1964.
J. S. Bell, "On the problem of hidden variables in quantum mechanics," *Rev. Mod. Phys*, vol. 38, p. 447, 1966.
A. D. Aczel, Entanglement: The Greatest Mystery in Physics, Chichester: John Wiley & Sons Ltd, 2003.

CM

Equation 181: Phase Shifting Interferometry

Beautiful Equation 30 for *Interference* tells us that if we have two light beams superimposed on a detector, under the right conditions, we generate sinusoidal intensity signal with a phase equal to the difference $\Delta\theta$ in phase of the two original waves, like this:

$$I = I_A + I_B + 2\sqrt{I_A I_B}\cos(\Delta\theta)$$

Where I_A, I_B are the intensities of the light waves. If for example we change how far one of the light beams travels, then $\Delta\theta$ changes by a 2π radians for every half-wavelength of distance change. One-half wavelength is a fraction of a micron, which means that this is a pretty sensitive distance gage!

This is all fine in principle, but how do we measure the phase $\Delta\theta$? Mathematically, it looks as though we need to solve simultaneously for three quantities: I_A, I_B and $\Delta\theta$. The problem is made all the more difficult by having the phase locked inside the argument of a nonlinear, trigonometric function. For generations, the problem was solved by a visual interpretation of fringe patterns resulting from a small tilt between the light beams, like this:

This is equivalent to evaluating the interference equation at many *phase shifts*, since the interference fringes are the result of a varying path length across the visual field of the pattern. As long as I_A, I_B remain reasonably constant over this field, we can estimate $\Delta\theta$ from the locations of the location of the light and dark bands.

The basic idea of a phase shift can be moved to automated, time domain data processing by introducing a modulator that shifts the phase of one light beam with respect to the other. We then electronically sample several intensities as shown here:

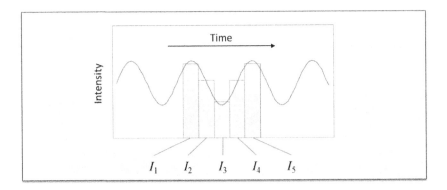

Once we have acquired the data samples, we calculate the phase $\Delta\theta$ from a least-squares fit of sines and cosines to the intensity data, in what amounts to a windowed, single-output digital Fourier analysis (see Eq.150). Usually there are only a small number of equally spaced data points, and the calculation simplifies to an algebraic ratio of weighted intensity values

$$\tan(\Delta\theta) = \sum_j s_j I_j \bigg/ \sum_j c_j I_j$$

where s_j, c_j are coefficients selected for their error-compensating properties. This is the principle of *phase shifting interferometry* or PSI. A well-known and still widely-used PSI algorithm employs five samples spaced by controlled, $\pi/2$ phase shifts:

$$\tan(\Delta\theta) = \frac{2I_2 - 2I_4}{-I_1 + 2I_3 - I_5}$$

Today phase shifting algorithms may have anywhere from three samples to seventeen or even more data frames, depending on the available data acquisition time, measurement speed and performance requirements.

The usual "first" reference to PSI is to the famous 1966 paper by P. Carré in Metrologia, describing the photoelectric and interference comparator of the BIPM for divided scales and end standards. However, the 1974 paper by John Bruning and coworkers is the true foundational document in the context of interferometric measurements of form and texture. The Bruning et al. paper includes the essential components of PSI instrumentation, including a piezoelectric phase shifter, a 32 X 32 element photodiode array and a PDP8/1 computer system with 8K of memory. Rapid developments in electronic cameras and personal computers lead to commercial PSI systems in the 1980's, first for the form and geometry testing of optical components and then to microscopic examinations of surface texture. The 1990's saw an expansion of PSI algorithm design to reduce certain systematic errors, while more recent work has concentrated on least-squares data analysis that goes well beyond the traditional PSI formula here, in part to address vibration sensitivity.

Today PSI enables the interferometric 3D analysis of millions of heights per second for surface areas ranging from a few tens of microns to a meter or more in diameter. The technique is fundamental to interferometry, fringe projection, deflectometry and other techniques that rely on this computationally-economical form of heterodyne detection.

A video for this equation is here: http://youtu.be/JpxzpFgt61c

PdG

Equation 182: The Affine Transform Matrix

$$\mathbf{p'} = \mathbf{Ap}$$

The new point $\mathbf{p'}$ is equal to the product of the transformation matrix, $\mathbf{A}$ and the old point, $\mathbf{p}$.

Globally the computer game industry is a huge business. In 2015 consumers around the world are expected to spend $91.5 Billion on computer games for phones, consoles and other devices. To put this figure into some perspective, that's around 50% more money than the UK's entire defence budget over the same period.

Computer graphics has come on a long way since early games. This is possible due to more computational power, the use of dedicated graphics hardware and new algorithms to, amongst other things, generate lighting effects and organic looking shapes. This means that the difference between Pong (1970) and modern gaming blockbusters is an incredible feat, yet they have a crucial piece of mathematics in common - the affine transform.

In mathematics, an affine transform is defined as any operation that preserves the properties of "colinearity" and "distance ratio" between the initial values and the transformed ones. In plain terms this means that applying an affine transform onto a line will always result in another line and the points that were at the beginning, middle and end of the old line will be at the beginning, middle and end of the new one.

For computer graphics, this allows us to translate a Pong paddle up and down a screen or rotate a complex 3D model of Batman to give the illusion of him turning around. Below are a number of examples of common affine transforms for computer graphics. For each example, there is an image based on performing that transform on a 2×2×2 cube centred on the origin

Translate: Move a point along an axis (see Figure 1)
Rotate: Rotate a point about an axis (see Figure 2)
Scale: Scale the lines to make them longer/shorter (see Figure 3)
Shear: Translate each point in the line along an axis by an amount that is proportional to its distance along another axis (see Figure 4)

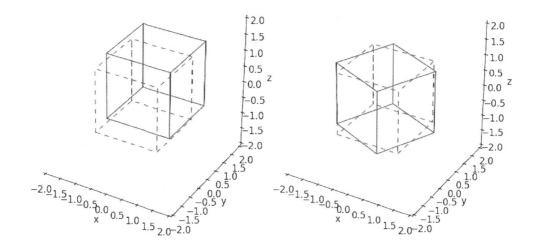

Figure 1. Left: The original cube (red dash) and the cube translated by 0.5 along the x axis, -0.2 along the y axis and 0.7 along the z axis (blue solid). Right: The original cube (red dash) and the cube rotated by 2 radians about the z axis (solid blue).

Matrices:

$$\begin{bmatrix} 1 & 0 & 0 & 0.5 \\ 0 & 1 & 0 & -0.2 \\ 0 & 0 & 1 & 0.7 \\ 0 & 0 & 0 & 1 \end{bmatrix} \text{ and } \begin{bmatrix} \cos(2) & \sin(2) & 0 & 0 \\ -\sin(2) & \cos(2) & 0 & 0 \\ 0 & 0 & 1 & 0 \\ 0 & 0 & 0 & 1 \end{bmatrix}$$

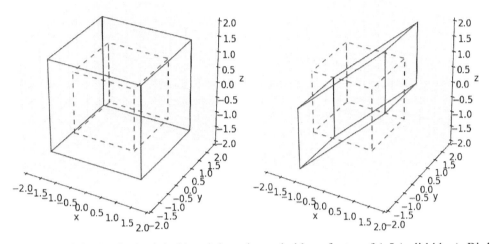

Figure 3. Left: The original cube (red dash) and the cube scaled by a factor of 1.5 (solid blue). Right: The original cube (red dash) and the cube sheared with respect to the yz plane by a factor of 0.8.

Matrices:

$$\begin{bmatrix} 1.5 & 0 & 0 & 0 \\ 0 & 1.5 & 0 & 0 \\ 0 & 0 & 1.5 & 0 \\ 0 & 0 & 0 & 1.5 \end{bmatrix} \text{ and } \begin{bmatrix} 1 & 0 & 0 & 0 \\ 0.8 & 1 & 0 & 0 \\ 0.8 & 0 & 1 & 0 \\ 0 & 0 & 0 & 1 \end{bmatrix}$$

Performing these transformations efficiently is vital to providing the realistic worlds that modern computer gamers are used to. A single character in a game may be constructed from upwards of 30000 polygons, each point requiring its new location to be recalculated for every frame, typically a minimum of 30 times per second to create the illusion of movement to the human eye. Dedicated hardware such as graphics cards can accelerate this process and key to this is the representation of the transforms.

As the equation above implies, the representation of choice for affine transforms is the use of matrix algebra, specifically matrix multiplication. A naive implementation of a 3D point would be in the style of a 1x3 column matrix representing the coordinates of the point $[x,y,z]^T$. However, this causes a problem. Rotation, scaling and shearing are linear in 3D space. I.e. they can be expressed as a simple matrix multiplication. However, translation is not linear. Moving a point that is near the origin e.g. $[0.1,0.2,0.1]$ by 2 along the x axis is a relatively large leap. Moving a point that is far away from the origin e.g. $[10000000,2000000,450000000]$ by 2 along the x axis is a relatively small leap. In addition, moving any point that is placed on the origin becomes impossible, as the product of $[0,0,0]$ with any matrix will be $[0,0,0]$. The effect of translation could be achieved by simply performing a matrix addition, but having a different rule for some operations than others becomes computationally difficult because of the sheer number of decisions that would have to be made per frame.

Instead computer graphics uses "homogeneous coordinates" to represent points. This means trivially adding a fourth dimension to the representation, which is assumed to always be equal to 1, i.e. every point becomes $[x,y,z,1]^T$. Now translation can be achieved by shearing along one axis, with respect to the fourth dimension. As shear is defined as moving points along one axis proportionally to their position on another, we move points along the axis of choice, proportionally to 1. This means that all of the useful affine transforms for computer graphics can be represented as 4x4 matrix multiplication. In fact, much of the functionality of a graphics card is the hardware to perform vast numbers of 4x4 matrix multiplications at high speed.

Other advantages of the use of a fourth dimension include the relatively easy application of perspective effects to add depth of field to the generated images, but this transform is outside the scope of this article.

This equation lies at the heart of all computer graphics, and is used hundreds of thousands of times per second to produce realistically animated vistas and fantastic creatures. In that sense, it truly is a beautiful equation.

RFO

Equation 183: Palermo Technical Impact Hazard Scale

$$P = \frac{\log pE^{0.8}}{0.03T}$$

Near-Earth Objects (NEOs) are asteroids and comets that have been nudged by the gravitational influence of nearby planets into paths that bring them into the vicinity of the Earth. The Palermo Technical Impact Hazard Scale is a convenient logarithmic scale employed to rate and prioritize the risks of these NEOs colliding with Earth. The scale has been derived using empirical data analysis, calculus and a bit of statistics. It encompasses a wide range of impact probabilities p, energies E and time intervals T and combines these parameters into a single "hazard" value P.

The formula above can be broken down as follows:

$$P = \log R$$

where R is the 'normalised risk' of an impact when the NEO under investigation nears the Earth and is given by

$$R = \frac{\text{Probability of impact when the NEO nears Earth at time } T}{\text{Number of impacts of energy } \geq E \text{ up until time } T}$$

$$= \frac{p}{f_B T}$$

Here, f_B is the background impact 'frequency' given by $f_B = 0.03E^{-0.8}$.

An alternative, though equivalent, interpretation of the normalised risk is given by:

$$R = \frac{\text{Expected impact energy when the NEO nears Earth at time } T}{\text{Cumulative energy of all impacts each of energy } \approx E \text{ up until time } T}$$

The denominator for both expressions of R, i.e. the threat posed by from the entire asteroid and comet population averaged over a very long time span, is referred to as the 'background hazard'. A rating of 0 on the Palermo scale (i.e. $R = 1$) means the hazard is as likely as the background hazard. A scale rating of +2 would indicate the hazard is 100 times more likely than a random background impact by an object at least as large before the date of the potential impact in question. Scale values less than −2 reflect events that are only 1% as likely as a random background event occurring in the intervening years. Potential impacts with positive Palermo Scale values will generally indicate situations that merit concern.

Example: (35396) 1997 XF11 is a Mars-crosser (an asteroid whose orbit crosses that of Mars) which is predicted to make an exceptionally close approach to Earth on 26 October 2028, at a distance of about 2.4 2.8 km in diameter.

The primary reference for the Palermo Technical Scale is a scientific paper entitled "Quantifying the risk posed by potential Earth impacts" by Chesley et al. (Icarus 159, 423-432 (2002)).

A scale that is similar to the Palermo one but less complex and less reliable is the Torino Scale, which tends to be used in the non-scientific media.

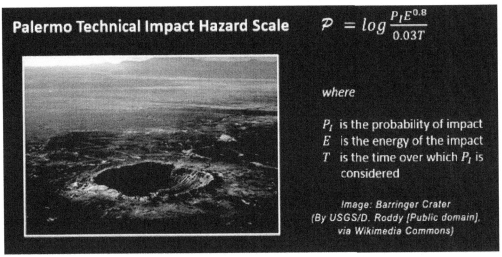

Palermo Technical Impact Hazard Scale

$$\mathcal{P} = log \frac{P_I E^{0.8}}{0.03T}$$

where

P_I is the probability of impact
E is the energy of the impact
T is the time over which P_I is considered

*Image: Barringer Crater
(By USGS/D. Roddy [Public domain],
via Wikimedia Commons)*

POSTSCRIPT

Asteroid Day is a global awareness movement that holds annual events on the anniversaries of the 1908 Siberian Tunguska blast. The 2015 event was hosted by astrophysicist and Queen guitarist Brian May and film-maker Grigorij Richters in San Francisco and London. Brian Cox, Martin Rees, Carolyn Shoemaker, Richard Dawkins and many other notable scientists are signatories to the movement. The aim is raise awareness of the potentially catastrophic risk of an impact and to campaign for vastly increased funds. However, some take a different view of the worth of directing so much money to hunting down NEOs. "The asteroid impact threat is very easy to overstate and misunderstand," says Eric Christensen of the University of Arizona in Tucson. "You could accomplish a lot of the same task with a couple of 4-metre dedicated telescopes from the ground, and that would not be half a billion dollars."

NEO scientists Clark Chapman and David Morrison estimated the chances of an individual dying from selected causes in the US (Nature, Vol. 367, page 39 (1994)). The chance of dying as a result of a NEO impact (1 in 40 000) is less than that of dying in a flood or a passenger air crash, though more than that of dying in a tornado or through a venomous bite or sting.

FR

Equation 184: The Raman Effect

In 1928, during his physics professorship at the University of Calcutta, Chandrasekhara Venkata ("C.V.") Raman discovered that some of the scattered light through a medium had a small shift in frequency (and hence wavelength). The discovery arose after several years of studying the scattering of light in transparent liquids, in which a significant role was played by Raman's colleague Kariamanickam Srinivasa Krishnan. Filters were used to produce the "monochromatic" light needed to observe the shifts in frequency. Due to its extremely low scattering efficiency (approximately 1 part in 10^{10}), the Raman effect did not come into common use until powerful lasers became available in the 1960s. Raman spectroscopy is now, as a complementary method to infrared absorption spectroscopy, a widely-used approach to the study of molecular vibrations and structure. The Raman effect is an example of the quantum nature of light and matter interactions, and Raman was awarded the Nobel Prize in physics for this work in 1930, the year directly following Louis de Broglie's famous prize for the discovery of quantum matter waves (BE102). Krishnan was omitted from the prize but received acknowledgement for his role in Raman's Nobel lecture.

This effect reveals an additional mechanism of light-matter interaction beyond elastic or Rayleigh scattering (BE28) - in which there is no net change in energy - and absorption, in which a photon of a certain frequency is used as energy transferred to a particle state (e.g. induced excitation of electronic states in the UV-visible region of the spectrum; induced molecular vibrations or rotations in the infrared).

Here we will focus on the range of electromagnetic energies associated with molecular vibrations, which is the main application of Raman spectroscopy. Molecular vibrations relate to the infrared region of the electromagnetic spectrum. In the case of infrared absorption and Raman scattering, the energy differences are the same for a particular molecular vibration as they result in vibrational states above the ground electronic state, but are caused by different mechanisms. In the case of infrared absorption (BE168) a photon is absorbed if it is the exact energy required to cause the resulting molecular vibration. This is a resonance effect whereby the photon frequency is matched to a change in electric dipole moment. The alternating electric field of the incident photon interacts with fluctuations in the electric dipole moment. This is the reason why symmetric diatomic molecules (eg H2, N2) do not absorb in the infrared (but are active in Raman). The dipole moment $\boldsymbol{\mu}$ is a vector that depends on the charge symmetry and size of a molecule (or molecular group in a larger compound). It is a measure of a molecule's polarity, relating to two opposing charges ($\pm q$) and the displacement $\boldsymbol{d}$ between them:

$$\boldsymbol{\mu} = q\boldsymbol{d} \tag{1}$$

The Raman effect is an inelastic scattering effect (which had actually been predicted in 1923 by Adolf Smekal) that arises due to the ability of a molecule to acquire an induced dipole moment from an applied electric field of much higher energy (typically in the visible region of the spectrum which relates to electronic states). It is not a resonant effect and its occurrence is independent of the excitation frequency. The ease with which the electron cloud around an atom or molecule can be distorted by an applied field is called the **polarizability** and is defined as

$$\mu_i = \alpha E \tag{2}$$

where α is the polarizability, μ_i is the induced dipole moment and E is the applied electric field. The units of α are Cm^2V^{-1}. It is described here as a scalar quantity as it is assumed that the induced dipole moment is in the same direction as the applied field. (The representation becomes rather more complicated if it depends on other directions.) In the simplest case of an atom, the polarizability is isotropic (same in all directions). In Raman scattering, the incident photons act as the applied field. The instantaneous polarization is dependent on displacement of the vibrational state which, ignoring the higher derivatives of the Taylor expansion, can be approximated by

$$\alpha = \alpha_0 + \left(\frac{\partial \alpha}{\partial r}\right)\Delta r \tag{3}$$

where r is the nuclear separation. In the case of vibrational states of frequency ν_m the displacement of the nuclei is sinusoidal

211

$$\Delta r = r_0 \cos(2\pi v_m t) \tag{4}$$

as is the strength of the applied electric field (excitation photons of much higher frequency v_0)

$$E = E_0 \cos(2\pi v_0 t) \tag{5}$$

Substituting (4) in (3) and then (5) and (3) into (2):

$$\mu_i = \alpha_0 E_0 \cos(2\pi v_0 t) + E_0 r_0 \left(\frac{\partial \alpha}{\partial r}\right) \cos(2\pi v_0 t) \cos(2\pi v_m t) \tag{6}$$

Using the trigonometric identity $\cos a \cos b = \frac{1}{2}[\cos(a+b) + \cos(a-b)]$ the two frequency difference terms are arrived at

$$\mu_i = \alpha_0 E_0 \cos(2\pi v_0 t) + \frac{E_0 r_0}{2}\left(\frac{\partial \alpha}{\partial r}\right)\{\cos[2\pi t(v_0 - v_m)] + \cos[2\pi t(v_0 + v_m)]\} \tag{7}$$

The scattering processes typically involve an intermediate "virtual" state that relates to the ground electronic state plus the energy of the incident photon, whereas in absorption the molecule goes directly from the ground electronic state to the first vibrational states (fundamental and overtones of $h v_m$ where h is Planck's constant). The first sinusoidal term is Rayleigh scattering with no change in frequency (see the diagram below). This is statistically much more likely as the atoms return to the ground state and the coupling between electronic and vibrational states is weak (nuclei are much larger than electrons). The second and third terms at lower and higher frequencies are Raman scattering. The second term is the Stokes line (energy loss) and the third is the anti-Stokes line (energy gain). These are named after George Stokes who had observed similar changes in frequency relating to fluorescence in 1852. The anti-Stokes line occurs with much lower probability than the Stokes line at ambient temperature because it has to start from an excited ground electronic state: Raman spectrometers typically measure the Stokes line, with the difference frequency relating to the molecular vibrations described as the Raman Shift, often measured in wavenumbers $\bar{v}$ (cm^{-1}) for convenience and direct comparison with infrared absorption spectra. Note that if the derivative term in the equation is zero there is no Raman effect, which requires a change in polarizability with respect to variation in nuclear separation. This gives us a selection rule for Raman scattering, and explains why some molecular configurations are Raman "active" and others are not. Raman scattering is typically weak for strongly polar molecular groups (e.g. O-H), because the electron clouds are more constrained and less easily polarizable. Large atoms are more easily polarizable and Raman is particularly useful for metallic groups, and is often used in mineralogy. The above derivation is essentially a classical approach which illustrates the different mechanisms. In the case of Rayleigh scattering, the classical and quantum approaches give the same results because there is no net change in vibrational quantum number. For a full explanation of the Raman energy states, however, a quantum approach is required. Modern Raman spectrometers can be made fairly compactly and cheaply, using a stabilised semiconductor laser, often at 532 or 785 nm but also at many other wavelengths depending on the application. 532 nm is preferred for signal strength. From scattering theory (small sphere approximation) it can be shown that the intensity of the scattered light is inversely proportional to the fourth power of the wavelength – this strong dependency results in much stronger signal in blue compared with red laser light, for example. However, the likelihood of fluorescence (a relatively strong absorption/emission process) interfering with the Raman signal is much more likely in the blue region, so both have their advantages. Notch filtering in the returning optics is quite precise and can allow frequencies close to the laser line ("Rayleigh" line), e.g. low molecular vibrations below 200 cm^{-1}, to be detected. A typical wavenumber range for a low-cost instrument is 200-3500 cm^{-1} which covers most fundamental vibrations and deformations and some of their overtones. The simplified "Jablonski diagram" (after the Polish physicist Aleksander Jablonski) shown below with the final equation (7) illustrates the different scattering effects compared with infrared absorption. Note that E_g is the ground energy state of the molecule and not to be confused with E_0 in the equation which is the electric field strength amplitude of the incident photon.

$$\mu_i = \alpha_0 E_0 \cos(2\pi \nu_0 t) + \frac{E_0 r_0}{2}\left(\frac{\partial \alpha}{\partial r}\right)\{\cos[2\pi t(\nu_0 - \nu_m)] + \cos[2\pi t(\nu_0 + \nu_m)]\}$$

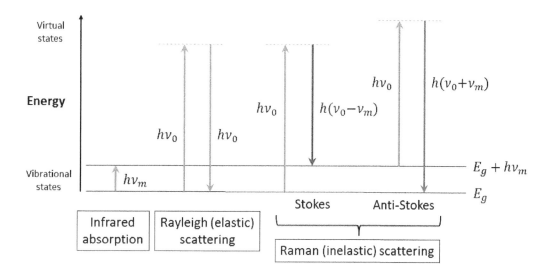

PB

Equation 185: Gravitational Lensing

A normal lens works by bending light passing through it using a process known as refraction to focus the light somewhere. Gravitational lensing, an effect of Einstein's theory of general relativity, is a process where mass bends light.

Isaac Newton speculated that massive objects should deflect light, but was unable to describe this as he only considered light to act as a wave. It wasn't until 1804 when Johann Soldner (1776 to 1833), a German physicist, mathematician and astronomer, predicted that a light ray passing close to the Sun would be deflected by an angle $\tilde{\alpha} = 0.84$ arcsec. Over a century later, and unknown to him, Albert Einstein calculated a value of $\tilde{\alpha} = 0.83$ arcsec. This however was before he had finished his General Theory of Relativity.

Once Einstein's theory was complete, he derived the correct formula for the deflection angle of light passing a massive object.

$$\tilde{\alpha} = \frac{4GM}{c^2 r}$$

where $\tilde{\alpha}$ is the deflection angle of light passing a massive object, G is the gravitational constant, M is the mass of the body, C is the speed of light, r is the distance the light passes from the object.

Since the Schwarzschild radius (BE 4) is defined as $r_s = \frac{2GM}{c^2}$, the angle of deflection can be expressed simply as

$$\tilde{\alpha} = 2\frac{r_s}{r}$$

Having derived this, Einstein was now able to calculate the angle of deflection for a light beam passing close to the Sun to be $\tilde{\alpha} = 1.74$ arcsec, and this was verified experimentally to within 20% by Arthur Eddington during the total solar eclipse of 1919.

As stated above, mass bends light, or more accurately, mass bends space time and the path in which the light travels. The more massive the object, the greater the gravitational field it produces. Probably the most famous examples of gravitational lensing are the Einstein Cross and an Einstein Ring. The Einstein Cross (shown in the first figure) is a gravitationally lensed quasar sitting directly behind a large galaxy (Huchra's lens). Due to the gravitational lensing effect, four images of the same quasar appear in the foreground. The quasar is located about 8 billion light years from Earth, while the lensing galaxy is about 400 million light years from earth. An Einstein Ring (shown in the second figure) is where light from a star or galaxy is distorted into a ring by an object with a very large mass (such as a black hole or massive galaxy). This occurs when source, lensing mass and earth are all aligned.

Gravitational lensing has recently caught the public imagination through Christopher Nolan's film "*Interstellar*". Special effects company, Double Negative Visual Effects, along with physicist Kip Thorne solved the equations for light propagation through the curved space time of a spinning (Kerr) black hole to produce IMAX-quality smoothness without flickering. This was a major departure from the image generation techniques used by physicists and indeed the techniques currently used in the CGI industry.

O. James, E. von Tunzelmann, P. Franklin and K. S. Thorne, "Gravitational lensing by spinning black holes in astrophysics, and in the movie Interstellar," Class. Quantum Grav., vol. 32, 2015.

CM

Equation 186: Cornu Spiral

The Cornu Spiral is a graph that is used to calculate an integral that appears in the Fresnel diffraction theory:

$$\int_0^u e^{i\frac{\pi}{2}u^2}\, du = \int_0^u \cos\left(\frac{\pi}{2}\right) u^2\, du + i \int_0^u \sin\left(\frac{\pi}{2}\right) u^2\, du = C(u) + iS(u)$$

$C(u)$ and $S(u)$ are known as the Fresnel integrals. The complex integral (1) can be represented graphically in the complex plane by plotting $S(u)$ against $C(u)$. The figure this gives is known as the Cornu Spiral.

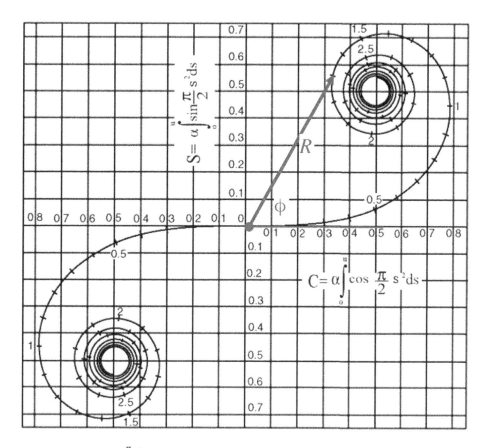

The value of the integral $\int_{u_1}^{u_2} e^{i\frac{\pi}{2}u^2}\, du = R \cdot e^{i\phi}$ is given by the straight line connecting the two points on the Cornu Spiral specified by the limits u_1 and u_2. The length of the line is the magnitude R of this complex integral; the angle the line makes with the real axis in the phase ϕ. If it is only the diffraction intensity, proportial to R^2, that is of interest, only the length of the chord between the two values u need to be read from the graph.

Alfred Cornu (1841 – 1902) was a French physicist who contributed significantly to the wave theory of light. Except for the spiral he is also named in the Cornu depolarizer. The Cornu spiral is used in diffraction theory as mentioned, but also in diverse areas such as violin and road design.

HH

Equation 187: Electric Field of A Point Charge

The presence of a point charge q in vacuum results in an electric field being generated. The magnitude of the electric field is inversely related to the square of the radial distance r from the point charge and is given by

$$E = \frac{q}{4\pi\epsilon_o r^2}$$

where ϵ_o is the permittivity of free space, which describes the resistance encountered in the forming of an electric field. The direction of the electric field is either radially outwards (in the presence of a positive charge) or radially inwards (in the presence of a negative charge).

215

The concept that the electric field of a point charge has an inverse-square behavior was first documented in the 18th century by Daniel Bernoulli and Alessandro Volta. In 1785, French physicist Charles-Augustin de Coulomb published results, in which he stated the inverse-square law of electrical force between two charges [1].

[1] Coulomb's Law - Wikipedia, The Free Encyclopedia, https://en.wikipedia.org/wiki/Coulomb's_law

MF

Equation 188: The Cayley-Hamilton Theorem

In Beautiful Equation 172 (the Generalized Eigenvalue Problem), it was mentioned that the eigenvalues of an $n{\times}n$ matrix $\mathbf{K}$ with respect to another $n{\times}n$ matrix $\mathbf{M}$ are obtained by solving the equation

$$\det(\mathbf{K} - \lambda\mathbf{M}) = 0$$

which is an nth-order polynomial in λ.

For the special case when $\mathbf{M} = \mathbf{I}$ (the identity matrix), the λs obtained by solving the above equation are simply called the eigenvalues of $\mathbf{A}$ and the nth-order polynomial is called the characteristic polynomial of $\mathbf{A}$. Thus, the eigenvalues of $\mathbf{A}$ satisfy the equation

$$\det(\mathbf{K} - \lambda\mathbf{I}) = 0$$

In terms of λ, the above can be written as the following polynomial $p(\lambda)$:

$$p(\lambda) = \lambda^n + c_{n-1}\lambda^{n-1} + c_{n-2}\lambda^{n-2} + \cdots + c_1\lambda + c_0 = 0$$

Here the coefficients c_i involve various invariants of the matrix $\mathbf{K}$.

The Cayley-Hamilton theorem states that every square matrix $\mathbf{K}$ satisfies its own characteristic equation, i.e., the matrix $\mathbf{K}$ satisfies the equation

$$p(\mathbf{A}) = \mathbf{A}^n + c_{n-1}\mathbf{A}^{n-1} + c_{n-2}\mathbf{A}^{n-2} + \cdots + c_1\mathbf{A} + c_0\mathbf{I} = \mathbf{0}$$

Note that the zero on the right-hand side is the zero square matrix of size n. This, in this author's opinion, is one of the most beautiful equations in the entire field of Matrix Computations. The theorem is named after Arthur Cayley (a British lawyer and mathematician) and William Hamilton (an Irish astronomer and mathematician). Hamilton, in 1853, proved the theorem for a special case in the context of quaternions. However, it was Cayley who stated it more formally in 1858.

As an example, consider the 3×3 matrix:

$$\mathbf{K} = \begin{pmatrix} 3 & 5 & 6 \\ 5 & 5 & 4 \\ 6 & 4 & 7 \end{pmatrix}$$

The characteristic equation of this matrix is the third-order polynomial in λ given by

$$\lambda^3 - 15\lambda^2 - 6\lambda + 58 = 0$$

The Cayley-Hamilon theorem states that the matrix $\mathbf{K}$ satisfies the equation

216

$$\mathbf{K}^3 - 15\mathbf{K}^2 - 6\mathbf{K} + 58\mathbf{I} = 0$$

The above can easily be verified to be equal to zero. This equation can also be used to illustrate an interesting application of the Cayley-Hamilton theorem in finding the inverse of a non-singular matrix $\mathbf{K}^{-1}$. Multiplying throughout by $\mathbf{K}^{-1}$ and solving for $\mathbf{K}^{-1}$, we find

$$\mathbf{K}^{-1} = -\frac{1}{58}(\mathbf{K}^2 - 15\mathbf{K} - 6\mathbf{I})$$

The above also allows us to find any negative integral power of $\mathbf{K}$. Other applications of the theorem include reducing the order of a polynomial in $\mathbf{A}$ and the calculation of the matrix exponential.

HC

Equation 189: Tupper's Inequality

Almost a mathematical conjuring trick, Tupper's inequality is a formula which is capable of generating an image of itself. It is:

$$\frac{1}{2} < \left\lfloor \mathrm{mod}\left(\left\lfloor \frac{y}{17} \right\rfloor 2^{-17\lfloor x \rfloor - \mathrm{mod}(\lfloor y \rfloor, 17)}, 2 \right) \right\rfloor$$

$\lfloor \cdot \rfloor$ is the floor function: for a sequence of numbers, the floor function maps a number to the largest previous number (so floor$(x) = \lfloor x \rfloor$ is the largest integer not greater than x).

mod is the modulo operation; it outputs the remainder following a division operation.

To use the formula, it is calculated for x-values from 0 to 105 and y-values from k to k+17, where k =

960,939,379,918,958,884,971,672,962,127,852,754,715,004,339,660,129,306,651,505,519,271,702,802,3
95,266,424,689,642,842,174,350,718,121,267,153,782,770,623,355,993,237,280,874,144,307,891,325,96
3,941,337,723,487,857,735,749,823,926,629,715,517,173,716,995,165,232,890,538,221,612,403,238,855,
866,184,013,235,585,136,048,828,693,337,902,491,454,229,288,667,081,096,184,496,091,705,183,454,0
67,827,731,551,705,405,381,627,380,967,602,565,625,016,981,482,083,418,783,163,849,115,590,225,61
0,003,652,351,370,343,874,461,848,378,737,238,198,224,849,863,465,033,159,410,054,974,700,593,138,
339,226,497,249,461,751,545,728,366,702,369,745,461,014,655,997,933,798,537,483,143,786,841,806,5
93,422,227,898,388,722,980,000,748,404,719

The results are then plotted using x and y as coordinates and reversing the axes such that x increases to the left and y increases downward. A dot is printed at each coordinate for which the inequality is satisfied.

The result is:

217

Jeff Tupper is a Canadian computer scientist, who introduced the inequality in a paper intended to assist computer scientists in writing graphical software for mathematicians.

MJG

Equation 190: Polarization of Light and the Jones Matrix

The propagation of light in space is characterized as a transverse wave, which has two oscillating field components (the electric and the magnetic) that are perpendicular to the direction of travelling. It is a convention in optics to describe the light wave by its electrical vector, whose direction of vibration with respect to the axis of travelling is known as the polarization state of the light. Polarized waves have a particular direction of vibration throughout its propagation. Natural light coming from the Sun or an incandescent bulb are broad spectrum and unpolarized because each is a mixture of multiple waves with random amplitudes, frequencies, polarizations, and phases.

In the 1940s R. Clark Jones (1916-2004), an American physicist published a series of papers in the Journal of the Optical Society of America explaining his method for calculating the polarization of light (known as the Jones calculus) as it propagates through different components of an optical system.

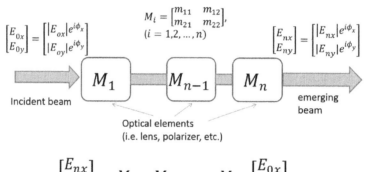

$$\begin{bmatrix} E_{nx} \\ E_{ny} \end{bmatrix} = M_n \cdot M_{n-1} \cdot \cdots M_1 \cdot \begin{bmatrix} E_{0x} \\ E_{0y} \end{bmatrix}$$

Jones started his notation for the complex pair of amplitudes of the electrical field in x and y directions by a vector:

$$\begin{bmatrix} E_{0x} \\ E_{0y} \end{bmatrix} = \begin{bmatrix} |E_{ox}|e^{i\phi_x} \\ |E_{oy}|e^{i\phi_y} \end{bmatrix}$$

such that when the incident light (with the subscript of 0) passes through one or a sequence of optical elements leads to the emerging light described by:

$$\begin{bmatrix} E_{nx} \\ E_{ny} \end{bmatrix} = M_n \cdot M_{n-1} \cdot \ldots M_1 \cdot \begin{bmatrix} E_{0x} \\ E_{0y} \end{bmatrix}$$

where the two by two complex matrix M_n is the Jones matrix for the nth optical element.

For example, a linear polarizer can be described as:

$M = \begin{bmatrix} 1 & 0 \\ 0 & 0 \end{bmatrix}$; a quarter-wave plate can be described as: $M = \frac{1}{\sqrt{2}} \begin{bmatrix} 1 & i \\ i & 1 \end{bmatrix}$. The Jones matrix can be normalized for energy considerations without any effect on the state of polarization of interest.

The primary application of Jones calculus is to calculate the state of polarization by describing the functionality of certain optical elements (polarizer, wave plate, phase retarder etc.) in its matrix form. For example, this matrix analysis approach is used for analyzing the errors sources (both mechanical and optoelectronic) in high precision displacement measuring interferometers. A further generalization of Jones matrix is the Mueller matrix, which takes the unpolarized or partially polarized light sources into consideration.

References:

[1]. R. Clark Jones, "A new calculus for the treatment of optical systems, I. Description and Discussion of the Calculus". Journal of the Optical Society of America 31 (7): 488-493, 1941
[2]. G. R. Fowles, "Introduction to Modern Optics", 2nd edition, 1975, ISBN: 0-486-65957-7
[3]. P. de Groot, "Jones matrix analysis of high-precision displacement measuring interferometer", ODIMAP II, 9-14, 1999

KN

Equation 191: Conductivity of a Semiconductor

$$\sigma = e(n\mu_e + p\mu_h)$$

where e is the electron charge, n and p are the number of electrons and holes respectively, and μ_e, μ_h are the mobility of electrons and holes.

Semiconductors are materials whose electric properties are in between a perfect conductor and a perfect insulator (See Equation (161)). The number of charge carries in a semiconductor can be modulated by adding other atoms (impurities) in the lattice structure. Through the so-called doping process it is possible to tune the available charge carriers. Impurities can be elements from the 5th or 3rd column of the periodic table which have 5 or 3 valence electrons available for bonding. Since only 4 electrons bond with adjacent molecules, there is either an electron more free to move, or an empty hole where an electron can move into. The mobility of the charge carriers takes into account how easy it is to move those charges in the semiconductor.

Typical electron mobility for SI at room temperature (300 K) is 1400 cm^2/(V·s) and the hole mobility is around 450 cm^2/ (V·s). Very high mobility has been found in several low-dimensional systems, such as two-dimensional electron gases (35,000,000 cm^2/(V·s) at low temperature, carbon nanotubes (100,000 cm^2/(V·s) at room temperature) and more recently, graphene (200,000 cm^2/ V·s at low temperature). Organic semiconductors developed so far have carrier mobilities below 10 cm^2/(V·s), and usually much lower. The available charge density can be tuned by doping the semiconductor or by applying a voltage to create a local abundance of charges. The latter phenomenon is employed in CMOS transistor to modulate the conductivity of the channel by applying a voltage to the gate (see equation 55).

GM

Equation 192: Depth of field

In the words of the song; oh what a picture what a photograph. But what makes a good photograph? Well from a technical point of view it can be about making good use of depth of field. Take the picture of the nude below; no there isn't one but that got your attention. OK back to the boring stuff. Before calculating depth of field we have to define the circle of confusion (sounds like something Geoff Boycott would say in a test match commentary).

Circle of confusion (disk of confusion, circle of indistinctness, blur circle) is an optical spot caused by a cone of light rays from a lens not coming to a perfect focus when imaging a point source [Wikipedia]

Hope this hasn't confused you too much but for modern digital cameras it's one of two numbers; for most of us with APS/APS-C sized sensors its 0.019948, for full sized sensors and 35 mm film its 0.02501.

OK, an equation now. We are going to calculate the hyperfocal distance (the closest distance at which a lens can be focussed while keeping objects at infinity acceptably sharp). This distance has the maximum depth of field. If you focus your camera at this distance, objects from half the hyperfocal distance to infinity will be acceptably sharp. Right find the focal length of your lens and here we go

$$H = \frac{f^2}{Nc} + f \approx \frac{f^2}{Nc}$$

where H is the hyperfocal distance, f is the focal length, N is the aperture (f-number), and c is the circle of confusion. Next we need to calculate the near point NP; the closest distance that will be in focus given the distance between the camera and the subject (lens focal distance - d).

$$NP = \frac{Hd}{(H + (d - f))}$$

Finally, we need to calculate the far point FP

$$FP = \frac{Hd}{H - (d - f)}$$

OK and now for the beautiful equation

$$TotalDof = FP - NP$$

I'll leave it as an exercise for the reader to calculate the depth of field for a Conon 30D with APS sensor and a 50 mm lens set at f/2.8. You can check your answer using an app such as PhotoCalc for iPhone.

The History of this equation is not that clear but I refer you to Sutton and Dawson (1867) who define focal range – what we now call hyperfocal distance, Abney (1881), Taylor (1892), Piper (1901), Derr (1906) and Johnson 1909. More recently, Kingslake (1951) makes the distinction between the two definitions of hyperfocal distance.

DF

Equation 193: Bertrand's Theorem

$$f(r) = -\frac{k}{r^{3-\beta^2}}$$
$$\beta^2(1 - \beta^2)(4 - \beta^2) = 0$$

where $f(r)$ is the force law between two rigid bodies, k is a constant of the system that includes angular momentum, mass, and a physical constant. r is the distance between the centers of mass of the two bodies. And β is a quantity that relates to the orbital frequency and, thereby, the nature of the orbital motion between the two celestial bodies.

It is well known that Newton's laws relating gravitational attractions between bodies in space can be used to predict interplanetary motions with remarkable accuracy. That these Laws provide such precise predictive power is a measure of the extent to which they embody the physics of the Universe in a fundamental way. An obvious question might be whether or not these are the only Laws or maybe just one of many equally viable mathematical models? A definitive answer to this question might never be known. However, we know that any other mathematical model must incorporate the same predictive power.

Finding solutions to particular questions helps to illuminate the power of this model. One such question relates to the ability of planets to form stable cyclic orbits based on arbitrary initial conditions (i.e. distance apart from each other and relative velocities).Newton ably demonstrated that consideration of many planets (or other astronomical objects) would result in mathematical formulations that do not admit of solutions in the form of a simple set of explicit equations. However, a reduced universe comprising only two bodies does provide considerable insight into the so-called 'many body problem'. This is not to say that the mathematics is straightforward.

Considering a relatively empty universe comprising two bodies that do not collide, a number of simplifications arise. Firstly, the Euler-Lagrange equation (beautiful equation 6) reveals that, independent of how the bodies move relative to each other, their center of mass is either stationary or moving at constant velocity in a straight line. This is great in that it enables us to ignore it altogether and plant a coordinate system at the center of mass and think of this as a stationary point. Even better, it is possible to lump the masses of the two bodies together and determine a complete solution from an equation that considers only a single mass (called the 'reduced mass') for which the angular momentum is always a constant. These result in substantial simplification of the mathematics leaving only a single, second order, differential equation to solve.

From considerations of solutions to the orbital equation, it is possible to determine constraints upon β. Firstly, if the orbits are to close, it is necessary after a fixed number of rotations the orbit returns to the same state (position and velocity). A second condition that the force law be capable of providing stable circular orbits implies $\beta^2 > 0$. However, observations of planetary motions indicate that deviations from a circular orbit are common. It might be reasonably assumed that these observed stable orbits have occurred due to celestial bodies coming into proximity with one another with varying initial velocities and trajectories. Imposing these conditions, stable cyclic orbital motions are only possible if β satisfies the second of the beautiful equations given above. In this case the stable orbital force law is of either forms

$$f(r) = -kr, \qquad \beta^2 = 4$$
$$f(r) = -\frac{k}{r^2}, \qquad \beta^2 = 1$$

The first of these is Hooke's law (Beautiful equation 18). The second of these recovers Newton's famous inverse square law (Beautiful equation 67) as well as Coulombs law for the force between two charged particles (Beautiful equation 187). This inverse square law also shows up in general field theories in which forces arise through localized gradients of potential in fields containing fluxes of conserved quantities (electric, gravitational, and fluid fields). In fact, both of these cases result in elliptical orbits and were studied extensively by Newton in propositions X and XI in his Principia.

While others had worked on this problem, in particular the French mathematician Jacques Charles Francois Sturm (1803 – 1855) in 1841, the above conclusions were first presented in their entirety in 1873 by the French mathematician Joseph Louis Francios Bertrand (1822 – 1900) for whom the theorem is now named.

References

J. Bertrand, 1873, Theoreme relatif au movement d'un point attire vers un centre fixe, *Comptes Rendus*, **77**, 849 – 853.
H. Goldstein, 1980, Classical mechanics, Addison Wesley, ISBN 0-201-02969-3, Chapter 3 and Appendix A.

F.C. Santos, V. Soares, and A.C. Tort, 2011, An English translation of Bertrand's theorem, *Lat. Am. J. Phys. Educ.*, **5**(4), 694 – 696.
Sturm, J. C. F., 1841, *Comptes Rendus*, **13**, 1046, 1841

SS

Equation 194: Wave Shift Function

The wave shift function is used in grinding to produce an optimal grinding wheel v work piece speed ratio so that regenerative grinding forces are minimized and surface finish is improved.

$$\text{Wave Shift} = \text{Gw Revs} / \text{Wp Revs} - \text{the integer number of the result of Gw Revs} / \text{Wp Revs}$$

$$\text{Surface speed} = \left((\varnothing\text{Gw} \cdot \pi) \cdot \text{GwV}\right) - \left((\varnothing\text{Wp} \cdot \pi) \cdot \text{WpV}\right)$$

where Gw is the grinding wheel diameter (M), Gw*V* is the grinding wheel Velocity (RPM), Wp is the work piece and Wpv is the work piece velocity

Note, this assumes that both surfaces are moving in the same direction, for different directions the work piece velocity would simply be a negative velocity. This is the difference between conventional and climb grinding.

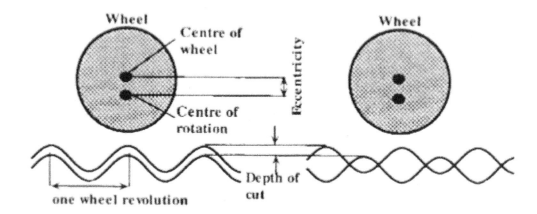

The wave shift function is effectively a phase shift of a cycle but is expressed as fractions of a revolution rather than in degrees. Optimum numbers are not easily divisible into a small number. Wave shift values such as 0.11111, 0.33333, 0.66666 etc. There are more optimal numbers however aiming to achieve such numbers would be a good choice. The goal is to produce the number of n per rev interactions whether it be from eccentricity, error motions or other vibrations.

Many commercial grinding processes are highly tuned and require an optimum surface speed to produce the most efficient process. This could mean maximum material removal rates or best surface finish.

The grinding wheel and work piece speed is dictated by the desired surface speed, being the interaction of the two speeds. As grinding occurs wheel and work piece diameters change and so speeds are adjusted to maintain surface speed. This gives a varying wave shift function and can allow regenerative grinding forces to effect the process, going in and out of good speed ranges. The effect can be quite significant on part quality.

Research in this field has been extensive and related studies are numerous. I am citing here the specific work of Trimal & Holesovsky (2001).

By taking the wave shift into account and selecting and transitioning through optimum wave shift values the effects of regenerative grinding can be mitigated giving a better quality manufacturing process.

Reference:
2001 - Trimal & Holesovsky - Wave-shift and its effect on surface quality in super-abrasive grinding
International Journal of Machine Tools & Manufacture 41 (2001) 979–989

STS

Equation 195: Pick's Theorem

$$\text{Area}(P) = I + (B/2) - 1$$

Where Area(P) is the polygon area, I is the number of grid points inside the polygon and B is the number of grid point on the polygon boundary

Since the earliest days of mathematics there has been a fascination with shapes and their area. Common geometric shapes all have their specific formulae for calculating their area, perhaps the most famous being Archimedes' formula for the area of a circle (circa 250BC). However it wasn't until 1899 that a quite remarkable discovery was made by Geory Pick relating the area of all 2 dimensional polygons, whose corners lie on a square grid, to just the grid points within the shape and those on its edges.

Pick's formula requires simply counting the number of grid points that lie on the interior of the shape, I and multiplying this by half the grid points that lie exactly on the boundary (corners and edges), B then subtracting 1. Amazingly, this always gives the exact area of the polygon. The proof is non-trivial but can be constructed by demonstrating that the equation holds true for any type of triangle then for any triangle combined with another triangle and finally any triangle combined with any number of triangles as any polygon can be deconstructed into triangles.

The formula can also be extended to include polygons containing hole by simply adding on the number of holes, H.

$$\text{Area}(P) = I + (B/2) - 1 + H$$

With the advent of digital measurement this formula has found many applications as any x,y based digital measurement will always place the vertices on a regular grid to satisfy Pick's criteria.

Although efforts were made to extend this theorem to 3-dimensional space in 1957 John Reeve showed this would never be possible. He constructed a family of irregular tetrahedra that contained no interior grid points and only 4 boundary grid point however their volumes were different.

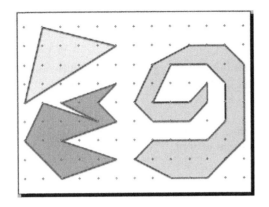
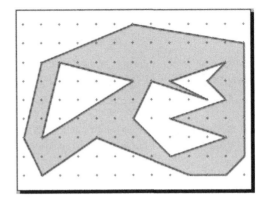

ILB

Equation 196: Linear Thermal Expansion

$$\frac{\Delta L}{L} = \alpha \Delta T$$

where ΔL is the change in length, L is the original length, ΔT is the change in temperature and α is the linear coefficient of thermal expansion.

Most solids expand when they are heated. This expansion tends to be linear, where the constant of proportionality is called the coefficient of linear thermal expansion (or some other permutation of these words). But why do most solids expand when heated? We can model a solid as a bunch of molecules spaced on a regular lattice. But molecules are never stationary – they vibrate. So, we can extend our solid model by imagining springs between the molecules. When the solid is in thermal equilibrium (all the same temperature and not changing), then the molecules will all vibrate with the same frequency and amplitude. Raising the temperature of the solid raises the kinetic energy in the solid, which raises the amplitude of the molecules' vibrations. This raising of the amplitude results in an overall increase in the size of the atom and; if we measure this in one direction, we will measure the linear expansion. Of course, the solid expands on all other directions, so there are also the concepts of area and volume expansion.

Very few solids contract on heating and this is usually only over very restricted temperature ranges. The coefficient of expansion (CTE) of water actually drops to zero at 3.983 °C and is negative thereafter. Highly pure silicon has a negative CTE for temperatures between about 18 K and 120 K.

In my field (metrology), we often want to minimise or at least control thermal expansion, as it is often a source or error (e.g. if I am measuring the length of a block, the expansion will clearly have an effect on my measurement – see BEs 05 and 19). Therefore, metrology labs are often temperature controlled and materials with well-known CTE are used. As an example of the degree of potential error, a 1 m bar will expand by approximately 11 μm for a 1 °C temperature rise (i.e. it has a CTE of 11×10^{-6} K^{-1}). This effect also needs to be considered when designing precision instruments. There are a number of low-expansion materials available these days, for example Invar is a nickel-steel alloy with a CTE of approximately 1×10^{-6} K^{-1} and the glass-ceramic Zerodur (also used to make optics) with a CTE of almost zero at 20 °C.

It has proved difficult to find any definitive history for the discovery of linear thermal expansion. Certainly, the ancients would have been aware of its effects and it played a significant role in the design and operation of clocks. Classic examples of products still in use that take advantage of the effect are the bimetallic strip and the mercury in glass thermometer (although this is of course liquid expansion).

RL

Equation 197: Relationship Between Half Value and Mean Value for Decreasing Exponential Functions

$$x_{\frac{1}{2}} = \langle x \rangle \ln(2)$$

where, for a function $f(x) = f(0)e^{-\lambda x}$, $x_{\frac{1}{2}}$ denotes the *half value* of x, i.e. its value when $\frac{f(x)}{f(0)} = \frac{1}{2}$, $\langle x \rangle$ denotes the *mean value* of x, which works out to be equal to $\frac{1}{\lambda}$, and coincides with the value of x when $\frac{f(x)}{f(0)} = \frac{1}{e} \approx 0.37$, λ is a parameter typically referred to as the 'exponential decay constant', or the 'linear attenuation coefficient', or the 'heat transfer coefficient', depending on context.

An approximation to the above formula is given by

$$x_{\frac{1}{2}} \approx 0.693 \langle x \rangle$$

'X' Referring To Distance:

The Beer-Lambert Law suggests that the intensity of radiation propagating inside a uniform material falls off exponentially with distance travelled. The quantity $x_{\frac{1}{2}}$ is referred to as the 'half thickness' or the 'half value layer (HVL)' of the medium for that type of wave, and $\langle x \rangle$ (often denoted as λ) is called the 'attenuation length' or the 'penetration depth'.

'X' Referring To Time:

Using the symbol 't' instead of 'x', then $t_{\frac{1}{2}}$ is referred to as the 'half-time' or 'half-life' of the process in question and T is often used instead $\langle t \rangle$, being referred to as the 'mean lifetime' or the 'time constant', depending on context.

Thus in the case of radioactive decay, the half-life $t_{\frac{1}{2}}$ is the mean time taken for the number of radioactive nuclei of the sample in question to decay to half the original amount. In the case of a first order chemical reaction, the half-life $t_{\frac{1}{2}}$ is the mean time taken for the concentration of the reactant to reduce to half the original value.

In diving physiology, different tissue types have different take-up and release 'half times' $t_{\frac{1}{2}}$ for a given inert gas and a given depth.

In a circuit containing resistors and capacitors and/or inductors, the time constant T is given by RC/L. In neuroscience, when there is an action potential (or passive signal spread) in a neuron, the time constant T is similarly given by rc, where r is the resistance across the membrane and c is its capacitance.

Thermal systems modelled as 'lumped capacitance systems', a simple example of which is a system conforming to Newtons Law of Cooling, also feature thermal time constants similar to the above. Here, $T = \frac{\rho c V}{hS}$, where ρ is the density of body in question, c is specific heat capacity, V is volume, h is thermal transfer coefficient and S is surface area.

FR

Equation 198: Spherical Harmonics

Spherical harmonics are encountered as the angular part of the solution to the Laplace equation in spherical polar coordinates. The solutions to the Laplace equation were known as harmonics, and by extension, functions that are solutions defined over the surface of a sphere are called spherical harmonics. Part of the angular problem can be formulated as the Legendre equation and hence part of the solution is conveniently represented by the associated Legendre polynomials.

Pierre-Simon de Laplace's treatment of the Newtonian gravitational potential and Adrien-Marie Legendre's discovery of the polynomials (which turned out to be a special case of the spherical harmonics, also investigated in relation to expansion of the Newtonian potential) both date to around 1782. The spherical harmonics, a more general form of the harmonics introduced by Laplace and Legendre, were described by Lord Kelvin and Peter Tait in their Treatise on Natural Philosophy in 1867 as solutions to the Laplace equation.

The Laplace equation, the simplest example of an elliptic partial differential equation, is

$$\nabla^2 \psi = 0$$

where ψ can be a scalar potential and the Laplacian is the sum of the spatial partial derivatives

$$\nabla^2 \psi = \frac{\partial^2 \psi}{\partial x^2} + \frac{\partial^2 \psi}{\partial y^2} + \frac{\partial^2 \psi}{\partial z^2}$$

The solutions to this type of equation are found in potential theory, of which there are numerous examples in classical physics areas such as astronomy, electromagnetism, fluid dynamics etc. The applications also extend to quantum theory of atomic orbitals. Spherical coordinates are shown below where r is the radial coordinate, θ the polar (colatitudinal) coordinate (ranging 0 to π) and ϕ the azimuthal (longitudinal) coordinate (ranging 0 to 2π).

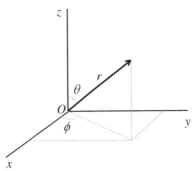

The spatial variables are transformed as

$$x = r \sin\theta \cos\phi; \quad y = r \sin\theta \sin\phi; \quad z = r \cos\theta$$

The Laplacian needs to be expanded into a form dependent on the variables r, θ, ϕ, for example starting with

$$\frac{\partial \psi}{\partial x} = \frac{\partial \psi}{\partial r} \cdot \frac{\partial r}{\partial x} + \frac{\partial \psi}{\partial \theta} \cdot \frac{\partial \theta}{\partial x} + \frac{\partial \psi}{\partial \phi} \cdot \frac{\partial \phi}{\partial x}$$

using the differentiation chain rule. This is quite a lengthy process and requires evaluating all the required partial derivatives $\frac{\partial r}{\partial x}, \frac{\partial r}{\partial y}, \frac{\partial \theta}{\partial x}, \ldots$etc.

The second partial derivatives are constructed from these which, after cancelling many terms, results in the following expression for the Laplacian in spherical coordinates

$$\nabla^2 \psi = \frac{1}{r^2} \frac{\partial}{\partial r} \left(r^2 \frac{\partial \psi}{\partial r} \right) + \frac{1}{r^2 \sin \theta} \frac{\partial}{\partial \theta} \left(\sin \theta \frac{\partial \psi}{\partial \theta} \right) + \frac{1}{r^2 \sin^2 \theta} \frac{\partial^2 \psi}{\partial \phi^2}$$

The next step uses separation of variables to split the differential equation into radial and angular equations

$$\psi(r, \theta, \phi) = R(r) Y(\theta, \phi)$$

This type of "product solution" is known to exist provided that the differential equation is linear and homogeneous (and so are its boundary conditions), which is typically the case. The above results in a radial equation in R and angular equation in Y, whose solutions can be multiplied together at the end to arrive at the complete solution. After cancelling out $\frac{1}{r^2}$, substitution for ψ in the Laplacian and dividing by RY to separate the functions gives two expressions equated to eigenvalues which are equal and opposite constants $\lambda, -\lambda$ (since the Laplacian = 0)

$$\frac{1}{R} \frac{d}{dr} \left(r^2 \frac{dR}{dr} \right) = \lambda; \quad \frac{1}{Y} \frac{1}{\sin \theta} \frac{\partial}{\partial \theta} \left(\sin \theta \frac{\partial Y}{\partial \theta} \right) + \frac{1}{Y} \frac{1}{\sin^2 \theta} \frac{\partial^2 Y}{\partial \phi^2} = -\lambda$$

Now that the radial component is removed as an ordinary differential equation, the angular component (spherical harmonic) can be further separated into polar and azimuthal components as

$$Y(\theta, \phi) = \Theta(\theta) \Phi(\phi)$$

with similar processing and multiplying through by $\sin^2 \theta$ resulting in

$$\frac{1}{\Phi} \frac{d^2 \Phi}{d\phi^2} = -m^2; \quad \lambda \sin^2 \theta + \frac{\sin \theta}{\Theta} \frac{d}{d\theta} \left(\sin \theta \frac{d\Theta}{d\theta} \right) = m^2$$

In this variable separation the constant m^2 is expressed as a squared integer assuming a simple periodic solution in ϕ of the form $\Phi = e^{\pm im\phi}$, and it is convenient to represent λ as $l(l+1)$ where l is an integer such that $l \geq |m|$. This is required so that $Y(\theta, \phi)$ is a regular function at the poles and enables the polar equation to take the form of the Legendre equation in the variable $\cos \theta$. The Legendre equation has the general form

$$\frac{d}{dx} \left[(1 - x^2) \frac{d}{dx} P_l^m(x) \right] + \left[l(l+1) - \frac{m^2}{1 - x^2} \right] P_l^m(x) = 0$$

where $P_l^m(x)$ are the associated Legendre polynomials of degree l and order m, a special case of Legendre functions which have many useful properties, including orthogonality and symmetry.

The resulting spherical harmonics are

$$Y_l^m(\theta, \phi) = \sqrt{\frac{(2l+1)}{4\pi} \frac{(l-m)!}{(l+m)!}} P_l^m(\cos \theta) e^{im\phi}$$

where the square-root term is a normalization factor. The spherical harmonics may be defined in slightly different ways (normalizations) depending on their application, such as in seismology, electromagnetics or quantum mechanics, and depending on the exact definition of the associated Legendre polynomials.

227

The first few examples are:

$$Y_0^0(\theta, \phi) = \frac{1}{2}\frac{1}{\sqrt{\pi}}; \quad Y_1^0(\theta, \phi) = \frac{1}{2}\sqrt{\frac{3}{\pi}}\cos\theta; \quad Y_1^1(\theta, \phi) = -\frac{1}{2}\sqrt{\frac{3}{2\pi}}\sin\theta\, e^{i\phi}; \quad \dots$$

Please refer to other texts such as G. B. Arfken: Mathematical Methods for Physicists (6th ed, 2005) for a more detailed analysis and description of the Legendre functions and their properties and applications, and a full treatment of the use of spherical harmonics in, for example, describing the eigenfunctions that represent atomic orbitals in quantum theory. Another useful text for this is A. M. Rae: Quantum Mechanics (4th ed. 2002).

PB

Equation 199: The Carothers Equations

American chemist, William Hume Carothers, was working with Paul John Flory at DuPont in 1930, when he and his colleague developed a series of equations that radically revolutionised the polymer chemistry and the world in general. These equations, known as Carothers equations, are the building blocks of a reaction model, the *step-growth polymerisation*, that designed the synthesis of a series of important polymers including, polyesters, polyamides and polyurethanes. This polymerisation process is also the first chemical reaction to be predicted by a scientific theory. Carothers and Flory designed the reaction with the purpose of creating polyesters with high molecular weight, a large step beyond the production of the first polymeric material, Bakelite, in 1907.

The basic Carothers equation, which is valid for linear polymers formed with two monomers in equimolar quantities, is:

$$\bar{X}_n = \frac{1}{1-p}$$

Where $\bar{X}_n$ is the *degree of polymerisation*, and p is the *fractional monomer conversion*, also indicating the extent of the step-growth polymerisation reaction.

The fractional monomer conversion is expressed as:

$$p = \frac{N_0 - N}{N_0}$$

Where, N_0 is the number of monomers at the beginning of the reaction, and N is the number of remaining monomers at a certain time during which the polymerisation reaction proceeded. It follows that $0 < p < 1$.

A high degree of polymerisation (high $\bar{X}_n$) corresponds to the consumption of a large number of monomers at a certain time, bonding with each other to form a long linear chain. This process implies that N is very small, and p very close to 1. For example, a fractional conversion of 98% ($p = 0.98$) corresponds to $\bar{X}_n = 50$, while a conversion of 99%, corresponds to $\bar{X}_n = 100$.

The equation for a linear polymer, with one of the two monomers in excess is:

$$\bar{X}_n = \frac{1+r}{1+r-2rp}$$

228

Where r is the *stoichiometry ratio* of the two monomers, with the concentration of the monomer in excess as denominator ($r \leq 1$). The effect of the excess reactant is to reduce the degree of polymerisation. In fact, in the limit of complete conversion of the monomers in the polymeric chain ($p \to 1$), $\bar{X}_n \to \infty$ in the first equation, while $\bar{X}_n \to \frac{1+r}{1-r}$ in the second equation. In the case of a monomer in excess of 1% ($r = 0.99$), $\bar{X}_n \to 199$.

DEGREE OF POLYMERISATION

The *degree of polymerisation* $\bar{X}_n$ is a very important factor to control in polymer science, as it is related to the ration of the number-average molecular weight of the polymer M_n and the weight of the molecular weight of the monomer unit M_0. A large value of $\bar{X}_n$ means a high M_n. The latter determines the physical properties of the polymer, in particular its melting temperature and its mechanical properties. Higher $\bar{X}_n$ correlates with higher melting temperature and mechanical strength. For most industrial applications, $\bar{X}_n$ values of thousands or tens of thousands are desired. Molecular weights higher or lower than the desired weight are equally undesirable. Polymer with identical chemical composition, but different in total molecular weights may exhibit different physical properties.

STEP-GROWTH POLYMERISATION

The Carothers equations are derived from the kinetics of the *step-growth polymerisation* model, in which $\bar{X}_n$ is a specific function of the reaction time. This is a type of polymerisation mechanism, in which monomer units bind together (through functional groups) to form first *dimers*, then *trimers*, hence longer *oligomers* and eventually long chain polymers (Figure 1).

LINEAR POLYMERS

The units may consist of two identical monomers, like in the case of polyethylene (PE), in which the monomer is an ethylene molecule) or two different monomers like in the case of polyesters, such as polyethylene terephthalate (PET - monomers: ethylene glycol and terephthalic acid) or polyamides, such as Nylon 6,6 – monomers: hexamethylenediamne and adipic acid), as illustrated in Figure 2. Nylon was synthesised by Carothers in 1935 at DuPont in the US; PET was patented in 1941 by John R. Whinfield and James T. Dickson at Calico Printers' Association in the UK.

The step-growth polymerisation mechanism itself presents different types of chemical reactions. The two typical step-growth chemical reactions are:
addition polymerisation, which producers only one polymer
condensation polymerisation, which produces a polymer and a molecule with low molecular weight (H_2O, NaCl, etc..) as sub-product.
An alternative mechanism to step-growth polymerisation is called *chain-growth polymerisation*, which obeys different kinetics laws.

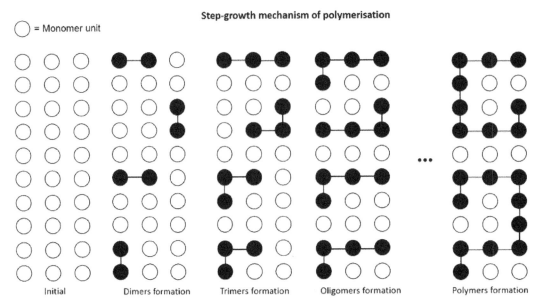

Figure 1. Schematic illustration of Step-growth mechanism of polymerisation.

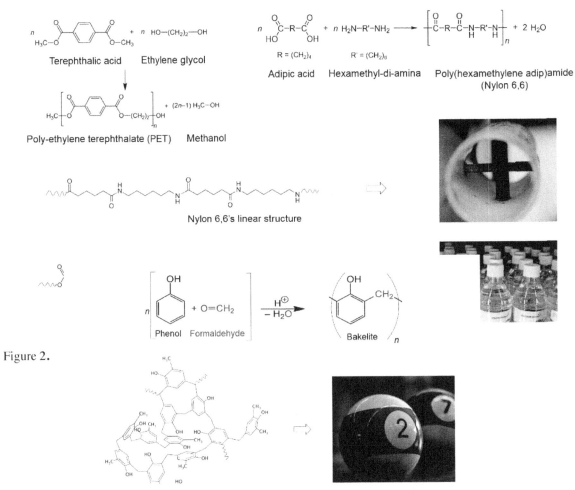

Figure 2.

Figure 3. Branched polymers: Bakelite (Sources: Wikipedia, Flickr Commons).

The polymerisation of monomers with 2 lateral functional groups will always result in a linear polymer. However, the polymerisation of monomers with 3 or more functional groups can produce branched polymers, which form cross-inked macrostructure or networks, with completely different chemical and physical properties compared with linear polymers. Bakelite is an example of branched polymer, which is made from phenol and formaldehyde as monomers (Figure 3).

The *modified Carothers equation* for branched polymers is:

$$\bar{X}_n = \frac{2}{2 - pf_{av}}$$

Where $p = \frac{2(N_0 - N)}{N_0 f_{av}}$ is the fractional monomer conversion, and

$$f_{av} = \frac{\sum N_i \cdot f_i}{\sum N_i}$$

Is the *average functionality* per monomer unit. The *functionality* f_i of a monomer is the number of functional groups of the monomer participating in the polymerisation.

AT

Equation 200: Legendre Polynomials

Legendre polynomials $P_n(x)$ are polynomials that have fixed combinations of coefficients in such a way that they are orthogonal over the interval [-1:1] (or any other interval by scaling). It is common to normalize them so that $P_n(1) = 1$.

The first few polynomials are:

$$P_0(x) = 1, \qquad P_1(x) = x, \qquad P_2(x) = \frac{(3x^2 - 1)}{2}, \qquad P_3(x) = \frac{(5x^3 - 3x)}{2}$$

Once two previous polynomials are known, the next one can be derived from

$$(n + 1)P_{n+1}(x) = (2n + 1)xP_n(x) - nP_{n-1}(x)$$

The first 10 polynomials are depicted in the figure below:

231

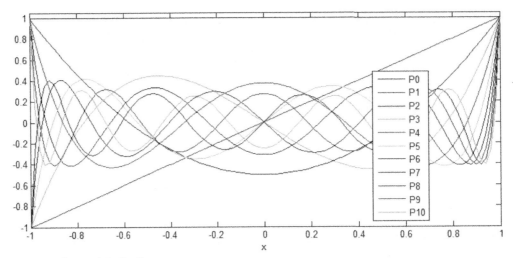

Figure: Legendre Polynomials $P_0..P_{10}$

So what is beautiful about these polynomials ? In the first place any function can be described as a linear combination of these polynomials:

$$f(x) = \sum_{n=0}^{\infty} a_n P_n(x)$$

That on its own is maybe not yet so special/beautiful; the real beauty is that this is combined with the independence of the polynomials, that is also called their orthogonality:

$$\int_{-1}^{1} P_m(x) \cdot P_n(x) \, dx = \frac{\delta_{nm}}{n + 0.5}$$

with $\delta_{nm} = 1$ for $n = m$ and $\delta_{nm} = 0$ otherwise.

An example is that all coefficients remain the same if a least-squares line P_1 is removed from a function, something that cannot be said of the Fourier analysis, although it is common practice to remove a least-squares line before doing a Fourier analysis.

This is a difference with the Fourier analysis, where there are many aspects in common: the independence of coefficients and completeness of the description. The relatively scarcer use in numerical analysis relative to the Fourier analysis is in the numerical instability of very high powers, where the Fourier analysis can well deal with rapidly oscillating signals. On the other hand the Legendre polynomials have little problems with wavelengths that do not precisely fit in the interval.

In physics, Legendre polynomials are widely used, e.g. in multipole expansions and in solutions of Laplace equations.

Adrien-Marie Legendre (1752 – 1833) was a French mathematician who made numerous contributions to mathematics.

HH

232

Equation 201: Magnetic Flux

Magnetic flux Φ_B is a quantity used to describe the magnetic field through a given surface. The general equation for magnetic flux is the following.

$$\Phi_B = BA$$

where B is the magnetic field magnitude and A is the surface area through which the magnetic flux is calculated. In the equation's general form, there are two assumptions being made. The first is that the magnitude of the magnetic field is constant throughout the surface. The second is that the surface is always normal to the direction of the magnetic field. In practice, these assumptions are not satisfied and the components of the magnetic flux equation must be accordingly modified.

The calculation of magnetic flux is very useful in the field of power generation. In particular, knowing the magnetic flux through a coil allows for the estimation of the electrical current generated through electromotive force.

In 1831, Michael Faraday first introduced the concept of electromagnetic induction, in which a changing magnetic field generates an encircling electric field [1]. This relationship between magnetism and electricity is pivotal to power generation, both early and modern.

[1] Magnetic Field – Wikipedia, The Free Encyclopedia, https://en.wikipedia.org/wiki/Magnetic_field

MF

Equation 202: The Cayley-Hamilton Theorem

In Beautiful Equation 172 (the Generalized Eigenvalue Problem), it was mentioned that the eigenvalues of an $n \times n$ matrix K with respect to another $n \times n$ matrix M are obtained by solving the equation

$$\det(K - \lambda M) = 0$$

which is an nth-order polynomial in λ.

For the special case when $M = I$ (the identity matrix), the λ values obtained by solving the above equation are simply called the eigenvalues of A and the nth-order polynomial is called the characteristic polynomial of A. Thus, the eigenvalues of A satisfy the equation

$$\det(K - \lambda I) = 0.$$

In terms of λ, the above can be written as the following polynomial $p(\lambda)$:

$$p(\lambda) = \lambda^n + c_{n-1}\lambda^{n-1} + c_{n-2}\lambda^{n-2} + \cdots + c_1\lambda + c_0 = 0.$$

Here the coefficients c_i involve various invariants of the matrix K.

The Cayley-Hamilton theorem states that every square matrix K satisfies its own characteristic equation, i.e., the matrix K satisfies the equation

$$p(A) = A^n + c_{n-1}A^{n-1} + c_{n-2}A^{n-2} + \cdots + c_1A + c_0I = 0$$

233

Note that the zero on the right-hand side is the zero square matrix of size n. This, in this author's opinion, is one of the most beautiful equations in the entire field of Matrix Computations. The theorem is named after Arthur Cayley (a British lawyer and mathematician) and William Hamilton (an Irish astronomer and mathematician). Hamilton, in 1853, proved the theorem for a special case in the context of quaternions. However, it was Cayley who stated it more formally in 1858.

As an example, consider the 3×3 matrix:

$$K = \begin{bmatrix} 3 & 5 & 6 \\ 5 & 5 & 4 \\ 6 & 4 & 7 \end{bmatrix}.$$

The characteristic equation of this matrix is the third-order polynomial in λ given by

$$\lambda^3 - 15\lambda^2 - 6\lambda + 58 = 0.$$

The Cayley-Hamilon theorem states that the matrix K satisfies the equation

$$K^3 - 15K^2 - 6K + 58I = 0.$$

The above can easily be verified to be equal to zero. This equation can also be used to illustrate an interesting application of the Cayley-Hamilton theorem in finding the inverse of a non-singular matrix K^{-1}. Multiplying throughout by K^{-1} and solving for K^{-1}, we find

$$K^{-1} = -\frac{1}{58}(K^2 - 15K - 6I).$$

The above also allows us to find any negative integral power of K. Other applications of the theorem include reducing the order of a polynomial in A and the calculation of the matrix exponential.

HC

Equation 203: The Ideal Rocket Equation

The fundamental principle of rocket science is that a craft that expels some of its mass in one direction at high speed will be projected in the opposite direction (as required both by conservation of momentum or, equivalently, by Newton's Third Law of Motion). The rocket equation relates the velocity and amount of the expelled mass to the change in velocity of the craft.

$$\Delta V = V_e \ln \frac{M_0}{M_1}$$

M_0 is the initial total mass = rocket mass + fuel mass, M_1 is the final total mass = rocket mass, V_e is the exhaust velocity, ΔV ("delta vee") is the change of velocity of the craft.

The equation is "ideal" in the sense that it is valid only in the absence of applied forces. For a rocket rising from the Earth, aerodynamic drag is a very significant such force. Nevertheless, the rocket equation is the most important of all equations related to spaceflight.

In order to entirely overcome the gravitational attraction of a planet of other body, a speed called the escape velocity must be achieved. However "entirely overcome" means that the craft will continue endlessly through space (barring accidents!) without ever falling back to its origin. This is not a necessary condition for reaching nearby destinations such as space stations or the Moon. In fact, no human has ever travelled at the Earth's escape velocity: Moon travellers need not escape the Earth's gravity, since the Moon itself has not.

So, in practice, the velocity required for a space journey depends both on the start and the end-point. However, this velocity is not necessarily the same as the Delta-V of the rocket equation, due to the fact that, if a rocket is falling into a gravitational field, any thrust that it exerts in its direction of fall is converted. (This is due to the fact that work is the product of force and distance; the force on the rocket depends on its exhaust velocity and the distance depends on its speed - so, the faster the rocket is moving, the more work a given exhaust velocity will do on it). Most actual space journeys exploit this so-called Oberth Effect to increase efficiency. Figure 2 gives some examples of Delta-V values (in kilometres per second) required for different journeys, including the Oberth Effect where applicable.

Earth	Low Earth Orbit	9.7
Low Earth Orbit	Moon	5.93
Moon	Low Earth Orbit	2.74
Low Earth Orbit	Mars transfer orbit	4.3
Mars transfer orbit	Low Mars orbit	2.7
Low Mars orbit	Mars	4.1

This velocity is the Delta-V of the rocket equation, and values for selected journeys are in Figure 2.

The ideal rocket equation was developed by William Moore, a British engineer, and published in 1813 in his *Treatise on the Motion of Rockets*. Moore also made contributions to the mathematical theory of gunnery and to calculus but, ironically, most of his other work was destroyed by a V2 rocket attack on London in the Second World War. The ideal rocket equation was rediscovered by Russian engineer Konstantin Tsiolkovsky and often bears his name.

Further reading
Understanding Space, an Introduction to Astronautics (3rd edition 2006), Jerry Jon Sellers, William J. Astore, Robert B. Giffen and, Wiley J Larson.

MJG

Equation 204: Homogenous Transformation Matrix (Htm)

Physical quantities like displacement, velocity and acceleration are widely used in daily life as well as scientific community to describe the motion of objects. The concept of coordinate system and the coordinate values of an object is introduced by French philosopher and mathematician Rene Descartes (1596-1650) in the 17th century. Named after him, the Cartesian coordinate system is used to specify a geometrical point (with no volume and mass) with a unique pair of numbers (x, y, z) called the coordinates of a point. It is the birth of analytic geometry providing a mathematical framework to explore problems and solutions for the manipulation of geometrical objects.

The position and orientation of an object can be determined using any number of different coordinate systems. The idea of the coordinate frame is essential. It is the general mathematical structure of describing

235

targets of interest and is not limited to the representation of static objects like a point, vector, plane or sphere. It can also be used for describing kinematic motions of further coordinate frames like translations and rotations. When displaying and manipulating parameters that represent these motions a matrix formalization is preferred. The difference between transforming an element and transforming a coordinate system should be kept in mind in applications.

In three dimensional space, a common Homogeneous Transformation Matrix (HTM) is a four by four matrix in which both the translation and the rotation can be prescribed. The homogeneous transformation belongs to a more general classification in geometry, called the affine transformation (beautiful equation 182).

It is a widely adopted method to describe the location of a target in a positioning system, where multiple coordinate systems are attributed to various mechanical components like sliders, axes, rotary table etc. The following equation in green text box demonstrate that the coordinate of a point in a new coordinate system is calculated by multiplying its coordinates (in an old coordinate system) to a HTM. In comparison the HTM based method uses multiplication for both rotation and translation while the vector method (in red text box) has to use summation to describe translation.

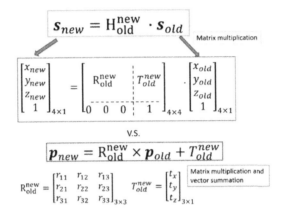

where s_{new} is the homogeneous coordinates of the point in the new coordinate system augmented from the 3D p_{new} position vector of $(x_{new}, y_{new}, z_{new})$. s_{old} is the homogeneous coordinates of the point in the old coordinate system augmented from the 3D p_{old} position vector of $(x_{old}, y_{old}, z_{old})$. H_{old}^{new} is the 4-by-4 homogeneous transform matrix. R_{old}^{new} is the 3-by-3 matrix describing the rotation of the new coordinate system with respect to the old coordinate system. T_{old}^{new} is the 3-by-1 vector describing the linear translation of the origin of the old coordinate system with respect to that of the new coordinate system.

KN

Equation 205: Bernoulli Probability Distribution

$$f(k; n, p) = \Pr(X = k) = \binom{n}{k} p^k (1 - p)^{n-k}$$

where p is the probability of success, n is the number of repeated trials, k is the number of successes and $\binom{n}{k}$ is the binomial coefficient whose factorial formula is hereby reported

$$\binom{n}{k} = \frac{n!}{k!(n-k)!}$$

for

$$0 \le k \le n.$$

The Binomial distribution is a discrete probability distribution useful to describe repeated success/failure phenomenon or experiment, also called a Bernoulli process. A single success/failure experiment can be described by a random variable which has success with probability p, and fails with probability 1-p (also called the Bernoulli distribution). For a value of $p = 0.5$ it is the outcome of the fair coin toss.

The binomial distribution describes the probability of having k successes of a repeated success/failure phenomenon n times. The factor $\binom{n}{k}$ is the binomial coefficient (equals to $\frac{n!}{k!(n-k)!}$) and takes into account that in n repeated experiments, k successes can happen in any sequence.

The distribution was explained to me in those terms: let's imagine a drunk sailor walking down the pier, each step forward is also a bit wonky towards right or left with a probability of 0.5. What is the probability of reaching his boat without falling into the water?

The distribution was described first by Jacob Bernoulli, in 1685 just before he became professor at the Basel university in 1687.

GM

Equation 206: Exhaust Tuning

An exhaust on a car is just a pipe that collects the waste gases from the cylinders and dumps them in to the atmosphere isn't it. It's just a pipe of flowing gases isn't it?

Well not quite. Firstly, we want the gas pressure in the exhaust to be lower than that at the cylinder head to assist scavenging (removing gases from the cylinders as quickly as possible) through gas inertia. Secondly we don't want the exhaust gas of one cylinder to pressurise another cylinder. If we ignore the silencer the lengths of the primaries, secondary's and collector affect the pulse tuning of the exhaust system. So how do we calculate the optimum length and diameters of the primary and secondary components?

If we alter the length and bore of the primaries and secondary's we can ensure that the negative pressure component of each exhaust pulse reaches the cylinder head when the exhaust valve is open, thereby further assisting cylinder scavenging. All this will depend upon engine speed and the valve opening time, *i.e.* exhaust valve duration. To give an example, on a four-cylinder engine, we can use the negative pressure pulse from number one cylinder to assist the exhaust scavenging of number four cylinder. Cylinders that are 180 degrees apart are paired. This assumes a wide valve overlap, *i.e.* both exhaust and inlet valves are open at the same time (hence they 'overlap').

In addition, not only does the negative pulse from number one cylinder assist the scavenging of number four cylinder, but, because of the negative pressure and the fact that number four cylinder's exhaust and inlet valves are both open, this negative pulse will actually assist in sucking the new inlet charge into the cylinder. Hence, gains in power and torque.

As we have said this will only work perfectly at a given rpm. Tuning an exhaust for maximum power will inevitably reduce the torque lower down and 'close up' the engine's 'power band'. Ever wondered why race engines idle badly with associated popping and farting and lumpy idle rpm.

Similarly, engines with a wide torque spread produce less peak brake horse power. Compromises have to be made.

So tuning an exhaust is all about getting the negative (and, hence, scavenging) pressure pulse to arrive at the exhaust valve as it is opening. To do this we have to set the pipe lengths and diameters correctly.

First equation, the Primary pipe length is

$$P = \frac{850ED}{RPM} - 3$$

where *RPM* (revolutions per minute) is the engine speed to which the exhaust is being tuned. *ED* is 180 degrees plus the number of degrees the exhaust valve opens before bottom dead centre. *P* is the Primary pipe length (on a 4-1 manifold), or Primary pipe length plus Secondary pipe length (on a 4-2-1 manifold), in inches. The numbers refer to the exhaust pipe layout.

For a road going vehicle the manifold will be tuned to the rpm that gives max torque whereas race engines will be tuned to work either at the rpm giving max bhp (brake horse power) or a speed midway between the max bhp rpm and max torque rpm.

Note that 4-1 manifolds (all four exhaust pipes merge in to one) restrict the power band, whereas 4-2-1 manifolds (four pipes that merge to two pipes that merge to one pipe) give better mid-range power but sacrifice top end power by as much as 5-7%.

Making the assumption that with a 4-2-1 manifold the starting point for Primary pipe length (*P*) is 15 inches, the Secondary pipe length is *P* - 15 inches. Changing the length of the Primary pipe tends to rock the power curve around the point of maximum torque. Shorter Primaries gives more top end power but less mid-range, and *vice-versa*. There is, however, little change in the peak torque or the rpm where this occurs.

Ideally, the Primary pipes should come off the cylinder head in a straight line for about four inches before any turns occur.

The equation for the inside diameter of the pipe given the cylinder volume in cc and the primary length in inches is,

$$ID = 2.1\sqrt{\frac{V}{25(P + 3)}}$$

where *V* is the cylinder volume in cc and *P* is the Primary pipe length in inches.

For a 4-2-1 system then, Primary pipe diameter is calculated as above. Secondary pipe diameter is

$$ID_s = 0.93\sqrt{ID^2}$$

where *ID* is the calculated inside diameter of the primary pipes.

A change in pipe diameter can be used to change the peak torque rpm and a reduction in pipe diameter of 0.125 inches will drop the peak torque rpm by 500-600 rpm in engines over 2 litres and by 650-800 rpm in smaller engines. Increasing the pipe diameter by 0.125 rpm has approximately the opposite effect.

The total length of the collector and tailpipe (measured to the front of the silencer) should be equal to *P* + 3 inches (or any full multiple of *P* + 3 for a road car).

Tailpipe internal diameter is given by:

$$ID_T = 2\sqrt{\frac{2V}{25(P+3)}}$$

where P is calculated as above.

Collector length is given by

$$CL = \frac{ID_2 + ID_3}{2}\cot A$$

where ID_2 is the diameter of Collector inlet, ID_3 is the diameter of Collector outlet and $\cot A$ is the cotangent of angle of Collector taper.

This article is based on the thread posted here in 2008. http://sideways-technologies.co.uk/.../1395-exhaust-tuning-t.../

Further reading PH Smith's "Scientific Design of Exhaust & Intake Systems".

DF

Equation 207: The Gibbs-Appell Equation

$$\frac{\partial \mathfrak{I}}{\partial \ddot{q}_r} = Q_r$$

$$r = 1,2,3,\dots k$$

$$\dot{q}_r = \sum_{s=1}^{k} \beta_{rs}\dot{q}_s + \beta_r$$

$$r = k+1, k+2, \dots, n$$

where k is the number of freedoms of the system best considered as the number of independent coordinates needed to locate all components of the system. l are the number of constraints that need to be added in the case of non-holonomic systems (examples of non-holonomic systems being the bicycle or a ball rolling on a surface). p are the number of quasi coordinates that are chosen to more conveniently represent system coordinates in differential form. $n = k + l + p$ is the total number of coordinates necessary to describe motion of all components of our system. $\mathfrak{I}$ is the Gibbs function expressible as a sort of 'kinetic energy of accelerations' and equal to $\frac{1}{2}\sum_{r=1}^{3N} m_r \ddot{x}_r^2$ for a system of N point masses with the x coordinates representing the three components of a Cartesian coordinate system for each of the masses. β_{rs} represents geometric constraints of the system.

Early in the development of mechanics, using the principle of virtual work, D'Alembert was able to reduce Newton's laws of motion for rigid particles to a single equation. Later Lagrange would expand this to enable systems, particularly holonomic systems, to be specified in terms of any coordinate system again with all of this mathematics expressible as a single equation. In many ways the Euler-Lagrange equation was considered a pinnacle in dynamics, particularly if Rayleigh's dissipation function is included. It was therefore not surprising that a new formulation presented by the brilliant American scientist J. Willard Gibbs (1839 – 1903) in 1879 was mostly ignored. One possible reason is the terse nature of his presentation of this new approach to dynamics and an absence of discussion relating to the limitations of the new formulation.

One reason for the absence, other than the mathematical barriers to implementation, is that there really aren't any limitations! Another reason for the lack of recognition might have been that his other great achievements in thermodynamics (the Gibbs free energy), vector algebra, optics, and signal processing (Gibbs phenomena) might have been more readily appreciated for their direct application. Even in today's historical studies, this contribution commands only a brief mention. Not being aware of the work of Gibbs, twenty years later the French Mathematician Paul Appell (1855 – 1930) published the first edition of his treatise on 'Rational Mechanics' in 1896 deriving the same equation. The problem with this derivation was that it used an incorrect formulation of Lagrange's equation in its application to non-holonomic systems. Realizing this mistake, he further worked on this and presented a complete set of equations that would apply to holonomic and non-holonomic systems and readily admits the use of 'quasi-coordinates'. Quasi-coordinates are often necessary to incorporate constraints such as rolling or sliding on a surface. For those of a mathematical inclination, the Gibbs-Appell equations are rigorously derived and discussed in the treatise of Pars referenced herein.

So, how does this equation work? First it is necessary to identify the motion constraints of any particular system. This is a point in the formulation of the problem where a little mathematical and mechanical skill is required. The outcome of this first step is the identification of the degrees of freedom of the system as well as constraint equations and any convenient quasi-coordinates that incorporate constraints that cannot be expressed as exact differentials. Take, for example, the two degree of freedom system (i.e. $k = 2$) represented by motion of a particle of mass m that can slide freely (i.e. no friction) but is constrained to move on a plane surface. It would be easy to use any two coordinates (such as polar coordinates r, θ or linear coordinates x, y) to determine the position on the surface. However, in this case a simple radial coordinate r plus a quasi-coordinate dq representing half of the area swept out by the moving particle will be chosen. Notice that this quasi-coordinate can only be expressed in differential form. Steps in the solution of this problem are shown in Table 1. Typically based on mathematical convenience, any of the coordinates or quasi coordinates can be chosen to represent the k degrees of freedom of the system. Only these chosen coordinates can appear in $\mathfrak{J}$. Once obtained express the 'energy of acceleration' in terms of the accelerations in the chosen coordinates. Because of the partial derivative in the Gibbs-Appell equation any terms not containing accelerations can be 'thrown away', an easy simplification of the formulism. Generalized forces can be obtained in the usual manner by considering the work done in a virtual displacement in the generalized coordinates. Once formulated, the Gibbs-Appell equations are often very simple to solve to determine the equations governing motion of the system. These equations plus the geometric equations of the non-holonomic system provide the n differential equations to solve for the dynamics of any system.

References

Appell P., 1899, Sur les mouvements de roulement; equations analogues à celles de Lagrange, *Comptes rendus*, 129 , 317-320

Appell P., 1899, Sur une forme générale des équations de la dynamique, *Comptes rendus*, 129, 423-427 & 459-460.

Appell P., 1904, *Mecanique rationale*, second edition (note; the first edition in 1896 contained a serious error in the derivation of the Lagrangian used in the Euler-Lagrange equation. A third edition came out in 1911)

Gibbs J.W., 1879, On the fundamental formula of dynamics, *Amer. J. Math.*, **II**, 49 – 64. See also *The scientific papers of J. Willard Gibbs*, Dover Publications, volume II, 1 – 16.

Pars L.A., 1965, *A treatise on analytical dynamics*, J. Wiley, chapters XII & XIII; -reprinted Oxbow Press, 1979, ISBN 0-918024-07-2. There is a concise discussion of the development of these equations on pages 631 – 632. Chapter XIII is a series of applications of the Gibbs-Appell equations mainly for deriving the equations governing motion of non-holonomic systems

Table 1: Example of steps in solving dynamical equations using the Gibbs-Appell equation for a system comprising motion of a particle in a plane (adapted from Pars, 1964)

Steps in a solution	Equations
Select coordinates	$r, dq = x\,dy - y\,dx$
Express the acceleration energy in terms of coordinates	$\ddot{x}^2 + \ddot{y}^2 = \dfrac{1}{r^2}\left(\ddot{q}^2 + \left(r\ddot{r} - \dfrac{\dot{q}^2}{r^2}\right)^2\right)$
'Throw away' terms not involving accelerations to determine $\mathfrak{J}$	$\mathfrak{J} = \dfrac{1}{2}m\left(\ddot{r}^2 - \dfrac{2}{r^3}\dot{q}^2\ddot{r} + \dfrac{1}{r^2}\ddot{q}^2\right)$
Given radial and tangential forces, F_r, F_θ determine virtual work δW	$\delta W = F_r\delta r + F_e\delta\theta = F_r\delta r + \dfrac{F_e}{r}\delta q$
Solve to give the Gibbs-Appell equations of motion	$m\left(\ddot{r} - \dfrac{\dot{q}^2}{r^3}\right) = R_r,\ m\ddot{q} = rF_\theta$

SS

Equation 208: Interfacial Area Ratio

The Sdr areal surface parameter named the developed interfacial area ratio or the surface area ratio parameter is a useful metric for understanding how much texture or structure that there is on a measured topography.

$$Sdr = (TSA - CSA)/CSA$$

where TSA is the Texture Surface Area and CSA is the Cross Sectional Surface Area.

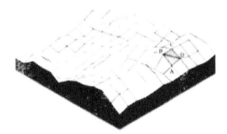

The Sdr parameter takes the three dimensional surface area of a topography (TSA) subtracted by the two dimensional plane (CSA) and divides it by the two dimensional surface area (CSA) of the plane of the measured sample. This gives a ratio of the amount of texture on a surface. Sdr is part of the ISO 25178 surface metrology standards. It is considered a hybrid parameter because it contains both amplitude and spatial characteristics. It can be useful for indicating the nature of the surface for many applications but lends itself to applications where adhesion and lubrication are important. (ISO 25179-2:2012)

Ref: Michigan Metrology gives a very nice overview of the Sdr parameter
http://www.michmet.com/3d_s_hybrid_parameters_sdr.htm

STS

Equation 209: Molar Ellipticity Equation

In nature, there are molecules which are able to rotate linearly polarised light -more precisely the

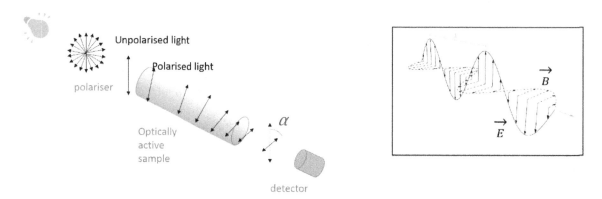

Figure 1. Rotatory effect of an optically active sample on polarised light (Source: Wikipedia).

electromagnetic field of light – by a specific angle α (Figure 1).

Such a type of molecules, and in general a sample (often a solution) composed of these molecules, is said *optically active*. Its *optical activity* is measured through a polarometer and it depends on:

9. the concentration (c, in *g/mol*)
10. the path length of the polarised electromagnetic wave (l, measured in *dm*) through the cell containing the sample
11. the temperature (T, measured in °C), and
12. the wavelength (λ in nm) of the electromagnetic wave.

The equation that defines the optical activity is

$$[\alpha]_{T,\lambda} = \alpha \cdot l \cdot c$$

In the case of *circularly polarised light*, another observed phenomenon, known as circular dichroism (CD), occurs and consists in the ability of the optically active sample to modify a circularly polarised light, by selectively absorbing electric field components of the light (Figure 2).

CD also depends on the optical activity of the molecules in the sample. This results in a variation from circular to elliptical polarisation of light. The ability of the medium to cause this transition is called *molar ellipticity* and is expressed as

$$[\theta] = \frac{100 \cdot \theta}{l \cdot c}$$

where l is the path length of the cell containing the sample (in *cm*); c is the concentration of the optically active species in the sample (in *g/L*); and θ is the angle defining the *ellipticity* of the light beam. Another expression for the *molar ellipticity* is

$$[\theta] = 3298.2 \cdot \Delta\varepsilon$$

where $\Delta\epsilon$ is the difference between the molar extinction coefficients of the left and right circularly polarised light, respectively indicated as ϵ_L and ϵ_R. This is expressed as

$$\Delta\varepsilon = \varepsilon_L - \varepsilon_R = \frac{A_L - A_R}{l \cdot c}$$

where A defines the *absorbance* of the circularly polarised light by the sample.

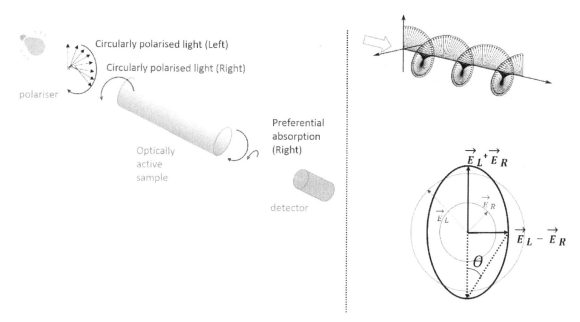

Figure 2. Rotatory effect of an optically active sample on circularly polarised light – *circular dichroism* (Source: Wikipedia).

OPTICAL ISOMERS AND CHIRALITY

Optically active molecules are called optical isomers, or enantiomers. They can be either dextrorotatory enantiomer – indicated with the sign (+), as they rotate a linearly polarised light towards the right – or levorotatory enantiomers – indicated with the sign (-) as they rotate a linearly polarised light towards the left. In the case of circularly polarised light, an enantiomer (+) will preferentially absorb E_R while an enantiomer (-) will absorb E_L component of the electric field of the wave.

Molecular enantiomers are also called *chiral molecules*. These molecules are characterised by having the same chemical elements, but with different spatial distribution around a central atom. In particular, a chiral molecules and its mirror image cannot superimpose (Figure 3, left). This property is also valid for objects such as the human hand and circularly polarised light. Chiral objects do not have any symmetry axis. In the case of molecules, an atom bonded to four different chemical groups is called *chiral centre*. Even different isotopes of an element in the same molecules can determine its *chirality* (Figure 3, right). Enantiomers have very similar physical properties, including melting and boiling points and they behave identically to each other – that is to say, they have identical NMR and IR spectra and identical migration in thin-layer chromatography. However, they interact in different ways with other chiral molecules or objects such as circularly polarised light or chiral chromatographic media (for example, quartz).

Some enantiomers of organic molecules have completely different effects, due to different interactions with chiral molecules in proteins. For example, (+) ethambutol treats tuberculosis, but (-) ethambutol causes blindness. Enzymes, molecules which play a crucial role in regulating the physiological activity of living organisms, are also chiral molecules. Interestingly, most amino acids are L-form, while most sugars are D-form on Earth. The reason is still unknown.

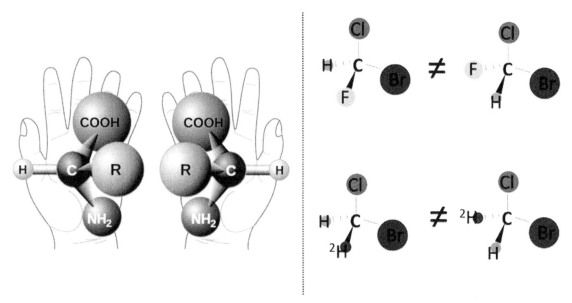

Figure 3. Examples of chiral molecules. An atom with four different substituting groups (including isotopes) is a chiral centre (Source: Wikipedia).

Optical activity was first observed by the French physicist Jean-Baptiste Biot, as well as Augustin Fresnel and Aime Cotton, in the first half of the 19th century. Louis Pasteur, who supported Biot's work by observing the rotation of polarised light by crystals of tartaric acid (an acid commonly present in wine), was the first scientist to show the existence of chiral molecules. Pasteur also observed that a mixture of two enantiomers rotated the polarised light either clockwise or anti-clockwise, depending on which enantiomer was in excess. In the case of a 50-50 mixture of the two enantiomers (also called a *racemic mixture*) the light was not rotated as the two polarising effects cancelled each other. The definition of chirality, however, was first introduced by Lord Kelvin in 1884.

APPLICATION OF CD IN BIOLOGY

CD-based spectroscopy is a powerful technique to study the secondary structure of proteins. Biological molecules, including enzymes, amino acids and proteins, have dextrorotatory and levorotatory components, and for this reason they exhibit optical activity. For example, an alpha-helix and a double-helix structure in proteins will have distinctive CD spectral signatures (Figure 4).

Therefore, CD spectroscopy can reveal important features of several biological molecules, such as their geometrical conformations and relative proportion in a solution. This technique is also useful to study metal-protein interactions, which play a fundamental role in a number of biological processes.

Despite giving less specific structural information when compared with X-ray crystallography or NMR spectroscopy, CD spectroscopy is a quick method which does not require a large amount of proteins and extensive data processing. For this reason, this technique is particularly suitable for the survey of a large number of solvents in a wide variety of conditions, including temperature, salinity and pH.

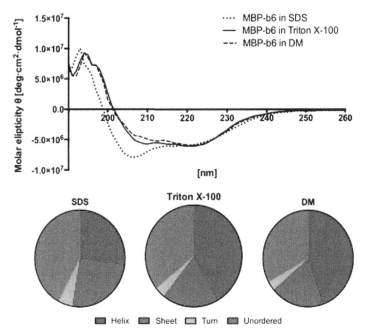

Figure 4. Example of CD spectra of Maltose Bonding Protein (MBP) in different media. Qualification of relative proportion of its configurations in pie charts (Source: Wikipedia).

AT

Equation 210: The Sellmeier Equation

The Sellmeier equation was first proposed in 1871 by Wilhelm Sellmeier and was a further development on Cauchy's equation for modelling dispersion.

The Sellmeier equation gives the relationship between refractive index n and wavelength λ for a particular transparent medium. The equation is used to determine the dispersion of light in the medium. However, it is only valid at room temperature.

The Sellmeier equation and coefficients for silicon is given and is only applicable to the wavelength region where absorption is negligible.

$$n^2(\lambda) = 1 + \frac{B_1\lambda^2}{\lambda^2 - C_1} + \frac{B_2\lambda^2}{\lambda^2 - C_2} + \frac{B_3\lambda^2}{\lambda^2 - C_3}$$

where n is the refractive index, λ is the optical wavelength (given in μm), B_i and C_i are the Sellmeier coefficients given in the table.

Sellmeier coefficients	
B_1	10.66842933
B_2	0.003043475
B_3	1.54133408
C_1	0.3015116485
C_2	1.13475115
C_3	1104

The Sellmeier coefficients are normally obtained through a least-square fitting procedure, applied to refractive indices measured in a wide wavelength range. If we consider a tunable laser, producing light in the wavelength range 1530 nm to 1610 nm incident on a thin silicon sample, there is a change in refractive index of the sample. Plotting the Sellmeier equation over this range demonstrates the wavelength dependant nature of the refractive index. The Sellmeier equation for silicon at wavelength range 1530 nm to 1610 nm and at room temperature is plotted in the figure below

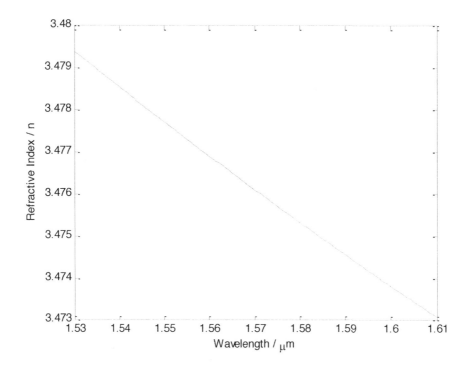

The refractive index of a material is one of the most important properties in any optical system. The refractive index is also a fundamental physical property of a material, and so can be used to determine a material type or determine purity.

B. Taitian, "Fitting refractive index data with the Sellmeier dispersion formula," Appl. Opt., vol. 23, p. 4477, 1984.

CM

Equation 211: Dirac Delta Function

This bizarre but useful function is arguably not a function at all: The Dirac delta $\delta(v - v_0)$ is infinitely tall but at the same time infinitely thin. It is so strange that it really only exists as part of an integral transform:

$$f(v_0) = \int_{-\infty}^{\infty} \delta(v - v_0) f(v) \ dt.$$

$$\delta(v) = \begin{cases} 0 & v \neq 0 \\ "\infty" & v = 0 \end{cases} \qquad 1 = \int_{-\infty}^{\infty} \delta(v)\ dt$$

Infinitely thin
Infinitely tall

Area = 1

Practically, the delta function is a way to extract the value of the function $f(v)$ at a specific value v_0, using a multiplying factor $\delta(v)$ that is only nonzero when $v = v_0$. If we let $f(v) = 1$, then clearly the integral of $\delta(v - v_0)$ equals one, and therefore the area of the delta function equals one.

So why is this useful? Suppose we are working with Fourier Transforms (Beautiful Equation 150), one of the most common integral transforms in engineering. Here the Dirac delta $\delta(v - v_0)$ may be the distribution of frequencies that results in a pure sinusoidal oscillation in time, at a frequency exactly equal to v_0. As an inverse Fourier cosine Transform, this reads as

$$\cos(2\pi v_0 t) = \int_{-\infty}^{\infty} \delta(v - v_0) \cos(2\pi v t)\ dv.$$

The forward Fourier transform then readily provides a definition of sorts for the Dirac delta function itself:

$$\delta(v - v_0) = \int_{-\infty}^{\infty} \cos[2\pi(v - v_0)t]\ dt$$

This last equation shows that the "function" is zero everywhere because of the oscillatory nature of the cosine, except when $v = v_0$, where the integral becomes infinite. And yet, its integral must *by definition* be equal to 1. Weird! Another hint that the delta function is really not quite a function is that it can be defined in entirely different ways, for example, as a limit of a rectangle centred on v_0 of width Δv and height $1/\Delta v$ as $\Delta v \rightarrow 0$. It can also be described as a Gaussian with zero standard deviation! Conceptually, it is a very sharp spike at the frequency v_0. Alternatively, if you have a really short pulse in the time domain, you get a sinusoidal frequency distribution representing a boat-load of equally-spaced frequencies, like in a mode-locked or "comb frequency" laser.

As a mathematical idea, the first appearance of the delta function dates back to the 1878 invention of the Fourier Transform; even though at that time the name and the symbol had not yet been fixed. Paul Dirac introduced the now familiar notation in his 1930 book *Principles of Quantum Mechanics*. The Dirac delta function is a beautiful equation because it is at the same time an intriguing mathematical concept and an extremely useful tool for modelling and analysing real systems.

PdG

Equation 212: The Welch-Satterwaite Equation

$$v_{eff} = \frac{u_c^4(y)}{\sum_{i=1}^{N} \frac{u_i^4(y)}{v_i}}$$

When we calculate measurement uncertainty, we use an equation to find the quadrature sum of the various uncertainty contributions, i.e. we find the square root of the sum of the squares of the various contributions (assuming they are all variances – called standard uncertainties). This process was described in Equation 40 – please refer to this equation before continuing. The result of this calculation is called the combined

247

uncertainty. But, with uncertainty calculation, there is still one step to go. We need to express the final uncertainty as a confidence interval at a given percentage, i.e. we are confident that $X\%$ of the measurement values fall within a stated interval, usually expressed as the measurement result ± the combined uncertainty multiplied by a coverage factor. To find the coverage factor, we need to know two things: 1. The statistical distribution of the combined uncertainty – usually this is a normal distribution, 2. The degrees of freedom of the result. With these two bits of information, it is possible to use standard statistical tables to look up the coverage factor for $X\%$.

But, what are degrees of freedom (dof)? The number of dof is the number of values in the final calculation of a statistic that are free to vary. As an example, when we take N repeated measurements and find the mean of those measurements, the number of dof is $N - 1$. It is not N because we have to take off the number of parameters used as intermediate steps in the estimation of the parameter itself, and in this example, the parameter used as an intermediate step is the mean. If we also had to use, e.g. the standard deviation (this is the case when calculating the degrees of freedom of a linear fit), the there would be $N - 2$ dof.

Now, when we estimate relatively complex combined uncertainties, we often have parameters that are not simple statistics, e.g. we may be importing the value from a calibration certificate or relying on a manufacturer's specification. It is not such a simple matter to calculate the dof with these parameters. There are ways to estimate the dof (which we won't cover here), but we need some way to combine all the information about the dof for each term in the calculation of the combined uncertainty – this is where the Welch-Satterwaite equation comes to our rescue. The effective dof is given by the equation shown, where $U_c(y)$ is the value for the combined uncertainty, $U_i(y)$ is the ith standard uncertainty and V_i is the dof associated with the ith standard uncertainty.

The equation was developed in two papers, the first published by Satterwaite in 1946 and the other by Welch in 1947. Note that there are international attempts ongoing to use a Bayesian approach to uncertainty calculation – this approach would negate the need for dof and today's equation.

RL

Equation 213: Motion of an Electron Accelerated Through a Cylindrical, Electric or Magnetic, Lens Field

For an electrostatic lens

$$r'' = \frac{(1 + r'^2)}{2V(r,z)}\left[\frac{\partial V(r,z)}{\partial r} - r'\frac{\partial V(r,z)}{\partial z}\right]$$

For a magnetic lens

$$r'' = -\frac{r}{8}\frac{q}{mV}B_z^2$$

$$\frac{d\phi}{dz} = B_z\sqrt{\frac{q}{8mV}}$$

where r is the radial distance from the axis of the cylindrical field, the dashes represent differentiation along the axis of the cylindrical field, z is the distance along the axis of the field, ϕ represents rotation of the particle around the axis of the cylindrical field, $V(r,z)$ is the potential in the cylindrical region, B_z is the magnetic field along the axis of the magnetic lens system, q is the charge of the particle (typically a single electron charge $e = 1.60217657 \times 10^{-19}$ coulombs), m is the mass of the particle (typically either an electron, $m_e = 9.10938291 \times 10^{-31}$ kg, or ion) and V is the potential difference between electrodes used to accelerate the particle through the magnetic lens.

The two sets of equations above are significant to the development and understanding of globally transformative technologies spanning the previous century and continuing into the present.

Before discussing the impact of particle beam technology it is important to discuss the information contained in the above equations. Equation (1) can be used to compute the curvature, and subsequently the trajectory, of an electron that has been accelerated through an electrostatic lens as illustrated in Figure 1. This lens has been produced by applying a difference in electrical potential (called a voltage) to two cylinders (potential V_a and V_b applied to cylinders A and B respectively), see Figure 1. Electric fields can readily exist in space (it would not be possible for light to get to us from the Sun otherwise) and therefore choosing any arbitrary path in space from one to the other, there will be a continuous variation in the potential. For example, between the end faces of the two cylinders, the voltage will change linearly from 0 V at A to 8 V at cylinder B. Midway between these two cylinders there will be a potential of 4 V and so on. Solving Maxwell's equations for this configuration it is possible to map the potentials throughout the space encompassed by the cylindrical boundary as shown in the figure. It is then possible to plot a contour map of constant potential values in exactly the same way that it is possible to plot a contour map of a hill (the height in this case represents gravitational potential). Again analogous to the contour map of a hill, an object placed at one of these potentials will experience a force perpendicular to the 'equipotential' contours. This produces lines of force labeled the electric field in the figure. A quick glance at the shape of the equipotentials implies that it might (and does) act on a particle in a manner analogous to how light is acted upon by a traditional optical lens. It also turns out that a solenoidal coil can be wrapped around a pair of magnetically permeable cylinders to create a similar field for which the electric potential and field can be replaced by a magnetic potential and magnetic field.

Any charged particle in an electric field will experience a force given by the Lorentz equation. Combining this with Maxwell's equations, it is possible to determine the trajectory of a particle carrying a charge q as is progresses axially along the cylinder. Equation (1) expresses the rate at which the charged particle curves towards the axis and is interesting because it is independent of the mass and charge of the particle (really the mass to charge ratio). Basically this predicts that a particle initially moving in an axial direction along the cylinder will be bent toward the axis when it passes through the region of the lens. At axial distances remote from the lens the particle experiences no radial force and will continue in a straight line in accordance with Newton's laws (with, of course, modifications at high acceleration voltages where particle velocities can approach that of light and relativistic effects become significant). Being independent of mass makes this is a preferred method for focusing heavy particles such as atoms, for which the mass comes from the relatively heavy protons (electrons are around 1/2000 of the mass of a proton). A drawback of these types of lens is the large voltages that can cause breakdown (sparks) upon undesired circumstances.

On the other hand, equations (2) and (3) indicate that the deflection due to a 'magnetic lens' will be a function of mass and charge. Additionally, the particle will rotate in a tangential path around the axis. When passing through the lens, the charged particle will spiral towards the axis but again follows a straight path outside of this region. Hence both systems can be characterized by the simple model shown in Figure 1. The mass, charge, and voltage dependence of the magnetic lens means that particles having different masses and charges will deflect through different angles. While this is a problem for focussing of charged particles, this has been exploited in mass spectrometers for identifying different atoms and their isotopes enabling, for example, forensic analysis and isotope (carbon) dating techniques.

The ability to focus and deflect particle beams has, indirectly, had an enormous impact on society and culture. Electron beams were extensively used in early studies of atomic structure using cathode rays that were adapted to produce television. Focused electrons can be used to impact a surface and 'knock out' secondary electrons that are measured to provide spectacular scanning electron microscope images of physical and biological systems down to atomic scales. These same focused electrons are today being used to write nanometers scale patterns on masks for making modern computers. Focused ion beams are now readily used to machine specimens to further explore the nature of physical processes at nanometer scales. More recently focused helium ions have been used instead of electrons in the helium ion microscope, see Ward et al., 2006.

Early in the 20th century there was extensive research in the field of electron optics and this expanded considerably with the advent of vacuum tube amplifiers. Possibly because much of this research took place within secretive industrial laboratories, the origins of these equations are surprisingly difficult to trace. Early theoretical papers appeared in the 1930's and are to be found in textbooks and monographs from these early day to the present. More detailed discussions of the equations presented here can be found in the books of Gewartowski and Watson, 1965, and Harris L.A., 1962.

References

Bertram S., 1940, Determination of the axial potential distribution in axially symmetric fields, *Proc. IRE*, **28**, 418-420.
Bertram S., 1942, Calculation of axially symmetric fields, *J. Appl. Physics*, **13**, 496 – 502.
Gewartowski J.W. and Watson H.A., 1965, Principles of electron tubes, D. van Nostrand Company, Inc. Chapter 3.
Harris L.A., 1962, Introduction to electron beam technology, Bakish R., editor, John Wiley and Sons, available free on line.
Ward B, Notte J, Economou N, 2006, Helium ion microscope: a new tool for nanoscale microscopy and metrology, *J Vac Sci Technol B*, 24(6), 2871–2875.

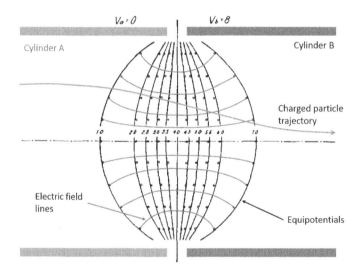

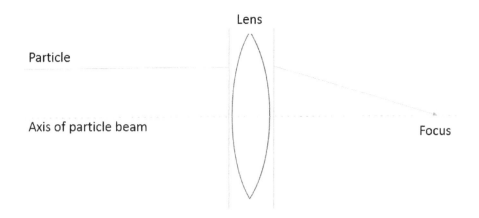

Electric field in a cylindrical electrostatic lens plus the trajectory of an electron accelerated through this field (substantially modified from the paper of Bertram, 1942). The figure below shows a typical 'black box' approach to modeling of this system.

SS

Equation 214: The Markov Property

$$\Pr(X_{n+1} = x \mid X_1 = x_1, X_2 = x_2, ..., X_n = x_n) = \Pr(X_{n+1} = x \mid X_n = x_n),$$

$$\Pr(X_1 = x_1, X_2 = x_2, ..., X_n = x_n) > 0$$

A sequence of random variables $X_1, X_2, X_3, ...$ is said to have the 'Markov property' if each variable X_{n+1} in the sequence is dependent only on the variable X_n preceding it. Thus it is independent of those variables in the sequence prior to that, i.e. $X_{n-1}, X_{n-2}, ..., X_1$. This property of a stochastic sequence is sometimes loosely alluded to as 'memorylessness'.

A sequence characterised by the Markov property is referred to as a 'Markov chain'. The possible values of X_n form a countable set called the 'state space' of the chain.

The term 'Markov process' is sometimes used instead of Markov chain, but is often reserved for a discrete-time Markov chain (DTMC).

Markov processes are named after the Russian Mathematician Andrey Markov (1856-1922).

A Markov chain is often described using a 'state transition diagram', specifically a directed graph, where the state transitions are labelled by the probabilities of going from one state at instance n to the other states at instance n+1. The same information can be represented as a state transition matrix.

A state transition diagram and matrix for a simple example are illustrated. The states represent whether a hypothetical stock market is exhibiting a bull market, bear market, or mixed market trend during a week. According to the figure, a bull week is followed by another bull week 90% of the time, a bear week 7.5% of the time, and a mixed week the other 2.5% of the time.

The MARKOV PROPERTY

$$Pr\,(X_{n+1} = x \mid X_1 = x_1, \, X_2 = x_2, \, ..., \, X_n = x_n) \;=\; Pr\,(X_{n+1} = x \mid X_n = x_n)\,,$$

$$Pr\,(X_1 = x_1, \, X_2 = x_2, \, ..., \, X_n = x_n) > 0$$

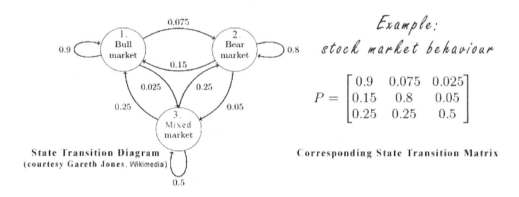

Example:
stock market behaviour

$$P = \begin{bmatrix} 0.9 & 0.075 & 0.025 \\ 0.15 & 0.8 & 0.05 \\ 0.25 & 0.25 & 0.5 \end{bmatrix}$$

State Transition Diagram
(courtesy Gareth Jones, Wikimedia)

Corresponding State Transition Matrix

Markov chains have many applications as statistical models of real-world processes. Markovian systems appear extensively in the fields of thermodynamics and statistical mechanics. In chemistry they can be used to model enzyme activity, copolymers and some cases of crystallisation. They are used in biology, social sciences, games, music, information sciences, queuing theory and the internet (Google's PageRank is apparently defined by a Markov chain - at the time of writing).

FR

Equation 215: The Planck Length (and Related Units)

The Planck length, together with the Planck time and Planck mass and other related units, is an example of a Planck unit that can be defined using three universal constants: the gravitational constant G, the speed of light c, and the reduced Planck constant $\hbar$ – and hence be dependent only on natural constants which are themselves ideally independent of experimental parameters. It turns out that these representations are particularly useful in unified theories.

The Planck length is defined using three universal physical constants as

$$l_p = \sqrt{\frac{\hbar G}{c^3}} \approx 1.6 \times 10^{-35} \mathrm{m}$$

where $\hbar = \dfrac{h}{2\pi}$ is the reduced Planck constant, G is the gravitational constant and c is the speed of light. It is an extremely small and immeasurable quantity. The German physicist Max Planck derived the Planck length and similar transformed units shortly after his discovery of his Planck constant h, the universal constant of action which appears everywhere in the quantum world. The relationship above, together with similar expressions for a Planck time and Planck mass represented by the same three constants, was derived by Planck in 1899, two years before his revolutionary paper on black body radiation and quantisation of energy. The only way Planck could derive this at the time was using dimensional analysis, but with the hindsight of quantum mechanics and relativity it can be arrived at in other interesting ways. Although essentially devised as a rearrangement of units and scale, the Planck length and related units have profound implications in fundamental physics. The Planck length can be related to limiting dimensional parameters such as the Schwarzschild radius (BE4) and Compton wavelength.

Using dimensional analysis the three universal constants can be broken down into the fundamental physical units of mass (M), length (L) and time (T), using their definitions (BE67 & 73):

$$G = \frac{Fr^2}{m^2}: \qquad \rightarrow \qquad M^{-1}L^3T^{-2}$$

$$c: \qquad \rightarrow \qquad LT^{-1}$$

$$\hbar = \frac{E}{\omega}: \qquad \rightarrow \qquad ML^2T^{-1}$$

where E is energy, ω is angular frequency, F is force, r is distance and m is mass.

If we assume a general form for the Planck length in terms of the universal constants

$$l_p = G^x c^y \hbar^z$$

we find that

$$l_p = M^{-x+z} . L^{3x+y+2z} . T^{-2x-y-z}$$

Because we are looking for a length, we know that the power term of M is 0, that of L is 1 and that of T is 0. This results in three simultaneous equations for the power terms, from which the relation for Planck length can be derived. Similar expressions for time and mass are obtained in the same way. The Planck length is an example of a base or natural unit. When quantities are expressed in these terms, the constants are eliminated by a normalisation, e.g.

$$F_p = G \frac{m_p{}^2}{l_p{}^2} \equiv \frac{\left(\frac{m_1}{m_p}\right)\left(\frac{m_2}{m_p}\right)}{\left(\frac{r}{l_p}\right)^2}$$

In which the gravitational constant has disappeared and can be redefined (within this system of units) as 1. In a similar way, c can be factored out so that $G = c = 1$. There is no reason why the usual set of SI units should be preferred other than familiarity and convenience – although base unit transformations should not necessarily be applied to standard measurements, they are valid mathematical operations that provide insight into fundamental physical theories.

The Planck length is considered to be the shortest distance with any physical meaning, or the smallest uncertainty in a measurement of position. It could be comparable, for example, with the size of a superstring or the size of the universe at which the fundamental forces began to separate. The Planck length and Planck mass can be shown to be related to the scale at which the Schwarzchild radius (the maximum size a particular mass can be without collapsing into a black hole) and the Compton wavelength (the length scale at which quantum field theory becomes significant for explaining the behaviour of a particular mass) are similar, and is therefore a useful concept in explaining theories of quantum gravity - the scale at which gravity is expected to become significant at a quantum level, such as one might find in the vicinity of a black hole or the very early stages of the big bang.

PB

Equation 216: Photostriction

A very basic definition of photostriction is the generation of mechanical strain by the irradiation of light. The photostrictive effect has been observed in a variety of materials. Figielski (1961) demonstrated the effect in germanium while Gauster and Habing demonstrated the effect in silicon (1967).

When light from a laser source in incident on a material, heat is produced on the surface and this radiates into the material, depending on the material's properties. This is the mechanism used in the optical generation of ultrasound, where local heating causes a thermal expansion, leading to a mechanical strain. If the laser source is pulsed, the rapid heating and cooling in the material produces elastic waves that radiate out from the heat source. However, in photostrictive materials, while the same outcome is observed, no heating takes place to produce the elastic waves.

Light is absorbed into a semiconducting material and penetrates to a depth dictated by the absorption coefficient. In metals, for example, this leads to a thermal expansion, but in silicon, the light also generates free charge carriers (electron-hole pairs), producing a local mechanical strain. The mechanical strain is a result of a contraction in silicon up to a depth equal to the skin depth. The contraction is caused by an increase in electron concentration which also increases the pressure of the electron gas. The volume changes to minimise the energy. The pressure dependence of the bandgap, given by $\frac{d\epsilon_g}{dP}$ is negative in silicon. Charge carriers also form much faster than the thermal stresses so the contraction will appear sooner than the thermal expansion stress wave. Light incident on the silicon surface must meet the condition $\lambda < \lambda_{\epsilon_g}$, where λ is the wavelength of incident light and λ_{ϵ_g} corresponds to the wavelength of the bandgap energy. The pressure dependence of the bandgap in germanium is positive and so results in an expansion of the material.

This effect is illustrated in the figure below. Light of wavelength 532 nm was incident on a silicon wafer, half of which was covered with a 118 nm aluminium film. The detection beam was incident on the opposite side of the wafer, in effect detecting the bulk wave through the silicon wafer. Two readings were taken. The first (dotted line) was where the light source is incident on the aluminium coated section of the wafer. The second (solid line) was where the light source is incident on the silicon. Offsets are for presentation purposes.

We can see the onset of the thermally induced stress waves in both the aluminium coated silicon measurement and the silicon measurement at 0.22 μs. However, a contraction (of opposite sign) is visible in the silicon sample from 0.13 μs to 0.24 μs. No thermal effects contribute to this, concluding that this is purely an electronic effect.

The value for the contraction can be calculated from the equation

$$x_3 = \frac{(1-v)l^2}{\delta} \frac{d\epsilon_g}{dP} \Delta n$$

where x_3 is the value of the contraction, v is Poisson's ratio, l is the length of the source, δ is the skin depth, $\frac{d\epsilon_g}{dP}$ is the pressure dependence of the bandgap, ϵ_g is the bandgap energy, P is pressure and Δn is the excess charge carriers.

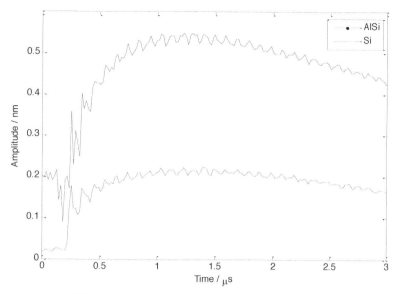

While this is an interesting subject in terms of elastic wave generation, it is of no real benefit, compared to laser (thermal) based techniques. However, photo-actuators, used extensively in MEMS research, work through the process of photostriction.

References

T. Figielski, "Photostriction effect in Germanium," Physica Status Solidi (B), vol. 1, no. 4, pp. 306-316, 1961.
W. B. Gauster and D. H. Habing, "Electronic volume effect in Silicon," Phys. Rev. Lett., vol. 18, pp. 1058-1061, 1967.

CM

Equation 217: Method of Exact Fractions

The calibration of gauge blocks by interferometric means is one of the cornerstones of dimensional metrology. As an example the Kösters-gauge block interferometer is schematically sketched in the figure (left). The gauge block to be calibrated is wrung on a baseplate and this assembly is illuminated by one of the lines of a spectral lamp that acts as a light source. The observer sees the interferogram with fringes as in the figure 1 (right). The method to calculate the gauge block length from interferograms like in figure 1 (right) is called the method of exact fractions.

For every vacuum wavelength λ_n a fraction f_n is measured. The gauge block length L can be expressed as

$$L = \frac{\lambda_1}{2n(\lambda_1)} \cdot (N_1 + f_1) = \frac{\lambda_2}{2n(\lambda_2)} \cdot (N_2 + f_2) = \frac{\lambda_3}{2n(\lambda_3)} \cdot (N_3 + f_3) = \cdots$$

where n is the air refractive index and N_n is the integral fringe order which is unknown, except for the fact that it is an integer.

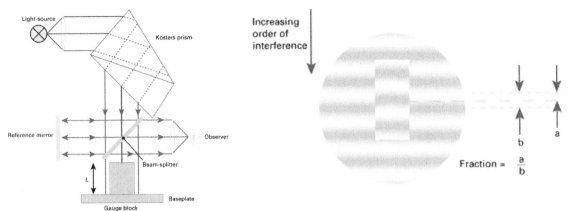

Schematic of Kösters inteferometer with Gauge block (left) and interference pattern (right)

The method of exact fractions works as following:

- For one wavelength, λ_1, the integral order $N_{1,0}$ is estimated from the nominal gauge block length, which is commonly well known within 1 μm from its specification or a previous calibration.
- A number of possible lengths around this estimated nominal value $N_{1,0}$ is taken, e.g. $L_{-2}, L_{-1}, L_0, L_1, L_2$ corresponding to $N_{1,0}-2, N_{1,0}-1, N_{1,0}, N_{1,-2} N_{1,-2}$.
- for all of these possible lengths, the fractions f corresponding to the other wavelengths are calculated, and compared to the measured fractions
- As final value for the gauge block length, the value of L_n is taken for which the calculated values of all fractions agree with the measured fractions to within the experimental uncertainty.

In the pre-computer and even pre-calculator age this was the employed method. Its practical implementation involved tables with nominal values and even dedicated sliding rules for specific spectral lamps.

The method was first described in 1898 by Jean-Renée Benoît, a French physicist who coöperated with Michelson, Fabry and Perot in relating wavelengths to physical lengths.

HH

Equation 218: Biot-Savart Law

In the presence of moving electrons, e.g. electric current flowing through a wire, a magnetic field is generated. The Biot-Savart Law provides a mathematical relationship between the current flowing through a wire as well as the size and spatial distribution of the wire and the magnitude and direction of the resulting magnetic field at a certain distance away from the wire. The following form of the Biot-Savart Law applies to the magnetic field generated by a steady flow of electrons through an infinitesimally thin wire along a closed path and in vacuum. The equation consists of a line integral since the contribution to the magnetic field is calculated for infinitesimally small segments along the wire.

$$\vec{B} = \frac{\mu_o}{4\pi} \int \frac{I d\vec{l} \times \hat{r}}{|\vec{r}|^2}$$

where μ_o is magnetic permeability in vacuum, l is the current flowing through the wire, $d\vec{l}$ is an infinitesimal length section along the wire, and $\vec{r}$ is the displacement vector from the segment of the wire to the point at which the magnetic field is calculated. $\hat{r}$ is the unit displacement vector. In the case of electric current flowing through a straight segment of wire, the nature of the magnetic field is shown in the figure below.

The direction is given by the right-hand screw rule, which corresponds to the cross-product in the above equation.

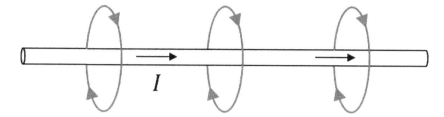

The Biot-Savart Law is named after two French scientists, Jean-Baptiste Biot and Felix Savart, who in 1820 postulated the mathematical relationship between current and the resulting magnetic field.

MF

Equation 219: Transmissibility

$$\frac{F_t}{F_0} = \left|\frac{x}{y}\right| = \sqrt{\frac{1 + 4\xi^2 \frac{\omega^2}{\omega_n^2}}{\left(1 - \frac{\omega^2}{\omega_n^2}\right) + 4\xi^2 \frac{\omega^2}{\omega_n^2}}}$$

$$\theta = \tan^{-1}\left(2\xi \frac{\omega}{\omega_n}\right) - \tan^{-1}\left[\frac{2\xi \frac{\omega}{\omega_n}}{1 - \frac{\omega^2}{\omega_n^2}}\right]$$

where F_t is the force transmitted to the frame supporting a solid body of mass m that is attached through a linear spring of stiffness k plus a damper of value b. F_0 is a dynamically varying force applied to, or generated by, the solid mass. x is the displacement of the solid mass. y is the displacement of the support measured in the same direction as that of the solid mass. ω is the frequency of the dynamically varying applied force or the frequency of the support motion. $\omega_n = \sqrt{k/m}$ and is the natural frequency of the oscillation of the mass if the damper is removed. ξ is the damping ration calculated from $\xi = b/2\sqrt{km}$. θ is the phase shift between input and output of the system.

During operation, an internal combustion engine comprises a large number of dynamically moving parts, pistons, valves, crankshafts, etc. Essentially, it is a solid container full of parts moving cyclically with high energy. Because the engine has to be attached to the chassis, this must result in vibrations that will have to be constrained somehow. Even the most perfectly balanced 6 cylinder engine produces vibrations that are harmonics of the crank speed (this being the limit to which most of the forces due to the crank motions can be arranged to cancel each other out). So why do these vibrations not get transmitted into the chassis of the vehicle and, by physical connection, into all other components such as the steering column and the body panels that would, in turn, act like loudspeakers amplifying the noise inside the car?

Looking at the simplified sketch of an engine and chassis, it is apparent that there are a number of soft rubber interfaces between the two (labeled as the 'engine mount'). Modern engine mounts are more complex, often containing chambers of fluids and complex flow channels. However, the net result is that these provide a relatively low stiffness support and will dissipate energy when subject to oscillating forces. At first, a soft spring supporting a shaking engine would appear to create more problems.

It turns out that useful information about the dynamics of this, and many other, systems can be inferred from a simple 'lumped' model such as that shown in the figure. The forces producing shaking of the engine have been replaced by an externally applied force, F_0, (not an accurate model but the conclusions will be similar) and the force that is transmitted to the chassis (labeled 'Base' in the figure), F_t. Solving for the applied forces considered in terms of their frequency content (we can do this using the tools provided by the Fourier series and transforms, BE's 12 and 150) results in the plot shown.

The vertical axis on the plot is the ratio of the force transmitted to the chassis to that applied while the horizontal axis represents the ratio of the frequency of the shaking to that of the natural frequency of oscillation of the engine on the mounts. There are three regions of interest in this plot. At low frequency, any forces generated at the mass will be transmitted directly to the base with the same value. In this case the system 'looks' like a spring. As the frequency of applied forces approaches resonance it turns out that the spring force becomes balanced by the inertial force of the mass. Consequently, the motion of the mass (engine) can build up resulting in an amplification of the forces transmitted to the base (chassis). At the resonant frequency, the only forces that will oppose motion of engine are due to the damper In this case, it is seen that the **higher** the value of the damping ratio, the lower will be the forces transmitted. Finally, at higher frequencies, two effects are apparent. Firstly, the amplitude is attenuated. Secondly, the attenuation is greater for the system having the **lower** damping ratio. This attenuation starts at a frequency that is the square root of 2 times the natural frequency, a point labeled in the figure. Generally, at high frequency the response is dominated by the inertial resistance to motion of the mass.

For the engine, as well as other similar processes, it is generally desired to attenuate force transmission. Fortunately, the frequencies that an engine will generate will generally be equal to, or higher than, the rotational speed of the crankshaft. Hence given that idle 'might' represent a minimum the springs can be made sufficiently soft that the natural frequency of the suspension is well below that of the idle. Typically, if you push an engine and let go, it will wobble at frequency of one to a few times a second. Typical idle speeds are around a 120 or so rpm corresponding to a couple of Hertz. This is comparable to that of the engine and therefore the forces are likely to be transmitted into the chassis with only little attenuation. However, while it might be possible to feel the engine vibration through the steering column, the frequencies at idle are too low to be audible. When moving, the engine speed is considerably higher and for which the forces are considerably attenuated.

With some other physical processes, the mass of the system is shaken by the movement of the base. Examples include building on a foundation during an earthquake, a car travelling at uniform speed on a rough road, or a delicate scientific instrument on a table attached to the ground. In these cases it is desired to somehow isolate them from these unwanted vibrations, in particular the high frequency vibrations that contain more energy. In this case, we are interested in how ground borne vibrations, y, are transferred into motion of the mass, x. It turns out that the exact same equation can be used to model the response. Hence to increase the ability of a building to withstand earthquakes it is common to put them on a more flexible foundation so that the building resonates at frequencies considerably lower than damaging seismic motions. Similarly, delicate scientific experiments usually take place with the instruments being mounted on very large tables supported by soft springs with very little damping.

Generally, solutions of the equation from this simple model show us that, to avoid the effects of shaking the answer is often not to try to 'clamp' things down but to 'loosen' them up.

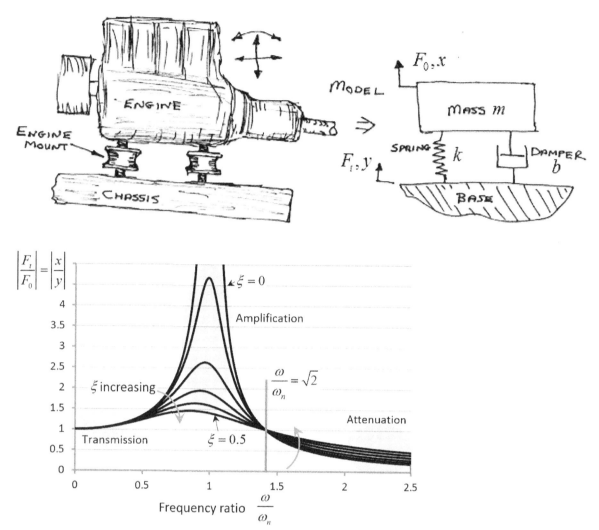

Figure 4: Simplified model of an engine mounted onto a chassis (top left), lumped model (top right), and magnitude frequency response (transmissibility) (bottom). The shaded regions show where responses are dominated by the stiffness (yellow region), the damper (blue), and the mass (green).

SS

Equation 220: The Verhulst Equation

Thomas Malthus' 1798 "Essay on the Principle of Population" proposed that catastrophe must result from unrestrained geometrical population growth combined with arithmetical increase in food supplies. The Essay prompted responses from many disciplines, from politics to philosophy, and from theology to biology. Its mathematical impact, though less well-publicised, was also profound.

Pierre-Francois Verhulst was made Professor of Mathematics at the Free University of Brussels in 1835, and was inspired that same year to explore the mathematics of Malthusianism through the publication, by Adolphe Quetelet, of a theory that the geometrical progression of population could be treated like a moving object, and would encounter a natural resistance proportional to the square of its rate.

This bizarre theory led Verhulst to suggest an equation, still rather arbitrary, which is now called the Verhulst Equation, or the Logistic Function:

$$P(t) = \frac{P(0)e^{rt}}{1 + P(0)(e^{rt} - 1)/K}$$

where P is population, t is time and r and K are arbitrary constants.

The theory was published in 1838 in "An Note on the Theory of Population Growth."

Despite its arbitary basis, the equation proved highly successful in modelling the populations of France, Belgium and the county of Essex over several decades.

In words, the equation states, roughly, that a population will initially grow near-exponentially before the rate of growth slows and tails off.

Visually, the equation is a sigmoid function:

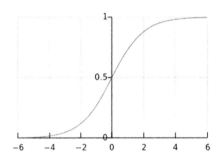

Application of the Verhulst equation is very wide, applying as it does to practically any type of self-limiting natural process.

MJG

Equation 221: The Q Factor

In physics and engineering, the Q factor is a figure of merit to distinguish the performance of an oscillatory system. First defined by American electrical engineer K. S. Johnson around 1924 in "Transmission Circuits for Telephone Communication" when he was working on the design and evaluation of passive circuits at the American Western Electric Company. The letter Q was chosen simply because Mr. Johnson noticed it was not widely used in technical publications, but it turned out to be a widely accepted measure to the "quality" of a system.

The Q-factor is defined in the context of resonance, a phenomenon that occurs when a given system (electrical, mechanical and optical etc.) is driven by another vibrating source to oscillate with a maximum amplitude at a specific frequency. In the figure, three resonance peaks are plotted representing systems (same resonance frequency) with Q factor of 5, 10 and 50 respectively. When the amplitude of the system is plotted against the frequency of driving source, a sharper peak of resonance is obtained compared to that of the lower Q systems.

Mathematically, it is the ratio of the system's resonance frequency (f_r) to the 3 dB bandwidth (Δf).

$$Q = \frac{f_r}{\Delta f} = \frac{1}{2\zeta}$$

Additionally, the denominator of the equation 2ζ is the twice the damping ratio of a system. The damping ratio gives system designers an insight of how well the system is damped. A high Q system means that, if an oscillated is applied, after this source is removed its oscillatory behavior is sustained for a long time while a low one indicates the amplitude of the system decays after a few cycles of oscillation.

Q factor can also be interpreted in terms of energy loss, it is proportional to the ratio of the energy stored in a system to the energy dissipated per oscillation. So the higher the Q factor is the less the energy is lost for each cycle of oscillation. The figure shows mechanical systems whose Q values are typically around 5 to 100. Tuning forks (typical of quartz oscillators used for timing in watches and computers) are in the regions of hundreds to thousands in air and can reach 10's of thousands in vacuum. Comparatively, an optical resonator has Q factor value around 10^7 to 10^{10} in various applications.

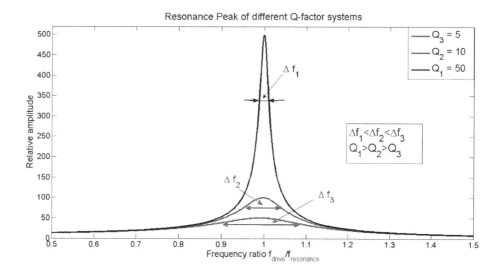

KN

Equation 222: Poisson Probability Distribution

$$f(k;\lambda) = \Pr(X = k) = \frac{\lambda^k e^{-\lambda}}{k!}$$

where λ is the mean value of the even occurrences, k is the number of occurrences happening and $k!$ is the factorial of k ($i.e. \ k * (k-1) * (k-2) \dots 3 * 2 * 1$).

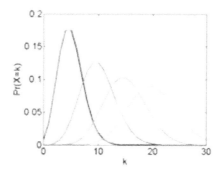

The Poisson distribution is a discrete probability distribution which expresses the probability of a given number of events occurring in a fixed interval of time and/or space (if these events occur with a constant rate per time unit and independently of the time since the last event). The graphs shows example of the probability distribution for mean (and variance) of 5,10,15 and 20. Notice that the mean and the variance have the same value in a Poisson distribution. Examples of phenomenon that can be modelled as Poisson distribution are: The number of soldiers killed by horse-kicks each year in each corps in the Prussian cavalry, the number of goals in sports involving two competing teams, the number of mutations in a given stretch of DNA after a certain amount of radiation, the targeting of V-1 flying bombs on London during World War II, the arrival of photons on a pixel circuit at a given illumination and over a given time period.

The last example is relative to noise in picture taken from a digital photo camera: due to the quantum nature of light, the photons arriving to the camera pixels will be the same in mean value, but they will vary from adjacent pixels according to a Poisson distribution. This source of noise in pictures is called shot noise and is particularly relevant in telescope where images are recorded where only a few photons are collected, and therefore the shot noise intensity might be above the level of the stars images.

The distribution was named after the French mathematician Siméon Denis Poisson, who published it in the 1827 in his work Connaissances des temps.

GM

Equation 223: The Flack Parameter

Today I am writing about an equation from my namesake H. D. Flack (no relation). The parameter is used in x-ray crystallography. X-ray crystallography is a tool used for identifying the atomic and molecular structure of a crystal. Crystalline atoms cause a beam of incident X-rays to diffract into many specific directions. By measuring the angles and intensities of these diffracted beams, a three-dimensional picture of the density of electrons within the crystal can be produced. From this electron density information on the mean positions of the atoms in the crystal can be determined, as well as knowledge of chemical bonds, their disorder and various other pieces of information. [See equation 87]

The Flack parameter, introduced in the 1980s by H. D. Flack became one of a standard set of values being checked for structures with non-centrosymmetric space groups. (In crystallography, a point group which contains an inversion centre (centre of symmetry) as one of its symmetry elements is centrosymmetric).

OK, let's hang on a minute we need to define some terms.

In geometry, a point group is a group of geometric symmetries (isometries) that keep at least one point fixed. If we think in 2D for a moment, there are only two one-dimensional point groups, the identity group and the reflection group. In 2-D we have cyclic groups – n-fold rotations and dihedral: cyclic with reflections. Beyond that it gets more complicated and I refer the reader to the Wikipedia article on point groups.

In crystallography, a point group which contains an inversion centre as one of its symmetry elements is centrosymmetric. In such a point group, for every point (x, y, z) in the unit cell there is an indistinguishable point $(-x, -y, -z)$. Such point groups are also said to have inversion. Point reflection is a similar term used in geometry. Crystals with an inversion centre cannot display certain properties, such as the piezoelectric effect. Point groups lacking an inversion centre (non-centrosymmetric) are further divided into polar and chiral types. A chiral point group is one without any rotoinversion symmetry elements (combinations of rotation with a centre of symmetry perform the symmetry operation of rotoinversion). Piezoelectric, and pyroelectric, crystals are non-centrosymmetric. The polarization properties of non-centrosymmetric crystals produce a broad range of interesting optical properties.

In short a chiral molecule can occur as one of an otherwise identical pair which have the exact same composition, functional structure, but shapes which are mirror images of the other and so display a definite "handedness". These left–right handed pairs even have identical physical and chemical properties unless they are interacting with some other chiral species or environment.

In X-ray crystallography, the Flack parameter is a factor used to estimate the spatial arrangement of the atoms of a chiral molecular entity that is to say determine the absolute structure of a non-centrosymmetric crystal. The processes uses the anomalous dispersion effect. (The 'anomalous' dispersion corrections, which are not in fact anomalous, take into account the effect of absorption in the scattering of phonons by electrons.)

There are several ways to determine the absolute structure by X-ray crystallography. One of the more powerful and simple approaches is using the Flack parameter, because this single parameter clearly indicates the absolute structure.

The Flack parameter is calculated during the structural refinement using the equation given below

$$I(hkl) - (1 - x)|F(hkl)|^2 + x\ |F(-h - k - l)|^2$$

where x is the Flack parameter, I is the square of the scaled observed structure factor and F is the calculated structure factor.

By determining x for all data, x is usually found to be between 0 and 1. If the value is near 0, with a small standard uncertainty, the absolute structure given by the structure refinement is likely correct, and if the value is near 1, then the inverted structure is likely correct. If the value is near 0.5, the crystal may be racemic or twinned. The technique is most effective when the crystal contains both lighter and heavier atoms. Light atoms usually show only a small anomalous dispersion effect. In this case, the Flack parameter can refine to a physically unrealistic value (less than 0 or greater than 1) and has no meaning.

There is a long-standing interest in finding ways to improve the precision of the Flack parameter in light-atom structures.

DF

Equation 224: The Risk Equation

$$Risk\ =\ probability\ x\ impact$$

A businesswoman wants to know if she should invest in a company. A general is determining if an assault is viable. A schoolteacher needs to show that a class trip is safe. All of these people are attempting to make a decision that involves some element of risk.

The reality is that all people are exposed to risk every day. Nearly every action (and inaction) has potential hazards that could result in harmful consequences. Despite how ubiquitous the concept of risk is, it is often not well understood nor well defined. Colloquially "risk" has become almost synonymous with "likelihood" but this doesn't offer a complete picture when assessing decisions. In almost all arenas where people have attempted to formalise their treatment of risk they use the above equation (or a variant) that factors in the negative impact of the risk coming to pass.

The direct effect that the risk equation has on our daily lives is highlighted by its use in safety engineering. When safety engineers are determining which hazards need to be mitigated, they use the risk equation to calculate the risk associated with those hazards, using estimates of the probability and impact of the accidents that those hazards could become. When discussing risk, the design principle of "As Low As Reasonably Practicable" (commonly abbreviated to "ALARP") is commonly cited by safety engineers and health and safety specialists alike. The UK Health and Safety Executive interprets ALARP as demanding that risks are weighed against "the trouble, time and money" needed to control them. As a result, ALARP (and by extension the risk equation) informs the level to which risks in the workplace and public spaces are controlled. In safety engineering ALARP is applied after a system has been designed to specification and drives organisations to go further than "acceptable risk", and to do everything they can to reduce risk as much as they can with the resources they have.

In other areas, such as cyber security, quantitative probability is problematic. This is typically because of a lack of sound data. When assessing cyber security vulnerabilities many potential attacks have not been seen before, and certainly haven't been seen enough to form an estimate of probability, but they still need to be considered to produce an effective defence. In these cases, probability is replaced with a qualitative measure: likelihood. The likelihood of a cyber-attack is, in reality, the likelihood that an attacker will achieve their goal, factoring in the technical capability of the attacker, how difficult a vulnerability is to exploit and the motivations of different types of attacker.

Impact is also used as both a qualitative and quantitative measure. The most common quantitative measure for risk is cost. i.e. the financial cost of a particular event occurring. Whilst numerically accurate, it is ethically and practically difficult to place a cost on things like a human life, or the quality of an environment. It is of note that legal frameworks, such as health and safety legislation, do serve to ensure that people wishing to rely on quantitative assessments of loss of life or environmental harm, have a discouragingly high figure associated with breaches. Practically, it is often better to describe impact using qualitative terms such as 'low', 'medium' and 'high'.

When using the equation, it is important to state the scope of the system being assessed. One action can involve multiple people/organisations and systems so it is important to know whose risk is being assessed (an assessment can be of the risk to some or all of the parties involved). In addition, it is important to determine the time window that is of interest. The risk posed by the ultimate heat death of the universe is huge, as the probability is 1 and the impact is the end of all life, but the risk becomes so diminishingly small over the lifetime of a human, or even over the likely era of humanity, that it rarely appears in a risk assessment! In fact, minimising "time at risk" is a recognised mitigation within safety engineering as reducing the window of opportunity for an accident, pushes the probability of an incident occurring down to an acceptable level.

When qualitative factors are introduced, the equation can no longer be evaluated as a simple product. Instead, organisations, government bodies, and entire industrial sectors generate look-up tables that convert the qualitative likelihood and impact measures into similarly qualitative measures of risk.

Measuring risk is only really useful within a wider risk-management framework. There are a number of competing standards, but in broad terms risk management is about systematically identifying risks, measuring the level of risk associated with them and comparing that with the risk-appetite (level of acceptable risk) of the organisation or person making the decision. Once the risk has been assessed, one of the four following options are available:
- 'accept' - conclude the risk is worth taking and carry on as intended
- 'avoid' - simply stop performing the action being assessed
- 'transfer' - typically a legal action, this could take the form of either taking out insurance against the risk occurring or getting another party to take ownership of the risk (personal responsibility waivers or similar)
- 'treat' - re-engineer the system so that the risk is removed or reduced

The level of acceptable risk not only varies from organisation to organisation, but can be highly context dependent for a given individual. In business terms it is common to not only calculate the risk associated with an action, but also calculate the ratio of the risk to the benefit of the action (assuming that the risk does not manifest). This implicitly alters the acceptable risk level proportionally to the possible reward.

Perhaps ironically, there is an inherent risk with the use of the risk equation. Its simplicity can make the act of performing a risk assessment seem easy and can be used to quickly produce a convincing looking argument that a system or decision is safe. However, it should not be relied upon too heavily, nor used to trump common-sense. In particular, in the domains of safety and security, it is important to consider other factors such as an organisation's culture, the capability and competence of the people interacting with a system and the maturity of the processes those people follow. Without considering those factors, this equation can leave dangerous gaps in the resilience of systems and cause massive harm. Another common criticism of the risk equation is that it oversimplifies the complex, multifaceted notion of risk into a single measurement. This is particularly misleading when using the qualitative form, as it is extremely difficult to unpick the bias introduced by the designers of the lookup tables or the individual who selected the input labels to begin with.

Despite its short-comings, the risk equation is an extremely useful way to talk about risk in a way that allows us to engineer the plethora of complex systems that we trust with our lives from planes and cars through to medical devices and food handling systems.

RFO

Equation 225: Straightness Reversal Technique

INTRODUCTION

In precision machine metrology a straightness reversal technique can be applied which measured an artefact (or material measured) in two opposite directionally sensitive positions to determine the straightness of a linear axis.

Two probes (A & B) measure either a known good straight edge or a reference component with fiducial markings that can be referenced and offset with respect to the measured position.

Probe A & B shall have the same direction sensitivity where there is a bump or straightness deviation on the component, however they shall be arranged such that they give an opposite directionally sensitive measurement where the axis they are mounted on has a bump or straightness deviation.

A simple equation can be used to determine the straightness of the linear axis and also to tell the profile of the artefact. In fact the reversal nature of this technique first measured and then drops out the artefact from the straightness of the axis in the same calculation.

EQUATION & PARAMETERS

$$A(n) + B(n) / 2 = C(n) \ (artefact)$$

where n is the measured sample number or more typically a linear position where the measurement occurs

$$A(n) - C(n) = \text{Straightness}$$

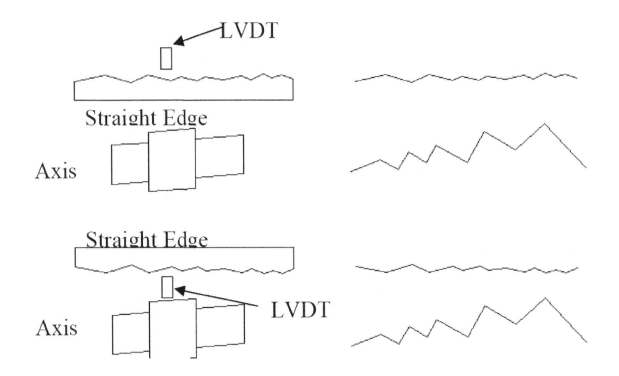

USES AND APPLICATION
In precision machine tools a straight-edge straightness measurement or an artefact reversal method are often carried out to measure a machine tools straightness of linear axis motion and then to verify that an error compensation has been correctly applied.

ADAPTATIONS
Adaptations of this include

- Measuring the back side of a turned component on a lathe - the straightness error is half the deviations measured. This method is a coarse check only and does not account for several other parameters that may influence the measurement.
- Using a known good artefact and ignoring the artefact contribution can be a much quicker test where less accuracy is required. This also involves removing the slope of the artefact from the measurement.

HISTORY
The original origins of the reversal principles can be traced back to ancient egyptian times.

Early attempts at measuring straightness used a probe at the cutting point to measure the generated surface, which usually turns out to be a very small error because you are measuring with the error you just made. This mistake is surprisingly common and gives a false reading.

Evans, Hocken and Estler, 1996 produced a test which they referred to as a nulling test which is based on the reversal principle, this was used in manufacturing to drive errors as close to 0 as possible.

SUMMARY
A reversal measurement is a useful metrology technique that allows you to get highly accurate measurements and calibrations with less than perfect artefacts on which to measure and calibrate.

This technique is really very useful as it is very practical and gives high levels of accuracy, traceability and confidence. As with any measurement it is limited by uncertainty and improved statistically by increased sampling and averages of measurements.

TS

Equation 226: Harmonic Series

HARMONICS
The Harmonic Series is a sequence of all multiples of a base frequency, as its name suggests, it is derived from the concept of overtones, or harmonics in music as many pitched musical instruments use a string or column of air which approximates a harmonic oscillator.

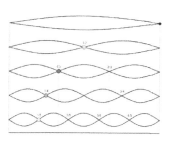

In these instruments resonant frequencies travel in both directions along the string or air column and interfere constructively or destructively with each other to form a standing wave. This standing wave interacting with the surrounding air creates the sound waves heard by the listener. The musical pitch of the note is usually perceived as the lowest frequency, the fundamental frequency, whilst the relative strengths of each harmonic give the instrument its timbre which distinguishes different instruments even when playing the same note. Eg Guitar and Piano

HARMONIC SERIES
In mathematics the harmonic series is the divergent infinite series of the form

$$\sum_{n=1}^{\infty} \frac{1}{n}$$

$$\frac{1}{1} + \frac{1}{2} + \frac{1}{3} + \frac{1}{4} + \frac{1}{5} + \frac{1}{6} + \frac{1}{7} + \frac{1}{8} + \frac{1}{9} + \cdots = \infty$$

Every term of the series after the first is the harmonic mean of the neighbouring terms; the phrase harmonic mean is also derived from music.

There are now many accepted proofs that this series is divergent however it was first demonstrated by a French monk, Nicole d'Oresme (ca. 1350). Nicole's proof is considered by many mathematicians to be the pinnacle of medieval mathematics and is still the standard proof taught in schools today. His proof uses the comparison test…

$$\frac{1}{1} + \frac{1}{2} + \frac{1}{3} + \frac{1}{4} + \frac{1}{5} + \frac{1}{6} + \frac{1}{7} + \cdots$$

or

$$1 + \frac{1}{2} + \left(\frac{1}{3} + \frac{1}{4}\right) + \left(\frac{1}{5} + \frac{1}{6} + \frac{1}{7} + \frac{1}{8}\right) + \left(\frac{1}{9} + \frac{1}{10} + \frac{1}{11} + \frac{1}{12} + \frac{1}{13} + \frac{1}{14} + \frac{1}{15} + \frac{1}{16}\right) + \cdots$$

is greater than…

$$1 + \frac{1}{2} + \left(\frac{1}{4} + \frac{1}{4}\right) + \left(\frac{1}{8} + \frac{1}{8} + \frac{1}{8} + \frac{1}{8}\right) + \left(\frac{1}{16} + \frac{1}{16} + \frac{1}{16} + \frac{1}{16} + \frac{1}{16} + \frac{1}{16} + \frac{1}{16} + \frac{1}{16}\right) + \cdots$$

which equals

$$1 + \frac{1}{2} + \frac{1}{2} + \frac{1}{2} + \frac{1}{2} + \cdots$$

which clearly diverges.

Therefore the Harmonic Series also diverges.

An interesting point about the harmonic series is that for many people the fact it is divergent is very counterintuitive. This is because as n increases to infinity the nth term goes to zero. Therefore the harmonic series diverges very slowly, in fact so slowly that although the partial sum will exceed 10 after 12,367 terms more than 1.509×10^{43} terms are needed to exceed 100!

A total of 15 092 688 622 113 788 323 693 563 264 538 101 449 859 497 terms to be exact.

The counterintuitive nature of the harmonic series also creates some interesting apparent paradoxes when applied to practical problems.

DOMINO STACKING

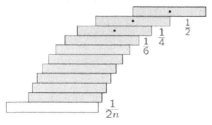

Stacking identical dominos to overhang an edge follows a harmonic series because to maintain balance the center of gravity of the complete stack must never extend beyond the edge. This is achieved by adding successive dominoes with progressively smaller and smaller overhang distances. For the optimal stack each successive overhang will be 1/2n where n in the total number of dominoes. This then means that the greatest overhang that can be achieved is infinite as long as you have enough dominos.

SPEED OF LIGHT SWIMMER

If a swimmer starts swimming at 1 Km / h across a pool which is 1 Km / 60 (approx. 16.667m) wide then it will take 1 minute to cross the pool. If every time the pool is crossed the swimmer pushes off the edge and increases their speed by 1 Km / h the second crossing will take 0.5 minutes. If there is no limit to the maximum speed of the swimmer how long will it take for them to be swimming at the speed of light? Well, the number of times the pool needs to be crossed is very large at over 1,000 million times. However, it will only take 22 minutes!

Very similar in form to the harmonic series (which is the sum of reciprocals of the natural numbers) is the series formed by the sum of the squares of the reciprocals of the natural numbers. That is to say

$$\sum_{n=1}^{\infty} \frac{1}{n^2}$$

However, this series *is* convergent and turned out to be considerably more challenging to solve. It became known as the Basel Problem and will be the topic of a future 'Beautiful Equation'

ILB

Equation 227: Coulomb's Law

$$F = \frac{kq_1q_2}{r^2}$$

268

In the 1780's, Charles Augustin de Coulomb did a series of experiments that allowed him to quantify the force of repulsion between two electrically charged spheres. He could quantify the strength of repulsive force as a function of separation using a highly sensitive torsion balance. See https://www.youtube.com/watch?v=_5VpIje-R54 for a video.

The result of his experiments was today's equation, where q_1 and q_2 are the strength of the two charges and r is their separation. k is a constant of proportionality and in SI units is equal to 1 divided by $(4\pi\varepsilon_0)$, where ε_0 is known as the permittivity of free space, and is equal to 8.854×10^{-12} F m^{-1} (1 F is a 1 farad and is equivalent to 1 C^2 N^{-1} m^{-1}). Note that the k value given here is if the two charges are in a vacuum. If a material is present, then a further constant has to be applied and is called the relative permittivity.

$$P = \frac{\pi^2 B_p^2 d^2 f^2}{6k\rho D} \quad B_p^2 \ d \ f \ k \ k \ = \ 1 \ k \ = \ 2 \ \rho \ D$$

RL

Equation 228: Allometric Scaling

$$\log y = \alpha \log x + \log b$$

The term 'allometry', used in biology, originally referred to the scaling relationship between the size of a particular body part and the size of the body as a whole. It has since been extended to refer to almost any co-varying biological measurements, be it for morphological, physiological or ecological traits. Thus allometry describes how traits or processes scale with one another - and how they relate to evolution

The term was coined by Julian Huxley and Georges Tessier in 1936. Huxley had been studying the extraordinarily large claw (or chela) of the male fiddler crab, Uca pugnax. When he plotted the body size and chela size of crabs at different developmental stages using a log-log scale - see diagram (from one by A. W. Shingleton, Nature Education 2010), he got a straight line whose slope was greater than 1. This indicated that, through time, for each unit increase in body size there was a proportionally larger increase in chela size. Huxley concluded that the reason the chela was exaggerated in this crab was because it was growing at a faster rate than the body as a whole.

Huxley and Tessier recognized that many scaling relationships were linear when plotted on a log-log scale. As a result, these relationships could all be described using the simple linear equation:

$$log \ y = \alpha \ log \ x + log \ b$$

... or, in exponential form, ...

$$y = bx^\alpha$$

where x is body size, y is organ size, α is the line's gradient (also known as the allometric coefficient) and $log \ b$ is the y-intercept. α captures the differential growth ratio between the organ itself and the body as a whole. $\alpha > 1$ (eg. the male fiddler crab's chela) is called positive allometry or hyperallometry. $\alpha < 1$ (e.g. the human head) is called negative allometry or hypoallometry. For an organ that grows at the same rate as the body as a whole, $\alpha = 1$, a condition called isometry.

269

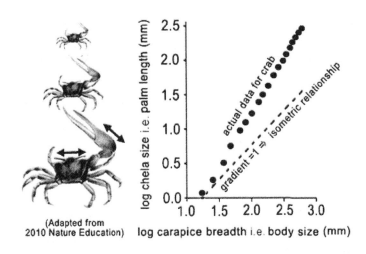

(Adapted from
2010 Nature Education)

The great progress made over the last 100 years investigating biological scaling relationships among myriad traits has made clear that many traits are genetically, physiologically, developmentally, and functionally integrated: changes in the scale of one trait alter the scale of another - though the mechanisms underlying this integration are far from obvious.

PB

Equation 229: Radiocarbon and other Radiometric Dating

Radiometric or radioactive dating was first suggested as a technique for determining the age of the earth by Ernest Rutherford in 1905. Radiocarbon dating was discovered by the American physical chemist Willard Libby in the 1940s after work that started in the Radiation Laboratories at Berkeley during WWII. He was looking for elements with a sufficiently long radioactive half-life for use in biomedical research. This involved the discovery that carbon-14 can be produced via a slow neutron interaction with nitrogen (previously thought to be via deuterons and carbon-13). Recent work by Danworth and Korff in 1939 had predicted the production of carbon-14 in the atmosphere by the action of cosmic rays that had been slowed down by collisions. Libby published a paper in 1946, then working at Chicago, that indicated the presence of carbon-14 in living matter compared with inanimate matter, and subsequent work demonstrated the principle on methane gas from sewage compared with petroleum. Eventually in 1949, wood samples from an ancient Egyptian tomb were dated with unprecedented accuracy. Willard Libby received the Nobel prize in Chemistry in 1960.

The most common form of carbon is carbon-12 (^{12}C). Approximately 1% of carbon exists as carbon-13 (^{13}C), a stable isotope. Approximately 1 part in 10^{12} of carbon exists as the unstable isotope carbon-14 (^{14}C).

^{14}C is continually produced in the lower stratosphere and upper troposphere by the action of cosmic rays. This is readily converted to CO_2 which is absorbed by the oceans, consumed by plants in photosynthesis and subsequently consumed by animals eating the plants. In this way it is completely distributed throughout the biosphere. Due to the continuous exchange of carbon with the environment in living matter through respiration and nutrition, the amount of ^{14}C in living matter is more or less in equilibrium with the atmosphere.

$$n + {}^{14}_{7}N \longrightarrow {}^{14}_{6}C + p \quad \text{(atmosphere)}$$

The beautiful equation given here is the decay of ^{14}C to stable nitrogen. Once a living organism dies, the carbon exchange no longer continues and the residual ^{14}C continues to decay but is no longer replenished by ^{14}C from its surroundings. The decay of ^{14}C therefore provides a dating mechanism for any living organic matter that has deceased.

270

$$^{14}_{6}C \longrightarrow {}^{14}_{7}N + e^- + \bar{\nu}_e \text{ (decay)}$$

The decay is described by the following

$$N = N_0 e^{-\lambda t}$$

where time $t = 0$ is the time of death, λ is the decay constant (the reciprocal of the mean-life, the average time that an atom will take to decay), N is the number of atoms left after time t and N_0 is the number of atoms at time $t = 0$.

A more common measure of time that is usually quoted is the half-life ($t_{1/2}$), which is the time for the radioactive isotope to convert half of its material to the stable isotope. The half-life of ^{14}C is about 5730 years. This means that the method is ideal for archaeology, but the limit of age that can be accurately measured is about 70,000 years. Beyond this, the levels of ^{14}C left in the material are too low to be measured accurately (for example using mass spectroscopy). So ^{14}C dating is ideal for determining the age of ancient human artefacts and corpses, ice-age mammals, Neanderthals etc, but nothing much prior to that.

So how are ancient fossils, which are millions of years old, dated? Fortunately there are many other radioactive materials that can be measured, some of which have a much longer half-life than carbon. Ancient fossils are incorporated into rock material and therefore the age of the surrounding rock is important. Many rock minerals contain potassium, for example. One of the main processes used here is the decay of radioactive potassium (^{40}K) to stable argon (^{40}Ar), K-Ar dating.

In the case of potassium-40, the half-life is approximately 1.25×10^9 years. The stable form of potassium most commonly encountered (approximately 93%) is ^{39}K and about 7% is ^{41}K. A small fraction (approximately 0.01%) is ^{40}K.

The most important factor is the stage at which the "radiometric clock" begins. In the case of fossils imbedded in rock it is often the point at which molten lava cooled and became igneous rock. If fossils are surrounded by this then K-Ar dating can be applied to the igneous rock above and below the fossil and a fairly accurate date range can be established for the organism. Another commonly-used radioactive transition is uranium to lead. In the case of uranium-235 which decays to lead-207 the half-life is about 700 million years; for uranium-238 to lead-206 the half-life is about 4.5 billion years. Different mechanisms are used for scale and accuracy. Before the radioactive techniques were available, stratigraphy (comparing layers of sedimentary rock) was the only way to estimate the age of fossils. This was a relative method; however, the radioactive methods are not entirely reliable because of the assumptions of trapping within minerals, gaseous release and absorption and events occurring before and after the radiometric clock event, and usually have to be checked with other dating methods depending on the time scale (examples are fission track, paleomagnetism, luminescence and electron spin resonance).

Wikipedia is a good starting point for those interested in the various techniques used for dating in geology, archaeology and paleontology: https://en.wikipedia.org/wiki/Radiometric_dating

Willard Libby wrote a book "Radiocarbon Dating", second edition published 1955.

PB

Equation 230: The Scherrer Equation

The Scherrer equation provides the relationship between the average size of sub-micro particles in a crystals – also known as crystallites – and the broadening of the peak in an X-ray diffraction powder pattern.

$$B(2\theta) = \frac{K\lambda}{<L> \cos\theta}$$

where B is peak broadening (sometimes indicated with β); θ is Bragg's angle of diffraction (see Equation 87); $<L>$ is the average size of the crystallites; λ is wavelength of the X-ray radiation and K is a dimensionless shape factor.

THE SHAPE FACTOR
K depends on different factors including:

13. The method to calculate the peak width
14. The shape of the crystallites
15. The size distribution of the crystallites

For $K = 0.94$ for spherical crystallites when calculating peak width as FWHM; for $K = 0.89$ for spherical crystallites when calculating peak width with Integral breadth method. Sometimes, K value is approximated to 1 for both the aforementioned methods. In general, K varies from 0.62 to 2.08.

According to this equation, the peak width (B) is inversely proportional to the crystallite average size, $<L>$.

The equation, published by the Swiss physicist Paul Scherrer in 1918, is used to determine the average size of solid, nano-scale crystallites in the form of powders, corresponding to average dimensions of 0.1- 0.2 micrometres. It is not easily applicable to large grains in metals and alloys, which contain linear defects such as twinning centres and dislocations.

Scientists sometimes erroneously refer to this equation as the Debye-Scherrer equation. Among a multitude number of applications, X-ray powder spectra are often used to collect information about the chemical composition of unknown samples and to study phase transitions in solids.

HISTORY
Between 1915 and 1917, Paul Scherrer and his supervisor, Peter Debye, worked together to develop methods for the analysis of crystal structures in the form of fine powders, in order to avoid the problem of growing large single crystals to extract the structural information from the method developed by Max von Laue, based on the Bragg's law (see equation 87).

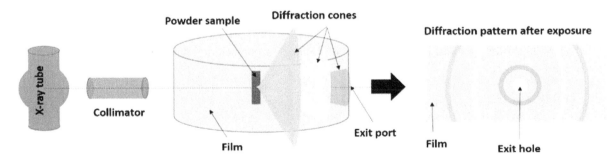

Figure 1. Schematic illustration of the Debye-Scherrer experiment on powder diffraction.

This method, which was also independently developed by Albert Hull at the General Electric Research Laboratory, became known as the Debye-Scherrer method. It consists of analysing two X-ray films in a powder diffraction detector. The X-rays pass through a collimator and strike a sample consisting of randomly oriented tiny crystals with uniform distribution. The diffracted beams form a cone projected on a film surrounding the specimen (Figure 1, right) after it has been exposed to the scattered radiation. The pattern on the film consists arcs of cones. The radiation that is not scattered leaves the system through an exit port.

The of arcs of cones on the film can be more intuitively understood by considering a single crystal rotating about the axis of the system and other two axes perpendicular to this axis, for a given Bragg's diffraction angle (theta). This is the same pattern that would be observed with a powder of tiny crystals randomly oriented in all directions. On the contrary, in the case of a non-rotating single crystal, the X-ray exposed film typically consist of dots, corresponding to diffracting crystalline planes (Figure 2: X-ray pattern of a single crystal of an enzyme. X-ray diffraction patterns of two different samples collected from the Martian surface by NASA's Curiosity rover).

Regarding X-ray spectra, the peaks of a diffracting crystal are very "narrow", almost corresponding to straight line, while those of a powder pattern present a certain amount of broadening.

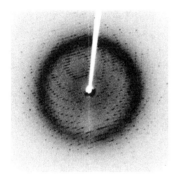

X-ray pattern of a single
crystal of enzymes

X-ray powder pattern of Mars' rock
samples collected by NASA's Curiosity rover

Figure 2. Examples of an X-ray single crystal pattern and an X-ray powder pattern (Source: Wikipedia).

Following the development of this method, Scherrer continued his studies on the relationship between the average size of the crystallites within a powdered sample and the broadening of the peaks in its corresponding spectrum. This work was developed without Debye's collaboration and eventually lead to the formulation of the Scherrer's equation in 1918.

FACTORS CAUSING PEAK BROADENING

There are several factors which affect the broadening of the peak in an X-ray powder pattern, including:

- Instrument and peak profile
- Crystallite size
- Microstrain: this is due to the presence of linear defects such as non-uniform lattice distortions, dislocations, twinnings, domain boundaries, stack faulting
- Solid solution inhomogeneity
- Temperature

The contributions from microstrain and solid solution inhomogeneity are typically present in crystalline grains of metals and alloys. The peak profile is therefore the convolution of the individual profiles generated by all of these contributions.

CALCULATION OF THE PEAK WIDTH

For a sample in which the microstrain contribution to the peak profile can be neglected, it is still important to separate the sample and instrument contributions from the peak width. Once this operation is performed, the peak width can be evaluated with two methods:

- Full Width at Half Minimum (FWHM), as used by Scherrer in his work. This is the width of the peak in radians, at a height half-way between the background spectrum and the peak maximum (Figure 3, left).
- Integral Breadth (IB), a method developed by Stokes and Wilson. This method calculates the area under the peak and divides it by the peak height (Figure 3, right). When compared with the FWHM, this method has the advantage of not depending on the distribution of size and shape of the crystallites. However, it requires a careful evaluation of the tales of the peak in the spectrum.

Another technique, known as the Williamson-Hall method, was developed in 1953 to calculate the average size of crystallites, when the microstrain contribution cannot be neglected. This method was further improved to quantify also anisotropic effects in the peak profile in strain induced by dislocations and other line defects.

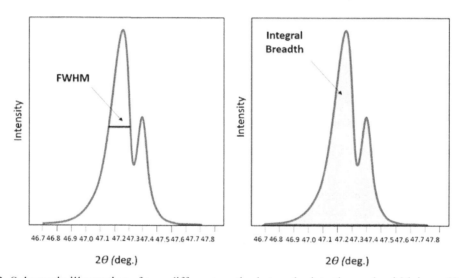

Figure 3. Schematic illustration of two different methods to calculate the peak width in an X-ray diffraction spectrum.

AT

Equation 231: Spatial Carrier Fringe Method

In equation 30 it was explained how a surface can be depicted in an interferogram as a kind of height map. The inverse problem: how to derive a surface height from a given interferogram is less straightforward, especially when only a single interferogram is available. However, when the surface is given a fair tilt, the surface topography can be derived from a single interferogram by a method called the "Spacial Carrier Fringe"- method.

274

For simplicity the derivation is done along 1 dimension (x), it works similarly for a 2-D interferogram.

When a surface with profile $h(x)$ is tilted with an angle α, its interferogram can be written as

$$I(x) = a(x) + b(x)\cos [\varphi(x) + 2\pi f_0 x]$$

with

$$\varphi(x) = \frac{2\pi h(x)}{\lambda}$$

and

$$f_0 = \frac{2\alpha}{\lambda}$$

here $a(x)$ and $b(x)$ are slowly varying functions describing the intensity and contrast, λ is the wavelength.

The trick is now to write the cosine function as $cos(x) = (e^{ix} + e^{-ix})/2$ and consider the Fourier transform

$$g(x) = a(x) + 0.5 \cdot b(x)e^{i\varphi(x)}e^{i2\pi f_0 x} + 0.5 \cdot b(x)e^{-i\varphi(x)}e^{-i2\pi f_0 x}$$
$$= a(x) + c(x)e^{i2\pi f_0 x} + c^*(x)e^{-i2\pi f_0 x}$$

with

$$c(x) = \frac{b(x)}{2}e^{i\varphi(x)}$$

The Fourier transform of the intensity $g(x)$, $G(f)$ can be described as the sum of the Fourier transform of the 3 terms

$$G(f_x) = \int g(x)e^{-i2\pi f x}dx = C^*(f_x + f_0) + A(f_x) + C(f_x - f_0)$$

The Fourier spectrum now consists of 3 parts: a central part consisting of $A(f)$ with on both sides the shifted transform of $c(x)$.

The 'beautiful equation' consists of isolating the peak in the spectrum that represents $C(f - f0)$ and take the inverse transform.

$$g'(x) = \int C(f - f_0)e^{2\pi i f x}dx = c(x) = \frac{b(x)}{2}e^{i(\varphi(x)+2\pi f_0 x)}$$

$$\varphi(x) + 2\pi f_0 x = arctan\frac{Im(c(x))}{Re(c(x))}$$

This means that the phase $\varphi(x)$, and with that the height $h(x)$ can be found from the phase of the inverse Fourier transform of the area around f_0 in the spectrum.

This phase needs to be unwrapped*, then the tilt can be removed and $h(x)$ can be calculated. An alternative for this is shifting the frequency axis so that the remaining spectrum is centred around zero, before taking the inverse transform.

This method has its limitations and ambiguities in sign, gradient and spatial frequencies, but it is certainly an elegant method that makes a 3-D topography appear from a flat black-and-white interferogram.

The method was described first by Takeda in 1982: Takeda, M, H. Ina, and S. Kobayashi, "Fourier-Transform Method of Fring-Pattern Analysis for Computer-Based Topography and Interferometry". J. Opt. Soc. Am, **72**, 156-160 (1982)

*unwrapping: as the phase is defined modulo $2 \cdot \pi$, jumps of $2 \cdot \pi$ may appear if the phase has a gradient. Adding or subtracting 2π at these jumps so that a continuous phase function appears is called unwrapping. In 1-D this is no problem, in 2-D this is a field of its own.

Graphical illustration of the carrier fringe method:

Interferogram of a groove in a surface:

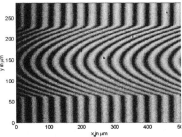

Taking the Fourier transform gives as amplitude spectrum:

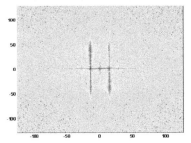

Now select single area and zero the rest:

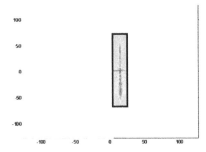

Take inverse Fourier transform and consider the phase of the result

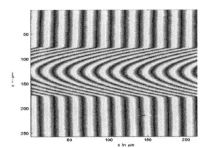

Unwrap, scale and level surface:

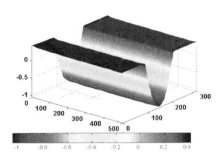

HH

Equation 232: Power Dissipation of Eddy Currents

When a conductor is placed in a changing magnetic field, small currents are generated within the conductor. The currents are formed in circular loops, hence the name 'eddy currents'. An interesting phenomenon is the generation of a secondary magnetic field induced by the eddy currents, which opposes the original magnetic field. However, resistivity in the conductor means that some of the power will be dissipated in the form of heat.

A very specific equation for the power lost due to eddy currents in a conductor is presented below. This equation applies to a thin sheet or wire of uniform conducting material placed in a uniform magnetic field; other simplifications apply and can be found in reference [1].

The power dissipation per unit mass due to eddy currents is given by

$$P = \frac{\pi^2 B_p^2 d^2 f^2}{6k\rho D}$$

where B_p^2 is the peak magnetic field, d is the thickness of the sheet or diameter of the wire, f is the frequency at which the magnetic field is changed, k is a constant the value of which depends on the geometry of the conductor (in this case, $k = 1$ for a thin sheet and $k = 2$ for a wire), ρ is the resistivity of the conductor material, and D is the material's density.

The phenomenon of power dissipation due to eddy currents is instrumental to the exploitation of eddy currents to slow down moving objects. More can be read about eddy current brakes in reference [2].

The discovery of eddy currents is credited to French physicist and baguette baker Léon Foucault who, in 1855, observed that rotating copper disk required more power when the rotation was performed near magnets. Foucault also observed the heating of the disk in the presence of the magnetic field [1].

[1] Eddy currents – Wikipedia, The Free Encyclopedia, https://en.wikipedia.org/wiki/Eddy_current
[2] Eddy current brake – Wikipedia, The Free Encyclopedia,

MF

Equation 233: Cauchy's Stress Theorem

The concept of stress is important in understanding material and structural failure due to mechanical, thermal or any other type of load. Stress is a vector quantity that represents the force per unit area (intensity of the force) at a given point in a body. Consider a body B subjected to some external forces. Then, internal forces develop within the body due to the action of these external forces. Suppose we isolate a part P of this body.

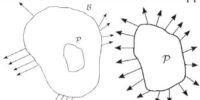

Figure 1: (Left). A body B subjected to external forces. (Right). The traction or stress vector acting on a part P isolated from B.

Then, the forces acting on the surface of this part are due to the action of the rest of the body on this part. The force per unit area acting on the surface of this part is what is called the stress vector or the traction vector and denoted by t. Clearly, this is a function of position x of a point on the surface of the part and time t. Furthermore, based on physical grounds, it is also assumed to be a function of the orientation of the surface, i.e., the unit outward normal n to the surface at x. Thus, the stress vector at a point x on the part surface at time t is denoted by $t(x, t; n)$.

An infinite number of surfaces (with distinct normal vectors n) can pass through a given point and therefore, the stress vector magnitude and direction can change as n changes. A beautiful theorem due to Augustin-Louis Cauchy states that the dependence of the stress vector on the normal vector is linear. The stress vector t at point A, on an arbitrary plane/surface passing through A is given by

$$t = Tn$$

where n is the outward unit normal to the plane/surface at point A. T is the second-order stress tensor that has the matrix form

$$T = \begin{bmatrix} T_{11} & T_{12} & T_{13} \\ T_{12} & T_{22} & T_{23} \\ T_{13} & T_{23} & T_{33} \end{bmatrix}$$

when a rectangular coordinate frame is used. Column i in the above matrix represents the components of the stress vector t_i on a plane perpendicular to the ith coordinate plane.

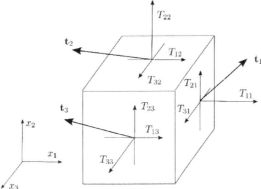

Figure 2: Traction vectors acting on three mutually perpendicular planes passing through a point. Note that $T_{ij} = T_{ji}$.

Thus, in matrix form, the above linear dependence of the stress vector on $\boldsymbol{n}$ can be written as

$$\begin{Bmatrix} t_1 \\ t_2 \\ t_3 \end{Bmatrix} = \begin{bmatrix} T_{11} & T_{12} & T_{13} \\ T_{12} & T_{22} & T_{23} \\ T_{13} & T_{23} & T_{33} \end{bmatrix} \begin{Bmatrix} n_1 \\ n_2 \\ n_3 \end{Bmatrix}$$

The Cauchy's stress theorem is a powerful theorem that says that the stress vector at a point can be determined for arbitrary $\boldsymbol{n}$ once the stress vectors on three mutually perpendicular planes (typically taken to be parallel to the coordinate planes) are known. These components are usually determined by solving the balance of linear momentum equations together with the constitutive laws that describe material response to applied loads and the boundary/initial conditions.

For example, consider a square plate formed by two angular plates welded together along one of the diagonals.

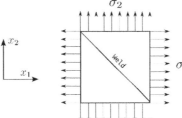

Figure 3: A square plate with a weld along a diagonal and subjected to biaxial loading.

Suppose that the plate is subjected to biaxial loading as shown below. Then, the stress-state in the plate can be shown to be given by $T_{11} = \sigma_1$, $T_{22} = \sigma_2$ and all other $T_{ij} = 0$.

Here, σ_1 and σ_2 are applied loads (per unit area) on the edges of the plate as shown in the figure. The stress vector on the diagonal welded plane can be obtained using the Cauchy stress theorem. The outward normal vector to the diagonal can be taken as $[n_1 \quad n_2 \quad n_3]^T = \frac{1}{\sqrt{2}}[1 \quad 1 \quad 0]^T$ and the stress vector on the weld plane is given by

$$\begin{Bmatrix} t_1 \\ t_2 \\ t_3 \end{Bmatrix} = \begin{bmatrix} \sigma_1 & 0 & 0 \\ 0 & \sigma_2 & 0 \\ 0 & 0 & 0 \end{bmatrix} \begin{Bmatrix} 1/\sqrt{2} \\ 1/\sqrt{2} \\ 0 \end{Bmatrix} = \frac{1}{\sqrt{2}} \begin{Bmatrix} \sigma_1 \\ \sigma_2 \\ 0 \end{Bmatrix}$$

Based on the strength of the weld, the above result can be used to determine the allowable applied loads (i.e., σ_1 and σ_2).

HC

Equation 234: Henry's Law

To construct the Eads Bridge (above) over the Mississippi in the 1870s, the workers had to work on the bed of the river. This was made possible by lowering metal cylinders onto the bed. Air was pumped into them, so that the water was forced out and a working environment made. But the price paid was in the health - and in twelve cases (fifteen according to some courses) the lives - of the workers.

Though "caisson disease" was known at the time, its cause was not. But it should have been, since the phenomenon was an example of Henry's law, formulated by English chemist William Henry in 1803.

Henry's Law states that

"At a constant temperature, the amount of a given gas that dissolves in a given type and volume of liquid is directly proportional to the partial pressure of that gas in equilibrium with that liquid."

Henry's Law in the form of an equation is

$$p = ck_H$$

where p is the partial pressure of the gaseous solute above the solution, c is the concentration of the dissolved gas and k_H is the Henry's law constant, and depends on the solute, solvent and temperature.

Caisson disease, now known as the bends or decompression sickness, is caused by dissolved nitrogen bubbling out of the blood when the air pressure is rapidly lowered, as it was when the caissons were decompressed. Henry's law is also the cause of the fizz when a can or bottle of effervescent drink is opened; in that case it is carbon dioxide that comes rapidly out of solution when the pressure on the drink is reduced by opening its container.

MJG

Equation 235: P-Norm and Unit "Circle" in 2D Space

In physics and engineering, the concept of "length" (or "distance") is ubiquitous. From a rigorous mathematical perspective, it makes a lot of difference how the dimensional quantity is defined. It is a commonsense to quantify the size of an object or a how far away a point is from one location to another in a 3D Euclidean framework. But more profoundly, the concept of length can go far beyond its geometrical interpretation to an abstract notion of "norm". A norm is defined on a space that consists of points, vectors or even functions attributed with a series of properties. It is a function that assigns a positive value of length to the mathematical objects of interest. How the Norm is calculated depends on the way it is defined.

The beautiful equation as follow presents a general treatment of various definitions of norm, called the p-Norm, where the p value can be specified arbitrarily. For $p \geq 1$, $x = (x_1, x_2, ..., x_n)$, the p-Norm of x is defined as

$$\|x\|_p = \left(\sum_{i=1}^{n} |x_i|^p \right)^{1/p}$$

In practice, the most common values for the p are $1, 2$ and infinity, which are known as the "absolute norm", the "Euclidean norm" and the "infinity norm" respectively. Imagine in a two dimensional space (i.e. $n = 2$), a point is located at the coordinate of $(3, 4)$ for x_1 and x_2 respectively. Then, the distances (D) from the point to origin $(0, 0)$ calculated by the p norm are

$$D_{p=1} = |3| + |4| = 7,$$
$$D_p = \sqrt[2]{|3|^2 + |4|^2} = 5$$

and

$$D_{p=infinity} = max(3,4) = 4$$

for $p = 1, 2$ and infinity respectively.

For example, a circle is a simple shape defined as a set of points in a plane (2D space) that are at a constant distance from a given point. The figure attached describes the shape of a unit "circle" when the distance is defined by of norms with four different p values! As the p value increases, the shape of the "circle" is evolving from a diamond to a curved line and to a fully expanded square boundary. It is astonishingly beautiful even it is anti-intuition.

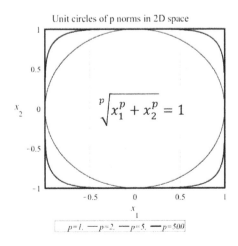

$$\sqrt[p]{x_1^p + x_2^p} = 1$$

Unit circles of p norms in 2D space

Application of norms is frequently used in the theory of functional analysis. The Polish mathematician Stefan Banach (1892-1945) is considered one of the founders of the branch in mathematics, whose name is used for defining the complete normed space commonly referred to as a Banach space.

Reference:

Erwin Kreyszig, Introductory functional analysis with applications, 1978, ISBN 0-471-50731-8

KN

Equation 236: Pareto Probability Distribution

$$\bar{F}(x) = \Pr(X > x) = \begin{cases} \left(\dfrac{x_m}{x}\right)^{\alpha} & x \geq x_m \\ 1 & x \leq x_m \end{cases}$$

where α defines the shape of the probability distribution and must be $\alpha > 0$ and x_m is the scale parameter and which is also the mode of the distribution. It must also be $x_m > 0$.

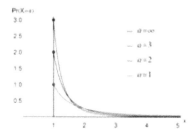

The Pareto distribution is a probability law which was originally used by Vilfredo Pareto to describe the distribution of wealth in society: it seemed that a very large portion of the wealth of any society is owned by a very small percentage of the people in that society. Pareto probability distribution was published in 1896, in his first paper "Cours d'économie politique'', where British income data indicates that the richest 30% of the population had about 70% of the income. Globally, those numbers are moving towards a wider separation between rich and poor: a 2014 report on wealth distribution from Credit Suisse highlights that 1% of the world population own 48% of the wealth; 10% own 87% and bottom 50% own less than 1% [1].

Pareto developed the principle by observing that 20% of the peapods in his garden contained 80% of the peas. The case 20-80 is a special case of the Pareto probability distribution for alpha equal 1.16, and it is widely used in trading and business. Other examples of variables having a Pareto probability distribution are:

- The sizes of human settlements (few cities, many hamlets/villages);
- The values of oil reserves in oil fields (a few large fields, many small fields);
- The numbers of software crashes and the line of code responsible for it (few lines of code responsible for most of the crashes)

Another and last example of how the Pareto probability distribution affects everyone lives is attributed to Woody Allen: "*80 percent of success is showing up* ".
https://publications.credit-suisse.com/tasks/render/file/?fileID=5521F296-D460-2B88-081889DB12817E02

GM

Equation 237: Bend It Like Beckham

A classic moment when England captain David Beckham scores with a sensational 30-yard free kick, three minutes into injury-time. Its two and a half minutes in to injury time. England are 2-1 down against Greece and are not going to the World Cup finals. England get a free kick outside the area and up steps David Beckham with a curving free kick to score.

Ball games have universal appeal because of their simplicity with the pleasure lying in simply kicking a ball as fast as possible or striking it sweetly with a bat, racket or club. But as games became more competitive, players began to realise that the ball's flight could be modified by hitting or kicking it in a particular way. This might involve hooking the ball around an obstacle in golf or swerving it over the defensive wall in a football match.

So although Beckham's ability to score in that match was the result of skill and hours of practice, what is the underlying Physics here?

Isaac Newton (see earlier equations) was the first scientist to record that a tennis ball would swerve when it was struck with an oblique racket (sliced shot) when trying to explain refraction. However, very little science was applied to the subject of ball flight until the mid-19th century when a Scottish mathematician, Peter Guthrie Tait, turned his attention to golf, his favourite game. Tait modelled a golf ball's flight using the familiar forces of gravity and air resistance and produced a range and time-of-flight that were much too short. His great insight was to realise that a spinning ball would generate aerodynamic lift, which would prolong the flight considerably.

This third force, only present when the ball is both spinning and moving forwards, is nowadays called the Magnus force after its discoverer, the German physicist Heinrich Magnus.

The beautiful thing about the Magnus force is its direction. The force direction is always at right angles to the plane containing the velocity vector and the spin axis. So for a ball spinning about a horizontal axis (that is, pure backspin) the force will be vertical, just as Tait had envisaged. But tilt the spin axis and the deflecting force follows suit. Therefore, if you control the inclination of the spin axis, and hence the direction of the Magnus force, you modify the flight of the ball, something that was discovered by a number of gifted Brazilian footballers in the 1950s.

Both the drag and the Magnus forces follow a squared relationship of the form

$$F = \frac{1}{2} \times air\ density \times ball\ cross\ sectional\ area \times ball\ speed^2 \times C$$

where C is a dimensionless number that scales the strength of the drag or Magnus force and is usually subscripted d or m so that we know which force we are considering.

$$\frac{d^2x}{dt^2} = -vk\left(C_d\frac{dx}{dt} + C_m\frac{dy}{dt}\right)$$

$$\frac{d^2y}{dt^2} = -vk\left(C_d\frac{dy}{dt} + C_m\frac{dx}{dt}\right)$$

$$\frac{d^2z}{dt^2} = -g - vkC_d\frac{dz}{dt}$$

where

$$v = \sqrt{\left(\frac{dx}{dt}\right)^2 + \left(\frac{dy}{dt}\right)^2 + \left(\frac{dz}{dt}\right)^2},$$

$$k = \frac{\rho A}{2M}$$

and M is the mass of the ball.

Solving these differential equations is not a trivial task. There are no known solutions such that the ball's co-ordinates, x, y and z can be expressed as exact formulae. We therefore need to model the process using a personal computer. The other problem is determining the values of C and this is usually done by mounting the football in a wind tunnel. An alternative method is to measure the actual trajectory of a football using high speed cameras.

The human brain solves all these equations in an instant. However, the ability to do so relies on hours of practice on the football field when the brain works out which combinations work and which don't. It's easy isn't it? Well, actually no, only 10% of free kicks in English football's premier league are actually successful. If you look at the process in more detail it's actually chaotic. Future research in this field will look much more closely at the ball's specific aerodynamic properties, its surface structure and panel pattern.

All of the methods described, from Tait's rudimentary approach, to the complexity of computational fluid dynamics, derive ultimately from the Navier-Stokes equations. Solve the Navier-Stokes equations and you can claim a prize of $1 million. A fortune for most mathematicians but hardly enough to pay a middle ranking footballer for a year.

Reference
Modelling the flight of a soccer ball in a direct free kick by Ken Bray and David Kerwin in the *Journal of Sports Sciences*, Vol 21, pp 75-85, February 2003.)

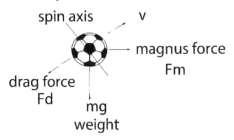

DF

Equation 238: The Column Buckling Formula

Consider a long, slender bar such as a straight ruler. If a compressive load is gradually applied at the two ends of the ruler, the ruler decreases in length while remaining straight initially. However, as soon as the applied load reaches a critical value, the ruler bends sharply with a very large lateral deflection. Such a deformation due to compressive axial load is termed buckling.

Buckling is an important consideration in structures containing members in compression. For example, in the truss structure shown below, depending on the magnitudes and directions of the loads, some members may be in compression and hence can buckle when the forces acting in them approach a critical value.

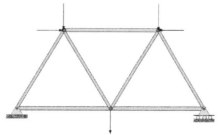

Figure 1: A simple truss with members in possible compression due to the applied external loads.

Each of the members of the truss can be thought of as a two-force member (bar) with the ends being pinned. For such a member, the critical load for buckling was obtained by the mathematician Leonhard Euler in 1757 (under the assumption of elastic behaviour) as

$$P_{cr} = \frac{\pi^2 E I}{L^2}$$

Here, E is the elastic (Young's) modulus of the bar, L is the length of the bar, I is the second moment of the cross-sectional area about an axis about which the bar bends (buckles). The above equation is valid for a pinned-pinned bar as shown

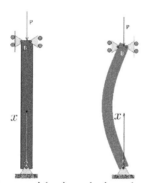

Figure 2: An example of buckling: a column with pinned-pinned ends.

in Figure 2 and by determining the conditions under which a lateral deflection is possible due to the axially applied compressive load.

The buckling load equation has several interesting features:

- The critical load for buckling is inversely proportional to the square of the length of the bar. Thus, a bar of length $2L$ has a critical load that is one fourth of the critical load for a bar of length L.
- If we consider two columns (bars) with the same cross-sectional area, the one with the lower I will have the lower critical load for buckling. For example, the I for a circular cross-section is smaller than the I for a square cross-section with the same area. Therefore, a column with the circular cross-sectional area will have a lower critical load for buckling.
- Another interesting conclusion is that for a given cross-section, buckling will take place about that axis with respect to which the second moment of area is the smallest. For example, if we consider a rectangular cross-section with a base b and height h with $b < h$, then, then I is smaller when the moment is taken about 2-2 axis instead of 1-1. Therefore, the bar will buckle about the 2-2 plane.

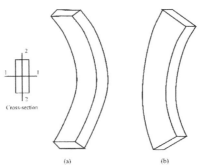

Figure 3. Under compressive loads, a rectangular bar will buckle in the mode shown in (a) and not (b) since the second moment of the cross-section about 2-2 axis is smaller than about 1-1 axis.

16. Given a circular cross-section and a tube, both with the same cross-sectional areas, the tube will have a higher I and therefore, the critical buckling load will be higher for a tubular column. For this reason, tubular columns are preferred over solid circular columns when failure by buckling is of concern.

When the end conditions are different from pinned-pinned, the critical load for buckling is multiplied by a factor k that depends on the nature of the boundary conditions. For example, if a column is fixed at one end and pinned at the other end, the factor k is approximately 2.05.

KN

Equation 239: Strehl Ratio

The Strehl ratio is a number between zero to one that gives a representation of the amount of aberration of an image from its point source compared to an ideal optical system.

The Strehl ratio is the ratio of the peak abberated image intensity from a point source compared to the maximum theoretical intensity from a "prefect" system.

The Strehl ratio is expressed as S given by

$$S = \left|\langle e^{i\phi}\rangle\right|^2 = \left|\langle e^{i2\pi\delta/\lambda}\rangle\right|^2$$

where i is the imaginary unit and the phase error over the aperture is expressed as (see second equation)

$$\phi = 2\pi\delta/\lambda$$

The Strehl ratio is evaluated over the average area of the aperture expressed as $A(x, y)$ and is typically compared to the point source peak intensity on axis given by $A(0,0)$.

The Strehl ratio is commonly used in astronomy where atmospheric aberration can play a significant part in imaging (interesting aside, read about adaptive optics and correction of atmospheric aberrations - very interesting).

The Strehl ratio is also a useful real world performance metric by which to evaluate an optical system. Perfection is hard to come by but is often the target, it is important to note that the reliability and effectiveness of the Strehl ratio diminishes as the quality of the optical system is reduced, therefore it is of little use in standard photography applications and is usually reserved for high end optical applications.

286

The Strehl ratio is named after Karl Strehl(1864-1940) who initially proposed it. Adaptations and real world improvements have been made by applying various Fourier techniques and also by a statistical phase evaluation that simplifies the equation by Mahajan's formulas.

Reference
https://en.wikipedia.org/wiki/Strehl_ratio

STS

Equation 240: The Born-Lande' Equation

The Born-Lande' equation calculates the lattice energy of an ionic crystal as a function of the distance between two adjacent ions (r_0). The equation is the sum of two contribution, an attractive term and a repulsive term.

$$E(r_0) = -\frac{M N_A z^2 e^2}{4\pi \varepsilon_0 r_0}\left(1 - \frac{1}{n}\right)$$

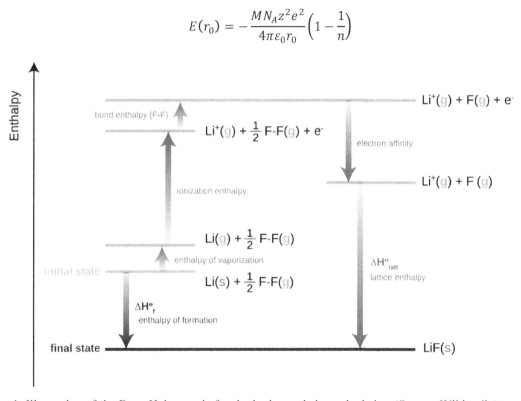

Figure 1. Illustration of the Born-Haber cycle for the lattice enthalpy calculation (Source: Wikipedia)

where z is electric point charge of the ions (z^+ and z^-), e is the electron elementary charge ($1.6022\cdot10^{-19}$ C); ε_0 is the dielectric permittivity in vacuum ($1.6022\cdot10^{-19}$ C); N_A is Avogadro's number ($6.022\cdot10^{23}$); M is Madelung constant; n is integer number, which varies between 5 and 12.

The Born-Lande' equation assumes a simple model of the ionic crystal, in which the ions can be approximate to hard spheres with point charges.

The lattice energy of an ionic crystal corresponds to the variation of enthalpy that is needed to break the bonds of a solid ionic crystal and bring the ions in gas phases. This enthalpy cannot be directly evaluated through the transformation of a crystalline solid to a plasma state, as the latter state is unstable and requires high energy for its formation. The lattice enthalpy is usually derived indirectly, by evaluating all the energy terms involved in the reactions within the Born-Haber cycle (Figure 1).

The value of this enthalpy can be compared to the energy calculated through the Born-Lande' equation, and can be used to quantify the distance between two ions. If the radius of an ion is known, this distance can be used to evaluate the atomic radius of the other ion.

The equation was proposed by the German physicists Max Born and Alfred Lande' in 1918, who worked together at the University of Goettingen. Among other physicists who worked with Lande' and Born were Edmund Landau and Carl Runge, while also Niels Bohr frequently visited the university.

REPULSIONS AND ATTRACTIONS

The attractive term of the equation, also called Madelung energy, is the result of the Coulombic electrostatic potential between pairs of ions of equal and opposite charges $z-$ and $z+$, with r being the distance between the two ionic species.

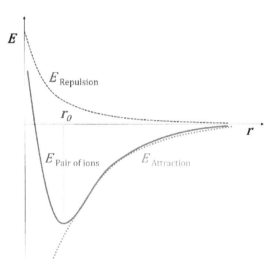

$$E_M(r_0) = -\frac{M N_A z^2 e^2}{4\pi\varepsilon_0 r_0}$$

The repulsive term in the Born-Lande' equation is due to the repulsive interactions between two ions when at short distance from each other. This term is not only the result of the electrostatic repulsion between two ions of same charge, but also of a repulsive force between ions of opposite charge caused by the repulsive electron – electron repulsions of the atomic shells.

Figure 2. Schematic trend of the energy of ions as a function of their relative distance.

$$E_R(r_0) = -\frac{N_A B}{r_0{}^n}$$

The exponent (n) is derived experimentally by measuring the compressibility of solids. B is a constant that scales the strength of the repulsive interaction. It is derived by the condition:

The repulsive term becomes dominant for ion-ion distances smaller than the equilibrium distance r0. A typical curve of the potential energy for a pair of ions is shown in Figure 2.

MADELUNG CONSTANT

An important factor of the Born-Lande' equation is the Madelung constant, M. This parameter takes into account the electrostatic potential experienced by a single ion from the whole crystal. As this potential depends on the crystal structure and its symmetry, this factor ultimately account for the crystal structure contribution and varies according to the different crystalline geometries and parameters such as lattice constants. M is a numerical value to which a mathematical series converges. For example, for an ionic solid AB (e.g. NaCl, LiF, ZnS, MgO, etc...), an A+ ion is surrounded by 6 anions B+ at closest distance $r = a/2$ (Figure 3, green), where a is the size of the cubic unit cell of the crystal. The next closer species are 12 cations A^+ at a distance $r = a/\sqrt{2}$ (Figure 3, red). The next nearer species are 8 anions B^- at a distance $r = a/\sqrt{3}$ (Figure 3, blue), and so on.

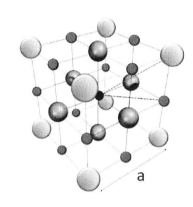

Figure 3. Schematic representation of the closer neighbours to a central ion (black) in an AB type ionic solid structure (Source: Wiki Commons)

The Madelung constant for this type of structure will be:

$$-6 + 12/\sqrt{2} - 8/\sqrt{3} + 6/\sqrt{4} - 24/\sqrt{5}...$$

This series converges to the numerical value of -1.748. This method of defining the sum is called "expanding sphere", because the contributions to the sum are evaluated by exploring the surrounding environment of a central ion by moving radially away from the species.

In 1951, the German mathematician O. Emersleben demonstrated that the above series is only conditionally convergent, depending on how the terms are grouped and truncated. Today, there are other methods nowadays to calculate Madelung's constants for different crystals avoiding the problem of conditionally convergent sums, for example the Evjen method or the Ewald method.

AT

Equation 241: Campbell Equation

The Campbell equation is the expression for the propagation constant of a loaded line. The propagation constant is a measure of the amplitude of a wave as it propagates in a given direction.

Developed by George Campbell (27th November 1870 – 10th November 1954), an American Electrical Engineer, Campbell developed mathematical methods to solve the problems of long-distance telegraphy and telephony, with his most important contribution in the use of loading coils.

Loading coils are inductors used to prevent signal distortion over long distance transmission cables or in radio antennas. Originally proposed by Oliver Heaviside in the 1860s, while studying slow signal speeds in the first transatlantic telegraph cable, he concluded that amplitude and time delay distortion of a transmitted signal could be prevented by adding additional inductance to the line.

Campbell, working for AT&T at the time, proposed distributing loading coils along a transmission line and this was tested successfully on a telephone cable for the first time in Boston in 1899. Loading coils were put into public service, again in Boston the next year.

The Campbell equation, which can be used to calculate the separation distance of loading coils on a line, is given in the equation.

$$\cosh(\gamma'd) = \cosh(\gamma d) + \frac{Z}{2Z_0}\sinh(\gamma d)$$

where γ is the propagation constant of the unloaded line, γ' is the propagation constant of the loaded line, d is the interval between coils on the loaded line, Z is the impedance of a loading coil and Z_0 is the characteristic impedance of the unloaded line.

Campbell's work later provided the basis for his work on filters, which became very important for frequency division multiplexing, where the total available bandwidth is divided into a series of non-overlapping frequency sub-bands, each used to carry a separate signal.

CM

Equation 242: Principal Planes of a "Thick" Lens

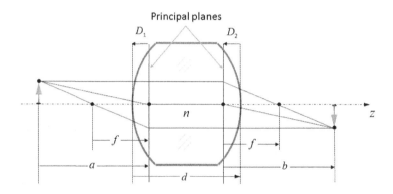

In beautiful equation 115, we learned that "thin" lenses have a very simple imaging formula:

$$\frac{1}{a} + \frac{1}{b} = \frac{1}{f}$$

where a is the object distance, b is the image distance and f is the focal length measured from the centre of the lens. A thin lens has the peculiar definition that it has no thickness at all. When tracing rays from object to image, the refraction is imagined to occur abruptly at a single plane bisecting the lens.

Real lenses have curved surfaces and some thickness to them, which makes the ray paths discouragingly hard to draw and to calculate when compared to the imagined thin lens. It can be done, of course, and in the magical paraxial approximation where trigonometry is linear, matrix methods take us from one surface to another. But the simplicity of the thin lens equation is lost.

To recover the simplicity of the thin lens, we simply declare that the thick lens *must* behave like a thin lens. We insist upon it. To make this happen, we replace the single imaginary plane of the thin lens with *two* planes - the *principal planes* - one for the object side and the other for the image side, and allow the positions of these planes to be adjustable to satisfy the maths. To our joy this actually works, as long as we position these two planes from the apex of each of the two lens surfaces like this:

$$D_1 = -\frac{fd}{f_2 n}$$

$$D_2 = -\frac{fd}{f_1 n}$$

where f_1 and f_2 are the focal lengths of the two refracting surfaces considered independently, n is the refractive index, d is the lens thickness, and the effective focal length of the complete lens is

$$f = \frac{f_1 f_2}{f_1 + f_2 - \dfrac{d}{n}}$$

This last equation is the lens-makers formula, Beautiful Equation 23, in disguise. Equipped with the principal planes, the thin-lens formula works just as before, as long as we measure the focal length f and the distances a and b from the planes D_1 and D_2 respectively. This also rescues the geometric tricks that allow us to trace out the locations of images using chief rays and so on. The concept of principal planes is even more powerful than this, in that we can use these two planes to simplify not just thick lenses, but entire optical systems with multiple lenses on axis. Really cool. Beautiful, actually.

As a historical note, C. F. Gauss proposed principal planes or *Hauptebenen* dates already in 1840, in his *Dioptrische Untersuchungen*. For this reason, analysis of optical systems in the paraxial approximation is referred to as Gaussian optics.

PdG

Equation 243: Planck's Law

Today's equation describes the spectral energy distribution of the radiation emitted by a blackbody. In physics, a "blackbody" is a hypothetical entity that completely absorbs all radiant energy incident upon it, reaches some equilibrium temperature, and then reemits that energy at the same rate as it absorbs it. In the equation, U is the spectral energy distribution, λ is the wavelength of the emitted radiation and T is the absolute temperature. The other terms are well-known universal constants: h is Planck's constant (6.626×10^{-34} J s), c is the speed of light in a vacuum ($299\ 792\ 458$ m s^{-1}) and k is the Boltzmann constant (1.38×10^{-23} m^2 kg s^{-2} K^{-1}). The figure is a plot of E against λ and basically shows that as one applies radiant energy to the body, i.e. heats it up, the body will respond by emitting different colours which are dependent on the amount of heat. We have all seen a metal bar rod glow when heated, but maybe not appreciated that the colour of the glow varies with heat. Most of the "glow" will actually be in the form of infrared, then as more heat is applied will shift down to red and then towards blue – this spectrum will be dependent on the material.

The equation was derived by Max Planck, a German physicist, in 1900. Planck was treating the body as being made up of atoms and the only way he could get the maths to work, was to assume that they emitted radiation at a set of discrete wavelengths, with nothing in between. This, along with insights from Albert Einstein, led to quantum mechanics, a theory that would revolutionise physics and is still the most proven theory ever. For me, this equation is one of the most beautiful ever: it describes fundamental behaviour of radiation and atoms, it contains three of the most basic universal constants and led to one of the cornerstones in physical theory. What a guy!

$$U = \frac{8\pi hc}{\lambda^5} \frac{1}{\exp\left(\dfrac{hc}{\lambda kT}\right) - 1}$$

291

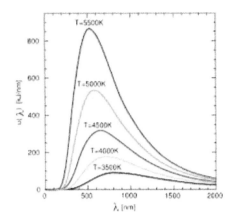

RL

Equation 244: De Morgan's Laws

$$\neg (A \wedge B) = \neg A \vee \neg B$$

$$\neg (A \vee B) = \neg A \wedge \neg B$$

In a grand house, a chandelier hangs from a ceiling via a two-link chain. If either of the links breaks, the chandelier shall fall to the floor. Put formally, the statement, "the chandelier is falling" becomes true when either of the following statements becomes false: "Link A is secure" or "Link B is secure".

Boolean algebra is a shorthand for expressing true/false statements. The description above can be turned into the following:

$$Q = \neg A \vee \neg B$$

where Q is the statement "the chandelier is falling", A is the statement "Link A is secure", B is the statement "Link B is secure", $\neg$ is the logical NOT operator (i.e. $\neg X$ is NOT X or the logical inverse of X) and $\vee$ is the logical OR operator (described in Table 1).

Table 1: The Truth Table for a Logical OR Operation where X and Y are inputs and O is the output. A gate that is true so long as either, or both, of it inputs are true.

X	Y	O
0	0	0
0	1	1
1	0	1
1	1	1

Another way of expressing the same situation is that for the chandelier to fall is to say that the statement "Both Link A and Link B are secure" must become false.

$$Q = \neg (A \wedge B)$$

292

where ∧ is the logical AND operator (described in table 2).

Table 2: The Truth Table for a Logical AND Operation where X and Y are inputs and O is the output. A gate that is only true if both of its inputs are true.

X	Y	O
0	0	0
0	1	0
1	0	0
1	1	1

This relationship demonstrates that the two equations are related, i.e.

$$\neg (A \wedge B) == \neg A \vee \neg B$$

This is the essence of the first part of De Morgan's Laws, that describe the relationship between the logical OR function and the inversion of the logical AND function (the inversion of "AND", is equivalent to "NOT AND", and is usually abbreviated to "NAND").

The second half of De Morgan's Laws can be illustrated by another example in our grand house. The house has an electrical system with two generators, one primary generator and a backup. For the statement "the house has no power" to be true, the statements "Generator A is functioning" and "Generator B is functioning" must both be false.

$$Q = \neg A \wedge \neg B$$

Equally, the system can be said to have failed if neither system A nor system B is functioning

$$Q = \neg (A \vee B)$$

i.e.

$$\neg (A \vee B) == \neg A \wedge \neg B$$

Table 3: The Truth Table for the statements in the previous examples

A	B	¬A ∨ ¬B	¬(A ∧ B)	¬A ∧ ¬B	¬(A ∨ B)
0	0	1	1	1	1
0	1	1	1	0	0
1	0	1	1	0	0
1	1	0	0	0	0

Table 3 is a truth table (a table that compares every possible logical input with the corresponding output(s)) comparing the logical expressions used in the previous examples. Note how the final 2 columns are identical and the 3rd and 4th columns are identical. In these examples, demonstrating that these relationships hold may seem trivial, but De Morgan's Laws have had a significant impact on modern society, through their effects on the cost of affordable electronics.

Before the advent of affordable microprocessors, household electrical devices such as washing machines and other white goods that required a low-level of 'intelligent operation' had their 'intelligence' implemented using combinations of logic gates. The microchips that implemented logic gates would be clusters of a particular gate type, for example you could buy 4 NAND gates on a single microchip and 4 NOR gates on another. The microchips themselves were, at that time, significantly more expensive than they are today and the space they occupied also necessitated larger, more expensive circuit boards. Layout constraints also drove cost, as connecting multiple microchips could introduce timing errors between them. This meant that complex logic using multiple logical operations was expensive and potentially unreliable. To make matters worse, these collections of gates on a single microchip meant that there were often many unused gates on the board (imagine a circuit that uses only one AND and one OR, this would leave you with six unused gates if using two microchips, implementing four gates of each type).

De Morgan's Laws are a formalism that allow circuit designers to rephrase logical expressions in a way that optimises microchip use. In fact, combined with another logical law that states that cross-linking the inputs to a NAND gate turns it into a NOT gate ($\neg(A \wedge A) = \neg A$ in Boolean algebra), De Morgan's Laws allow any logic circuit to be rephrased as only NAND gates, cutting back on the number of microchip types and allowing designers to reuse the gates on a single microchip. This in turn reduced the number of microchip to microchip connections and improved reliability. Not only were manufacturers able to reuse gates on a single microchip, they could also employ a greater economy of scale, ordering hundreds of thousands of NAND gates instead of tens of thousands of multiple microchip types.

This huge impact on the affordability of consumer electronics was a major driver for the digital revolution that we see today. An achievement of note for a family of equations first formalised in the Nineteenth Century by the British mathematician Augustus De Morgan, whom the laws are named after.

RFO

Equation 245: The Compton Effect

$$\Delta\lambda = \frac{h}{mc}(1 - cos\theta)$$

INTRODUCTION
Compton scattering is the inelastic scattering of an X or γ ray photon by a (virtually) free charged particle (usually an electron), in which some of the photon's energy is transferred to the recoiling particle. (If the scattered photon has enough energy left after the first interaction, the process may be repeated.)

The term 'Compton effect' is virtually synonymous with 'Compton scattering'; perhaps one might feel inclined to use the former term when alluding mainly to the wavelength change aspect (the amount by which the wavelength changes is called the 'Compton shift'), and the latter term (scattering) when alluding mainly to the photon's change of path.

Inverse Compton scattering (important in astronomy) also exists, in which it is the charged particle that transfers part of its energy to the photon.

PHOTONS IMPINGING ON ATOMS

294

Although nuclear Compton scattering (where protons are the scattering particles) exists, the term Compton scattering mostly refers to interactions involving the atomic electrons. X-ray photon energies are very much larger than the binding energy of loosely bound electrons, so the latter can be treated as being free.

• A lower energy photon - a few eV to a few keV, i.e. visible light through soft X-rays - will eject an electron from its host atom (the photoelectric effect), instead of undergoing Compton scattering.

• A photon of energy comparable to the electron's rest energy of 511 keV, i.e. within the hard X-ray range, may lose part of its energy to the electron, making it recoil (Compton scattering). Compton scattering is the most probable type of interaction of high energy photons with atoms.

• A higher energy photon of 1.022 MeV or more, i.e. gamma rays or super hard X-rays, may cause the formation of an electron plus a positron (pair production).

HISTORY
The Compton Effect was discovered in 1923 by Arthur Holly Compton, who earned the Nobel Prize in physics for this work.

The effect is historically important because it corroborated existing evidence that light cannot be explained purely as a wave phenomenon.

APPLICATIONS
Compton scattering is used in, amongst other things, radiobiology, radiation therapy, material physics and gamma spectroscopy.

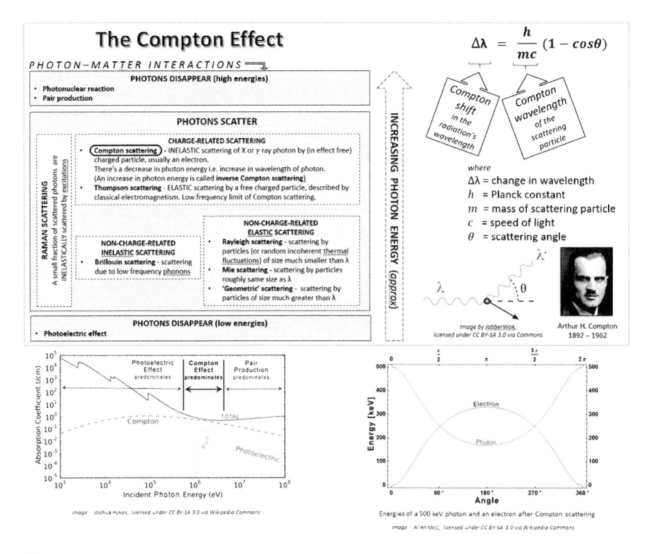

The Compton Effect

$$\Delta\lambda = \frac{h}{mc}(1 - cos\theta)$$

Compton shift in the radiation's wavelength

Compton wavelength of the scattering particle

where

$\Delta\lambda$ = change in wavelength
h = Planck constant
m = mass of scattering particle
c = speed of light
θ = scattering angle

Image by JabberWok, licensed under CC BY-SA 3.0 via Commons

Arthur H. Compton
1892 – 1962

PHOTON–MATTER INTERACTIONS

PHOTONS DISAPPEAR (high energies)
- Photonuclear reaction
- Pair production

PHOTONS SCATTER

CHARGE-RELATED SCATTERING
- (Compton scattering) - INELASTIC scattering of X or γ ray photon by (in effect free) charged particle, usually an electron. There's a decrease in photon energy i.e. increase in wavelength of photon. (An increase in photon energy is called **inverse Compton scattering**)
- **Thompson scattering** - ELASTIC scattering by a free charged particle, described by classical electromagnetism. Low frequency limit of Compton scattering.

RAMAN SCATTERING
A small fraction of scattered photons are INELASTICALLY scattered by excitations

NON-CHARGE-RELATED INELASTIC SCATTERING
- **Brillouin scattering** - scattering due to low frequency phonons

NON-CHARGE-RELATED ELASTIC SCATTERING
- **Rayleigh scattering** - scattering by particles (or random incoherent thermal fluctuations) of size much smaller than λ
- **Mie scattering** - scattering by particles roughly same size as λ
- **'Geometric' scattering** - scattering by particles of size much greater than λ

PHOTONS DISAPPEAR (low energies)
- Photoelectric effect

INCREASING PHOTON ENERGY (approx)

Image - Joshua Hykes, licensed under CC BY-SA 3.0 via Wikipedia Commons

Energies of a 500 keV photon and an electron after Compton scattering

Image - Allen McC, licensed under CC BY-SA 3.0 via Wikipedia Commons

FR

Equation 246: Conic Section

A conic section (or "conic") is a two-dimensional curve that arises from the intersection of a right (symmetric) cone with a plane. In the case of conic sections the cone is taken to be a double cone, which is a quadratic surface of revolution that can be constructed by tracing a line at a fixed angle about an axis which intersects it. The curves generated by the intersecting planes are second-degree plane algebraic curves which are often encountered in physical situations, such as those involving relative motion. Depending on the angle of the intersecting plane relative to the cone axis, on the whole two curve types of variable proportions result: ellipse and hyperbola. There are two special cases: a circle (ellipse with equal major and minor axes, in which the angle of the intersecting plane is perpendicular to the cone axis of symmetry) and a parabola (the unique curve that occurs between ellipses and hyperbolae in which the intersecting plane is oriented parallel to the cone edge). The general form of a conic section can be described in Cartesian coordinates by the bivariate quadratic equation

$$Ax^2 + Bxy + Cy^2 + Dx + Ey + F = 0$$

which reduces to the well-known forms of equations for the circle (beautiful equation 157), the ellipses (beautiful equations 92 & 171), the parabola and the hyperbolae, and can be classified by the value of the discriminant $B^2 - 4AC$ is represented by the powers of x and y (provided the conic is not degenerate, see below). If the discriminant is less than 0 the curve is an ellipse; when it is equal to 0 the curve is a parabola; when it is greater than 0 the curve is a hyperbola. This distinction is convenient for the classification of linear second-order partial differential equations (PDEs) encountered in physics, in which the order of derivative (space or time) is in a characteristic equation similar to that above. This is the basis for most of the common PDEs describing systems that change in space and/or time coordinates, and the nature of their solutions is closely related to their "conic" classification. The three classic forms are represented by the Laplace equation (elliptic, steady state); the heat or diffusion equation (parabolic, evolves in time but generally decays in spatial coordinates); the wave equation (hyperbolic, evolves in both space and time coordinates). The Laplace equation and the wave equation have been covered in beautiful equations 48 and 198. Going back to the geometric representation, examples of the plane intersections generating the curves are shown below.

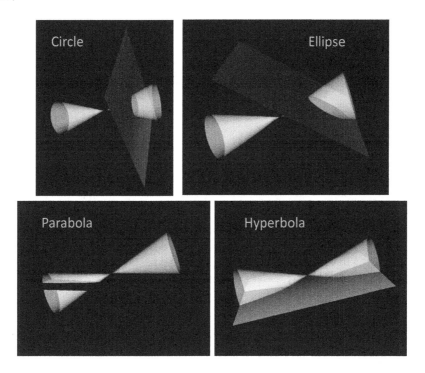

Note that only in the case of a hyperbola does the plane intersect both cones and hence the curve is separated into two curves that are always mirror images of one another, regardless of the intersection angle.

The curves produced by conic sections have a defining property called eccentricity ε, (previously described for an ellipse in beautiful equation 92) which varies depending on the curve type relating to the angle of plane intersection. The eccentricity can be related to a point inside the curve called the focus and a line outside it called the directrix. The ratio of the distance of a point on the curve to the focus and its perpendicular distance to the directrix line is the same for all points on the curve. In the case of the parabola (a unique curve), the ratio is 1. The focus of a parabola is very important in optics for example: if a parabola is extended by rotating about its axis of symmetry in three dimensions into a paraboloid, a mirror with a source at the focus will result in a collimated (parallel) beam. For ellipses the ratio is less than 1 (and for a similar reason, an ellipsoidal mirror may be used as a focusing mirror), in the special case of a circle being 0; for hyperbolae it is greater than 1.

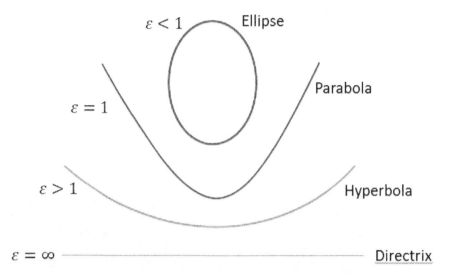

In the examples shown in the first picture the curves were generated by intersecting the plane through a section of the cone that is away from the apex (the point at the centre). An interesting limiting case occurs when the plane passes through the apex – instead of ellipses there is a continuum of angles of intersection that generate a point only (effectively the same point), which instantaneously becomes a line when the angle of intersection reaches the cone angle, and then a "X"-shaped object (two intersecting lines) for the angles that would otherwise generate a hyperbola. These are examples of "degenerate" conics.

The ancient Greeks were aware of conics, and an understanding of them probably started with Menaechmus in the 4th century BC, followed by Euclid and Archimedes, although there are no surviving accounts from them. A very extensive treatment of conics was given by Apollonius around 200 BC and covered much of the definition that is still used today. A large amount of Apollonius's work only survived via its translation into Arabic by the Persian mathematician Omar Khayyam in the 11th century AD. Later on, conics were further advanced by the likes of Kepler and Pascal, and eventually it was Descartes who was central to developing the theory into its modern algebraic form.

PB

Equation 247: Plasma Frequency

The optical properties of metals are determined by their conduction electrons. These electrons move freely through the ion lattice and can be thought of as a plasma. Plasma oscillations, or Langmuir waves as they are sometimes known, are rapid oscillations of the electron density in metals and were discovered by Irving Langmuir (31 January 1881 – 16 August 1957), an American chemist and physicist.

The relationship between dielectric constant, refractive index and extinction coefficient are well known from the Drude theory and the reflectivity of a metal is a function of both refractive index and extinction coefficient. A consequence of the Drude theory is that for longer wavelengths, materials with high numbers of free carriers (e.g. metals) are almost perfect reflectors.

When radiation is incident on a material, energy is transferred to the lattice through absorption. The electrons can be treated as a plasma and as such, respond to electromagnetic radiation depending on frequency. Metals have a characteristic frequency, known as the plasma frequency, ω_p. This plasma frequency relates electron density to optical properties and determines the point at which the electrons cannot respond fast enough to the frequency of the incident radiation, screening the electric field from the material. In other words, it determines the point at which the metal becomes transparent to the incoming radiation.

The plasma frequency is given in the equation

$$\omega_p = \sqrt{\frac{N_e q^2}{m_e \varepsilon_0}}$$

where N_e is the material's electron density, q is the electronic charge, m_e is the mass of the electron and ε_0 is the permittivity of free space.

The plasma frequency is the point at which the square of the refractive index n, is equal to the square of the extinction coefficient k, or when the real part of the dielectric constant, ε_1 is equal to zero. This is illustrated in the figure below, which shows the point where $n^2 = k^2$ and ε_1 crosses the zero axis in copper and equates to a wavelength value of 155 nm. Below the plasma frequency, the real part of the dielectric constant is negative.

The plasma frequency can be thought of as dividing the frequency range into two parts. The first is the low frequency (increasing wavelength) part where the complex refractive index has an imaginary part and waves are attenuated. This occurs above 155 nm (or 1.93×10^{15} Hz) in copper. The second part is where the refractive index is real and the material becomes transparent (below 155 nm in copper). For metals, the plasma frequency occurs in the ultraviolet range of the electromagnetic spectrum and the gradient of the drop in reflectivity is related to the relaxation time.

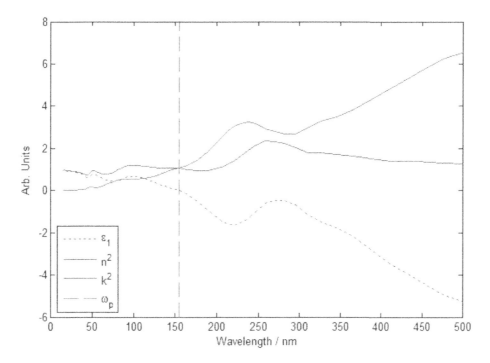

D. R. Lide, Ed., CRC Handbook of Chemistry and Physics, CRC Press, 2005.

CM

299

Equation 248: Polygon Calibration

A polygon with nominal equal angles is one of the basic angular material standards. It can be calibrated by two autocollimators, using the fact that all angles sum up to a multiple of 360°. Figure 1 depicts a polygon and gives a schematic of the calibration set-up.

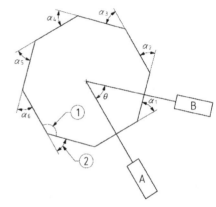

Figure 1. Left: picture of 12-sided polygon. Right: schematic of polygon with 1: prism angles and 2: α_i: polygon angles. A and B are autocollimators

Taking autocollimator A as a reference, the n-sided polygon is rotated n times over its polygon angle until it is in the original position again. The reference-autocollimator should give a reading '0', with small differences that are accounted for. The reading of the second autocollimator is a measure for the polygon angle. For a perfect polygon it should also be zero or at least constant. A measurement M is the difference in reading between autocollimators B and A. By rotating the polygon, a set of measurements $M_i = B_i - A_i$ is obtained, with $i = 1 \dots n$ and n is the number of polygon angles. For position i it can be written that $\alpha_i = M_i + \theta$ (Eq 1), with θ the unknown constant angle between the autocollimators. θ is calculated from the summation over all measurements:

$$\sum_{j=1}^{n} \alpha_j = \sum_{j} M_j + \sum_{j} \theta$$

or

$$360° = \sum_{j} M_j + n \cdot \theta$$

which gives,

$$\theta = \frac{360° - \sum_{j=1}^{n} M_j}{n}$$

Substituting gives the beautiful equation that relates the polygon angles α_j to the measurements M_j:

$$\alpha_i = M_i - \frac{\sum_j M_j}{n} + \frac{360°}{n} = M_i \cdot \left(1 - \frac{1}{n}\right) - \frac{1}{n} \cdot \sum_{j=1}^{n} M_j + \frac{360°}{n}$$

The beauty is that the sensitivity of the autocollimator, that is typically a fraction of an arcsecond measurement, is transferred to the polygon angles that are far beyond its range of a few arcminutes. Also beautiful is the uncertainty in α_i, $u(\alpha_i)$ that is just a bit smaller than the uncertainty in M_i

$$u^2(\alpha_i) = \sum_j \left(\frac{\partial \alpha_i}{\partial M_j}\right)^2 \cdot u^2(M_j) = \left(1 - \frac{1}{n}\right)^2 \cdot u^2(M_i) + \left(\frac{n-1}{n^2}\right) \cdot u^2(M_i) = \left(1 - \frac{1}{n}\right) \cdot u^2(M_i)$$

And the uncertainty in the angle α, that is the sum of m polygon angles that can be derived from:

$$\alpha = \sum_{i=1}^{m} \alpha_i = \left(1 - \frac{m}{n}\right) \cdot \sum_{j=1}^{m} M_j - \frac{m}{n} \sum_{j=n-m}^{n} M_j + \frac{m \cdot 360°}{n}$$

This gives for the uncertainty in α:

$$u^2(\alpha) = m \cdot \left(1 - \frac{m}{n}\right)^2 \cdot u^2(M) + \left(\frac{m}{n}\right)^2 \cdot (n-m) \cdot u^2(M) = \left(m - \frac{m^2}{n}\right) \cdot u^2(M)$$

With as typical results a zero uncertainty for the trivial case that $m = n$, and a maximum uncertainty for $m = n/2$; that means for $\alpha = 180°$ when n is even: $u(\alpha) = 0.5 \cdot \sqrt{n \cdot u(M)}$

HH

Equation 249: The Yellow Change Formula

The duration of a yellow light at a traffic junction isn't the same for all traffic lights and it isn't arbitrary. There is, in fact, a formula that takes into consideration several factors related to the junction itself. The duration of a yellow light is determined to give drivers enough time to completely stop their vehicle or to clear a junction before the light turns red and is given by the following equation [1].

$$Y = t_p + \frac{1}{2}\left(\frac{v}{a + Gg}\right)$$

where Y is the duration of the yellow light, t_p is a constant corresponding to the time it takes for a human to react   given a visual signal (typically 1 second), v is the approach speed[1] of the vehicle (in m/s), a is a 'safe and comfortable' deceleration rate of a vehicle[2] (m/s²), G is the grade of the road (in %/100, downhill grade is negative), and g is the acceleration of gravity (9.8 m/s²).

The formula was first proposed in a 1960 publication [2] by three employees of the General Motors Company. The formula was modified in 1982 to account for the effects of gravity on a vehicle traveling either uphill or downhill (non-zero G).

[1] The value assigned to v can be equivalent to the speed limit of the road or another value deemed appropriate for the junction. If the approach speed is overestimated, the likelihood of a vehicle clearing the traffic junction before the light turns red increases.

[2] The value attributed to the 'safe and comfortable' deceleration rate can vary by organization, but is typically 3 m/s² or more.

[1] McGee, H., Moriarty, K., Eccles, K., Liu, M., Gates, T., and Retting, R. *NCHRP Report 731: Guidelines for timing yellow and all-red intervals at signalized intersections*, Transportation Research Board 2012.

[2] Gazis, D., R. Herman, and A. Mardudin. The Problem of the Amber Signal Light in Traffic Flow. Operations Research: Vol. 8, No. 1, 112–132, January–February 1960.

MF

Equation 250: Orthogonal Matrices

One of the most beautiful equations in all of Linear Algebra is the equation that defines orthogonal matrices (denoted here by Q):

$$QQ^T = Q^T Q = I$$

Here, the superscript T stands for transpose and I is the identity matrix. The above equation says that an orthogonal matrix is that matrix whose transpose is its inverse! Orthogonal matrices play an important role in linear algebra and in many applications such as numerical linear algebra, mechanics and computer graphics.

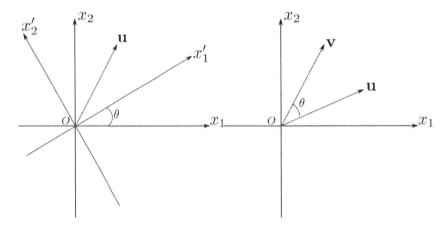

Figure 5: (Left). The primed coordinate system is obtained by rotating the unprimed system through an angle θ. (Right). The vector v is obtained by rotating the vector u through an angle θ.

Suppose that we have two Cartesian coordinate systems Ox_1x_2 and $Ox_1'x_2'$ with a common origin O. The second coordinate frame is obtained by rotating the first about the origin through an angle θ. Consider a vector u with components u_1 and u_2 in the unprimed coordinate system. Then, it can be shown that the components $[u_1' \quad u_2']^T$ of u in the primed coordinate system are given by the equation

$$\begin{Bmatrix} u_1' \\ u_2' \end{Bmatrix} = \begin{bmatrix} \cos\theta & \sin\theta \\ -\sin\theta & \cos\theta \end{bmatrix} \begin{Bmatrix} u_1 \\ u_2 \end{Bmatrix}$$

The 2×2 matrix, denoted by Q is an orthogonal matrix since

$$\begin{bmatrix} \cos\theta & \sin\theta \\ -\sin\theta & \cos\theta \end{bmatrix} \begin{bmatrix} \cos\theta & -\sin\theta \\ \sin\theta & \cos\theta \end{bmatrix} = \begin{bmatrix} \cos\theta & -\sin\theta \\ \sin\theta & \cos\theta \end{bmatrix} \begin{bmatrix} \cos\theta & \sin\theta \\ -\sin\theta & \cos\theta \end{bmatrix} = \begin{bmatrix} 1 & 0 \\ 0 & 1 \end{bmatrix}$$

If instead, the vector u is rotated through an angle v through the angle, then, the two vectors are related through

$$\begin{Bmatrix} v_1 \\ v_2 \end{Bmatrix} = \begin{bmatrix} \cos\theta & -\sin\theta \\ \sin\theta & \cos\theta \end{bmatrix} \begin{Bmatrix} u_1 \\ u_2 \end{Bmatrix}$$

302

Or

$$v = Q^T v$$

That is, the Q in the transformation matrix for the vector rotation case is just the transpose of the transformation matrix for the case where the vector is fixed but the coordinate axes are rotated. Similarly, in three dimensions, the components of a vector transform according to the following rule:

$$\begin{Bmatrix} u_1' \\ u_2' \\ u_3' \end{Bmatrix} = \begin{bmatrix} \lambda_{1,1} & \lambda_{1,2} & \lambda_{1,3} \\ \lambda_{2,1} & \lambda_{2,2} & \lambda_{2,3} \\ \lambda_{3,1} & \lambda_{3,2} & \lambda_{3,3} \end{bmatrix} \begin{Bmatrix} u_1 \\ u_2 \\ u_3 \end{Bmatrix}$$

Here, $\lambda_{i,j}$ is the cosine of the angle between x_i' axis and x_j axis. It can be shown that the coefficient matrix $[\lambda_{i,j}]$ is also an orthogonal matrix. It is worth pointing out that vectors are often defined using the above transformation rule.

In numerical linear algebra, orthogonal matrices occur in many places. For example, the well-known spectral theorem in linear algebra states that a symmetric, real matrix A of size n admits the spectral decomposition $A = UDU^T$ where D is the diagonal matrix consisting of the eigenvalues of A and U contains the eigenvectors of A. This theorem is exploited to obtain the eigenvalues of a given symmetric matrix A through a series of orthogonal similarity transformations (See beautiful equation 158). Another method to extract the eigenvalues of a matrix A is the QR algorithm which uses the QR decomposition of the matrix into an orthogonal matrix Q and an upper triangular matrix R such that $A = QR$.

HC

Equation 251: Hardy Weinberg Equation

Evolution does not occur continually. For evolutionary change to occur, a disturbing factor such as selection (perhaps through food shortage), mutation, the choosing of mates or the transfer of genes between individuals must be present.

In order to determine whether such factors are present and how effective they are, one must determine what the genetic variation within the population of interest actually is and compare it with what that variation would be in the absence of disturbing factors. The latter is called the equilibrium variation.

How do we know what the equilibrium variation is?

As Gregor Mendel showed, many traits are inherited in a way that can best be understood on the assumption that a particular trait (such as eye colour) occurs in two forms (brown or blue), and that every individual has a pair of alleles for each such trait. One member of each pair is inherited from the mother, one from the father. So, if B represents the allele that "causes" brown eyes and b is blue eyes, an individual may have either BB, Bb or bb allele pairs.

In every such trait, one form is dominant - in the case of eye colour B (brown) dominates. This means that an individual with at least one B allele will have brown eyes. Only if no B alleles are present will blue eyes form. So, of the three possible pairings ("genotypes") BB, Bb and bb, only the bb genotype gives rise to blue eyes.

Given any allele pair A and a, there will always be the three possible genotypes, AA, Aa and aa. If we know the A:a ratio in a particular population, then we can predict the proportions of the genotypes from the Hardy-Weinberg equation:

$$p^2 + 2pq + q^2 = 1$$

where p is the frequency of allele A, q the frequency of allele a, p^2 the frequency of genotype AA, q^2 the frequency of genotype aa and $2pq$ the frequency of genotype Aa.

The equation assumes equilibrium variation and so, by comparing its predictions of genotype frequencies with observed values, the degree of departure from equilibrium can be determined. If there is a significant such departure, then evolution is occurring.

The "Hardy" in the equation is G. H. Hardy, a British mathematician who thought (and wrote) that to use mathematics to deal with practical problems was to degrade what should be a pure subject. However, he was forced to denigrate himself in just this way when, while playing cricket with biologist Reginald Punnet, he heard of an ongoing debate among geneticists.

While Mendel's genetics papers - disregarded in his own time since he published them in an obscure journal - had been "rediscovered" in 1900, his theory was criticised because it was believed that recessive allele proportions could not sustain themselves over generations. The argument had become bogged down, thanks to the ignorance of mathematics among its proponents (it was not only Hardy who thought that "applied mathematics" was an unsavoury concept).

Hardy reluctantly formulated and proved the equation required (a very simple one indeed by modern standards) and published it in 1908, remarking acidly that " I should have expected the very simple point which I wish to make to have been familiar to biologists."

The "Weinburg" in the name of the equation is that of Wilhelm Weinburg, a medical doctor who independently derived it, also in 1908.

MJG

Equation 252: Singular Value Decomposition

The singular value decomposition (SVD) of a rectangular matrix A is the factorization of A into the product of three matrices $A = U\Sigma V^T$, where the matrix U and matrix V^T are orthogonal (see beautiful equation 250) and the square matrix Σ is diagonal with positive real elements. The factorization is illustrated in the first figure. The columns of U are called the left singular vectors (u_i) of A; the rows of V^T are called the right singular vectors (v_i) of A; the elements (σ_i) in matrix Σ is known as the singular values of A, where $\sigma_1 \geq \sigma_2 \geq \cdots \geq \sigma_r. r = \min(size\ of\ A)$.

The theoretical foundation of the SVD was developed by a group of mathematicians Eugenio Beltrami (1835-1899), Camille Jordan (1838-1921), James Sylvester (1814-1897), Erhard Schmidt (1876-1959) and Hermann Weyl (1885-1955) [1]. The first numerical algorithm for calculating the SVD was developed by Gene Golub [2], who is known as the "Professor SVD" [3].

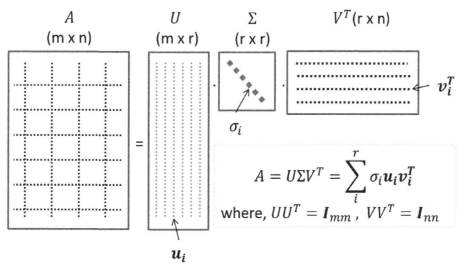

$$A = U\Sigma V^T = \sum_i^r \sigma_i \boldsymbol{u}_i \boldsymbol{v}_i^T$$

where, $UU^T = \boldsymbol{I}_{mm}$, $VV^T = \boldsymbol{I}_{nn}$

Figure 1: Graphic illustration of SVD

Applications of SVD occur in almost any fields for which matrix representation of data is used. Most analysis today utilizes a variant of the algorithm invented by Gene Golub and William Kahan in 1965 [4]. Many consider this achievement to be of comparable importance to the creation of the FFT algorithm (Fast Fourier Transform) for the implementation of the Fourier Transform in frequency domain analysis.

An example of analysing and reconstructing an image is shown in the second figure. The logo of the beautiful equation webpage in black and white is represented by a two-dimensional matrix (270 rows and 700 columns). The SVD computation outputs the 270 singular values (ranging from 0 to 75) of the original image in the upper right plot. It is interesting to see the reconstruction of original image using its first 135, 27 and 7 singular values as shown by the column of images on the left side of the second figure (also referred as rank-k approximation, k=135, 27 and 7). By using only the first 27 singular values and setting the rest to 0, the image can be reproduced in a fairly good quality! This is because that "information" of the original image/matrix is retained in the singular values and their associated left and right singular vectors.

The cumulative percent of each singular value with respect to the arithmetic sum of all the singular values is plotted on the bottom right of figure 2. One can view that it is a straightforward way of compressing an image by reconstructing it using a certain amount of singular values without deteriorating the overall quality of the original image. SVD provides a computationally easy foundation for other post processing of the original image/matrix because some near zero singular values are "tossed away" (i.e. by setting them to zero so the reconstructed image/matrix is comparatively sparse).

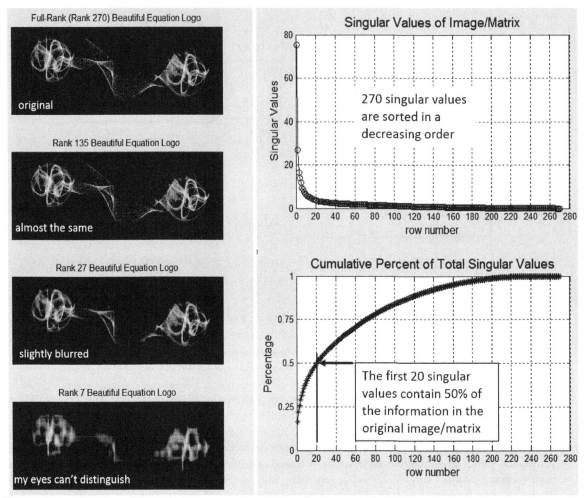

Figure 2: Application of SVD in image compressing and reconstruction

References:

[1]. P. Stewart, "On the Early History of the Singular Value Decomposition", SIAM Review, Vol. 35, Issue 4, 1993, pp. 551-566

[2]. G. Golub and W. Kahan, "Calculating the singular values and pseudo-inverse of a matrix", J. SIAM Number. Anal. 1965, Vol. 2, No. 2: pp. 205-224

[3]. C. Moler, "Professor SVD", http://www.mathworks.com/company/newsletters/articles/professor-svd.html

[4]. C. Martin and M. Porter, "The Extraordinary SVD", arXiv:1103.2338 [math.NA]

KN

Equation 253: Amicable Numbers

$$\sum_i divisor_i(f) = m \quad AND \quad \sum_i divisor_i(m) = f$$

Where f and m are an amicable pair, and the divisor is an operator that retrieves the i-th proper divisor of its argument.

Two numbers are defined amicable when the sum of the proper divisors of each is equal to the other number. For example, a pair of amicable numbers is (220, 284); the proper divisors of 220 are 1, 2, 4, 5, 10, 11, 20,

306

22, 44, 55 and 110, of which the sum is 284; and the proper divisors of 284 are 1, 2, 4, 71 and 142, of which the sum is 220.

Amicable numbers were known to the Pythagoreans, who credited them with many mystical properties. A general formula by which some of these numbers could be derived was invented circa 850 by the Iraqi mathematician Thābit ibn Qurra (826–901). Other Arab mathematicians who studied amicable numbers are al-Majriti (died 1007), al-Baghdadi (980–1037), and al-Fārisī (1260–1320). The Iranian mathematician Muhammad Bagir Yazdi (16th century) discovered the pair (9363584, 9437056).

Thābit ibn Qurra's formula was rediscovered by Fermat (1601–1665) and Descartes (1596–1650), and extended by Euler (1707–1783), and by Borho in 1972. The second smallest pair, (1184, 1210), was discovered in 1866 by a then teenage B. Nicolò I. Paganini (not to be confused with the composer and violinist), having been overlooked by earlier mathematicians.

The concept has been extended to sociable numbers, where the "amicability" property leads to a chain of numbers that finishes where it started. The length of the chain is also called the sociable set order. A list of all known sequences of amicable numbers can be found in [1]. A sociable set of order one, is also called a perfect number. The smallest known perfect number is 6 (1 + 2 + 3 = 6)

[1] Djm.cc, (2015). A LIST OF ALIQUOT CYCLES OF LENGTH GREATER THAN 2. [online] Available at: http://djm.cc/sociable.txt [Accessed 8 Sep. 2015].

GM

Equation 254: Mathematics and Railway Signalling

Whilst sitting on a train the other day delayed by a signal failure I started to wonder what beautiful equation governed railway signalling. It turns out that railway signalling makes use of a limited number of simple mathematical equations including the equations of motion covered earlier.

The main purpose of signalling is to keep trains separated so that each train has a length of track in front of it that is clear of other trains, enabling it to brake to rest without hitting the train in front.

The signalling has to be designed to cope with a certain number of trains per hour (the line capacity). The inverse of the number of trains per hour is called the headway. To give an example if we have 5 trains per hour the headway is 12 minutes. However, typically the design would allow for a headway of 8 or 9 minutes to allow some flexibility.

The actual definition of headway is slightly different and is defined as the time between following trains such that the second train runs under clear signals (Green).

Let's define

D_1 is an approach distance to the first signal to allow for the driver to observe a clear signal, $D_2 + D_3$ is the distance between the first signal and the first warning signal (signal 2), $D_4 + D_5$ is the distance to the signal protecting the first train (signal 3) D_6 in the UK is the overlap distance past the second signal (D6). This is a distance for a short overrun in the case of a misjudgement of braking, and D_7 is the length of the train so that the rear of the train is past the signal. Speed will be taken as v and time as t.

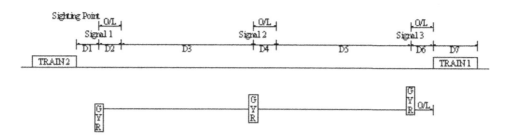

Figure 1

The headway is simply the time taken to cover the total distance, That is to say the sum of D1 to D7, at the speed of the train. (Speeds in the UK are stated as miles per hour with distances often in metres.)

$$t \ (in \ seconds) = (D_1 + D_2 + D_3 + D_4 + D_5 + D_6 + D_7) \times \frac{1000 \times 3600}{(v \times 1760 \times 36 \times 25.4)}$$

where v is the speed in mph. 25.4 converts inches to mm, 36 is the number of inches in a yard and 1760 is the number of yards in a mile. The number 3600 is the number of seconds in an hour. Finally, the 1000 coverts mm to metres. I am glad you like imperial units!

This equation is very much a simplification and assumes no speed changes and no stations. In practice, station stops, speed restrictions and gradients have to be taken into account. So let's introduce some speed restrictions.

The above equation is for constant speed (no speed restrictions). The distance between signals is taken as being equal to the braking distance from the maximum permitted line speed. For a line speed of 60mph the braking distance on level ground is taken as 1065 metres (this is a UK industry standard). The minimum distance between successive signals is, therefore, 1065 metres. The sighting distance is taken as a nominal 183 metres and the overlap is 180 metres. For a typical three coach passenger train, the train length is 69 metres.

The resulting headway is, therefore:

$$t \ (in \ seconds) = (183 + 1065 + 1065 + 180 + 69) \times \frac{1000 \times 3600}{(65 \times 1760 \times 36 \times 25.4)} = 96 \ seconds$$

This is a very straightforward model, but various modifications are required as speed limits and station stops are included. Now consider the case where the speed is restricted to 40mph over a distance of 200 metres starting 400 metres after signal 1. Figure 2 shows this.

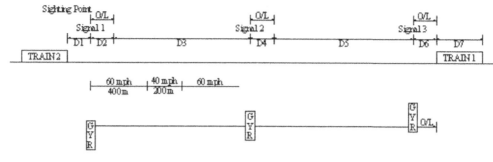

Figure 2

308

The model must include a realistic deceleration from 60 mph to 40 mph: 0.5ms^{-2}. (The m in mph is miles, the m in 0.5m/s^2 is metres – oh why do we persist with non-SI units)

We now have

$$u = 60 \times \left(\frac{1760 \times 36 \times 25.4}{(1000 \times 3600)} \right)$$

$$u = 40 \times \left(\frac{1760 \times 36 \times 25.4}{(1000 \times 3600)} \right)$$

$$a = -0.5$$

Remember all those conversions again!

When these values are substituted into

$$v^2 - u^2 = 2as$$

Note that a is acceleration and s is distance. From this we find

$$s_{60-40} = 400m$$

The length of the speed restriction is $s_{40} = 200m$

To return to the normal running speed for the line, the values for the acceleration will vary as shown below. (Typical values for a diesel multiple unit.)

$$40 - 45\text{mph } 0.240 \text{ m/s}^2$$
$$45 - 50\text{mph } 0.195 \text{ m/s}^2$$
$$50 - 55\text{mph } 0.156 \text{ m/s}^2$$
$$55 - 60\text{mph } 0.126 \text{ m/s}^2$$

Repeated applications of the above formula give
$$s40\text{-}45 = 177 \text{ metres}$$
$$s45\text{-}50 = 244 \text{ metres}$$
$$s50\text{-}55 = 337 \text{ metres}$$
$$s55\text{-}60 = 456 \text{ metres}$$

We also calculate the time through the braking section and each acceleration section, using the formula (we have seen this one before)

$$s = 0.5 \times (u + v) \times t$$

which can be rearranged to give

$$t = \frac{2s}{u + v}$$

From this we can obtain the values of t:

$$t_{60\text{-}40} = 17.9 \text{ seconds}$$

$$t_{40} = 11.2 \text{ seconds}$$
$$t_{40-45} = 9.3 \text{ seconds}$$
$$t_{45-50} = 11.5 \text{ seconds}$$
$$t_{50-55} = 14.3 \text{ seconds}$$
$$t_{55-60} = 17.7 \text{ seconds}$$

This gives us a total distance and total time for dealing with this restriction of

$$(400 + 200 + 177 + 244 + 337 + 456) = 1814 \ metres$$

and,

$$(17.9 + 11.2 + 9.3 + 11.5 + 14.3 + 17.7) = 81.9 \ seconds.$$

Subtracting this distance from the sum of D_1 to D_7 gives the length of track remaining that will be travelled at 60mph. In the example quoted, this is $2562 - 1814 = 748 \ metres$. The time for this portion of track is also needed. The new total time is the revised headway.

So next time you are delayed on a train have a thought for all the work that goes in to a safe signalling system.

The above examples and diagrams are taken from a document by Anthony Hoath of Hoath Enterprises Limited dated December 2001

DF

Equation 255: Inductance of a Circular Coil above a Flat Conducting Surface

$$L(z) = \frac{\pi \mu n^2 \bar{r}}{l^2 t^2} \int_0^\infty \frac{I^2(r_2, r_1)}{\alpha^5} \left[2l + \frac{1}{\alpha} \left\{ 2e^{-\alpha l} - 2 + \left[e^{-\alpha(l+z)} + e^{-\alpha z} - 2e^{-\alpha(l+2z)} \right] \left(\frac{\alpha - \alpha_1}{\alpha + \alpha_1} \right) \right\} \right] d\alpha$$

where, $L(z)$ is the inductance of the coil (H) that is a function of the height, z (m), of the coil above a conducting surface, n is the number of turns on the coil, μ is the permeability (H/m), l is the length of the coil (m), t is the thickness of the coil (m), $\bar{r}$ is the mean radius of the coil (m) and $I^2(r_2, r_1)$ is the Bessel function integrated over the outer and inner radius of the coil given by

$$I^2(r_2, r_1) = \int_{\alpha r_1}^{\alpha r_2} x J_1(x) \, dx = \left[\sum_{k=0}^{\infty} \frac{(-1)^k x^{2k+3}}{2^{2k+3}(2k + 3)k! \, (k + 1)!} \right]_{\alpha r_1}^{\alpha r_2}$$

J_1 is a Bessel function of the first kind of order 1 and $\alpha_1 = (\alpha^2 + j\omega\mu\sigma)^{\frac{1}{2}}$, with frequency ω (rad·s⁻¹), permeability μ and electrical conductivity σ ($\Omega^{-1} \cdot m^{-1}$).

Once an expression for the inductance of a coil is determined, the force can be solved by taking the derivative of the equation,

$$F = \frac{i^2}{2} \frac{\partial L}{\partial z} = \frac{i^2}{2} \frac{2\pi \mu n^2 \bar{r}}{l^2 t^2} \int_0^\infty \frac{I^2(r_2, r_1)}{\alpha^5} \left(e^{-2\alpha z}(e^{-2\alpha l} - 2e^{-\alpha l} + 1) \right) \left(\frac{\alpha - \alpha_1}{\alpha + \alpha_1} \right) d\alpha$$

Anyone with experience in the fields of eddy current inspection will know the paper published in 1968 by the American Physicists Caius Dodd and Ed Deeds. This paper presents mathematical equations for solving

310

a broad variety of axially symmetric eddy-current coil problems. The above equation provides a method to calculate the inductance of a circular coil that is placed above the surface of a flat conducting solid as shown in Figure 1, one of the many solutions to be found in this paper.

The information that is provided by their equations represents a complete description of the electromagnetic fields and subsequent currents that occur when the coil is provided with an oscillating current. This oscillating current produces a magnetic field that circulates around the wires of the coil so that there is a relatively large magnetic field flowing through the centre of the coil in the axial direction. This magnetic field emerges at one end and diverges to circulate around the outside of the coil and flow back into it from the other side (magnetic fields always circulate in closed loops). This oscillating magnetic field emerging from the coil permeates the conducting surface and induces an electric field that, in turn, cause a current to flow within the conductor. This is illustrated in Figure 2 that shows the three dimensional distribution of currents in the conducting material. All of these fields and the effects of them are solved by the theoretical models for this and other geometries.

While the geometries, being radially symmetric, are relatively simple, creating a mathematical framework

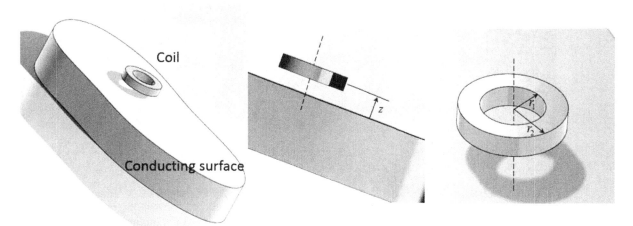

Figure 1: Illustration of a copper coil placed above a conducting surface assumed to be very thick. In practice this coil will be made up of a number of turns of copper wire.

that incorporates Maxwell's equations had eluded researchers until Dodd and Deeds identified a solution involving the determination of magnetic potentials in the regions between the coil and surface as well as inside the conducting surface itself. Having determined the magnetic potentials, a number of important physical effects can be calculated. These include the changing electrical properties of the coil as a function of its proximity to the surface, the change in voltage across the coil, the induced eddy-currents and the effects of a small flaws contained within the conducting material.

These solutions are particularly useful in the field of eddy-current detection for identifying buried flaws and/or the onset of fatigue cracks in aerospace and other structures for which failure can have catastrophic consequences. An example of a miniature eddy-current sensor is shown in figure 3 (described in more detail in Dogaru and Smith). This shows a small (less than 1 mm diameter) coil that has been attached to the surface of a giant magneto resistive (GMR) magnetic field sensor. This sensor has the unique property that it does not detect the field from the coil, neither does it detect the field from the oscillating current induced in the conducting material. However, if the field of the induced current has any axial asymmetry (caused by defects, cracks, or surface slopes) this is very sensitively picked up by the sensor. Hence moving this sensor over a copper surface (a US penny in the example shown) it is possible to map eddy-current asymmetries as shown in figure 3.

Surprisingly, the equations for inductance enable a simple formulation (shown above) for eddy-current levitation forces (these tend to be rather small) for a coil placed near to a conducting surface.

311

Another field of study for which this work directly applies is the development of displacement sensors based on the variation of the coil inductance with the separation z. Such eddy-current displacement sensors can be readily manufactured by placing the coil in a cylindrical housing that provide an output signal as it is moved close to a conducting surface as, again, can be modelled by the above equations. It is my pleasure to acknowledge Dr. Ed Deeds who provided background information for this short essay.

References:
Dodd C.V., and Deeds W.E., 1968, Analytic solutions to eddy-current probe-coil problems, *J. Appl. Phys.*, **39**(6), 2829 – 2838.
Dogaru T., and Smith S.T., 2001, A GMR based, eddy-current sensor, *IEEE Trans. On Magnetics*, **37**(5), 3831-3838.

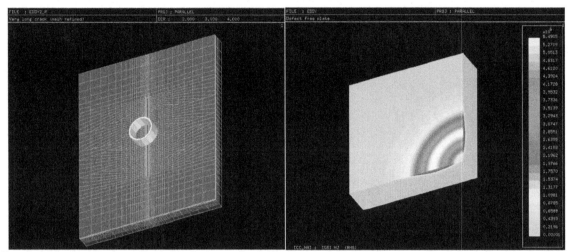

Figure 2: Computer numerical simulations of a coil (silver) placed above a conducting surface (red). When the coil is provided with an oscillating current is induces a current in the conducting surface. This current is shown in the figure on the left in which the conducting surface (now colored green) has been cut into a square with the lower right corner being coincident with the axis of the coil. It should be noticed that the induced current follows the shape and average radius of the coil.

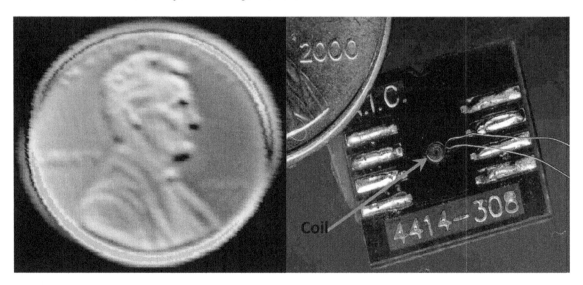

Figure 3: Map of eddy-current asymmetries when a miniature sensor is scanned over a US penny. The coil of the eddy-current probe is shown on the right (with a segment of a penny shown for scale) attached to the surface of a magnetic field sensor.

SS

Equation 256: Piezo Electric Capacitance

Piezoelectricity is a very interesting property and characteristic of certain materials. The piezoelectric properties of some materials can generate electricity from motion of can generate motion from electricity. It is applicable to solids such as ceramics and crystals as well as some other organic compounds. A key component of piezoelectric material is its capacitance. The equation for piezoelectric capacitance is

$$C = \frac{\varepsilon A}{t}$$

where C is the capacitance, ε is the electrical permittivity, A is the area and t is the thickness.

The capacitance can be seen as an electrical damping characteristic or the effective electrical mass of a circuit.

Piezoelectricity has many interesting uses in such things as high frequency buzzers, adaptive optics for ground based space telescopes and high force and low displacement applications in precision mechanics.

The Piezoelectric effect was actually first discovered due to further work on the pyroelectric effect which is the electrical potential generated by a change in temperature. This was first studied in circa 1850. Following on from these discoveries Pierre Curie and Jacques Curie demonstrated that an electrical charge could be created by a change in the crystal state of a material, therefore they had uncovered the piezoelectric effect which can generate electricity from motion or generate motion due to an electrical charge.

TS

Equation 257: The Basel Problem

$$\sum_{k=1}^{\infty} \frac{1}{k^2} = \frac{\pi^2}{6}$$

The Basel Problem was first posed in 1644 by Pietro Mengoli and remained open for 90 years. The problem deals with finding the exact sum of the infinite series of reciprocals of integers squared.

$$1 + \frac{1}{2^2} + \frac{1}{3^2} + \frac{1}{4^2} + \frac{1}{5^2} + \cdots \quad or \quad 1 + \frac{1}{4} + \frac{1}{9} + \frac{1}{16} + \frac{1}{25} + \cdots$$

In 1655, John Wallis, an English mathematician, communicated that he had found the sum to three decimal places. Jakob Bernoulli wrote about this problem in 1689 and it came to be known as the Basel problem, after the hometown of the Bernoulli's, one of the most prominent mathematical families of the time, as well as that of the eventual solver of the problem, Leonhard Euler.

It was known that the series converges but finding the exact sum proved to be remarkably difficult. Many prominent mathematicians, including Leibnitz, Mengoli and the Bernoulli brothers had tried their hand at solving it but had failed.

In 1721, a young Euler began studying mathematics under the mentorship of Johann Bernoulli (a leading mathematician at that time). It is likely that is was Johann who first told Euler about the Basel problem but it is unclear exactly when he did so but certainly by 1728 Euler was working on the problem. It was around that time that Daniel Bernoulli (Johann's son) wrote to Christian Goldbach that he had found an approximate value of the sum of the series (the value he gave was 8/5). Goldbach replied that he had found that the sum is between 41/35 and 5/3. One of Euler's earliest attempts was to find numerical approximations of some of the partial sums of the series but these were not too helpful. Indeed, the sum of the first thousand terms is 1.643937 but this is only accurate up to the first two digits (the problem being that the series converges extremely slowly).

The ingenuity of Euler

In 1734 Leonhard Euler solved the problem. The result was truly remarkable and even now many people who see it for the first time cannot help but be amazed. Euler's proof is a shining example of his ingenuity and mathematical prowess. In order to prove the result he used the well-known series expansion for sin x.

$$\sin x = x - \frac{x^3}{3!} + \frac{x^5}{5!} - \frac{x^7}{7!} + \frac{x^9}{9!} - \cdots$$

Next he considered the 'infinite polynomial'

$$P(x) = 1 - \frac{x^2}{3!} + \frac{x^4}{5!} - \frac{x^6}{7!} + \frac{x^8}{9!} - \cdots$$

Euler then multiplied P(x) by x/x

$$P(x) = x \left[\frac{1 - \frac{x^2}{3!} + \frac{x^4}{5!} - \frac{x^6}{7!} + \frac{x^8}{9!} - \cdots}{x} \right] = \left[\frac{x - \frac{x^3}{3!} + \frac{x^5}{5!} - \frac{x^7}{7!} + \frac{x^9}{9!} - \cdots}{x} \right] = \frac{\sin x}{x}$$

We can recognise that the numerator of this series is the expansion of sin x. In multiplying by x/x the series is equivalent to $\sin x/x$ which has zeros at $x = \pm n\pi$ for $n = 1,2, \ldots$ since these are the zeros of the function $\sin x$, we can now write $P(x)$ as an 'infinite product' and factor it, and set $P(x)$ equal to the equations below:

$$P(x) = 1 - \frac{x^2}{3!} + \frac{x^4}{5!} - \frac{x^6}{7!} + \cdots$$

$$P(x) = \left(1 - \frac{x}{\pi}\right)\left(1 - \frac{x}{-\pi}\right)\left(1 - \frac{x}{2\pi}\right)\left(1 - \frac{x}{-2\pi}\right)\left(1 - \frac{x}{3\pi}\right)\left(1 - \frac{x}{-3\pi}\right)\cdots$$

$$P(x) = \left[1 - \frac{x^2}{\pi^2}\right]\left[1 - \frac{x^2}{4\pi^2}\right]\left[1 - \frac{x^2}{9\pi^2}\right]\left[1 - \frac{x^2}{16\pi^2}\right]\cdots$$

The next step illustrates Euler's foresight and genius, for he expands the right-hand side as follows

$$1 - \frac{x^3}{3!} + \cdots = 1 - \left(\frac{1}{\pi^2} + \frac{1}{4\pi^2} + \frac{1}{9\pi^2} + \frac{1}{16\pi^2} + \cdots \right) x^2 + \cdots$$

where the remaining terms in the expansion on the right are not relevant to the problem at hand. Euler then compared the coefficients of x^2 in the above equation to get

$$-\frac{1}{3!} = -\left(\frac{1}{\pi^2} + \frac{1}{4\pi^2} + \frac{1}{9\pi^2} + \frac{1}{16\pi^2} + \cdots \right)$$

multiplying both sides by $-\pi^2$ we get

$$\frac{\pi^2}{6} = 1 + \frac{1}{4} + \frac{1}{9} + \frac{1}{16} + \cdots$$

This is a truly remarkable result. No one expected the value π, the ratio of the circumference of a circle to the diameter, to appear in the formula for the sum of reciprocals of integers squared. Euler went on to solve the general case for summing p-series with even exponents but a solution for odd exponents is still unknown.

Riemann-zeta function

Euler's ideas were exploited by Bernhard Riemann in the 19th century in studying the Riemann-zeta function in connection with his investigation of the distribution of primes. This will be explored in a future 'beautiful equation'.

ILB

Equation 258: The Duffing Equation

The Duffing Equation, often called the Duffing oscillator, is a second order non-autonomous non-linear differential equation that describes the motion of a forced harmonic oscillator. In the equation, x is displacement of the object being driven, t is time, δ is a coefficient that describes the degree of damping of the object, ω_o is a spring constant, β is a coefficient giving the degree of non-linearity in the equation, γ is the amplitude of the driving force and ω is the angular frequency of the driving force. If the x^3 term and the driving force term were not present, the equation would describe a simple harmonic oscillator. If just the x^3 term were not present, the equation would describe a very common second order (linear) driven oscillator, for example the pendulum on a clock. However, the x^3 term is what makes the equation so interesting – this term introduces the non-linearity.

The figure shows a solution to this equation, where the y in the figure is t in the equation.

As the figure suggests, this equation is chaotic. Due to the x^3 term, a very small perturbation in the initial conditions results in a wildly different solution – predictable, but complex. In my bachelor degree thesis, I built an analogue circuit (computer) to solve this equation (using integrators as differentiators which are highly subject to noise) and an analogue multiplier. It was fascinating to plot x against t on an oscilloscope. I would turn the circuit on and get a lovely spiral pattern. Turn it off, then on again, and get a completely different pattern. Why?

Because the sensitive dependence on initial conditions (the hallmark of a chaotic system) means that just a few electrons travelling down the wire slightly differently can cause a completely different solution. My famous party cry as a student was "Chaos!" but that's another story…

An example of another system described by the equation is a metal strip suspended vertically but with permanent magnets either side of its free end. If the strip is oscillated externally, it will move in a chaotic

315

manner described by the equation. Another is a drive double spring system. I also recall seeing papers describing machining mechanics and atomic force microscopy motion using this equation.

The equation was developed by German electrical engineer Georg Duffing in 1918.

$$\frac{d^2x}{dt^2} + \delta\frac{dx}{dt} - \omega_o^2 x + \beta x^3 = \gamma\cos(\omega t)$$

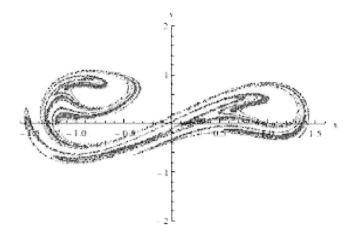

RL

Equation 259: Einstein's Photoelectric Law

$$K_{max} = hf - \varphi$$

Under the right circumstances, light or other electromagnetic radiation directed on to the surface of a material can cause electrons to be ejected from the material. This process is sometimes called '(electron) photoemission', along with an assortment of alternative names, in particular the (external) photoelectric effect.

In the above equation, K_{max} represents the maximum kinetic energy that a photoelectron (emitted from a particular material whose work function is φ and subject to radiation of frequency f) can have. h represents Planck's constant. The work function is the minimum energy needed to extract an electron within the material to a point in a vacuum immediately outside the surface. For metals, work functions range from 1.95 eV for caesium to 5.63 eV for platinum. A plot of K_{max} against frequency produces a straight line (Fig 3).

The most usual context in which this phenomenon is discussed is when light (UV or visible) is incident on a metal, with a single valence electron being ejected. Some non-metallic elements, notably metalloids, can demonstrate weak photoemission if the energy of the incident photons is high enough, i.e. in the X or gamma ray region.

When incident on a metal, these X or gamma rays can cause the emission of a core electron, with secondary effects ensuing. If the frequency is high enough, the nucleus itself may absorb the radiation, resulting in photofission or the nuclear photoelectric effect.

Electron photoemission can occur in a gas as well as a solid or liquid, with the electron being ejected from an atom or molecule.

Note that the photo effect can happen alongside other interactions in which a photon is incident on a material. These include both photon-absorbing and photon-scattering phenomena. (See e.g. Fig 4).

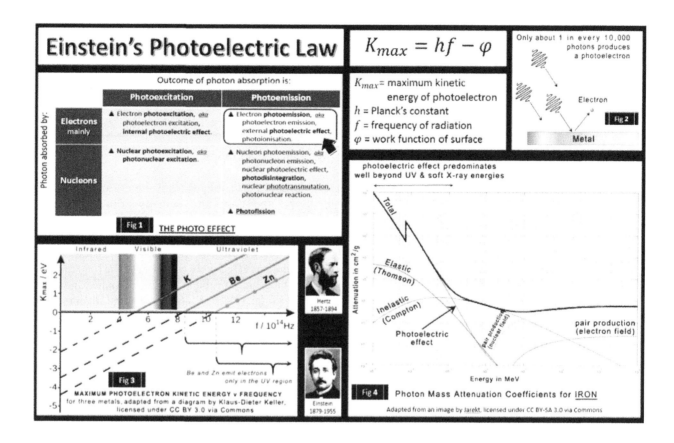

Einstein's Photoelectric Law

$$K_{max} = hf - \varphi$$

K_{max} = maximum kinetic energy of photoelectron
h = Planck's constant
f = frequency of radiation
φ = work function of surface

Only about 1 in every 10,000 photons produces a photoelectron

Fig 2

Outcome of photon absorption is:

	Photoexcitation	Photoemission
Electrons mainly	▲ Electron photoexcitation, *aka* photoelectron excitation, **internal photoelectric effect**.	▲ Electron **photoemission**, *aka* photoelectron emission, external **photoelectric effect**, photoionisation.
Nucleons	▲ Nuclear photoexcitation, *aka* **photonuclear excitation**.	▲ Nucleon photoemission, *aka* photonucleon emission, nuclear photoelectric effect, **photodisintegration**, nuclear phototransmutation, photonuclear reaction. ▲ Photofission

Fig 1   THE PHOTO EFFECT

Fig 3
MAXIMUM PHOTOELECTRON KINETIC ENERGY v FREQUENCY for three metals, adapted from a diagram by Klaus-Dieter Keller, licensed under CC BY 3.0 via Commons

Be and Zn emit electrons only in the UV region

Hertz 1857-1894

Einstein 1879-1955

photoelectric effect predominates well beyond UV & soft X-ray energies

Fig 4   Photon Mass Attenuation Coefficients for IRON
Adapted from an image by Jarekt, licensed under CC BY-SA 3.0 via Commons

MORE TERMINOLOGY

A photoemissive surface is one that can exhibit photoemission. The ejected electrons are, for convenience, referred to as photoelectrons. The photoelectric threshold is the point at which the energy of photons is just sufficient to cause photoemission. The threshold frequency and threshold wavelength of a material refer to the frequency and wavelength at this point. Most elements have threshold frequencies in ultraviolet and only a few dip into the visible region. The materials with the lowest threshold frequencies – even down in the infrared - are all specially manufactured semiconductors. The internal photoelectric effect is a phenomenon in which the absorption of a photon in a semiconductor results in an excitation, whereby an electron from the valence band jumps to the conduction band, rather than being ejected from the material entirely.

The photoelectric yield is the number of electrons ejected per photon incident on the surface. For most metals, this value is low (roughly 0.0001 near the photoelectric threshold - see Fig 2) because (a) their surface is strongly reflective; thus only a small fraction of the incident radiation penetrates the metal, and (b) only photoelectrons produced at a depth (the 'escape depth') of less than a few nanometres have enough energy left to escape.

HISTORY

Some time after Becquerel discovered the 'photovoltaic effect', Heinrich Hertz discovered the photoelectric effect in 1887 whilst investigating the effect of UV radiation on spark length. A number of scientists made contributions to knowledge about the effect, including Philipp Lenard, who observed that the energy of individual emitted electrons increased with the frequency of the light, in contravention of Maxwell's wave theory of light. In 1905, Einstein solved this seeming paradox by describing light as composed of discrete 'quanta', now called 'photons', rather than continuous waves. He theorised that the energy in each quantum of light was equal to the frequency multiplied by a constant (Planck's constant as it is now known). A photon above a threshold frequency has the necessary energy to eject a single electron, producing the photo effect.

This observation and theory led to the quantum revolution and earned Einstein the Nobel Prize in Physics in 1921.

APPLICATIONS
Photoemission is utilised in devices such as photoemissive cells & photo-multiplier tubes, photocopiers, automatic door openers, burglar alarms, light detectors and scintillators.

FR

Equation 260: Lagrangian Points

In any system with two rotating bodies, such as the Sun and Earth, it happens that there are five locations at which a third body, of negligible mass compared to the other two, can nominally sit in an orbit about the large central body (e.g. Sun) at a fixed position relative to the larger orbiting body (e.g. Earth). This has important implications for spacecrafts, particularly space-based astronomical observatories. The positions are called Lagrangian (or Lagrange, or libration) points. They represent positions where in a rotating frame of reference the gravitational potentials due to the large masses effectively cancel out. The Lagrangian points are examples of a restricted three-body problem. The three-body problem had been studied in detail since Newton and eventually proved by Bruns and Poincare in the 1880s to have no general algebraic solution (such a system is, in general, unrepeatable or chaotic). However, in many cases approximate solutions can be calculated and there are special cases such as these. Using a rotating (non-inertial) frame of reference relating to a two-body system and considering the resulting pseudoforces (such as centrifugal and Coriolis) an accurate solution can be obtained for a restricted three-body problem: a "stationary" solution in the non-inertial frame. Of the five Lagrangian points, three (typically notated L_1, L_2 and L_3) are located along the axis of rotation between the two larger masses. The other two, L_4 and L_5, are located at symmetric points in the plane of rotation which complete equilateral triangles with the Sun and Earth (see picture below, not to scale). By convention L_5 is travelling behind the orbiting mass and L_4 ahead of it.

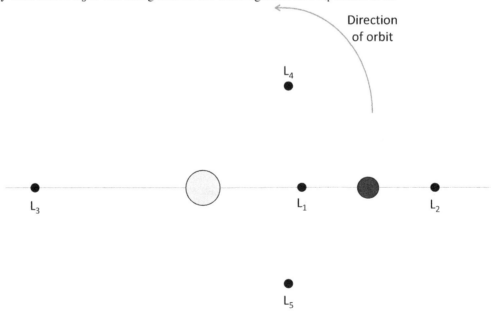

It turns out that in practice L_1, L_2 and L_3 are not stable – a slight perturbation can cause a spacecraft to quite easily move away from one of these points over a period of time – the location is critical, whereas L_4 and L_5 are stable, and the reason for this is that the gravitational contours are such that a movement away from the point generates a restoring Coriolis force. The Coriolis force is a turning force perpendicular to the direction of motion in the plane of rotation – if the rotating body has a velocity relative to the rotating frame, this couples with the angular velocity of the frame to generate an additional acceleration. In all cases the

318

forces to consider are the gravitational attraction of the two large bodies and three pseudoforces: the centrifugal force due to the motion of the small mass, a Coriolis force if the small mass starts to move with respect the already rotating reference frame, and a fifth force, the Euler force, which would need to be taken into account if an acceleration of the rotating reference frame was significant – for instance if the Earth was accelerating due to an eccentric orbit. In most cases, with near-circular and stable planetary orbits (minimal variation in angular momentum), this is negligible compared to the other effects.

For the example of L_1 shown below, its distance can be calculated.

Equating gravitational and centrifugal forces for the Sun and Earth, where m_s is the mass of the sun, m_e is the mass of the earth, v_e is the velocity of the Earth perpendicular to its axis of rotation and r is the distance between the Sun and Earth (the mass m_e cancels on both sides)

$$\frac{Gm_s}{R^2} = \frac{v_e{}^2}{R}$$

and $v_e = \frac{2\pi R}{T_e}$ where T_e is the period of the earth's rotation, keeping constant terms on the right hand side, we get

$$\frac{m_s}{R^3} = \frac{4\pi^2}{GT_e{}^2}$$

which is equivalent to Kepler's third law (beautiful equation 107) for an approximately circular orbit that relates the orbital period to the radius. Now we introduce the small spacecraft, with mass m_p, velocity v_p and period T_p. Taking into account the attraction of the Sun and Earth, the forces are equated again (this time m_p cancels out).

$$\frac{Gm_s}{(R-r)^2} - \frac{Gm_e}{r^2} = \frac{v_p{}^2}{R-r}$$

and using a similar substitution for the period as above this results in

$$\frac{m_s}{(R-r)^3} - \frac{m_e}{r^2(R-r)} = \frac{4\pi^2}{GT_p{}^2}$$

We now have similar expressions on the right hand side, and at the Lagrangian point L_1 (assuming the location of the craft in its vicinity is stable for now) the periods T_e and T_p are equal (the craft is at a stationary point in the rotating frame), so the left hand sides of the 2nd and 4th equations are equal. The constant terms are no longer required and the unique distance r is now a solution to the equation

$$\frac{m_s}{R^3} = \frac{m_s}{(R-r)^3} - \frac{m_e}{r^2(R-r)}$$

This equation can be solved analytically with the help of variable substitution and algebraic formulae to obtain a value for r which is, to a good approximation,

$$r(L_1, L_2) \approx R\left(\sqrt[3]{\frac{m_e}{3m_s}}\right)$$

For the Sun-Earth system this distance is about 1.5 million km from the centre of the Earth. The distance of L_2 on the far side of the Earth is almost exactly the same distance in the opposite direction, which is why it is referred to in the left hand side of the equation. Because the points L_1 and L_2 are unstable, any spacecraft sent to those positions require a small amount of regular adjustment ("station keeping" to prevent them from gradually drifting away from the location, with a critical period of about 23 days in this case). However, it only requires a small amount of fuel and these locations are very useful for astronomical observatories. Arthur C Clarke first suggested in 1950 that the L_2 point of the Earth-Moon system would be ideal for communications. In the 1960s, Robert Farquhar, a mission design specialist in NASA, examined potential orbits around Lunar L_2. Referring back to the Sun-Earth system, instead of being located at the points themselves, the craft are typically driven into a fairly large orbit (>100,000 km) around, or in the vicinity of, these nominally stable points. Such an orbit is called a halo orbit if it is periodic, and a Lissajous orbit if non-periodic. Orbits around L_1 are obviously chosen as the location for solar observatories like SOHO and Genesis, and the use of a large orbit reduces the impact of solar interference on earth communications. Orbits around L_2 were used for astronomical observatories WMAP and Planck, which have been used to study the cosmic microwave background (CMB), and will also be the location for the James Webb Telescope when it is hopefully launched in 2018. In this case the use of a large orbit ensures better illumination of the solar panels whilst having the benefit of being able to scan the Universe in a direction away from the Sun with the Earth behind, shielding most of the solar radiation. What about L_3? The location of L_3 is slightly less than the sun-earth distance on the opposite side of the sun (the slight difference relates to the barycentre of the Sun-Earth rotation, which is not exactly at the centre of the Sun but still well inside it). This location has been popularised as a place for a counter-Earth or Planet "X", which is of course impossible as it would be unstable for any significant mass and would for the most part find itself within view of the Earth. Due to its relative stability over a period of time, it could have been an ideal temporary location for a curious or hostile alien craft to hide from us for some years (at least before the time of space exploration). L_4 and L_5 are, as stated earlier, stable – and are the location of natural "trojan" objects, particularly in the Sun-Jupiter system, at which points exist stable orbits of a collection of about 2000 known asteroids outside the main belt. They are called trojans because the first ones discovered in the early 1900s were named after characters from Homer's Iliad. There are also some known trojans at the equivalent points for Neptune's orbit. Hypothetically others are expected to occur at the Lagrangian points for Saturn and Uranus but are yet to be found. At the L_4 and L_5 locations for the Sun-Earth system and Sun-Mars system there are smaller objects and collections of dust particles.

The points L_1, L_2 and L_3 were discovered by Leonhard Euler in 1760 (published in 1765) and the remaining two by Joseph-Louis Lagrange shortly after. The solutions to the five points were presented in a famous paper by Lagrange in 1772 ("Essay on the three-body problem"). Lagrange won an award (one of many) for this paper, which was well-deserved: in this work he was the first to introduce the method of a non-inertial rotating frame.

PB

Equation 261: The Hartree-Fock Equation

The Hartree-Fock (HF), (more precisely Hartree-Fock-Roothaan) equation is a matrix equation used to model the chemical and physical properties of multi-electron atoms (all atoms apart from hydrogen), molecules, nanocrystals and solids. It is also the base of the molecular orbitals (MO) models for molecules.

The equation in form of matrixes is

$$FC = SC\varepsilon$$

For one electron the Hartree-Fock equation is

$$\hat{f}_i \chi_i = \varepsilon_i \chi_i$$

where χ_i is the wave function of the electron, $\hat{f}_i$ is the Fock operator and ε_i is the energy level of the electron.

In the matrix equation, F is the Fock matrix. It is a matrix, which includes of all the individual Fock operators $\hat{f}$ for each electron in a given multi-electron system.

$$\hat{f}_i = [\hat{h}_i + \sum_{j \neq i} (J_j - K_j)] = -\frac{1}{2}\nabla_i^2 - \sum_{A=1}^{M} \frac{Z_A}{r_{iA}} + V_i(HF)$$

where, h_i is kinetic energy operator and potential energy of the nuclei on the electron I, J_j is the Coulomb interaction operator, and K_j is the exchange interaction operator.

C is the matrix of basis set coefficients for the wave function of the system. This wave function includes all the wave functions describing the individual electrons in the system. Each electron is described by a wave function χ_i expressed as a sum of specific functions $\tilde{\chi}_i$:

$$\chi_i = \sum_{\mu} C_{\mu i} \tilde{\chi}_i$$

All functions $\{\tilde{\chi}_i\}$ form the basis set for the electron wave function. $C_{\mu i}$ are the coefficients of the basis set for the individual electron i.

S is the overlap matrix, which includes all the mathematical overlaps $s_{\mu\nu}$ of the functions $\tilde{\chi}_i$ such as:

$$s_{\mu\nu} = \int \tilde{\chi}_\mu^* \tilde{\chi}_\nu \, d\boldsymbol{r}_i$$

The symbol $d\boldsymbol{r}_i$ indicates that the integral is over the whole space occupied by each electron.
If the functions $\tilde{\chi}_i$ are chosen in such a way that they are *orthonormal*, then the elements of overlap matrix are all equal to 1 and the Hartree-Fock equation reduces to an eigenvalue equation:

$$FC = \varepsilon C$$

ε is a vector that includes all the energy values $\{\varepsilon_i\}$ for all the electrons in the system. $\{\varepsilon_i\}$ are the eigenvalues of the Hartree-Fock equation with an orthonormal basis set.

HISTORY

321

The equation is the result of the application of quantum mechanics to multi-electron systems and it is rooted in the early 1920s, soon after the formulation of the Schroedinger equation for the hydrogen atom. Douglas Rayner Hartree, an English mathematician and physicist, elaborated the model after obtaining his PhD at the Cambridge University in 1926, in which he applied numerical analysis to the Bohr's atomic theory. Russian physicist Vladimir Fock contributed to the development of the method in the 1930, while also developing methods for geophysical explorations and helping develop general relativity.

FUNDAMENTAL APPROXIMATIONS
The Hartree-Fock method is built on two main approximations:

- Electrons of the atoms move around fixed nuclei. That is to say, the dynamics of the electrons and that of the nuclei for a given system of many atoms can be treated independently (Born-Oppenheimer approximation)
- The dynamics of each electron in the system is subject to the mean (electrostatic) field exercised by the other electrons. This approximation neglects the possibility of electron correlation between two electrons.

Hartree aimed to calculate the energy of a multi-electronic system in a very similar way to that used by Schroedinger to calculate the energy of the hydrogen atom, through the use of a wave function, whose square value gives the probability of finding the electron in a specific region of space (density distribution). Once the wave function is known, the application of an appropriate Hamiltonian operator will define a time-independent eigenvalue equation, the solution of which will provide the energy of the system. However, the three main problems faced by Hartree and Fock were: the construction of a wave function that includes all the electrons in the system; the definition of the appropriate operator acting on such a wave function and the solution of the eigenvalue equation for such a system.

In the mean field approximation, the wave function of a multielectron system can be defined as the product of the wave functions for the individual electrons. In order to satisfy the Pauli's principle, according to which two electrons in a given position cannot have the same spatial wave function and spin, the total wave function of a multi-electronic system must be antisymmetric. According to this property, American physicist John Clarke Slater defined the correct expression for the wave function of a multi-electronic system, known as Slater determinant. This function is a linear combination of all the different products of the wave functions for the individual electrons. This definition is equivalent to assuming that each electron moves independently of all the others and experiences the Coulomb repulsion due to the average positions of all electrons (the mean field approximation).

The Fock operator for a specific individual electron includes a kinetic term acting on a single electron, a coulombic term and an exchange term, with the last two acting on pairs of electrons. For each electron, these two terms are calculated from the wave functions of all the other electrons. For this reason, the Hartree-Fock equation is non-linear.

THE VARIATIONAL THEOREM
Slater and J.A. Gaunt showed independently in 1928, that Hartree's method could calculate the energy of a multi-electronic system by applying the "Variational theorem". This theorem states that, for a time-independent Hamiltonian operator, any trial wave function will have an energy expectation value that is greater than, or equal to, the true ground state wave function corresponding to the given Hamiltonian.

BASIS SET AND ITERATIVE ALGORITHM
An initial, hypothetical wave function is 'built' through a linear combination of mathematical functions, which together form the basis set of the wave function. The coefficients of the linear combination 'make the recipe' of the wave function and are allowed to vary. These are the elements of the C matrix of the Hartree-Fock equation. Hence, the Hartree-Fock method consists of searching the 'best coefficients' that form the

wave function corresponding to the 'best description' of the system in its ground state (lowest energy). The 'best coefficients' will hence correspond to the lowest energy of the system.

This search is performed through an iterative process called self-consistent field (SCF) to solve the Hartree-Fock equation:

- An initial wave function is built for a given position of nuclei, by using a hypothetical combination of basis set functions.
- The Fock operator (the V (HF) contribution) is calculated from this set of functions an equation is applied to minimise the energy levels corresponding to this wave function. This will produce new wave functions (or more precisely new coefficients).
- A new Fock operator (a new V (HF) contribution) can be calculated for this new set of functions (coefficients) and the process is repeated iteratively.

The iterative algorithm stops when the variation between the energy at the iteration N and the iteration N-1 is less than a specified threshold. The final energy and the electrons wave function of the system is calculated at the end of this iterative process. The electron density distribution for the system (Figure 1) is calculated from the wave function – that is to say, from the optimised coefficients of the basis set used to approximate the wave function.

POST HARTREE-FOCK

The Hartree-Fock equation is the starting point to calculate the physical properties of multi-electronic systems (atoms, molecules, crystals) ab initio. Hartree-Fock method is defined as ab initio, because it aims to calculate the properties of electronic systems, given the number of atoms and electrons in the systems, without relying on empirical parameters to be added in the model. However, this method presents its limits as it neglects important effects such as electron correlation.

Other methods have been developed to correct and improve the Hartree-Fock model in order to take into account these effects. One of the most effective methods in this direction is the Moeller-Plesset second order correction (MP2). These methods, which are known as Post Hartree-Fock methods, usually require expensive calculations in terms of costs and times and can be applied to systems with a relatively small number of electrons.

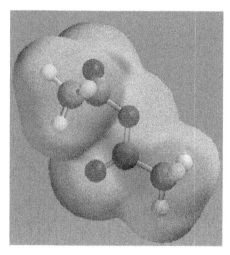

Figure 1. Electron density distribution of acetyl anhydride (Source: John Riemann Soong – Wikipedia)

APPLICATIONS

Initially, the Hartree–Fock (HF) method was applied exclusively to atoms, where the spherical symmetry of the system allowed one to greatly simplify the problem. Even so, solutions by hand of the Hartree–Fock equations for a medium-sized atom were laborious; small molecules required computational resources far beyond what was available before 1950. With the arrival powerful super-computers and parallel computing, the use of HF and Post HF methods became more common to study physical properties of molecules and solids. These methods are now part of a large number of software packages programmed for chemical and materials modelling, including GAUSSIAN, VASP, CP2K and many others.

Among the information provided by these models is:

- The equilibrium structures of molecules

323

- Reaction pathways and transition states
- Molecular electrical, magnetic and optical properties
- NMR and X-ray spectra
- Reaction mechanisms in chemistry and biochemistry
- Crystal structure and properties of solids

Intermolecular interactions to produce parameters for inter-atomic potentials used to study systems with a very large number of atoms (i.e. macromolecules, solvent effects, large crystals).

Pharmaceuticals and catalysis industries use these methods to design new drugs and predict/study catalytic behaviour of solid substrates. An alternative computational method widely used in materials and chemical modelling is the Density Functional Theory (DFT).

AT

Equation 262: Squareness and Diagonals

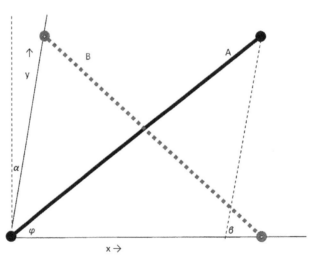

Ball bar or any other length measurement in a non-square coordinate system

Suppose a length bar with length L is measured over a diagonal in a 2-D coordinate system. All axes are perfect, just the x- and y axis have a squareness error α. The measurement is taken in diagonals; once in position A with an angle ϕ with the x-axis and once in position B. The difference in measured length is denoted by ΔL. For a square measurement with $\phi \approx 45°$ the following beautiful equation holds:

$$\alpha = \frac{\Delta L}{L}$$

The principle that for square axes the diagonals must have equal length is used for ages by carpenters who have to make doors and windows square. They know they have to adjust the axes so that along a diagonal bar half the difference between both diagonal lengths is indicated.

The equation can be derived using the cosine rule, that gives for the size L for the bar in position A:

$$L_A^2 = \Delta x^2 + \Delta y^2 - 2\Delta x \cdot \Delta y \cdot \cos \beta \approx \Delta x^2 + \Delta y^2 + 2 \cdot \Delta x \cdot \Delta y \cdot \alpha$$

This gives for the measured length

$$L_m : L_m^2 = \Delta x^2 + \Delta y^2 = L^2 - 2 \cdot \Delta x \cdot \Delta y \cdot \alpha(posA) = L^2 + 2 \cdot \Delta x \cdot \Delta y \cdot \alpha(posB)$$

With some approximations it can be derived that

$$\alpha = \frac{L \cdot \Delta L}{2 \Delta x \Delta y}$$

Here Δx and Δy are the difference in x- and y-coordinates between the balls – or end points - in figure 1. When the length bar makes an angle φ with the x-axis we have $\Delta x = L \cdot \cos \varphi$ and $\Delta y = L \cdot \sin \varphi$ so that

$$\alpha = \frac{L \cdot \Delta L}{2 \Delta x \Delta y} = \frac{\Delta L}{2L \sin \varphi \cos \varphi} = \frac{\Delta L}{L \sin 2\varphi}$$

For a diagonal at $\emptyset = 45°$ this gives the beatiful equation.

HH

Equation 263: The Lift Equation

The phenomenon that allows airplanes to take off from the ground and to stay there is called lift. Lift is defined as the upward (opposite the direction of gravity) force exerted on a body travelling through a medium. The equation below relates lift to various parameters of the body and the medium through which the medium is travelling [1].

$$L = (Cl) \left(\frac{\rho V^2}{2} \right) (A)$$

Where L is the lift exerted on the object (N), Cl is the 'lift coefficient', the value of which depends on body shape, inclination, air viscosity, and compressibility (unitless), ρ is the density of the medium (in Kg/m^3), V is the velocity of the body (m/s) and A is the surface area over which the medium flows (m^2).

The lift equation was initially proposed by the Wright brothers in 1900. The equation above is a modern iteration of the earlier equation. The difference between the two equations is in the reference for the value of the lift coefficient [2].

References:
[1] The Lift Equation, NASA Glenn Research Center, Editor: Nancy Hall
https://www.grc.nasa.gov/www/K-12/airplane/lifteq.html, Accessed: 12 September, 2015.
[2] Modern Lift Equation, NASA Glenn Research Center, Editor: Nancy Hall
http://wright.nasa.gov/airplane/lifteq.html, Accessed: 12 Septmeber, 2015

MF

Equation 264: Householder Matrices

Consider a vector **x** in a three-dimensional space. Let y be the reflection of this vector about a plane whose (unit) normal vector is **v** as shown in the figure below.

325

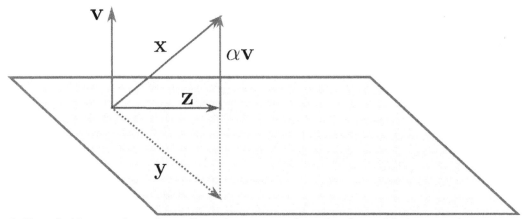

Figure1: Householder transformation reflects the vector **x** about the plane with the normal **v**.

Then, one can show that

$$\boldsymbol{y} = (\boldsymbol{I} - 2\mathbf{v}\mathbf{v}^T)\mathbf{x}$$

where $\boldsymbol{I}$ is the 3 by 3 identity matrix and the superscript $\boldsymbol{T}$ denotes the transpose of the unit vector **v**. The matrix multiplying the vector **x** in the above is called the Householder matrix. That is

$$\boldsymbol{H} = (\boldsymbol{I} - 2\mathbf{v}\mathbf{v}^T)$$

is called the Householder matrix. Although the matrix is introduced in the above in the context of a three-dimensional space, the above two equations are applicable to n-dimensional spaces with the reflection plane being a hyperplane in an $(n-1)$-dimensional space. Note that although the discussion here is limited only to matrices and vectors with real components, the Householder transformation can be defined for complex matrices and vectors as well.

The transformation $\boldsymbol{y} = \boldsymbol{H}\mathbf{x}$ is called the Householder transformation. The mathematician Alan S. Householder introduced the transformation in the 1950s. It was recognized as one of the top ten algorithms *with the greatest influence on the development and practice of science and engineering in the 20th century* [1]. The matrix $\boldsymbol{H}$ has some beautiful properties:

$$\boldsymbol{H} = \boldsymbol{H}^T$$

i.e. $\boldsymbol{H}$ is symmetric

For any vector **x** and $\boldsymbol{y} = \boldsymbol{H}\mathbf{x}, \boldsymbol{H}\boldsymbol{y} = \boldsymbol{H}^2\mathbf{x} = \mathbf{x}$, i.e. $\boldsymbol{H}^2 = \boldsymbol{I}$. This leads to the interesting result that $\boldsymbol{H}$ is non-singular, is its own inverse and is orthogonal!

For any two vectors $\mathbf{u}_1$ and $\mathbf{u}_2$, $\boldsymbol{H}\mathbf{u}_1 \cdot \boldsymbol{H}\mathbf{u}_2 = \mathbf{u}_1 \cdot \mathbf{u}_2$, i.e., the angle between any two vectors is preserved. This is a consequence of the fact that a Householder matrix is orthogonal.

$\boldsymbol{H}\mathbf{v} = -\mathbf{v}$. Thus -1 is an eigenvalue of H. $\boldsymbol{H}\mathbf{z} = \mathbf{z}$, therefore 1 is an eigenvalue of $\boldsymbol{H}$. In fact, since the reflection plane is an $(n-1)$-dimensional hyperplane, 1 is the eigenvalue of multiplicity $n-1$. The determinant of a Householder matrix is -1.

Note that the second property above implies that Householder transformation is an orthogonal transformation.

326

Householder matrices find applications in numerical linear algebra. The QR decomposition of a matrix A can be obtained using Householder transformations. To obtain the eigenvalues of a symmetric matrix, a sequence of Householder transformations can be used to first tridiagonalize the matrix. The eigenvalues of the tridiagonal matrix are then obtained using special-purpose fast algorithms developed for such matrices. Householder transformations can also be used to convert a matrix into an upper triangular matrix. This can be taken advantage of in the solution of systems of linear equations and in calculating the determinant of a matrix.

References
https://www.siam.org/pdf/news/637.pdf. Accessed at 14:34 EST on September 19, 2015.

HC

Equation 265: Isostacy (Pratt Equation)

Geodesy is the study of the measurement of the Earth, including its gravitational field, and its changes over time. British surveyor and engineer William Lambton proposed that geodesy, then a new concept, be applied to the whole of India, and on 10th April 1802, The Great Trigonometrical Survey of India began. One of its aims was the measurement of the heights of Everest and other Himalayan peaks. Though these were made successfully, it was found that the horizontal gravitational attraction of the Himalayas (as measured by plumb-line deflection, which is the angle which a freely hanging weight on a cord makes to the vertical) was much less than had been predicted based on the estimated mass of the mountains. Sir George Biddell Airy, the British Astronomer Royal, suggested that this might be caused by the presence of a mass of matter, of less than the average density, under the mountains; this explanation was further investigated by British Archdeacon and mathematician John Henry Pratt. Later (1892), the principle of isostacy was propounded and named by Major Clarence Edward Dutton, a US geologist: the weight of matter under any unit area of the earth's surface tends to become uniform, through the motion of underground materials which compensate for surface changes such as erosion. Today, isostacy is interpreted as the concept of light crustal materials floating on a denser fluid layer (called the mantle) - in other words, as the effect of buoyancy:

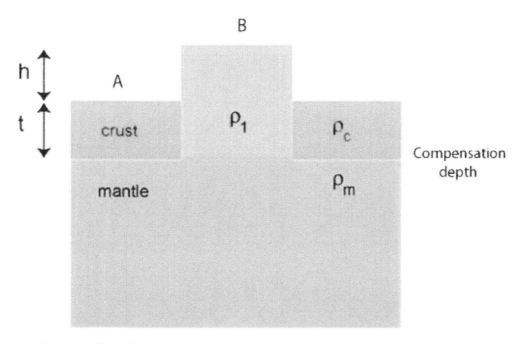

Pratt developed an equation of isostacy where

327

$$\rho_1 = \frac{\rho_c t}{(h+t)}$$

and then applied this to the Himalayas.

In fact, though the principle of isostacy is now well-accepted, the Himalayas are not in isostatic equilibrium - they are still rising. It is believed that they are being propped-up by the motion of the Indian continental plate.

MJG

Equation 266: Chebyshev Polynomials

Named after the Russian mathematician Pafnuty L. Chebyshev (1821- 1894), the Chebyshev polynomials (of the first kind) $T_n(x)$ are a family of orthogonal polynomials with respect to a certain weight function. This means that by multiplying different polynomials together and integrating the result the resulting integral will always be zero. The Chebyshev polynomials are defined over a horizontal interval spanning the range -1 to 1 by the recurrence relation as

$$T_0(x) = 1$$
$$T_1(x) = x$$
$$\dots$$
$$T_{n+1}(x) = 2xT_n(x) - T_{n-1}(x)$$

The letter of T was chosen by S. N. Bernstein, following the French translation of Chebyshev as Tchebischeff. An explicit definition of Chebyshev polynomials is

$$T_n(x) = \cos(n \cdot \arccos(x))$$

where n is the polynomial degree number. This explicit definition leads to the simple expression of Chebyshev roots (sometimes called Chebyshev nodes) for $T_n(x)$:

$$x_j = \cos\left(\frac{j\pi}{n}\right), 0 \le j \le n$$

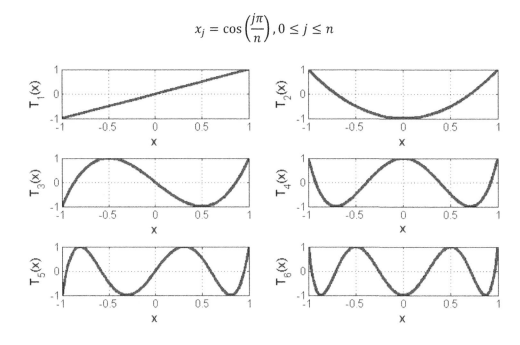

The first appearance of his polynomials was in 1854 [1], after Chebyshev's trip to western Europe in 1852 during which he met pioneers in modern mathematics history like Liouville, Hermite, Lebesgue and Dirichlet. Chebyshev was known as the founder of the Russian school of approximation theory. His visionary awareness of the concept and application of orthogonal polynomials has made a profound impact on today's computational algorithms for approximation of functions, numeric analysis and so on.

Similar to the popularity of using Fourier series (see beautiful equation 12) to approximate periodic functions, Chebyshev approximation play a vital role for choosing the best fitting algorithm for nonperiodic functions. For example, in the dimensional metrology community, a Chebyshev approximation is used for calculating the minimum zone fit of geometric features (like planes, circles, cylinders, etc.) from a measured data set. One reason for its popularity is that the Chebyshev approximation provides the optimal solution for finding minima and maxima for fitting equations to data, also known as the minimum zone fit. It is of importance in engineering when trying to compare measurements of the dimensions of manufactured components (that are not perfect) to the desired geometry (a practice often called Geometric dimensioning and tolerancing, or GD&T) [2]. A great online resource about approximation theory with numeric preference of using Chebyshev's method is Chebfun, which is software system built by researchers at the Oxford University [3] from 2004.

[1]. P. L. Chebyshev, Théorie des mécanismes connus sous le nom de parallélogrammes, Mém. Acad. Sci. Pétersb. 7 (1854), 539-568.
[2]. G Goch, K Lübke (2008), Tschebyscheff approximation for the calculation of maximum inscribed/minimum circumscribed geometry elements and form deviations. CIRP Annals – Manufacturing Technology 57 (1): 517-520.
[3]. http://www.chebfun.org/

KN

Equation 267: Murphy's Law

$$P_M = -K_M \left(e^{-\frac{I \times C \times U + F}{F_M}} - 1 \right)$$

where P_M is Murphy's probability of something going wrong, K_M is the Murphy's constant (and is equal to 1), F_M is the Murphy's factor (and is approximately equal to 0.01), I, C, U and F are constants that depends on the particular event conditions. I is the importance of the result or outcome, C is the complexity of the system, U is the urgency and F is the frequency. From these 4 parameters, virtually any event can be described.

Anything that can go wrong, will go wrong.

Edward A. Murphy, Jr. was one of the engineers on the rocket-sled experiments that were done by the U.S. Air Force in 1949 to test human acceleration tolerances (USAF project MX981). One experiment involved a set of 16 accelerometers mounted to different parts of the subject's body. There were two ways each sensor could be glued to its mount, and somebody methodically installed all 16 the wrong way around. Murphy then made the original form of his pronouncement, which the test subject (Major John Paul Stapp) quoted at a news conference a few days later [1].

For example considering an event whatsoever such as finishing and delivering a report or thesis the day before the deadline. For this event the constant I, C, U and F are specified as follow: I is the importance of the event and therefore equal to 8 (very important but not matter of life or death); C is the complexity of the system (laptop) and therefore equal to 7 (fairly complex but ordinary); U is the urgency and therefore equal to 10 (very urgent); finally F is the frequency of the event and therefore equal to 1 (not frequent). If we substitute in the Murphy equation we would obtain that the probability of your computer to start updating the operating system and the word processing software is equal to 1.

The English equivalent is the Sod's law which goes a step further: *if something can go wrong, it will, and it will happen at the worst possible time*.

The law is known in French as 'La loi d'emmerdement maximum', which translate to "the law of maximum shit-dipped". In italian is known as "la fortuna è cieca, ma la sfiga ci vede benissimo" which might be translated to: "Good luck is blind but bad luck sees us very well".

Reference
[1]. Scq.ubc.ca, (2007). *THE MURPHY'S LAW EQUATION | SCQ*. [online] Available at: http://www.scq.ubc.ca/the-murphys-law-equation/ [Accessed 9 Sep. 2015].

GM

Equation 268: Railway Adhesion

We are now in to autumn with the problem for railways of leaves on the line. However, this got me thinking. What does keep the wheels of a train on the tracks and what equations are involved?

Common belief is that the wheels are kept on the tracks by the flanges. However, if you look closely at a typical railway wheel you will see that the tread is burnished but the flange is not - the flanges rarely make contact with the rail and, when they do, most of the contact is sliding. The rubbing of a flange on the track dissipates large amounts of energy, mainly as heat but also including noise and, if sustained, would lead to excessive wheel wear.

The railway (whether horse drawn or drawn by steam locomotives) was probably the invention that had the most impact from the late 1700s onwards. It enabled materials to be moved more easily, fresh produce to be delivered to market and people to travel more widely.

Rail adhesion relies on the friction between a steel wheel and a steel rail. The term is particularly used when discussing conventional railways to distinguish from other forms of traction such as funicular or cog railway.

Adhesion of any kind is caused by friction and the maximum tangential force produced by a driving wheel before slipping given by:

$$F_{max} = coefficient\ of\ friction \times weight\ on\ the\ wheel$$

The force needed to start sliding is typically greater than that needed to continue sliding. The former is concerned with static friction, whilst the latter is called sliding friction.

When measured under laboratory conditions the coefficient of friction for steel on steel can be as high as 0.78 but typically on railways it is between 0.35 and 0.5, as low as 0.05 under extreme conditions.

Put another way a 100-tonne locomotive could have a tractive effort of 350 kN, under the ideal conditions), falling to a 50 kN under the worst conditions. We then have to consider that railways are not straight and have curves. Take a curve too fast and the outside wheel we begin to lift. Toppling will occur when the overturning moment due to the side force (centrifugal acceleration) is sufficient to cause the inner wheel to begin to lift off the rail. This may result in loss of adhesion - causing the train to slow or in the worst case the inertia may be sufficient to cause the train to continue to move at speed causing the vehicle to topple completely.

Let's do some physics. The downward force is the mass of the train multiplied by the acceleration due to gravity (mg). The sideward force is

$$F_s = \frac{mU^2}{R}$$

where F_s is the sideward force, U is the velocity, R is the track radius and m is the mass. If the centre of gravity is at height H then the train topples when R is equal to $\frac{2HU^2}{gd}$, d is the wheel separation or track gauge.

Going back to the original question as what keeps the wheels on the rail, centring is accomplished through the shaping of the wheel. The tread of the wheel is slightly tapered. When the train is in the centre of the track, the region of the wheels in contact with the rail, traces out a circle which has the same diameter for both wheels. The velocities of the two wheels are equal, so the train moves in a straight line.

However, the coning of the wheels can lead to a behaviour called hunting and manifests itself as a swaying of the train from side to side. The phenomenon of hunting was known by the end of the 19th century, although the cause was not fully understood until the 1920s and measures to eliminate it were not taken until the late 1960s. The instability of motion had become a limitation on the maximum speed achievable.

So today's equation is the one that describes this hunting behaviour.

A rudimentary analysis of the kinematics of the coning action yields an estimate of the wavelength of the lateral oscillation as

$$\lambda = 2\pi \sqrt{\frac{rd}{2k}}$$

where d is the wheel gauge, r is the nominal wheel radius and k is the taper of the treads.

For a given speed, the longer the wavelength and the lower the inertial forces will be, the more likely it is that the oscillation will be damped out. Note that the wavelength increases with reducing taper, thus increasing the critical speed requires the taper to be reduced, which implies a large minimum radius of turn.

A more complete analysis, taking account of the actual forces acting, yields the equation below for the critical speed of a wheelset, where W is the axle load for the wheelset, a is a shape factor related to the amount of wear on the wheel and rail, C is the moment of inertia of the wheelset perpendicular to the axle, m is the wheelset mass.

$$V^2 = \frac{Wrad^2}{k(4C + md^2)}$$

This equation is consistent with the kinematic result in that the critical speed depends inversely on the taper. This equation also implies that the weight of the rotating mass should be minimised compared to the weight of the vehicle. The wheel gauge implicitly appears in both the numerator and denominator and thus has only a second-order effect on the critical speed. However, life is never this simple and the true situation is much more complicated, as we need to take into account the response of the vehicle suspension. Restraining springs, opposing the yaw motion of the wheelset, and similar restraints on bogies, may be used to raise the critical speed further. However, in order to achieve the highest speeds without encountering instability, a significant reduction in wheel taper is necessary, so there is little prospect of reducing the turn radius of high speed trains much below the current value of 7 km.

Based on https://en.wikipedia.org/wiki/Adhesion_railway accessed 23/09/2015

DF

Equation 269: The Griffith Crack Criterion

$$\frac{\partial U}{\partial c} = 0$$

Where U is the total energy of the component (N·m) and c is the length of the crack (m).

At the turn of the 20th century, there was a growing body of research trying to understand the reason for failure of materials. Bolstered by demonstrable relationship between the strengths of chemical bonds and the stiffness of a material, much of this early work similarly focused on the strengths of chemical bonds. However, experimental measurement of the strengths of materials failed to show any clear correlations.

During this period Alan Arnold Griffith (1893 – 1963) working at the Royal Aircraft Establishment at Farnborough in England took a very different approach to the problem. For this approach, he sought to determine the free energy associated with an elastic solid containing a crack and stated that the rate of change of this free energy with crack length is zero in an equilibrium state, formally given by the above equation. This statement, based on simple thermodynamics of conservative systems (system that do not lose available energy) became the basis of all fracture mechanics. Of course, it would not be sufficient to present this hypothesis without demonstrating how this might be utilized for predicting the strength of a material.

Firstly, to apply this theory it is necessary to determine the energy in the region of a crack. For simple elastic deformation in the region of the crack, the energy can be considered to be comprised of elastic (mechanical) energy U_M, in the bulk of the material and the energy of the surfaces U_S, generated as the crack initiates and grows. Fortunately, a paper published in 1913 by the British engineer Charles Inglis provided a solution for the stresses in the vicinity of a sharp crack of length $2\,c$ for a material subject to a uniform stress field that would exist in the absence of the crack. Using this analysis, the elastic strain energy could be calculated by the relatively simple equation

$$U_E = -(1 - v^2)\frac{\pi c^2 \sigma^2}{E}$$

Where E [N·m^{-2}] is the elastic modulus (stiffness) of the material, σ [N·m^{-2}] is the uniform stress in the component at a distance remote from the crack which causes a localized increase in the stress particularly near to the crack tip (commonly referred to as a stress concentration), and v is the dimensionless Poisson's ratio of the material that is the ratio of how the stretching (strain) in one direction induces a corresponding strain at right angles to its direction, see BE 18. The surface energy is given by the product of the total area of the two crack surfaces and the surface energy, γ [N·m^{-1}], of the material. Hence the equilibrium condition for a stable crack becomes

$$\frac{\partial \left(4\gamma c - (1 - v^2)\frac{\pi c^2 \sigma^2}{E}\right)}{\partial c} = 4\gamma - 2(1 - v^2)\frac{\pi c \sigma^2}{E} = 0$$

If this value is greater than zero, Griffith hypothesized that the crack will grow indefinitely, leading, of course, to catastrophic 'failure'. Consequently, for a crack of length c_o the stress, σ_F at which failure is expected can be calculated from

$$\sigma_F = \sqrt{\frac{2\gamma E}{(1 - v^2)\pi c_o}}$$

Two of the three material properties E and v are constants (at constant temperature) and will be ignored. The surface energy however is going to be a function of the environment and can vary enormously from a clean surface (in a vacuum) to surfaces exposed to air and other conditions making it difficult to determine in many practical instances. Having profound implications, the appearance of the crack length in the denominator implies that as the crack gets smaller, the material can withstand higher stresses. Griffith's was concerned about this prediction and undertook a series of tests on glass fibres. The reasoning of these experiments was that glass is close to ideally brittle and is easily heated and drawn into thin fibres. Consequently, as the fibre diameters reduce so too must the sizes of flaws. Data from these experiments is plotted in figure 2 that indeed show a dramatic increase in strength with reduction in the fibre diameter.

The consequences of this equation also revolutionized the science of fracture mechanics and resulted in the development of a whole new approach to the science of strong materials. In terms of fracture mechanics, it is now normal to use a stress intensity factor that is typically given by $K_1 = Y\sigma\sqrt{\pi c}$ (N·m$^{-3/2}$) and a critical value of this K_{Ic} that is called the fracture toughness and is considered a material property. The parameter Y is dependent on the geometry of the crack and the factor π is a left-over from Griffith's original analysis upon which this new approach to fracture mechanics is broadly based.

And, as they say, the rest is history. Glass (and other) fibres are now used in many applications including planes, trains and automobiles and many high-tech structures are now designed based on the fact that there are cracks in the components of manufacture.

References
Gordon J.E., 1968, The new science of strong materials (or why you don't fall through the floor), Pelican Books, ISBN 978-0-14-192770-1.
Griffiths A.A., 1920, The phenomena of rupture and flow in solids, *Proc. Roy. Soc. Lond.*, **221**, 582 − 593.

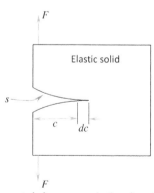

Figure 1: Illustration of an elastic body containing a crack that is subject to a vertical force that would tend to open the crack.

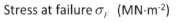

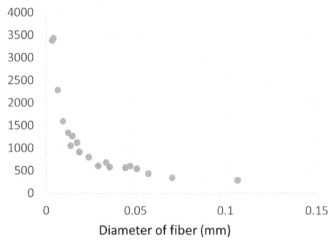

Figure 2: Plot of maximum strengths of glass fibres as a function of diameter. These fibres were heated and drawn using a gas flame and then left for 40 hours prior to testing. Data from Griffiths' original 1920 paper.

SS

Equation 270: Pythagorean Theorem

INTRODUCTION

Pythagoras is possibly one of the most famous Greek philosophers and early mathematicians. Pythagoras is also responsible for one of the most fundamental and common mathematical and geometric based equations the Pythagorean Theorem. The Pythagorean Theorem is used for determining the length of one side of a triangle Equation

$$a^2 + b^2 = c^2$$

where a^2 is typically the shorter of the two lengths either side of the right angle, b^2 is typically the longer of the two lengths either side of the right angle and c^2 is hypotenuse

The equation states that the sum of the squares of length a and b are equal to the square of the hypotenuse c. The formula can be rearranged to find the length of any side of a right angle triangle. For example $c^2 - a^2 = b^2$.

The proof for this equation was offered by Pythagoras as a graphical representation of rearrangement showing the constituent parts of the equation and why they worked. I think when an equation can be easily and simply visualized it qualifiers as beautiful equation.

USES

The practical real world uses of Pythagorean theory are numerous. Their use is complimentary to many trigonometry applications. There are also many geometric applications where the Pythagorean Theorem is applied.

HISTORY
Although the equation is attributed to Pythagoras (circa 550 BC) it is argued with some evidence that he was not the first to understand the equation although he is acknowledged to be the first person to identify with proof its validity.

TS

Equation 271: The Clausius-Mossotti Equation

The Clausius-Mossotti equation determines the relationship between dielectric constants of a medium with the polarizability of the atoms or molecules in the medium.

$$\frac{\varepsilon_r - 1}{\varepsilon_r + 2} = \frac{N_A \alpha}{3\varepsilon_0}$$

Where $\varepsilon_r = \frac{\varepsilon}{\varepsilon_0}$ is the relative dielectric constant of a substance; ε_0, the dielectric constant of the vacuum, N_A is the Avogadro's number, α is the molecular (or atomic) polarizability in SI units (C·m²/ V).

$$\varepsilon^* = \varepsilon + \frac{i\sigma}{\omega}$$

This is the complex form of the dielectric constant, in which σ is the conductivity of the medium and ω is the angular frequency of the applied electric field.

For a gas, the Clausius-Mossotti equation reduces to its simplest form:

$$\varepsilon_r = 1 + \frac{N_A \alpha}{\varepsilon_0}$$

The Clausius-Mossotti equation with the refractive index in place of the dielectric constant is:

$$\frac{n^2 - 1}{n^2 + 2} = \frac{N\alpha}{3}$$

Where n^2 is the refractive index, N is the number of particles per unit volume of the medium and α is the molecular (or atomic) polarizability.

The equation that relates the refractive index in place of the dielectric constant of a medium to its polarizability α is very important for two reasons:

- It allows to calculate (in principle) the dielectric constants of all materials, from the knowledge of its polarizability α.
- It relates the measure of microscopic properties such as polarizability of atoms and molecules to macroscopic properties such as dielectric constants and refractive index.

In fact, whilst the polarizability of a material is an intrinsic atomic property (microscopic), the dielectric constant depends on how the atoms assemble to form the crystal structure of a material (macroscopic).

In his Lectures on Physics published in 1964, American physicist Richard Feynman discusses atomic polarizability in these terms: when a sinusoidal electric field is applied to a material, it induces a dipole moment per unit volume on the atoms of the material (Figure 1), which is proportional to the electric field. The proportionality constant is the polarizability of the material α, that depends on the frequency of the electric field. However, this constant is in general a complex number, which means that it does not exactly follow the electric field but it may be shifted in its phase.

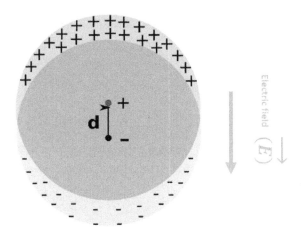

Figure 1. Schematic mechanism of the formation of an atomic dipole induced by an external electrical field (*E*).

The dipole moment induced on an atom by the presence of an electric field is proportional to the effect of the local electric field on the atomic orbitals. For non-spherical atomic orbitals, the polarizability will be not isotropic (the same in all directions), so that it will not be a number but a tensor. The local electric fields acting on a specific atom are generated by nearby atoms in a dense material. As these can vary in the different spatial directions, the final polarizability resulting from the overall contribution of all the local electric fields will be not isotropic, and will be mathematically expressed by a tensor instead of a number. This makes the polarizability of dense materials significantly different from that of a gas, in which it is isotropic. This is the reason why the Clausius-Mossotti equation for gases reduces to a simpler form.

HISTORY
The equation was the result of the studies of the Italian physicist Ottaviano-Fabrizio Mossotti, who investigated the relationship between the dielectric constants of two different media in 1850, and the studies of the German physicist Rudolf Clausius, who formulated the equation in 1879 not in terms of dielectric constants but of refractive indexes.

Mossotti, who exiled in Argentina for his liberal ideas and taught astronomy and physics at the University of Buenos Aires, developed a type of multiple-element lens correcting spherical aberration and coma.

Clausius was one of the main founders of thermodynamics. He introduced the name of Entropy and gave its first mathematical version in a landmark paper titled "The Mechanical Theory of Heat" in 1865. The paper ends with the following summary of the first and second laws of thermodynamics:

- "The energy of the universe is constant
- The entropy of the universes tends to a maximum."

APPLICATIONS
The equation finds a large number of applications to calculate dielectric properties of solids and liquids. Recent applications also include the evaluation of dispersion properties of optical fibres (Figure 2), in particular for structures having multi-component regions [1].

However, the equation does not apply well to polar liquids, such as water at room temperature, as it shows instabilities. Some attempts to improve the equation for these types of liquids were done by Onsager and Kirkwood. Some models argue that the instability in the Clausius-Mossotti equation for these liquids is a signature of a phase transition of the liquid into an ordered ferroelectric phase [2]. Recent experimental and theoretical studies of the frequency dependence of the Clausius-Mossotti equation for ash micro/nano particles, have suggested the possibility of using dielectrophoresis (DEP) to filter nanoparticles and purify exhausted combustion gases [3].

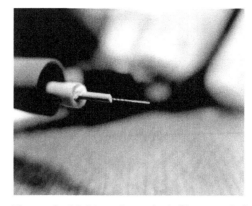

Figure 2. Multi-mode optical fibre used for short-distance communication (Source: Wikipedia)

References
[1] Melman P. and Davies, R.W. (1985), "Application of the Clausius-Mossotti equation to dispersion calculations in optical fibers", Journal of Lighwave Technology, 3 (5), 1123
[2] Sivasubramanian, S., Widom, A., Srivastava, Y.N. (2003), "The Clausius-Mossotti Phase Transition in Polar Liquids", Physica A: Statistical Mechanics and its Application, 345 (3-4), 356.
[3] Giugiulan, R., Malaescu, I., Lungu, M., Strambeanu, N., "The Clausius-Mossotti Factor in low Frequency Field of the Powders Resulted from Wastes Combustion", Romanian Journal of Physics, 59 (7-8), 862.

AT

Equation 272: Fibre Bragg Grating

A fibre Bragg grating is a short length of optical fibre that reflects a particular wavelength of light and transmits all others. This is achieved by creating a periodic variation in the refractive index of the fibre core.

The first fibre Bragg grating (FBG) was demonstrated in 1978 by Ken Hill at the Communications Research Centre, Ontario, Canada and later in 1989 Gerald Meltz (1934-2010) demonstrated the more common transverse grating pattern.

An FBG is created by "writing" a periodic variation of the refractive index onto the core of an optical fibre using a UV source. The grating that is created effectively becomes a wavelength selective mirror. When light, of a particular wavelength travels down the fibre, each variation within the grating partially reflects some of the light. The reflections interfere destructively at most wavelengths and the light continues down the fibre. At one narrow wavelength range, constructive interference occurs and light is returned up the fibre. A maximum in this reflected light occurs at the Bragg wavelength and is given in the first equation and shown in the first figure.

$$\lambda_B = 2n_{eff}\Lambda$$

where λ_B is the reflected wavelength or Bragg wavelength, n_{eff} is the effective refractive index of the grating and Λ is the grating period.

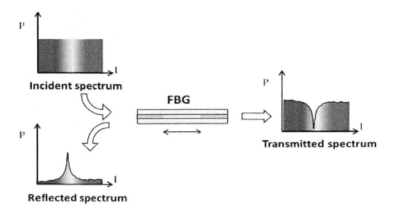

Reference FBGS (www.fbgs.com)

The Bragg wavelength is dependent on both the grating period and the effective refractive index. A change in the environmental conditions (i.e. a strain in the fibre or a change in temperature) will alter the Bragg wavelength and this is shown in the second equation.

$$\Delta\lambda_B = \lambda_B(1 - \rho_\alpha)\Delta\varepsilon + \lambda_B(\alpha - \xi)\Delta T$$

where $\Delta\lambda_B$ is the change in Bragg wavelength, ρ_α is the photoelastic coefficient, $\Delta\varepsilon$ is the change in strain, α is the coefficient of thermal expansion, ξ is the thermo-optic coefficient and ΔT is the change in temperature.

The Bragg wavelength is a function of both strain and temperature. If the FBG is strained, either through a mechanical change or a thermal change, the Bragg wavelength will also change. This is illustrated in the second figure.

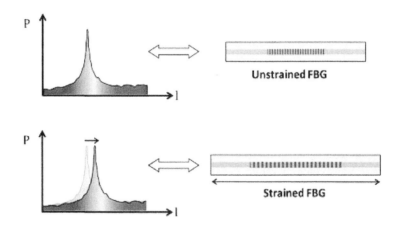

Reference FBGS (www.fbgs.com)

The FBG can be used as a temperature sensor by ensuring there is no mechanical strain and temperature compensated strain measurements are also possible, if the temperature is known or can be measured independently. FBGs can also be used to measure other parameters such as pressure, acceleration, displacement etc. by incorporating the FBG into a transducer. This sensor integration is increasingly becoming known as Smart Sensors.

K. O. Hill, Y. Fujii, D. C. Johnson and B. S. Kawasaki, "Photosensitivity in optical fiber waveguides : Application to reflection filterfabrication," Appl. Phys. Lett., vol. 32, pp. 647-649, 1978.
G. Meltz, W. W. Morey and W. H. Glenn, "Formation of Bragg gratings in optical fibers by a transverse holographic method," Opt. Lett, vol. 14, pp. 823-825, 1989.

CM

Equation 273: Magnification in a Compound Microscope

$$ m = \frac{D}{f_{EYE}} \cdot \frac{T}{f_{OBJ}} $$

The classical compound microscope dates from the 17th century, and is composed of two optical systems working together to magnify small specimens: The eyepiece, with a focal length f_{EYE} in the magnification equation above, and the objective, with a focal length f_{OBJ}. The traditional finite-conjugate geometry has the objective forming an intermediate real image at a distance T from the back focal point of the objective, which the eyepiece views and magnifies for visual inspection. The distance D is the near point for vision with the unaided eye. Microscopes are ubiquitous and are part of nearly everyone's basic science education. This makes it all the more interesting, and beautifully ironic, that the common expression above for microscope magnification m is so frequently misunderstood.

Starting with the eyepiece, the definition of the magnification is based on the eye's ability to focus on close-up objects. The conventional value is 250 mm, presumed to be typical for the average person. The proposition is that we can see objects sharply at one focal length f_{EYE} away when looking through the eyepiece with *relaxed vision* (focused on infinity), but no less than the near point D in *stressed vision* without the eyepiece. Inasmuch as 75% of adults use corrective lenses of some kind, the 250 mm distance is more of a choice than a physical fact. Another odd thing is that if a young person with good vision tries hard, the microscope magnification can be increased by simply refocusing—that is, the image size depends on how well the viewer can accommodate to different focus positions. This all adds up to an arbitrary definition for the eyepiece magnification. Not wrong, but not exactly rigorous math.

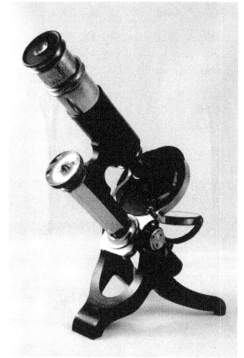

The second part of the magnification equation relates to the objective lens. Here the eye is not involved, so the magnification follows from simple geometry. One can readily show that the objective magnification is exactly equal to the ratio of the distance T to the objective focal length f_{OBJ}, where T measures from the back focal point of the objective to the intermediate image, which in relaxed vision is also the front focal point of the eyepiece. The distance T is known as the *optical tube length*. However, the optical tube length is not known *a priori* to the person using the microscope, so defining the magnification in this way is of limited practical use.

The commonly-accepted solution is to approximate, or rather confuse, the optical tube length with the *mechanical* tube length, which is the distance from the eyepiece flange to the mounting threads of the objective lens. The DIN (Deutsche Industrie Norm) standardizes the mechanical tube length to 160 mm and the part focal length from the objective threads to the sample to 45mm. The DIN also set the distance from the specimen to the intermediate image to 195mm. These standard mechanical dimensions make it possible for the objectives and eyepieces to be interchanged without a lot of refocusing. This is helpful generally and is essential for turreted microscopes. However, the *optical tube length* T *is not standardized* — it must vary to accommodate the mechanical constraints of the microscope body and the locations of the principal planes in the objective optics (see beautiful equations 242). Astonishingly, most optics textbooks state that it is the optical tube length T that is standardized to 160mm, which it most definitely is not. Setting $T = 160$mm makes for a simple calculation, but the results are wrong. We can easily see this by checking the specifications of commercial lenses: A DIN-standard Nikon Achromatic finite conjugate 4X objective has a focal length of 31 mm and an optical tube length of 124 mm, which is not even approximately equal to 160 mm.

The correct way to calculate the net magnification in a compound microscope is to multiply the eyepiece and objective magnifications stamped on their housings. How these magnifications are achieved, and what they mean, are interesting optical design questions not readily solved by the simple and often misunderstood equation at the beginning of this contribution.

For a beautifully correct description of the function of the compound microscope and many other topics, one can do no better than the exceptional three-volume series "Advanced Light Microscopy" by Maksymilian Pluta (Elsevier, 1989).

PdG

Equation 274: Mean, Median and Mode

When asked to provide the "average" of a set of numbers, we tend to compute the arithmetic mean. But what are we actually doing when we calculate that value? What are we looking for from an 'average'? In fact, this depends on what we're trying to achieve with the data set. Arithmetic Mean can be written as

$$\bar{x} = \frac{\sum_{i=0}^{n} x_i}{n}$$

where n is the number of samples and x_i is the ith element of the dataset.

We'll first look at the properties of the mean, median and mode, and then apply them to two activities that involve statistics, measuring something with a ruler and judging a hotel on a review site.

The arithmetic mean is an attempt to identify a value where the total distance between it and every other point in the set equals zero. For example consider a dataset A = {3,7,6,2,4} with an arithmetic mean of 4.4 as illustrated by Figure 1. The distance of each data point from the mean {-2.4, -1.4, 0.4, 1.6, 2.6} sums to 0. This is trivially proved for all cases by rearranging the equation for the mean. This technique is only a true indication of the 'average' if the frequency or magnitude of the values that are smaller than the average, are cancelled out by the frequency or magnitude of the values that are larger than the average.

The median value of a dataset is the 'central' point i.e. if the data is sorted into order of magnitude, the median is the point that is halfway down the sorted list. If there's an even number of points the median becomes the mean of the two middle points. For our dataset, sorting the list gives {2,3,4,6,7}, there are 5 values, so the median is the 3rd value in the list, i.e. 4.

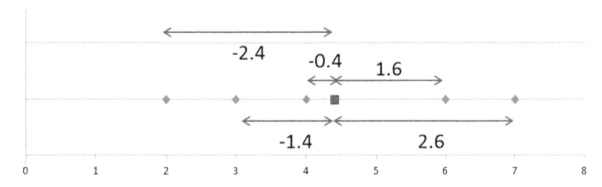

Figure 1. An illustration of an arithmetic mean. The data points are marked as blue diamonds and the mean is a red square. Numbers above red arrows illustrate the distance between each data point and the mean.

Median for a sorted dataset **A** the median value m is defined by

$$m = \{x_{(i+1)/2} \; for \; even(i) \; = false\}$$

$$\frac{x_{i/2} + x_{(i/2)+1}}{2} \; otherwise$$

where even(i) is a function returning true if a number is even and false if a number is odd.

The mode is simply the most commonly occurring value in the dataset. In this case every data point occurs only once, making it a 'multimodal' dataset, i.e. there are several modes. If we were to introduce a new value of 2 into the set, making it {2,2,3,4,6,7} the dataset becomes 'unimodal' i.e. 2 becomes the only mode. Mode for a dataset **A** the mode is the value(s) y that satisfy

$$\forall x, f(y, A) \; >= \; f(x, A)$$

where $f(x, A)$ is a function that returns the frequency of the value x in the dataset A. An important distinction between these averages is their differing resilience to anomalous data. The mode is best used for discrete data (i.e. integer values, labelled data such as days of the week etc.) and, by definition will always identify the most commonly occurring value in the dataset, ignoring other values. This makes the mode extremely resilient to many types of noise, but in datasets where the probability of the values is very similar (for example, if you recorded what day of the week a room full of people's birthdays occurred on) the mode can be volatile as new data is added. The varying effects of anomalous data on the mean and the median are best illustrated by observing the impact of introducing a new, deliberately misleading value into the data set. The table below illustrates the impact of introducing a new value into the data.

Table 1: The impact of introducing new values into the dataset. The top row is the original dataset for reference

New Value	New Mean	Difference from original Mean	New Median	Difference from original Median
-	4.4	0	4	0
0	3.67	0.73	3.5	0.5
-1000	-163	167.4	3.5	0.5
9	5.17	0.77	5	1
1000	170.33	165.93	5	1

For the mean, a single new value can have a significant impact on the 'average', which is proportional to the size of the new value. For the median, the important factor is really the number of new values introduced, not the size of a single new value. This makes the median less vulnerable to anomalous values. As alluded to at the beginning, the context of the 'average' is important so the mean's sensitivity is not necessarily a bad thing, as it makes it better for detecting shifts in the underlying data.

Let's consider our first example - measuring something with a ruler. When measuring things averages are used to estimate the "true" value of the thing being measured, eliminating any noise from the measurements. For the mean to be a good indicator of the true value, the noise needs to have a 'Gaussian' or 'normal' distribution. In other words the probability of being too big is roughly equal to the probability of being too small, and that extremely wrong values (big or small) are less likely than more correct ones. If we asked a number of people to measure the same line with a ruler, it is likely that variations in the answers would have this property and so the mean is a suitable average. The median is also likely to produce a good result, but can be misleading in the presence of a small dataset. Manually measuring things is not likely to be repeated many times, so the median is potentially a poor choice of average. The mode is particularly bad at real-world measurements. The dataset {1.1, 1.3, 1.31} is multimodal, so it's difficult to identify a good value. One solution to this problem is to group the data into "bins", for example, rounding to the nearest decimal place, which in this case would yield the value 1.3.

Moving on to our second example, picking a hotel, the mean can be a false friend. Review sites allow people to estimate the quality of a hotel using a star rating (for this example, we'll say marks from 1 to 10). The variability in people's opinions and hotel performance can result in these reviews taking on a Gaussian distribution. Let's consider a fictional hotel "The Dodgy Dump". It's reviews can be seen in Figure 2.

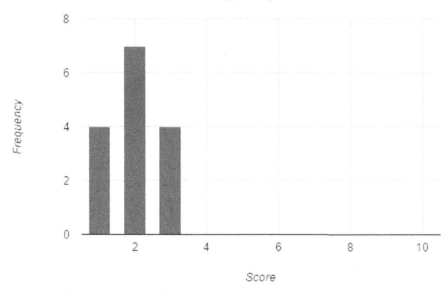

Figure 2. Reviews for "The Dodgy Dump"

Here the reviews have started to fall into an approximation of a Gaussian distribution (characterised by the fact that the plot of frequency against score starts to describe the shape of a bell), centred on the value 2. This is obviously not a great hotel. The review site knows that people don't like a lot of numbers and statistics so rather than showing the graph, it simply reports the average review score, calculated using the mean. The Dodgy Dump rightly has an "average" rating of 2 (which is incidentally also its median and mode review score).

Unfortunately, the owner of The Dodgy Dump is unscrupulous and hires a company to create accounts on the review site and add 10 star reviews. They generate ten such accounts, which shifts the average rating of the hotel on the website up to 5 (5.2 rounded). Not only is the median heavily skewed, the modal value becomes 10, which vastly over-estimates the hotel's quality. Had the review site used the median, this deception would be more difficult to pull off. Even in the presence of the new fake reviews, the new median value is still only 3 which is a much fairer indication of the hotel's quality.

As most large datasets collected from the real-world have the 'Gaussian' properties outlined earlier, it is easy to become used to using the mean as a quick and instructive way of identifying the 'average' value. In fact, a combination of its applicability and the ease with which it can be calculated has meant we have allowed the words 'average' and 'mean' to become synonymous in colloquial speech. However, with no knowledge of the nature of the data being analysed, it is the median that should be our default analytical tool, as it makes no assumptions about the distribution of the data. There are a number of different "averages" not covered by this article, including the harmonic average and the rms average. They all have particular properties designed to fit specific applications and each makes for a useful addition to an arsenal of analytical tools for any spreadsheet user or data scientist.

So next time you use the word 'average' try to think about what it is you are actually asking - do you want the most likely value to occur next, the 'true' underlying value that the data is estimating or a value that summarises the general consensus the data is describing? No matter what your application, select your beautiful equation wisely, and be sure that it is the answer to the question you are asking.

RFO

Equation 275: Pair Production

$$\gamma \rightarrow e^- + e^+$$

$$hf \approx 2m_oc^2 + K_{(-e)} + K_{(+e)}$$

The most well-known pair production process

Pair production is a phenomenon in which energy is converted into mass. The term usually refers the case of a spontaneous creation of an electron and, simultaneously, its antiparticle the positron, from a single high energy photon in the vicinity of a nucleus. (In a condensed medium, the positron quickly meets another electron with the result that the two are annihilated to produce another photon.)

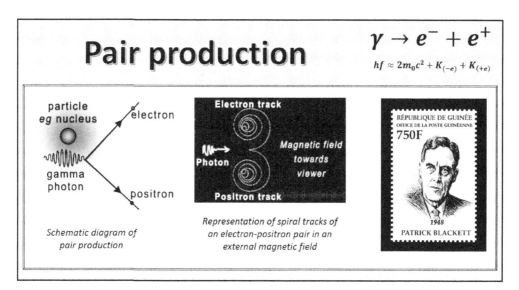

In order for pair production to occur, the photon's energy must exceed a certain threshold value, which equals the total rest mass energy of the particles produced – two times 0.511 MeV in the case of the electron-

343

positron pair. The probability of pair production occurring in photon-matter interactions increases with photon energy; it is also approximately proportional to the square of atomic number. This relationship means that pair production is most often observed near heavy nuclei such as those of lead or uranium.

Because of the momentum conservation principle, the creation of a pair of particles out of a single photon cannot occur in empty space. Something else must be around to absorb the photon's momentum. This something is usually a nearby nucleus, but it could be any particle – including an electron or a second photon. In the case of a nucleus of the order of 100,000 times more massive than the electron plus positron, the nucleus can absorb momentum whilst absorbing very little energy; in this situation, the energy-conservation relation, $hf \approx 2m_oc^2 + K_{(-e)} + K_{(+e)}$ is virtually an equality.

The fact that electrons and positrons are easily detected explains the use of pair production telescopes, which cover a very wide range of energies from about 30 MeV to 50 TeV – higher than that of Compton telescopes.

Less well-known pair production processes

The term 'pair production' is sometimes used in a more general sense, referring to the creation of any particle-antiparticle pair from an elementary neutral boson (or a field fluctuation) in a strong electromagnetic field. So the incoming boson might be a Z_0, say, rather than a photon. And the fermions produced might typically be, for example, a muon plus antimuon, a tau plus anti-tau, or a proton plus antiproton.

When the pair production process occurs in the electric field of an electron (this is sometimes called triplet production), the threshold energy is significantly higher than in the more usual interaction. On the other hand, in the strong magnetic fields near the surface of a neutron star, the threshold energy for the gamma rays producing the hypothesised e^+e^- pairs would be low. Theoretically, the pair produced might be gauge or scalar bosons, such as the production of Higgs particles from gluons. The term is also used when a particle plus its antiparticle are created as a result of a collision between two leptons or two hadrons.

The phrase 'internal pair production' refers to the process whereby an unstable nucleus that has at least 1.022 MeV of excess energy directly generates an electron-positron pair within its own electromagnetic field, without first producing a gamma photon. This process is a class of gamma decay.

HISTORY
It was in 1933 that experimental physicist Patrick Blackett discovered the now instantly recognisable opposing spiral tracks of positron/electron pair production, using a cloud chamber in which an incoming cosmic ray triggered a camera. In 1948 he was awarded the Nobel Prize in Physics.

FR

Equation 276: Parallax and Astronomical Distance

Parallax is a commonly used method to determine the distance of nearby stars (stars in our local region of the galaxy). It uses a reasonably straightforward triangulation concept, but requires the measurement of very small angles (typically much less than an arcsecond). Due to the earth's orbit around the sun, nearby stars appear to move slightly compared with the background of much more distant stars. This movement can be related to the angle traced by the earth's orbit from the point of view of the star being measured over the course of several months, during which the earth's location has moved from one side of its orbit to the other.

344

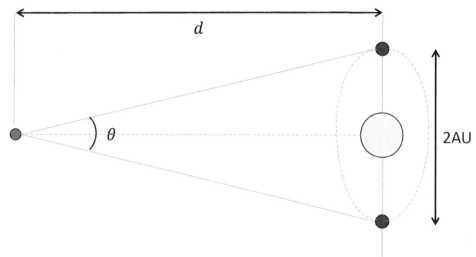

The picture above (not to scale) shows a schematic of the earth's orbit around the sun on the right (at an arbitrary angle to the plane of the paper) and a not-too-distant star on the left. The relatively fixed background of more distant stars is significantly further away on the left. Regardless of where the star is located in the celestial sphere, the parallax angle can still be symmetric about the line of sight between the star and the sun by adjusting the time of year at which the measurements are made (the relative location of the selected diameter of earth orbit). The parallax angle p is typically defined as the half-angle $\theta/2$ of the measurement giving the simple relationship for the stellar distance d in astronomical units (AU):

$$d = \frac{1}{\tan\left(\frac{\theta}{2}\right)} \cong \frac{1}{p}$$

At the tiny angles involved here, the tangent of the parallax angle is equivalent to the angle itself expressed in radians to above 10 decimal places. The AU is defined as the average distance between the earth and the sun, a very convenient standard for these measurements as the distance in AU can be determined directly from the parallax angle. Due to the large number of astronomical units in these stellar distances (even for the nearest star α Centauri, this is of the order of 10^6), a useful unit to define is the *parsec*, which is the distance of an imaginary star at which the parallax angle is one arcsecond. This allows simple scale measurements of a reasonable order of magnitude to be made for all the local stars. A parsec, which is about ¾ of the distance to our nearest star, is equivalent to approximately 206,265 AU (about 3.26 light years).

For stellar distance, parallax is sufficiently accurate to measure the distances of neighbouring stars and is still used as the standard for calibrating methods for measuring longer distances, such as the use of spectral red shift and standard "candles" (eg Cepheid variable stars, whose oscillation periods are accurately related to their absolute brightness and hence distance). For ground-based measurements the limit is around 100 parsecs; however, in 1989 an ESA space-based observatory called Hipparcos was launched and had a capability of measuring parallax to better than 1 milliarcsecond – this increases the maximum measurable distance to around 1000 parsecs, and over 4 years of observations this instrument measured the distances and proper motions (changes in position relative to the centre of the solar system) of more than 118,000 stars – providing a astrometric catalogue that far surpassed anything before it. The new successor to Hipparcos, together with other capabilities, is the ESA's Gaia observatory, launched in 2013 into an orbit around the Lagrange point L_2. The "Astro" instrument in Gaia's payload is capable of measuring parallax to an accuracy of better than 20 microarcseconds. This not only increases the achievable distance considerably but enables much more accurate measurements of stellar motions, and is anticipated to map the position, motion and luminosity of a billion objects in our galaxy. A density map of the galaxy was released this year and the first full catalogue of observations is expected in 2016.

Before the use of stellar parallax, solar parallax (observations of the sun from different locations on the earth) was used, in post-Copernican times (as it wouldn't mean anything without a heliocentric model of the solar system), to determine the distance of the sun from the earth, now defined as the astronomical unit. This was the first important step in calculating the distances of celestial objects. The first successful measurement of parallax of a star was taken by Friedrich Wilhelm Bessel in 1838, using an instrument called a heliometer (an instrument that uses split optics to generate two images of an object and measure their angular separation). He measured the parallax of the star 61 Cygni, roughly 11 light years distant, to a good accuracy (within better than 10% of the modern value) – this would have been extremely difficult to achieve for a star, the instrument was typically used to measure the variation in the sun's apparent diameter. The heliometer he used had been designed by Joseph von Fraunhofer at Königsberg Observatory, the first type to make use of an achromatic objective.

Parallax is derived from the Greek *parallaxis* which means "alteration". It is of course a familiar concept in earth-based applications such as binoculars, microscopes and rifle scopes – in fact binocular vision in general, and it can be observed easily enough with the naked eye at short enough distances (although for local measurement and compensation for the effect you would not necessarily be able to use the tangent-angle approximation that is so convenient for stellar parallax). The natural use of the phenomenon in the brain to generate depth perception is called stereopsis.

PB

Equation 277: Von Mises Stress

Von Mises stress or Von Mises yield criterion is considered the point at which a material begins to yield. Although first formulated by Maxwell in 1865, Huber published a paper in 1904 which received little attention. The Von Mises stress theory is mainly attributed to Richard von Mises (19 April 1883 – 14 July 1953), a German-American applied mathematician. At that point, the theory was no more than a mathematical equation without physical interpretation. It wasn't until 1924, that it was realised the theory related to deviatoric strain energy. Hencky suggested that yielding began when the elastic energy of distortion reaches a critical value and is related to the second deviatoric stress invariant by the shear modulus of the material.

The von Mises stress theory applies to materials which are normally isotropic or ductile i.e. metals and is used to predict the yielding of a material under any load condition The theory is based on the determination of the distortion energy in a material, that is, the energy associated with change of shape, not change of volume.

Mathematically, the von Mises condition is given in the equation.

$$f(J_2) = \sqrt{J_2} - k = 0$$

where J_2 is the second deviatoric stress invariant and k is the yield stress of the material in pure shear.

From the equation, the von Mises yield criterion for different stress conditions can be derived and are given in the table at the end of this article.

It is clear from the equation that yielding begins when $J_2 = k$ and at this point, the material makes the transition from the elastic to the plastic region. The simplest case is when a material is under uniaxial loading (the last entry in the table). In this case, when the stress in the direction of loading is equal to the von Mises stress, the material will yield.

346

The von Mises yield criterion is used extensively by mechanical design engineers who consider it to be a safety check of their designs; if the maximum Von Mises stress induced in a material is greater than the strength of the material, the design will fail.

W. N. Sharpe, Ed., Handbook of Experimental Solid Mechanics, Springer, 2008.

Load Scenario	Conditions	Von Mises Equation
General	None	σ_v $= \sqrt{\frac{1}{2}[(\sigma_{11} - \sigma_{22})^2 + (\sigma_{22} - \sigma_{33})^2 + (\sigma_{33} - \sigma_{11})^2 + 6(\sigma_{12}^2 + \sigma_{23}^2 + \sigma_{31}^2)]}$
Principle stress	None	$\sigma_v = \sqrt{\frac{1}{2}[(\sigma_1 - \sigma_2)^2 + (\sigma_2 - \sigma_3)^2 + (\sigma_3 - \sigma_1)^2]}$
General plane stress	$\sigma_3 = 0$ $\sigma_{31} = \sigma_{23} = 0$	$\sigma_v = \sqrt{\sigma_1^2 - \sigma_1\sigma_2 + \sigma_2^2 + 3\sigma_{12}^2}$
Principle plane stress	$\sigma_3 = 0$ $\sigma_{12} = \sigma_{31} = \sigma_{23} = 0$	$\sigma_v = \sqrt{\sigma_1^2 - \sigma_1\sigma_2 + \sigma_2^2}$
Pure shear	$\sigma_1 = \sigma_2 = \sigma_3 = 0$ $\sigma_{31} = \sigma_{23} = 0$	$\sigma_v = \sqrt{3}|\sigma_{12}|$
Uniaxial	$\sigma_2 = \sigma_3 = 0$ $\sigma_{12} = \sigma_{31} = \sigma_{23} = 0$	$\sigma_v = \sigma_1$
Notes	• Subscripts 1,2,3 can be replaced with any orthogonal coordinate system • Shear stress is denoted here as σ_{ij}; in practice it can also be given as τ_{ij}	

CM

Equation 278: E$_N$-Value

In order to make sure that measurement results are true and correct within their measurement uncertainty, and comparable world-wide, laboratories have to compare their results on artefacts or reference instruments that are circulated. Such comparisons are called "Interlaboratory comparisons (ILC's)", "Proficiency tests" or "Round robin tests". Whenever possible, one laboratory delivers a reference value, this is often a national laboratory where traceability can be best obtained as primary standards are realized there, so this measurements can have the lowest uncertainty.

To establish whether a laboratory's result X with its uncertainty $u(X)$ is consistent with the reference value R and its uncertainty $u(R)$ this is most conveniently done using the so-called normalized error, also called the E_N-value.

E_N is defined as:

$$E_N = \frac{X - R}{\sqrt{u^2(X) + u^2(R)}}$$

where X is the measured value, $u(X)$ is the uncertainty in measured value, R is the reference value and $u(R)$ is the uncertainty in reference value.

A measurement result is considered satisfactory when $\left| E_N \right| < 1$.

It is considered that there is a problem when $\left| E_N \right| > 1$. The cause can be that either the laboratory or the reference laboratory has underestimated its uncertainty (a rather trivial conclusion), more specifically it is common to consider that the reference value has been carefully checked and that the participating laboratory has done something wrong in its measurement method, evaluation and/or uncertainty calculation.

The equation follows directly from the uncertainty in the difference of two values X and R that each have their own, independent, uncertainty (see beautiful equation 40: Propagation of measurement uncertainty). When there is no specific hierarchy between the laboratories, i.e. when the uncertainties of all laboratories are comparable, the reference value can be taken as the weighted average of the different laboratories, including the laboratory under consideration itself. In that case the two values are not independent anymore and a correction must be made in such a way that every laboratories result is compared to the weighed mean and standard deviation of all others.

This equation is the basic verification tool to check if a laboratory can achieve the measurement uncertainty that it claims, next to documented calibration procedures and uncertainty evaluations. As such it is the basis of granting accreditation to calibration laboratories, or establishing confidence between national laboratories.

HH

Equation 279: The Day-length Equation

The duration of daylight depends on the day of the year and the geographical position on the Earth. Historically, the times of sunrise and sundown have been experimentally observed and stored in meteorological 'lookup' tables. Alternatively, a mathematical model can be used to calculate the duration of sunlight. One of the more recent models [1] was published in 1995 and claims to agree with experimentally observed data to within less than 1 minute for latitudes below 40° (both North and South). The maximum error for this model is 7 minutes and occurs at 60° latitude.

The mathematical equation for the model is given below and a detailed explanation of the various terms follows.

$$D = 24 - \frac{24}{\pi} \cos^{-1} \left[\frac{\sin\left(\frac{p\pi}{180}\right) + \sin\left(\frac{L\pi}{180}\right) \sin\varphi}{\cos\left(\frac{L\pi}{180}\right) \cos\varphi} \right]$$

where D is the day-length (hours); p is the 'day-length coefficient' (degrees), the value of which depends on how sunrise and sunset are defined; L is the latitude (degrees), which is positive for northern latitudes and negative for southern latitudes; θ is the Earth's revolution angle (radians), which is dependent on day of year J; and φ is the sun's declination angle (radians), which is dependent on the Earth's revolution angle θ. Two terms in the first equation are further described by the equations below.

348

$$\theta = 0.2163108 + 2\tan^{-1}[0.9671396\tan[0.00860(J - 186)]]$$

$$\varphi = \sin^{-1}[0.39795\cos\theta]$$

Day-length as defined by this model is plotted below as a function of day of the year for London, United Kingdom (L = 51°) and Reykjavik, Iceland (L = 64.1333°). The day-length coefficient p is 0.8333°, which corresponds to the US Government definition of sunrise and sunset: "when the top of the sun is apparently even with the horizon".

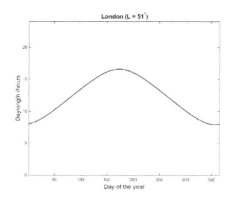 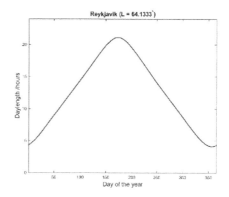

[1] Forsythe, W.C., Rykiel, E.J., Stahl, R.S., Wu, H., Schoolfield, R.M., A model comparison for daylength as a function of latitude and day of year, Ecological Modelling, vol. 80, pages 87-98 (1995).

MF

Equation 280: The Shape of the Surface of a Rotating Liquid

A cylindrical vessel of radius R is partially filled with a liquid. Suppose now the vessel is given a rotational speed of ω about its axis. Eventually, the liquid will rotate with the vessel at the same angular speed. We all know that the liquid surface is convex in shape with a dip in the middle and a rise near the edges. But what is the actual equation that governs the shape of the surface?

Since the fluid rotation is symmetric with respect to the rotational axis, the distance of the surface from the bottom face is a function only of the radial distance of a surface particle from the axis. We denote this by $h(r)$ where r is the radial coordinate of a fluid-particle in the vessel.

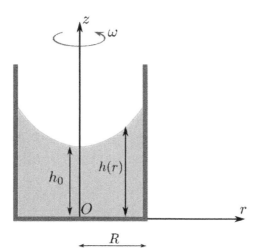

A schematic illustrating the nomenclature used in the text for the equation governing the surface of a rotating fluid in a rotating container.

Then, by ignoring the surface tension effects and assuming that the liquid is Newtonian and the rotation has reached steady-state, the shape of the liquid surface can be shown to be given by

$$h(r) = h_0 + \frac{\omega^2 r^2}{2g}$$

where h_0 is the distance of the surface from the bottom of the vessel along the axis of rotation and g is the acceleration due to gravity. This is the equation of a parabola in the $r - z$ plane and therefore, the liquid surface assumes the shape of a paraboloid of revolution. That the shape of the surface of a rotating liquid is paraboloid was noted by Newton in his monumental treatise Principia.

The above equation implies that the rise of the liquid near the edges is proportional to the square of the product of ω and R, the radius of the cylindrical vessel. The higher the angular speed, the sharper the rise. Due to the conservation of mass, this also implies that h_0 will decrease.

An interesting application of the equation given above is in the manufacture of liquid-mirror telescopes where the mirror is made of a reflective liquid such as mercury. The reflective liquid is placed in a container and spun at a prescribed speed so that the liquid surface assumes the paraboloid shape which has the interesting property that a parallel beam of light incident upon the parabolic surface is reflected to the focus of the surface. For such mirrors, the focal length is half the radius of curvature of the liquid surface at $r = 0$. From the equation given above for the liquid surface, it can be shown that the radius of curvature at $r = 0$ is $\rho = \frac{g}{2\omega^2}$. Therefore, the focal length f is given by

$$f = \frac{\rho}{2} = \frac{g}{2\omega^2}$$

Thus, the angular speed at which the liquid and the container need to be spun for a given focal length is given by

$$\omega = \left(\frac{g}{2f}\right)^{\frac{1}{2}}$$

This indicates that larger the focal length, the smaller the angular speed required. Furthermore, the angular speed required is directly proportional to the square root of g. Interestingly, many scientists have proposed

350

the possibility of building large liquid-mirror telescopes in the low gravity environment of the Moon (see References [1] and [2]). The largest liquid-mirror telescope currently in use is the Large Zenith telescope operated by the University of British Columbia [3]. The diameter of this mirror is 6 meters and has an effective focal length of 10 meters.

References
1. Borra, E. F., Seddiki, O., Angel, R., Eisenstein, D., Hickson, P., Seddon, K. R., & Worden, S. P. (2007). Deposition of metal films on an ionic liquid as a basis for a lunar telescope. Nature, 447(7147), 979-981.
2. Paul Hickson (2007). Liquid-Mirror Telescopes, American Scientist, May-June 2007, Volume 95, Number 3, Page: 216.
3. http://www.astro.ubc.ca/lmt/lzt/, accessed at 20:35 hours on October 5, 2015.

HC

Equation 281: Rossmo's Formula

Rossmo's formula makes quantitative two home truths: 1. don't shit in your own backyard, and 2. don't go further than you need. And it outputs - with luck - the best places to look for serial criminals.

The formula calculates the (x,y) coordinates of a (geographical) map, and a probability for each coordinate. The probabilities are those that the criminal being sought lives at that coordinate, and the data input to it are the coordinates of the locations of a number of crimes believed to have been committed by the same perpetrator.

Home truth 1. implies that a criminal will not draw attention to himself/herself by committing crimes close to home. In this case, there will be a buffer zone around that location free of his/her crimes (or, more precisely, that the probability of a crime being committed will increase with distance). The radius of the buffer zone depends on the recklessness/self-confidence of the criminal and so must be empirically determined.

Home truth 2. implies that the probability of a crime being committed by the criminal will decrease with distance. The rate of fall of probability with distance depends on the availability of victims and the transport mode of the criminal, so it too must be empirically determined.

Hence, both truths together imply that the criminal's residence will be surrounded by "hot zone", a circle (modified by local geography) along which crimes are most likely to be committed, with circles of progressively decreasing probability inside and outside it.

The formula is:

$$p_{i,j} = k \sum_{n=1}^{(total\ crimes)} \left[\frac{\emptyset_{ij}}{(|X_i - x_n| + |Y_j - y_n|)^f} + \frac{(1 - \emptyset_{ij})(B^{g-f})}{(2B - |X_i - x_n| - |Y_j - y_n|)^g} \right]$$

$$\emptyset_{ij} = \begin{cases} 1, & \text{if } (|X_i - x_n| + |Y_j - y_n|) > B \\ 0, & else \end{cases}$$

where k, f, g are empirical constants, selected by analyses of past crime data, the term $|X_i - x_n| + |Y_j - y_n|$ is a measure of the distance between the coordinates of the nth crime site (x_n, y_n) and of the location for which the probability is being calculated (X_i, Y_j), B is a characteristic radius of the buffer zone, and $\emptyset$ is a factor which weights the significances of the home truths.

The formula was developed by Kim Rossmo, a Vancouver police criminologist, in 1995. It was the subject of his Ph D thesis (he was the first Canadian police officer with a doctorate in criminology).

As the numerous user-set variables indicate, Rossmo's formula needs a great deal of experimental adjustment for each use. Nevertheless it has been successfully deployed on numerous occasions.

MJG

Equation 282: Process Capability

The production of goods in today's industrialized world must meet the ever demanding requirements of manufacturing at low cost to produce reliable processes and using sustainable methods. Quality, as one of the most important metrics for production started to play a quantitative role as early as the 1920's. American physicist, engineer and statistician Walter Shewhart (1891-1967) first applied statistics for the improvement of the quality in manufacturing processes when he worked at the Bell Telephone Laboratories. William Deming (1900-1993), who was inspired by Shewhart's work originated the concepts known collectively as statistical quality control (SQC) during WWII and promoted these statistical techniques resulting in them now being used throughout the world.

A core idea in SQC is to evaluate the variation of a functionally significant parameter on a product to indirectly understand what's going on in the manufacturing process by using a simple number called process capability C_p. The equation for calculating process capability is defined as:

$$C_p = \frac{USL - LSL}{6\sigma}$$

where, USL is the upper specification limit, LSL is the lower specification limit, σ is the standard deviation derived from measured sample parts. Ideally the value of C_p should be infinite assuming no variation is observed on the measured sample. But this is impossible in practical manufacturing, where disturbances from the shop floor environment, malfunction of the production tools, lack of experience of the machine operators, and the uncertainty of measurement on sample parts all come into play.

One typical use of C_p is to help quality engineers to identify possible sources causing the deviation of products during the manufacturing process. Once identified, the process can be investigated and the problem addressed to bring the manufactured components to within the allowed tolerances. The following graph is a "control chart" also known as Shewhart's chart, which illustrates the change of parameter from sampled parameter during a manufacturing process. The two black lines indicate the value of the supper and lower specification limits respectively symmetric to the mean μ of the parameter of interests. When a measurement falls within these two lines, a process is interpreted as "in control", otherwise "out of control".

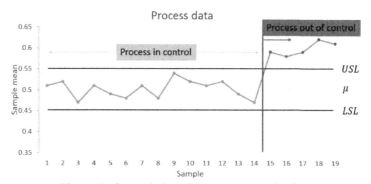

Figure 1: Control chart for process monitoring

Reference:
H. Kunzmann et al. "Productive Metrology – Adding Value to Manufacture", CIRP Annals-Manufacturing Technology, 2005, Vol. 54, 2, 155-168.

KN

Equation 283: Temperature of the Ocean at a Given Depth

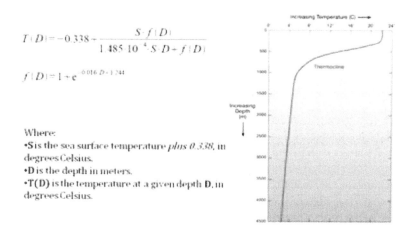

$$T(D) = -0.338 + \frac{S \cdot f(D)}{1.485 \cdot 10^{-4} \cdot S \cdot D + f(D)}$$

$$f(D) = 1 + e^{-0.016 D + 1.344}$$

Where:
- S is the sea surface temperature *plus 0.338*, in degrees Celsius.
- D is the depth in meters.
- T(D) is the temperature at a given depth D, in degrees Celsius.

The model of temperature of the ocean for a given depth has been derived by fitting to experimental data [1]. In the equation S is the temperature at the surface, D is the depth in meters. In the figure the temperature distribution is plotted as a function of the depth for a surface temperature of 22 degree Celsius.

An example of the temperature-depth curve is shown. 3 separate layers can be distinguish. The top layer also called the surface layer is where the ocean is heated by the sun, and wind and waves mixed it so also the heat gets transferred downwards. The second layer is called the thermocline and it is where the temperature gradient is the largest. Below the thermocline there is the bottom layer also called the deep ocean. The deep ocean is not very well mixed and it temperature is below 4 degree Celsius.

A note regarding the temperature of the ocean:

The Argos and BIOS program in charge for monitoring temperature and other parameters of the oceans have both warned that the ocean is warming [3]. Surface water temperatures obviously change from season to season and year to year, but the whole ocean has warmed 0.055 degree Celsius in the past 30-50 years [2]. This is an important amount considering the heat capacity of the oceans.

References:
[1] http://residualanalysis.blogspot.co.uk/2010/02/temperature-of-ocean-water-at-given.html
[2] http://www.windows2universe.org/earth/Water/temp.html
[3] http://www.windows2universe.org/earth/climate/cli_effects.html

GM

Equation 284: Duckworth-Lewis Method

There is nothing to beat sitting out in the sun and watching a day's cricket in jolly old Blighty. However, that ideal is infrequently enjoyed and one day games can be interrupted by the weather. So, when the overs available to the team batting second are reduced how do you set a target score? This is where the Duckworth-

Lewis method comes in, one of the most devious equations known to man. The Duckworth-Lewis method (or D/L method to its friends) was devised by two English Statisticians Frank Duckworth and Tony Lewis.

Setting an adjusted target for the team batting second is not as simple as reducing the run target proportionally. The reason for this is that a team with ten wickets in hand and 25 overs to bat will be expected to play more aggressively than if they had ten wickets and a full 50 overs and consequently will achieve a higher run rate. So Duckworth and Lewis devised a method to set a statistically fair target for the team batting second based on the score of the first team taking in to account their wickets lost and overs played into account.

So what is the equation, well the equation used in most first class cricket (*e.g.* internationals) is;

$$\text{Team 2's par square} = \text{Team 1's score}^* \left(\frac{\text{Team 2's resources}}{\text{Team 1's resources}}\right)$$

Team 2's par score is usually rounded up to the nearest integer. For example, if a rain delay means that Team 2 only has 90% of the resources that were available to Team 1, and Team 1 scored 254, then $254 \times 90\% = 228.6$, so Team 2's target is 229, and the score to tie is 228. Deceptively simple you might think. But oh no. This is where the fun starts. The actual resource values used in the Professional Edition are not publicly available and a computer software is used. It is the resource part that is statistical. Still with me?

OK, a team basically has two resources available to it to score the maximum number of runs; the number of overs they have to receive and the number of wickets they have in hand. At any point in the innings a team's ability to score depends on these two resources. What Duckworth and Lewis found was that there was a close correspondence between the availability of these to resources and a team's final score.

As with most non-trivial statistical derivations, the D/L method can produce results that are somewhat counter intuitive which adds to its mystical status. However, the announcement of the derived target score can provoke a good deal of debate amongst the crowd at the cricket ground. This can also be seen as one of the method's successes, adding interest to a "slow" rain-affected day of play.

Until 2003 a single version of the D/L method was used and it was fairly transparent as it used a single published reference table of total resource percentages remaining for all possible combinations of overs and wickets and some simple mathematical calculations. However, it had a known flaw in that it did not cope well with first innings scores of 350+.

After the 2003 world cup an improved version was introduced but this version relied on computer modelling rather than published tables and is less transparent. The original version was labelled the Standard Edition and the new version the Professional Edition.

So let's examine the standard edition. Here we will use the notation in the ICC Playing handbook where the First Team playing is Team 1 and there score S. The total resource available to team 1 for their innings is called $R1$. The team that bats second is Team 2 and their available resources $R2$.

For each reduction in overs, the loss in total resources available to the batting team is found using a published reference table, then Team 2's target score is changed as follows:

• If $R2 < R1$, reduce Team 2's target score in proportion to the reduction in total resources, i.e. $S \times R2/R1$.

• If $R2 = R1$, no adjustment to Team 2's target score is needed.

• If $R2 > R1$, increase Team 2's target score by the extra runs that could be expected to be scored on average with the extra total resource, i.e. $S + G50 \times (R2 - R1)/100$, where $G50$ is the average 50-over total. Team

2's target score is not simply increased in proportion to the increase in total resources, i.e. $S \times R2/R1$, as this 'could lead to some unrealistically high targets if Team 1 had achieved an early high rate of scoring [in the powerplay overs] and rain caused a drastic reduction in the overs for the match.] Instead, D/L Standard Edition requires average performance for Team 2's additional resource over Team 1.

The problem here is with $G50$. $G50$ will depend on the level of competition, ground and may not be invariant over time. To quote Duckworth and Lewis 'We accept that the value of $G50$, perhaps, should be different for each country, or even for each ground, and there is no reason why any cricket authority may not choose the value it believes to be the most appropriate. In fact it would be possible for the two captains to agree a value of $G50$ before the start of each match, taking account of all relevant factors. However, we do not believe that something that is only invoked if rain interferes with the game should impose itself on every game in this way. In any case, it should be realized that the value of $G50$ usually has very little effect on the revised target. If 250 were used, for instance, instead of 235, it is unlikely that the target would be more than two or three runs different.'

The Wikipedia article on the method gives examples of the tables, some examples of some calculations and how to deal with games where there are multiple interruptions.

DF

Equation 285: One Dimensional Capillary Wave

$$V^2 = \tanh(kl)\left(\frac{g}{k} + \frac{\gamma k}{\rho}\right) = \tanh(kl)\left(\frac{g\lambda}{2\pi} + \frac{2\pi\gamma}{\rho\lambda}\right)$$

where V is the velocity of the wave [m·s^{-1}], ρ is the density of the fluid [kg·m^{-3}], g is the gravity field [m·s^{-2}] (approximately 9.8 at the Earth's surface), λ is the wavelength [m], γ is the surface tension [N·m^{-1}], l is the depth of the fluid [m], $f = \frac{V}{\lambda}$ is the frequency of excitation [Hz] and $k = \frac{2\pi}{\lambda}$ is the wavenumber [m^{-1}].

For a fluid having a depth l that is many times larger than the wavelength λ, the hyperbolic tangent term $\tanh(kl)$ is close to a value of 1 and the above equation simplifies to

$$V^2 \approx \left(\frac{g\lambda}{2\pi} + \frac{2\pi\gamma}{\rho\lambda}\right)$$

or

$$\lambda \approx \frac{\pi V^2 \pm \pi\sqrt{V^4 - \frac{4g\gamma}{\rho}}}{g}$$

While somewhat difficult to realize in practice, the above equation represents a fluid that is being excited to produce ripples propagating out from a line source. Maybe imagine a wide infinity pool that is very long. At one end place a long ruler that spans the width of the pool and jiggle it up and down at a constant frequency. This will result in periodic ripples that will propagate along the pool with a velocity and wavelength that can be predicted from the above equations. The most interesting aspect of this equation is that the velocity is comprised from two terms, one of which contains surface tension (without a gravity term) while the other contains a gravity term (without a surface tension term). The gravity term is proportional to the wavelength while the surface tension term is divided by the wavelength. This means that for long wavelengths the gravity term dominates while for shorter wavelengths it is the surface tension that dominates the wave behavior. It is for this reason that these ripples are also called gravity-capillary waves.

The first of the above equations was presented by Lord Rayleigh in volume two of his seminal treatise on the theory of sound. An early formulation for infinitely deep pools was developed by Lord Kelvin and published in 1871.

In the case of water for which $\rho = 1000$, $\gamma = 0.076$, $g = 9.81$ Kelvins equation is plotted in figure 1 below. Because there are two possible values at each velocity, this can be a little confusing to read. It can be immediately seen that there is a minimum surface velocity corresponding to $V = 231$ mm s^{-1} at a frequency of 13.6 Hz. The minimum point at 231 mm s^{-1} represents a transition from gravity to capillary dominated waves. In the case of the excitation frequency, the capillary dominated waves are above this point at the higher frequencies while the opposite is true for the wavelength which is dominated by gravity waves at the lower frequencies. More discussion of this can be found in the second volume of Rayleigh's *Theory of Sound*.

References:
Strutt J.W.S. (Lord Rayleigh), 1896, *Theory of sound*, Volume II, The Macmillan Company, chapter XX. ISBN 486-60293-1
Thompson W.T. (Lord Kelvin), 1871, *Phil. Mag.*, XI(II), 375.

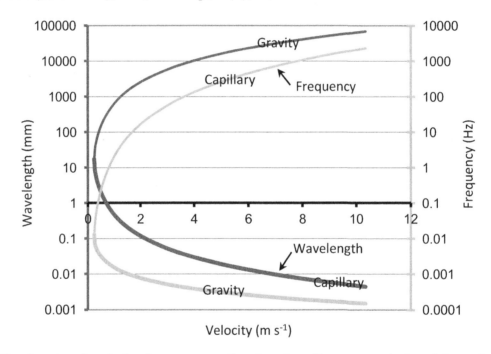

Figure 6: Wavelength and excitation frequency as a function of capillary wave velocity of the surface of a deep liquid.

SS

Equation 286: Drag Equation

INTRODUCTION
In fluid dynamics and particularly aerodynamics the drag equation is used to determine how much force a given object of size and shape will be exposed to due to its velocity. This is an equation based around the commonly specified drag coefficient which represents an object's impedance to passing through a fluid medium. The equation is

$$F_d = \frac{1}{2}(\rho u^2 * C_d * A)$$

Where F_d is the force due to the drag on the object, ρ is the density of the fluid, u is the velocity of the fluid flow with respect to the object, A is the cross sectional area of the object and C_d is the drag coefficient.

Both F_d and u assume that the fluid flow is normal to the direction of alignment of the object.
A assumes orthographic projection from a plane that represents the cross sectional area to the direction of fluid flow.

USES
The drag equation is commonly used in the automotive and aeronautic industries. It is common to see the coefficient of drag specified with a vehicle to give an indication as to how aerodynamically slippery a vehicle is. The lower the drag coefficient, the less the vehicle will be impeded due to the drag of the air, this means that a vehicle with a lower drag coefficient will require less power to travel at any given speed. This typically means it will have lower fuel consumption. Since the aerodynamic drag is a square of the velocity the drag coefficient and resulting force to move through a fluid becomes ever more significant the higher the velocity.

HISTORY
Lord Rayleigh (1842-1919) is credited with the equation. Lord Rayleigh originally used the equation with an L2 linear dimension parameter in place of the A parameter.

CONSIDERATIONS
It is important to note that the drag equation is a generalization and close approximation for many fluid dynamics systems, however there are special cases which require extra consideration dependant on the exact nature of the geometrics and dynamics in question and the application of the equation. For example, some systems are only valid if the Reynold's number is above a certain value or until certain other fluid dynamics effects take over.

TS

Equation 287: Goldbach Conjecture

$$p + q = 2n$$

where (p, q) are two prime numbers and n is a positive integer.

GOLDBACH CONJECTURE
In 1742 Christian Goldbach an amateur mathematician in correspondence with Leonhard Euler proposed a number of conjectures relating to how integers can be represented by sums of prime numbers. The simplest, most surprising and the one that continues to be the most challenging to prove is: every even integer greater than 2 can be written as the sum of two prime numbers. In response Euler wrote: "Dass … ein jeder numerus par eine summa duorum primorum sey, halte ich für ein ganz gewisses theorema, ungeachtet ich dasselbe nicht demonstriren kann." ("That … every even integer is a sum of two primes, I regard as a completely certain theorem, although I cannot prove it.")

Euler's confidence seems well placed even though there has been no full proof in the 270 years since its publication. In recent years Tomas Oliveira e Silva has been running a distributed computer project which has verified Goldbach's conjecture for every even number up to 1,609,000,000,000,000,000. Yet for all the attention this apparently simple problem has attracted a full proof remains a distant prospect.

Some simple examples:

$$4 = 2 + 2, 6 = 3 + 3, 8 = 3 + 5, 10 = 3 + 7, \ldots, 147 = 59 + 89, \ldots, 1969 = 19 + 1951$$

However, progress has been made over the years. For example, it has been proven that every even integer is the sum of at most six primes (Goldbach suggests two) and in 1966 Chen proved every sufficiently large even integer is the sum of a prime plus a number with no more than two prime factors (a $P2$). Vinogradov in 1937 showed that every sufficiently large odd integer can be written as the sum of at most three primes, and so every sufficiently large integer is the sum of at most four primes. One result of Vinogradov's work is that we know Goldbach's theorem holds for almost all even integers. Various speculative estimates, mostly based on probabilistic distribution of primes, are available for the number of solutions there should be to $2n = p + q$ and these grow in size with n. This suggests that for larger numbers the Goldbach Conjecture is increasingly likely.

When verifying the Goldbach Conjecture for n we quickly see that it is very easy to find many primes which add to n. In 1923 Hardy and Littlewood took the first major step toward the proof of the Goldbach conjectures using their circle method. Among other things, they conjectured that the number of ways of writing n as the sum of two primes, $G(n)$, is asymptotic to twice the twin prime constant times $n/(log\ n)2$ times the product of $(p-1)/(p-2)$ taken over the prime divisors p of n. That is a lot of solutions for a large n so there will be a solution with one of the prime quite small. For example, when checking all n up to 100,000,000,000,000, Richstein found the smaller prime is never larger than 5569 used in the following result: 389965026819938 = 5569 + 389965026814369.

Faber and Faber offered a \$1,000,000 prize to anyone who proved Goldbach's conjecture between March 20, 2000 and March 20, 2002, but the prize went unclaimed and the conjecture remains open.

ILB

Equation 288: Force on a Current in a Magnetic Field

$$F = BILsin\theta$$

The strength of a magnetic field is expressed in terms of the magnetic flux density of the field, often abbreviated by B. When a wire carrying a current is placed in a magnetic field, the wire experiences a force due to the interaction between the field and the moving electrical charges in the wire. The force on the wire is given by today's equation, where I is the electrical current in the wire, L is the length of the wire in the field and θ is the angle that the wire makes with the field lines. The greatest force on the wire will be when it is at right angles to the field. Fleming's Left Hand Rule gives direction of the force in comparison to the field lines and current direction (see figure). The force was first discovered by Faraday (1791 – 25 August 1867, English), and was preceded by work of Orsted (1777 – 1851, Danish) who discovered the magnetic field around a current-carrying conductor.

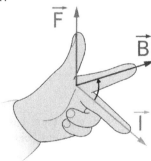

RL

Equation 289: Cross Section

$$\sigma = \frac{\mu}{n}$$

A handy concept in quantifying the likelihood of electromagnetic or particulate radiation interacting in a particular way with matter is an abstract one referred to as the 'cross section' for the event in question.

Examples of such interactions are collisions, or specific chemical/nuclear reactions. These involve some combination of beam particles (e.g. nuclei, elementary particles including photons) and target material (e.g. nuclei, atoms, gases, colloids).

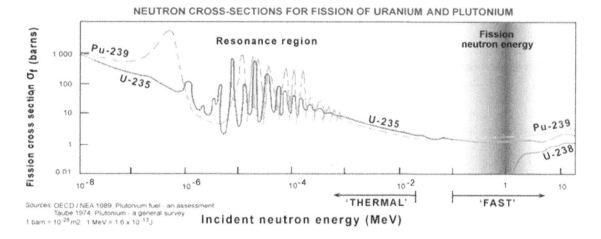

A real-life event is said to have a cross-section σ (which is conceptual) if its rate of occurrence is equal to the rate of collisions in an analogous and idealised classical scenario of a beam striking a target, where:
- the target particles are thought of as inert, impenetrable discs presenting an area σ (hence the name "cross-section") perpendicular to the beam.
- the beam is thought of as a stream of inert point-like particles;
- all other experimental variables are kept the same as in the real-life scenario.
-

    The cross section values described above are rarely the same as the geometric cross-sectional area of the target nucleus or particle.

Why talk in terms of 'cross section' rather than 'reaction rate' or 'probability'? The answer is that the likelihood of an event depends strongly on variables such as the density of the target material, beam intensity, or area of overlap between beam and target material. The cross section factors out these variables, which is what makes it so useful. So, for example, cross section values measured at one accelerator can be directly compared with those obtained at another, irrespective of performance differences in the accelerators.

The SI unit for cross sections is m^2, but in practice smaller units tend to be used. With visible light beams, the scattering cross-section is expressed in cm^2, whereas with X-rays, Å^2 is used (1Å, one angstrom, $= 10^{-10}$ m). In nuclear and particle physics, the conventional unit is the barn (b) or a derivative unit, where ($1b = 10^{-28}$ $m^2 = 100$ fm^2).

Cross section values depend on the energy of the bombarding particle and the type of reaction. Boron, when bombarded by neutrons travelling at 10 km/s, has a neutron-capture cross section of about 120 barns; for neutrons travelling at one tenth of that speed (1 km/s), the cross section increases by a factor of about ten to

359

roughly 1,200 barns. Hence so-called 'slow' neutrons are much more likely to be captured by boron, which is often used as a neutron absorber in nuclear reactors. (Note that boron's geometric cross-sectional area is only about 0.1 barn.)

In the formula $\sigma = \mu/n$, n is the number density of the target particles (SI units: m^{-3}); μ is the attenuation coefficient for this event (the fractional reduction in radiant flux per unit distance as the beam passes through a specific material; SI units: m^{-1}).

FR

Equation 290: Equal Temperament

Musical temperament is an adjustment made to the natural harmonics (pure intervals) between notes of an instrument or a collection of instruments for tuning purposes in order to make the collection of notes as consonant (in tune) as possible. This is done to correct for natural defects in the subdivisions of an octave, the most basic music interval which has a frequency ratio of 2:1. The midpoint of an octave in terms of frequency is the perfect fifth (ratio 3:2). Perfect intervals such as these are consonant and hence typically pleasing to the human ear. The history of tuning dates back to the time of the ancient Greeks and the four main types used at different times in musical history are known as Pythagorean tuning, just intonation, meantone temperament and equal temperament. There are alternatives described as "well" temperament which composers such as JS Bach experimented with in the early 18th century (his set of keyboard pieces called the "Well-tempered Clavier" being a famous example). Just intonation is the use of pure harmonics only. Pythagorean tuning is a form of this (quintic just intonation) that was widely used in medieval times and is based on a "cycle" of fifths. Let us first assume that there are 12 separate notes ("semitones") in an octave, as we see on a piano. Counting the keys of a piano, one may expect at first sight that a sequence of 12 fifths (each 7 semitones) is equivalent to 7 octaves (each 12 semitones). Although a fifth is actually a 7-note interval, the word "fifth" is based on the traditional major/minor scale in which it represents the fifth note of the scale, for the same reason the word octave relates to the 8th note.

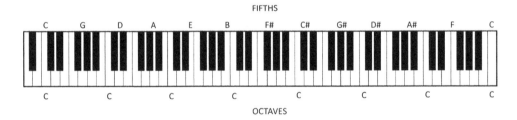

Starting at the lowest "C" we end up at the highest C in both cases – the same number of note intervals, $12 * 7 = 7 * 12 = 84$. This is the case on a piano, but only because it has been tuned by equal temperament for each key, in which the perfect fifths are slightly compromised. The reason for this is that in terms of pure harmonics (just intonation) these intervals are actually slightly different: the natural interval of 7 octaves has a frequency difference of 2^7; the natural interval of 12 fifths has a frequency difference of $\left(\frac{3}{2}\right)^{12}$; and

$$2^7 \neq \left(\frac{3}{2}\right)^{12}$$

These intervals are similar but not quite equal: the difference between a frequency ratio of 128 and 129.75 is about a fifth of a semitone – this error of dissonance was called the "Pythagorean comma". Traditionally pianos are tuned to octaves and violins to fifths (the intervals between open strings) and so this can cause a bit of a problem. Not only that, but the Pythagorean/just method means that throughout the scales we have certain intervals that agree (many fourths, tones and minor sevenths naturally fall into place) but others that do not, and the number of accurate notes depends on the key. This was acceptable in medieval times when harmonics were considered of prime importance and certain notes were simply not used. A particularly dissonant chord, the diminished 5th/augmented 4th ("tritone") was called the Devil's Interval and was generally forbidden, although by the Baroque period it became more commonly used. As an intermediate measure, meantime temperament was devised to have a preference for thirds and other intervals which became more popular – the fifths were slightly narrowed in their interval to achieve this – however this did not solve the problem of incommensurability of fifths and octaves and the resulting Pythagorean comma. The answer to the restrictions of just intonation was to develop equal temperament, which sets all octaves to exact (or as close as is physically manageable) ratios of 2:1, and then divide each octave into 12 equal semitones.

$$semitone\ (equal\ tempered) = \frac{f^{n+1}}{f^n} = \sqrt[12]{2} \approx 1.059$$

Note that this is a multiplier: the frequency of each successive note is a factor of 1.059 times the previous note, so the frequency difference increases with rising pitch – the intervals only appear equivalent on a logarithmic scale. This resolves many problems of dissonance by using a good approximation. Using this method of tuning, the error in important intervals such as tones, fourths, fifths and minor sevenths are reduced to a small fraction of a semitone (barely discernible to the human ear). Other intervals have larger errors (the largest being the tritone) but these are still small, and the important thing is that they are consistent throughout the whole range: every interval of the same span on a piano tuned with equal temperament, regardless of pitch, has an identical error (or none if it is an octave). For better agreement between musicians, the note we call an A (the A above middle C, often annotated A_4) is set by standard in concert pitch at 440 Hz. As a consequence of equal temperament, the pitch (frequency) of any note can be calculated by

$$P_n = P_a\left(\sqrt[12]{2}\right)^{n-a}$$

where P_n is the desired pitch, P_a is the reference pitch (usually 440 Hz) and n and a are integers referring to the semitone (key) number. Scientific pitch is a slightly different standard (not typically used in music) for investigation of acoustics, in which middle C is set to 256 Hz (which means that the lower octave C notes have integral numbers all the way down to 1 Hz, bottom C on the piano being 32 Hz. In concert pitch, middle C is 261.62 Hz. Although this definition of concert pitch eventually became an international standard in 1955, it is not adhered to universally and many orchestras around the world still prefer to set their pitch differently (typically slightly higher) – hence a Beethoven symphony played on one record compared with another may sound off-key.

Equal temperament was first defined in Western music in intervals of the twelfth root of 2 by the Flemish physicist/mathematician Simon Stevin in 1605, based on earlier work by Vincenzo Galilei (Galileo's father), who was a lutenist. It was later discovered that a Chinese prince, Zhu Zaiyu, had worked the same problem out a few years earlier in the 1580s and to a higher accuracy. Zhu Zaiyu used a system of 36 carefully carved bamboo pipes and a 12-string tuning instrument, and wrote about his discovery in much detail.

PB

Equation 291: Amdahl's Law

Amdahl's law estimates the maximum speed-up that can be achieved in computing, by parallelising a program to execute. Usually, not all parts of a program can be parallelised, with a certain number of instructions that can be executed by one processor only (serial).

$$S = \frac{N}{B \cdot N + (1 - B)}$$

Where S = Speed up, N = number of processors; B = fraction of the algorithm which is strictly serial (non-parallelised).

$$S(N) = \frac{T(1)}{T(N)}$$

Where $T(1)$ is the time spent to calculate the serial algorithm, and $T(N)$ the time spent to calculate the algorithm with a fraction B computed in parallel by N processors.

$$T(N) = T(1) \cdot [B + \frac{1}{N}(1 - B)]$$

The law was proposed at the Spring Joint Computer Conference in 1967 by the American computer architect, Gene Amdahl. It is based on the assumption that the size of the problem that needs to be computed remains the same when parallelised. When the parallelisation leads to larger problems or datasets, a better estimate of the parallelisation performance is given by Gustafson's law.

The law estimates the maximum expected improvement for an overall system when only one part of algorithm can be parallelised. The speed up of a program using multiple processors is limited by the time needed for the sequential fraction of the program (the fraction that is not parallelised).

For example, if a program can be solved in 20 hours using a single processor, and only 95% of this program can be parallelised (19 hours), then the minimum execution time cannot be less than 1 hour, which is the time needed to execute the portion of the program that cannot be parallelised. The maximum speed up is limited to at most 20 times, regardless the number of processors devoted to the parallelisation.

Amdahl's law estimates the speedup that can be achieved when we improve a portion P of the program computation, increasing the speed by S (equation 2). For example, if 30% of a program can be parallelised in such a way that its calculation will be twice as fast as that with a single processor, Amdahl's law will calculate the overall speed up as:

$1/[(1 - 0.3) + 0.3/2] = 1.176$ times the speed with one single processor.

In another example, let's consider a sequential program can be split in four parts, let's say $P1, P2, P3, P4$, contributing to the total time of the calculation for 11%, 18%, 23%, 48%, respectively. If $P1$ cannot be sped up (that is to say, not parallelised), whereas $P2, P3$ and $P4$ are sped up by 5 times, 20 times and 1.6 times, respectively, the total speedup calculated with Amdahl's law is:

$1/[P1/S1 + P2/S2 + P3/S3 + P4/S4] = 2.186$ (that means, only twice the previous time of calculation).

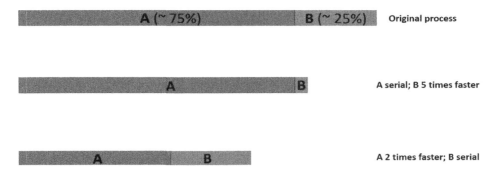

Figure 1. Graphic illustration of the effect of B parameter in Amdahl's Law.

The moderate effect of the parallelisation illustrated in this example, is because the major part of the program ($P4$, with 48%) is sped up by a factor of 1.6 only.

This example clearly shows that it is always more convenient to improve the speed of calculation a major part of a program, even if by a small amount, than speed up a minor part by many times. Figure 1 illustrates that a process consisting of two parts A and B, where is A is much bigger than B, a parallelisation of A which halves the time of calculation, is more convenient than parallelising B in such a way that its calculation is five times faster.

EFFICIENCY AND SPEED UP CURVE
The speed up S is defined as the ratio between the time spent in the execution of a program with one processor core only, $T(1)$, and the time spent in the execution of the parallelised program, $T(N)$, using n processors. The efficiency is then calculated as S/N, and it is a very important factor to consider for a company, due to the high cost of multiprocessor super computers.

A speedup curve is the graph speed up (S) *vs* number of processors (N), as illustrated in Figure 2. A curve with an angle of 45 degrees corresponds to the best speed we could hope for, that is $S = N$ (for 10 processors obtaining a speedup of 10 times). For a given fraction B, Amdahl's law yields a speedup curve that is logarithmic and remains below the line $S = N$. The graph shows that is the algorithm parallelisation and not the number of processor that determines the most effective speed up. As the curve begins to flatten out, the speed up is reduced despite a continuous increase in the number of processors. This trend corresponds to a reduction in efficiency.

PARALLELISATION IN COMPUTING AND ITS APPLICATIONS
Parallelisation is a technique that allows to divide an algorithm in a number of parts, each of which can be executed by different processors in parallel, at the same time. By doing this, each processor can execute a part of the algorithm without waiting for the result of the execution of another part. This technique, which can be achieved by using a supercomputer, which consists of a very large number of processors, allows computer calculations to be performed at high speed.

Parallel computing is vital in modelling systems that are characterised by a huge number of data, parameters, components, such as climate models, or multielectronic systems.

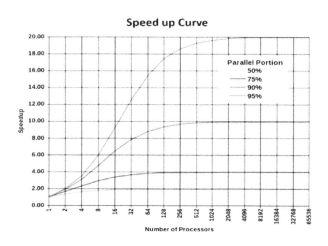

Figure 2. A typical speed up curve for different parallel portions (B) of a program (Source: Wikipedia).

It is also strategic in molecular dynamics of large systems, with a huge number of molecules and for long time and short time step. These simulations rely on solving equations for a huge number of variables (thousands or more) in iterative way, for a high number of steps (hundreds of thousands). Only through a correct parallelisation of the algorithm and effective super computers is possible to make these calculations feasible in terms of time.

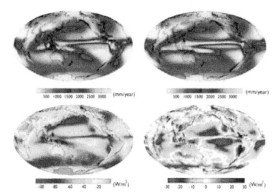

Figure 3. Climate models and data of rainfall (top left and right) and low-level clouds (bottom left and right). (Source: University of Washington).
AT

Equation 292: Kohlrausch Series of Mass Comparison

In equation 170 (equally arm balance) it was shown how a mass piece can be compared against a mass standard. In order to derive a set of mass pieces that can realize any mass, further comparisons are needed. A standard method to derive smaller masses from a reference mass, e.g. 1 kg, is called the Kohlrausch series and consists of comparing a set of masses of 500 g, 200 g (2 pieces, denoted A and B), and 100 g (2 pieces, detoted A and B) against a 1 kg reference and against each other. The measurement sequence can be such that also the un-equal arm is compensated. Imagine a balance with a left-side pan and a right-side pan. A reading y_i is the mass in the right pan minus the mass in the left pan. The mass pieces are weighed in the following sequence:

Weighing No	Left pan	Right pan	Nominal Mass in pan	Result
1	500+200+200+100(A)	1 kg	1 kg	y_1
2	1 kg	500+200+200+100(B)	1 kg	y_2
3	500	200+200+100(A)	500 g	y_3
4	200+200+100(B)	500	500 g	y_4
5	200(A)+100(A)	200(B)+100(B)	300 g	y_5
6	200(B)+100(B)	200(A)+100(A)	300 g	y_6
7	200(A)+100(B)	200(B)+100(A)	300 g	y_7
8	200(B)+100(A)	200(A)+100(B)	300 g	y_8
9	200(A)	100+100	200 g	y_9
10	100+100	200(A)	200 g	y_{10}
11	200(B)	100+100	200 g	y_{11}
12	100+100	200(B)	200 g	y_{12}

Using the least-squares method (see BE 152), it can be derived that the masses are given by:

$$M(500) = (y1 - y2 + y3 - y4 + 2mN))/4$$
$$M(200A) = (y1 - y2 - y3 + y4 + y5 - y6 + y7 - y8 + y9 - y10 + 2mN)/10$$
$$M(200B) = (y1 - y2 - y3 + y4 - y5 + y6 - y7 + y8 + y11 + y12 + 2mN)/10$$
$$M(100A) = (y1 - y3 + y5 - y6 - y7 + y8 - y9 + y10 - y11 + y12 + mN)/10$$

$$M(100B) = (y1 + y4 - y5 + y6 + y7 - y8 - y9 + y10 - y11 + y12 + mN)/10$$

Where $M(500..etc)$ is the nominal 500 g mass and mN is the reference mass of nominally 1 kg. With this set of masses, any nominal mass between 100 g and 1 kg can be realized in steps of 100 g. For lower masses, the 100 g mass piece can be used as reference, and can subsequently be used to calibrate a set of 10 g, 20 g en 50 g pieces, etc. In this way, reference standards can be derived in 1,2,5,10 sequence and masses in any amount of decimals can be realized. In this way, world-wide all masses are derived from the standard kilogram. The name 'Kohlrausch series' is derived from the textbook 'Practical Physics', published in 1870 by the German physicist Friedrich Wilhelm Georg Kohlrausch (1840 -1910) in which it was published. At that time it was probably already in practical use for some time.

HH

Equation 293: Peukert's Law

The time it takes for a battery to be discharged can be calculated using Peukert's Law, which relates the rated capacity and rated discharge time (for a rated discharge current) to the actual discharge time for a given discharge current. Often, a battery will indicate its capacity C in Ampere-hours and a rated discharge time H in hours. The rated discharge time is quoted for a discharge current given by the ratio of the capacity and the discharge time:

$$I_H = \frac{C}{H}$$

However, it is possible that the battery will be discharged at a current different than the one for which the discharge time is quoted. In that case, the discharge time t for an actual discharge current I can be determined using the following equation.

$$t = H\left(\frac{C}{IH}\right)^k$$

Where k is known as the Peukert constant and varies on the type of battery. This constant is sometimes provided with the battery. The equation indicates that a battery's capacity actually decreases when the discharge rate increases.

The equation was formulated by German scientist and hefeweizen brewer Wilhelm Peukert in 1897 [1]. Recently, it was found that Peukert's law does not predict actual discharge time of a battery unless the discharge current and operating temperature are kept constant [2].

Refrences:
[1] Peukert W (1897) Über die Abhängigkeit der Kapazität von der Entladestromstärke bei Bleiaakumulatoren, Elektrotechnische Zeitschrift 20

[2] D Doerffel, SA Sharkh (2006) A critical review of using the Peukert equation for determining the remaining capacity of lead-acid and lithium-ion batteries, Journal of Power Sources 155, pp 395-400

MF

Equation 294: The Binomial Series

Suppose that a, b and m are some real numbers. Then, the Binomial Series is the identity,

$$(a + b)^m = \sum_{n=0}^{\infty} \binom{m}{n} a^{m-n} b^n$$

In the above, the notation $\binom{m}{n}$ denote the binomial coefficients defined as follows:

$$\binom{m}{n} = \frac{m(m - 1)(m - 2) \ldots (m - n + 1)}{n!}$$

Note that the numerator is the product of a total of n terms. When $n = 0$, we assume that the binomial coefficient is equal to one. The binomial series converges when $\frac{|b|}{|a|} < 1$. The special case when m is a positive integer is known as the binomial expansion. It terminates at $n = m$ and thus, is given by

$$(a + b)^m = a^m + m a^{m-1} b + \frac{m(m - 1)}{2} a^{m-2} b^2 + \cdots + b^m$$

Note that there are a total of $m + 1$ terms on the right-hand side. The binomial expansion with $m = 2$ was known to the Greeks in the 4th century BC. The Hindu mathematicians in the 6th century appear to know the binomial expansion for $m = 3$. References to the binomial expansion for higher values of m appear in the texts from the 11th century Arabian mathematicians and the 13th century Chinese mathematicians. Isaac Newton in 1665 appears to have been the first one to generalize the series to include any real number m including negative numbers.

For $m = 2, 3,$ and 4, the binomial expansion of $(a + b)^m$ is given in the table below:

m	Binomial Expansion
2	$a^2 + 2ab + b^2$
3	$a^3 + 3a^2 b + 3ab^2 + b^3$
4	$a^4 + 4a^3 b + 6a^2 b^2 + 4ab^3 + b^4$

Some beautiful results follow from the general binomial series listed above. Suppose that $|x| < 1$. Then,

$$\frac{1}{1 + x} = (1 + x)^{-1} = 1 - x + x^2 - x^3 + \cdots = \sum_{n=0}^{\infty} (-1)^n x^n$$

These expansions have several important and practical applications in physics, chemistry, and mechanics. For example, in obtaining the solution to the stress waves propagating in a slender bar, Laplace Transforms are often used. The wave equation governs the stress waves propagating in the bar. Once the Laplace Transform of the wave equation is taken and the boundary conditions are applied, the solution in the transformed variable s contains terms such as $F(s)(1 \pm \exp(-2s\tau))$ where τ is a characteristic time for the bar and $F(s)$ is the Laplace transform of some function $f(t)$. With the help of the above expansions, these can be expressed as a series:

$$\frac{F(s)}{1 \pm \exp(-2s\tau)} = F(s)(1 \mp \exp(-2s\tau) + \exp(-4s\tau) \mp \cdots)$$

Using the second shifting theorem of Laplace Transforms, one can readily invert the right-hand side when $f(t)$ is known. The resulting solution will consist of an infinite series of terms that represent waves traversing the length of the bar in either direction and reflecting off of the boundaries at the left and right.

Another application of the binomial expansion is in the calculation of the electrostatic potential due to dipoles, quadrupoles etc. For example, in the case of dipoles, the binomial expansion of the potential leads to the conclusion that at distances r much bigger than the separation distance d of the charges, the dipole potential is inversely proportional to the square of r even though the potential due to individual charges is inversely proportional only to the first power of r.

HC

Equation 295: The Speed of Sound in Seawater

The speed of sound in seawater and most other liquids is given by an extremely simple equation:

$$C_{liquid} = \sqrt{\frac{k}{\rho}}$$

where k is the bulk modulus, which is the resistance of the liquid to a (uniform) compressional force, and ρ is the density.

In underwater acoustics, a knowledge of the speed of sound is of great importance, needed for applications including sonar, temperature measurement, signalling, hydrophone design, and mapping. Unfortunately, however the density of seawater varies enormously, being dependent on pressure (and hence depth and gravitational force), on temperature and on salinity.

Because of its great importance, equations to predict the speed of sound in a particular location (including depth) have long been sought, but a major (and perhaps surprising) stumbling block has been the practical impossibility of calculating the speed of sound accurately, even just as a function of pressure, even for pure water. Though early equations were regarded with scepticism, they could only be properly critiqued in 1999, when the first accurate measurements of the speed of sound in water (fresh and saline) over a sufficiently wide pressure range (up to 60 MPa) were made.

At the beginning of the 21st century, there was a bewildering variety of equations in use, some of which were highly accurate over narrow ranges of conditions and others which gave roughly correct values over wider ranges. Though the various formulae and data values could be gathered together, the resulting combined equations were of diabolical complexity, variable accuracy, required considerable expertise to use and plagued by transcription errors. And there were always some seas and oceans to which they did not apply.

The ideal was a single equation which provided accurate answers for locations in all the seas and oceans of the world, required the smallest number of input parameters and was as simple as possible. This was ultimately derived by Claude C. LeRoy, who pioneered a new generation of underwater measurement systems in the 1970s, and Stephen Robinson, Head of Underwater Acoustics at the UK's National Physical Laboratory:

$$C = 1402.5 + 5T - 5.44 * 10^{-2}T^2 + 2.1 * 10^{-4}T^3 + 1.33S - 1.23 * 10^{-2}ST + 8.7 * 10^{-5}ST^2 + 1.56 * 10^{-2}Z + 2.55 * 10^{-7}Z^2 - 7.3 * 10^{-12}Z^3 + 1.2 * 10^{-6}Z(\emptyset - 45) - 9.5 * 10^{-13}TZ^3 + 3 * 10^{-7}T^2Z + 1.43 * 10^{-5}SZ$$

where T is temperature in °C, S is salinity, Z is depth in metres, and $\emptyset$ is degrees latitude.

One of the great advantages of the equation is that, despite being long, it has a simple structure and so can easily be entered into spreadsheets, programmable calculators and mobile phones. It is accurate to +/- 0.2

metres per second and fails only in the vicinity of a small number of volcanically heated brine pools at the bottoms of some seas.

MJG

Equation 296: The Involute Equation of a Circle

In mathematics, an involute is a kind of curve that is associated with other types of curves such as the circle and ellipse. In 1673, Dutch mathematician Christiaan Huygens, who is known, among many other achievements, as the inventor of the pendulum clock, defined the involute in his book on the mathematical analysis of the motion of the pendulum [1].

The involute is always perpendicular to the tangent line to the given curve. For example, the involute of a circle is described by the following equation in a Cartesian coordinate system:

$$\begin{cases} x = r(\cos\theta + \theta\sin\theta) \\ y = r(\sin\theta - \theta\cos\theta) \end{cases}$$

Where, r is the radius of a given circle, θ is the angular parameter of generating the involute. Involutes of a curve can be constructed by attaching an imaginary taut string to a circle and tracing its free end as it is wound onto it as shown in the figure on the left side.

Perhaps, the most prevalent application of involute is in the design of gear profiles, which are almost invariably based on the involute of a circle. In 1760s, Leonhard Euler (1707-1783) first proposed the use of involute curve as the shape of gear teeth flank so the resistance and noise during the meshing of two gears can be minimized [2]. The figure the right side illustrates the meshing condition of a pair gears, where the contact between the gear teeth occurs at a single instantaneous point.

It is possible to use almost arbitrary shapes with cycloidal and epicycloid curves being an often discussed alternative (although almost never used except in superchargers in hot rods) gears. However, the involute has important properties such as a constant ratio of speeds between the two gears and a constant pressure angle at the contacting surfaces. It is these properties that enable multiple gear teeth to be in contact at any given time, thereby facilitating smooth, noise free, transfer of the loads as one gear leaves and new gears come into contact.

References:
[1] Christiaan Huygens, Horologium Oscillatorium sive de motu pendulorum ad horologia aptato demonstrations geometricae, Paris: F. Muguet, 1673.
[2] Leonard Euler, Novi Commentarii academiae scientiaeum Petropolitanae 5, 1760, pp 299-316
[3] Walter Gaustschi, Leonard Euler: His life, the man, and his works. 2008 SIAM Rev., 50(1), 3–33.

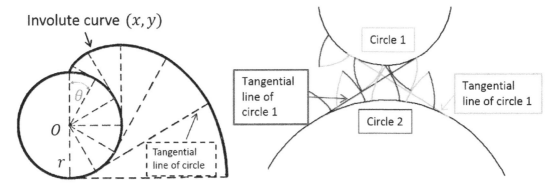

Involute curve (x, y)

Figure 7: Involute of a circle

Circle 1

Tangential line of circle 1

Tangential line of circle 1

Circle 2

Tangential line of circle

Figure 2: Euler's concept of involute gear [3]

KN

Equation 297: Signal to Noise Ratio of CCD Camera

In electronics, noise is a random fluctuation in an electrical signal. For example the intensity of light received on a camera pixel, observed under the same light condition, will not be a constant value but it oscillates around its mean value. The random oscillations constitutes the noise, which is an undesired disturbance of the useful signal. A measure of the quality of the signal it's the Signal to Noise ratio. In CCD (charged coupled device) cameras the observed signal is constituted by the number of generated electrons which is proportional to the number of photons incoming on the pixel (P). The parameter DQE (detector quantum efficiency) is the scale factor relating the number of photons to the numberof generated electrons.

The incoming photons have a Poisson statistical distribution (see equation 222) with an associated standard deviation, this variation in the number of incoming photons is called the shot noise.

$$\frac{S}{N} = \frac{D_{QE}\,P}{\sqrt{D_{QE}\,P + N_{dark} + \delta^2_{readout}}}$$

DQE: Detector quantum efficiency

P : average number of incoming photons

N_{dark} : thermally generated electrons

$\delta_{readout}$: read out noise

Another source of noise are the dark noise (Ndark) and the readout noise (δreadout). The dark noise is related to thermally generated electrons which do not correspond to an incoming photons. It is called dark noise because can be observed as intensity fluctuation of the pixels' intensity when the camera is not receiving any light. The readout noise is the noise introduced by the camera circuit when reading the number of electrons and converting this into a voltage to be transmitted. The average number of generated electrons divided by the quadrature sum of the electrons variance due to all the sources of noise is the SNR.

References:
1. http://www.andor.com/learning-academy/ccd-signal-to-noise-ratio-calculating-the-snr-of-a-ccd

GM

Equation 298: Euler's Numerical Integration

Students of pure mathematics will be familiar with the concepts of integration and differentiation. These opposing processes are used in the field of calculus to identify the gradient of a curve at a point or the area under a curve. For this article we'll focus on integration - calculating the area under a curve.

Here's a toy example:

A rocket's velocity is set to obey the following equation for the first 10 seconds of takeoff.

$$V = t^2$$

To work out how far the rocket has travelled we can calculate the area under the curve that describes the velocity. The integral of the velocity is called the displacement and can be calculated analytically to be:

$$d = \frac{t^3}{3}$$

So after 10 seconds, the rocket will be travelling at 100 meters per second and will have travelled 333.33 meters. This is illustrated in Figure 1.

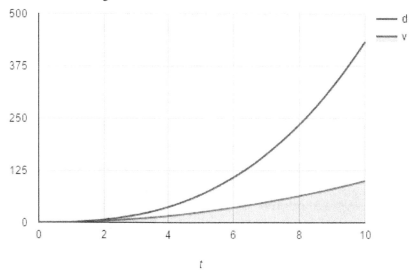

Figure 1: A rocket traveling with velocity profile V has moved d meters. At each point in time, t, the value of d is equal to the area of the blue shaded region from 0 to t.

In fact, integration is an incredibly powerful tool for engineering and science. Sadly it is not always easy to calculate the integral of a function. For complex functions, working out what the integral is analytically can be extremely challenging. For situations where the function is being measured dynamically (consider the velocity only being checked every so often by glancing at a speedometer) then how do you calculate the integral?

The answer comes from Euler's method of integration. In this method we can estimate the area under a curve by imagining the curve being broken up into a series of strips. In an extreme example, we'll assess the speed of our rocket every two seconds and build strips that are two seconds wide. The width of the strips is commonly referred to as Δt.

$$\widehat{f_n} = \widehat{f_{n-1}} + (x_n)\Delta t$$

In Figure 2 you can see how that estimate looks compared to the original area under the velocity graph. Table 1 compares the actual velocity with the estimated velocity.

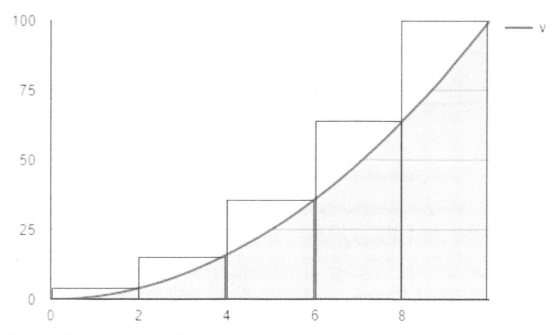

Figure 2: The area under the velocity graph (shaded) compared with the rectangles used to estimate it with Δt = 2

t in s	Real displacement in m	Estimated displacement in m	Error in m
0	0	0	0
2	2.67	8	5.33
4	21.33	40	18.67
6	72	112	40
8	170.67	240	69.33
10	333.33	440	106.67

Table 1. Estimates and errors for Δt = 2

There are two things to notice about this estimate, it isn't very good, and it gets worse as time goes on! However, Euler's Method becomes more accurate for smaller values of Δt. We can see this visually in Figure 3.

371

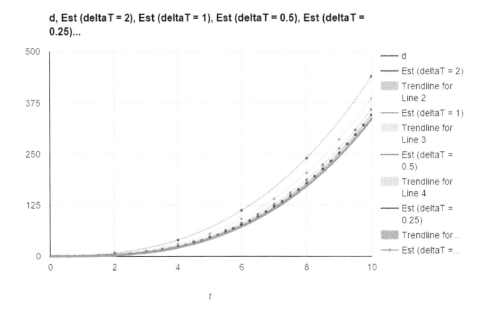

Figure 3: Different estimates of displacements using different values of Δt

As the value of Δt gets smaller the small pieces of rectangle that go over the curve become smaller too, making the estimate more accurate. The problem is that because the current estimate relies on the value of the last estimate, changing the value of Δt increases the amount of computational power required to calculate the estimate at each point. In our 10 second example, a window size of 2 seconds may be 107m off the mark, but we only had to make 5 calculations. To only have an error of 2.5m after 10 seconds, we need to estimate the displacement every 0.05s meaning a total of 200 calculations over 10 seconds.

Despite these problems, many simulations used for computer games and engineering applications use this method to calculate displacements and a variety of other properties. However, in areas where accuracy is vital, a set of equations, brought together using an algorithm called Runge-Kutta, are used - and will be discussed in Beautiful Equation 317.

RFO

Equation 299: The Fast Fourier Transform FFT

$$X_k = \frac{1}{2}\{Y_k + W^k Z_k\}$$

$$X_{k+N/2} = \frac{1}{2}\{Y_k - W^k Z_k\}$$

$$W^K = e^{-i\frac{2\pi k}{N}}$$

$$X_k = \frac{1}{N}\sum_{s=0}^{N-1} e^{-i\frac{2\pi ks}{N}} = A_k + iB_k \qquad\qquad k = 0,1,2\ldots(N-1)$$

$$\omega_k = \frac{2\pi k}{N}$$

Where X_k is the complex frequency component ω_k of the data array (i.e. the signal), N is the number of data points recorded and the values of the data points for each frequency, X_k is given by

$$x_k(s) = A_k \cos\left(\frac{2\pi ks}{N}\right) + B_k \sin\left(\frac{2\pi ks}{N}\right)$$

s is the location of the data point in the dataset and is an integer ranging from 0 to $N-1$. Matrices Y and Z will be explained in the text.

It is hard to find a definitive time at which the age of digital signals started. In terms of digital electronic transmission, this could be considered to have began with the telegraph that originated in the early 1800's. However, it was not until after the turn of the 1900's that digital signals started to attract research as a problem in dynamic signal processing associated with the limitations of bandwidth. One prominent researcher who contributed substantially to early theoretical development was the Swedish mathematician Harry Nyquist (1889 – 1976) working at Bell Laboratories in the US, see beautiful equation 44. Over the first half of the 20th Century a number of technologies were developed to convert a voltage into a numerical representation (i.e. an integer proportional to the real-world voltage). However, it was the arrival of the transistor and invention of the integrated circuit that eliminated other approaches and started the digital signal revolution. By the swinging 60's integrated circuits were capable of converting voltages to numbers with relatively high precision and conversion times fast enough to record signals having frequencies far beyond the audible range of humans, see Gordon and Colton, 1961.

After the arrival of the technology to convert signals into arrays of numbers, it was a natural progression to want to determine the frequency content contained therein. For real-world, continuous signals the mathematical procedures for doing this had already been established through the development of Fourier analysis resulting in equations to determine Fourier coefficients for periodic signals and spectral information in the form of the Fourier transform for many other signal types, see BE's 12, 88 and 150. When considering discrete numbers, the integrals of calculus must be replaced with suitably scaled sums of numbers. Hence equation (2) above represents a method for obtaining a numerical value (actually a complex number) that provides estimates of the amplitude of frequencies in a signal with frequency resolution/discrimination related to the number of data points recorded. The upper frequency that can be detected is typically limited by the need to record at least two values per period (the Nyquist frequency).

To understand the Fast Fourier Transform (FFT) it is informative to look at how equation (2) is computed. To do this, a very short set of only 8 data points will be chosen (for example, a high-resolution, 4 min recording of music could contain upward of 45 million data points). While not very practical, it will illustrate the mathematical procedures involved with these calculations. Table 1 shows the numerical values associated with the calculation of the first two frequency amplitudes out of a total of eight. The top two rows show the sample number s and the actual measured voltages x_s of each sample. Below this two rows illustrate the values of the first two cosine functions followed by the products of these and the first two sine functions with the measured voltages. Finally, to obtain the Fourier coefficients, A_k and B_k, these products are averaged resulting in the values on in the right hand column of this table. To determine all of the 8 pairs of coefficients, this table would have sixteen rows of calculations involving N^2 multiplications.

Table 2 lists all of the sine and cosine values necessary for these calculations. To help visualize the numbers in this table each different value is given a different color with equal values of opposite sign being given darker shades of the same color. A few things are immediately apparent. Firstly, ignoring signs, there are only three numbers in the entire table. Secondly, this table is symmetric about the $k = 4$ column for the cosine terms and symmetric with a multiplication of -1 (ultimately requiring addition of a -1 factor in the FFT algorithm that is often referred to as the 'fiddle' factor) for the sine terms. There is a corresponding symmetry about the $s = 4$ row. Finally, for this relatively small data set there are a lot of 1's and zeros that would not require multiplication. The symmetry of this matrix results in frequency amplitudes being mirrored about the central value (at the 'Nyquist' frequency) and this upper half can often be 'thrown away'.

'Bletchley Park' British mathematician Irving John Good (1916 – 2009) studied these types of calculation for which the square matrix (the array of 8 x 8 numbers in Table 2) has this dual symmetry and has a number of rows (and columns) that are an integer power of two (for example 2^3 =8 in our example) and presented a solution algorithm in a paper published in 1958. He also recognized that this would apply to Fourier coefficient calculations and that this would reduce the number of calculations from N^2 to less than a factor of $N \log(N)$. It was left to a pair of American mathematicians James William Cooley (1926) and John Wilder Tukey (1915 – 2000) to work out the details of an algorithm and determine that the number of required calculations would be less than $2N \log_2(N)$ all of which appeared in their seminal paper published in 1965. Table 2 uses these formulas to demonstrate the astonishing saving in the number of necessary calculations. While this is not great for small data sets, with around 16000 data points, this has reduced from 200 million calculations down to ½ million (1/400th). As a reference point to understand the size of modern digital signal data sets, modern digital sound recording will sample the pressure variations in air at between 44,000 and 192,000 times per second.

There is not enough space for a full explanation of this algorithm but it is worth a couple of sentences describing the interesting mechanics of how this works. Typically, this starts by splitting the original data into two new array one comprising the even number samples (i.e.$\{y\} = \{x_0, x_2, x_4, x_6\}$) and the odd samples ($\{z\} = \{x_1, x_3, x_5, x_7\}$). It is then possible to compute the discrete Fourier transform of these two reduced data sets Y_k and Z_k providing frequency data arrays of half the length of the original number of points. The key attribute of this algorithm is that it was recognized that these reduced datasets are related to the original frequencies by the 'Butterfly' relations given by equation (1). This procedure can be continued with by splitting this reduced data in half ultimately arriving at two data points after only n stages of splitting, where n is the integer power of 2 of the number of rows in the original matrix. The mechanics of this algorithm are illustrated for 8 data points in Figure 1 below.

The algorithm can, of course, be reversed to recover the original signal. Sometimes it might be desired to take a signal such as a recording of somebody singing out of key, convert it into its component frequencies and shift the slightly prior to reconstructing a modified signal (such as the singer now having perfect pitch). Applications of these Fourier techniques abound and are ubiquitous in almost all signal processing thereby influencing the sounds of music, our voices over phone transmissions, and all forms of communications including the internet.

References:
Cooley J.W., and Tukey J.W., 1965, An algorithm for the machine calculation of complex Fourier series, *Mathematics of Computation*, 19, 297 – 301.
Good I.R., 1958, The interaction algorithm and practical Fourier analysis, *J. Royal Statistical Society, Ser. B*, 20, 372 – 375.
Gordon B.M., and Evan T. Colton E.T., 1961, Signal Conversion Apparatus, U.S. Patent 2,997,704, filed February 24, 1958, issued August 22, 1961. (This disclosed the invention of a successive approximation method for converting voltages to number that would dominate design for the next 50 years and is commonplace today)
Newland D.E., 1993, Random vibrations, spectral and wavelet analysis, 3rd ed., John Wiley and sons (now available as a Dover publication), ISBN 0-470-22153-4, chapter 12: The fast Fourier transform. (provides a detailed discussion of this algorithm).

Table 2. The various calculations required to determine the Fourier transform coefficients A_k and B_k for the first two frequencies (i.e. $k = 1$ and $k = 2$) of this short signal.

	s	0	1	2	3	4	5	6	7	$\frac{1}{N}\sum$	
	x_s	4.000	10.000	-2.000	1.000	-7.000	-9.000	6.000	-1.000	0.250	A_0
$k=1$	$\cos\left(\dfrac{2\pi s}{N}\right)$	1.000	0.707	0.000	-0.707	-1.000	-0.707	0.000	0.707		
$k=2$	$\cos\left(\dfrac{4\pi s}{N}\right)$	1.000	0.000	-1.000	0.000	1.000	0.000	-1.000	0.000		
$k=1$	$x_s\cos\left(\dfrac{2\pi s}{N}\right)$	4.000	7.071	0.000	-0.707	7.000	6.364	0.000	-0.707	2.378	A_1
$k=2$	$x_s\cos\left(\dfrac{4\pi s}{N}\right)$	4.000	0.000	2.000	0.000	-7.000	0.000	-6.000	0.000	-1.375	A_2
$k=1$	$x_s\sin\left(\dfrac{2\pi s}{N}\right)$	0.000	7.071	-2.000	0.707	0.000	6.364	-6.000	0.707	-0.856	B_1
$k=2$	$x_s\sin\left(\dfrac{4\pi s}{N}\right)$	0.000	10.000	0.000	-1.000	0.000	-9.000	0.000	1.000	-1.000	B_2

Table 3. Values of the sine and cosine functions that will be multiplied with the measured signal x to determine the Fourier coefficients. The cosine and sine values are in the last 8 columns in this table (note that the first two, nonzero or 1, cosine values appear in rows 2 and 3 in Table 1).

s	Phase	cos0	cos1	cos2	cos3	cos4	cos5	cos6	cos7	cos8	
0	0	1	1	1	1	1	1	1	1	1	1
1	0.785398	1	0.7071	0	-0.7071	-1	-0.7071	0	0.7071	1	1
2	1.570796	1	0	-1	0	1	0	-1	0	1	1
3	2.356194	1	-0.7071	0	0.7071	-1	0.7071	0	-0.7071	1	1
4	3.141593	1	-1	1	-1	1	-1	1	-1	1	1
5	3.926991	1	-0.7071	0	0.7071	-1	0.7071	0	-0.7071	1	1
6	4.712389	1	0	-1	0	1	0	-1	0	1	1
7	5.497787	1	0.7071	0	-0.7071	-1	-0.7071	0	0.7071	1	1
	Phase	sin0	sin1	sin2	sin3	sin4	sin5	sin6	sin7	sin8	
	0	0	0	0	0	0	0	0	0	0	0
	0.785398	0	0.7071	1	0.7071	0	-0.7071	-1	-0.7071	0	0
	1.570796	0	1	0	-1	0	1	0	-1	0	0
	2.356194	0	0.7071	-1	0.7071	0	-0.7071	1	-0.7071	0	0
	3.141593	0	0	0	0	0	0	0	0	0	0
	3.926991	0	-0.7071	1	-0.7071	0	0.7071	-1	0.7071	0	0
	4.712389	0	-1	0	1	0	-1	0	1	0	0
	5.497787	0	-0.7071	-1	-0.7071	0	0.7071	1	0.7071	0	0

Table 4. Illustration of the reduction in the number of calculations using the Cooley Tukey (or FFT) algorithm rather than the discrete calculation referred to as a discrete Fourier transform or DFT.

n	$N = 2^n$	N^2	$2N\log_2(N)$
1	2	4	4
2	4	16	16
3	8	64	48
4	16	256	128
5	32	1024	320
6	64	4096	768
7	128	16384	1792
8	256	65536	4096
9	512	262144	9216
10	1024	1048576	20480
11	2048	4194304	45056
12	4096	16777216	98304
13	8192	67108864	212992
14	16384	2.68E+08	458752

An 8 Input Butterfly. Note, you double a 4 input butterfly, extend output lines, then connect the upper and lower butterflies together with diagonal lines.

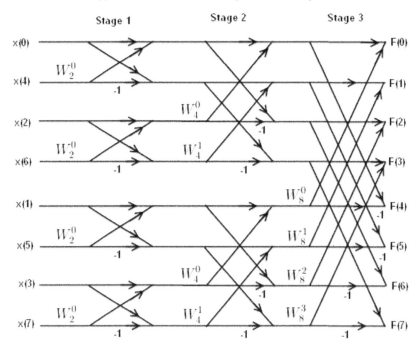

Figure 1. Butterfly diagram illustrating the 3 stages in computing the FFT of a signal comprising 8 data points (from http://alwayslearn.com/DFT%20and%20FFT%20Tutorial/DFTandFFT_BasicIdea.html). The symbols F on the right of this diagram represent the frequency components X_k in the above discussion.7

SS

376

Equation 300: Collatz Conjecture

The Collatz Conjecture is an interesting mathematical function that takes any natural number and if it is even half it, if it is odd then multiple it by 3 and add 1. The result of the function is then processed recursively by the same function until the solution of 1 is found. The Collatz Conjecture states that any natural number will resolve to 1in this manner.

This equation can be expressed as a function of n where n is the starting or iterated natural number.

$$f(n) = \begin{cases} n/2 & \text{if } n = 0 \pmod 2 \\ 3n + 1 & \text{if } n = 1 \pmod 2 \end{cases}$$

Take the result of $f(n)$ and iterate it i times.

$$a_i = \begin{cases} n & \text{for } i = 0 \\ f(a_i - 1) & \text{for } i > 0 \end{cases}$$

This equation can be reduced the following; assuming that the same function is applied and iterated each time until the result is 1.

$$a_i = f^i(n)$$

An example of this is; if $n = 6$, the sequence to one is $6, 3, 10, 5, 16, 8, 4, 2, 1$ (8 steps or iterations). There are many alternate names for this function or conjecture named after Lothar Collatz (1937) who first proposed this functional relationship. It has also been referred to as "oneness" and the hailstorm sequence.

The Collatz Conjecture when visualized graphically in many different forms follows natural and ordered patterns. Any problem or equation that can be visualized so readily qualifies as a beautiful equation.

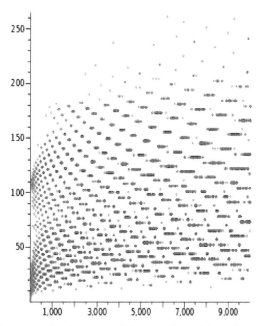

In this graph the integer n is shown in the X axis and the number of iterations termed "stopping distance" is shown in the Y axis.
TS

377

Equation 301: The Langmuir Isotherm

For a given equilibrium of adsorbed (A) – desorbed (AB) gas species on a solid surface (B) represented by the equation:

The Langmuir equation is:

$$A(g) + B(s) \rightleftharpoons AB$$

$$\theta = \frac{K.P}{1 + K.P} \qquad [1]$$

The equilibrium constant for the distribution of adsorbate between the solid surface and the gas phase is:

$$K = \frac{K_a}{K_b} = \frac{[AB]}{[A] \cdot [B]}$$

Where K_a and K_b are the adsorption and desorption constant rates, respectively.

θ is the fraction of adsorbed species on the surface and P is the pressure of the gas.

As the fraction of adsorbed species can be expressed as ratio of the volume of gas adsorbed (V) and the volume of gas adsorbed at high pressure conditions (V_{mono}, in the monolayer regime), the Langmuir equation can be also expressed as

$$\frac{P}{V} = \frac{P}{V_{mono}} + \frac{1}{K \cdot V_{mono}} \qquad [2]$$

The Langmuir equation, also known as Langmuir's isotherm, provides the number of adsorbed species on a heterogeneous surface as a function of the pressure of the gas.

In the case of low pressures, the $K \cdot P$ term is very small and the Langmuir equation approaches the linear regime:

$$\theta \sim K \cdot P$$

At high pressures, the $K \cdot P$ term is much larger than 1 and the Langmuir equation approaches a constant value:

$$\theta \sim 1$$

Likewise, the Clausius-Mossotti equation, also the Langmuir equation establishes a relationship between a microscopic quantity (the number of species adsorbed on a solid surface) and a macroscopic quantity (the pressure of the gas).

However, another form of the Langmuir equation (equation [2]) determines a relationship between two thermodynamic quantities, the volume and pressure of gas adsorbed on a surface.

HISTORY
Irving Langmuir, a New York-born metallurgical engineer and physical chemist, proposed this equation in 1916, while studying the properties of adsorbed films in vacuum. He was the first to observe very stable

mono-atomic films on tungsten and platinum and, after experiments with oil films on water, to formulate a general theory of adsorbed films. He also investigated the nature of electric discharges in high vacuum and in certain gases at low pressures. For his work on the surface chemistry, he won the Nobel Prize in Chemistry in 1932. He also studied the catalytic properties of adsorbed films. He discovered atomic hydrogen and used it to develop the atomic hydrogen welding process. Finally, he also contributed to the development of the Lewis theory of shared electrons in Chemistry.

CHEMISORPTION AND PHYSISORPTION

The adsorption of gases on a solid surface falls into two broad categories: physisorption and chemisorption. Physisorption is a process in which the adsorbate molecules loosely binds on the solid via van der Waals type interactions. This process allows for the formation of multi-layers of adsorbate species, which can be easily disrupted by increasing temperature. On the other hand, a chemisorption process takes place when the adsorbate species chemically bind to molecular or atomic species on the solid surface. In this case, only monolayer adsorption is possible. An interesting example of the occurrence of these two different adsorption regimes is given by the adsorption of nitrogen on iron. At - 190°C, nitrogen is physisorbed on iron as nitrogen molecules $N2$. At room temperature, iron does not adsorb nitrogen at all. At 500 °C, nitrogen is chemisorbed on the iron surface as nitrogen atoms (Figure 1). The Langmuir equation describes chemisorption processes, at the monolayer regime.

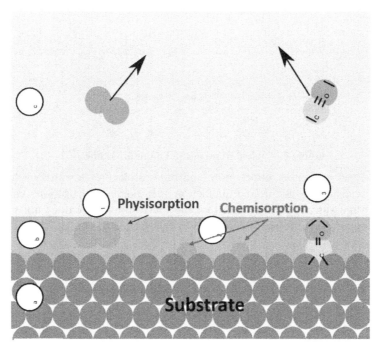

Figure 1. Schematic representation of a physisorption and a chemisorption process (Source: Wikipedia)

LANGMUIR ISOTHERMS ASSUMPTIONS AND LIMITATIONS

There are several assumptions within the Langmuir isotherm, which are summed up below:

- A fixed number of vacant sites are available on the solid surface
- All the vacant sites are equal in size and shape
- Each site can only hold one molecules/atom of adsorbate
- A constant amount of heat energy is released upon adsorption
- A dynamic equilibrium occurs between adsorbed gaseous molecules and free gaseous molecules, with a continuous exchange of adsorbed and desorbed species.

As it is shown in the graph of a Langmuir isotherm, as number of species vs pressure (Figure 2), there are two trends. A linear trend is present at very low pressures, that is to say, very low concentrations of adsorbed gases: the number of adsorbed species is directly proportional to the pressure of the adsorbed gas on the surface. On the other hand, at high pressures, the number of adsorbed species becomes constant as it reached the monolayer regime and all vacant sites have been occupied. No further species can be chemisorbed at the surface.

The Langmuir equation works as a model of gas adsorption at very low pressures, which is the condition under which monolayer regime exists. Langmuir's theory also assumes the sites on the surface to be equal, and that gas molecules do not interact with each other. However, solid surfaces are heterogeneous and weak forces of attraction exist among molecules of the same type.

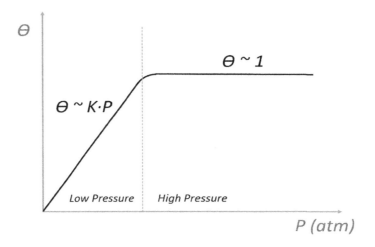

Figure 2. Typical trend of the Langmuir Isotherm.

At high pressures, gas molecules attract more and more molecules towards each other and a multilayer process is more realistic. In addition, liquefaction (capillary condensation) can take place upon gas adsorption at high pressure, a phenomenon that the Langmuir theory does not take into account. High-pressure adsorption processes are better described by the Brunauer, Emmett and Teller (BET) equation.

AT

Equation 302: Gurney Equations

In explosives engineering, the Gurney equations are a set of equations that describes how fast an explosive will accelerate the surrounding material when the explosive detonates. The equations were developed by Ronald Gurney in the 1940s. Ronald Gurney (1898 – 1953) was a British theoretical physicist and was at one point a research student of William Bragg.

When an explosive material, surrounded by a solid material detonates, the outer shell is accelerated by the detonation shockwave and by the expansion caused by the detonation gases contained within the shell. Gurney was the first to model the distribution of energy between the outer shell or sheet and the detonation gases produced from the explosion. From these models, he developed formulas to accurately describe the acceleration of outer material. In simple terms, he showed that the velocity gradient in the detonation gases was linear.

The table gives the Gurney equations for some simple explosive types and the following quantities are used in each equation, where C is the mass of the explosive charge, M is the mass of the accelerated shell or sheet

380

of material, V is the velocity of accelerated material after explosive detonation and $\sqrt{2E}$ is the Gurney Constant for a given explosive.

The Gurney constant is expressed in units of velocity and compares the relative accelerated material velocity produced by different explosives materials.

CM

Explosive Type	Equation	Applications	Notes
Cylindrical charge	$$\frac{V}{\sqrt{2E}} = \left(\frac{M}{C} + \frac{1}{2}\right)^{-\frac{1}{2}}$$	Military explosives (artillery shells, bombs & missile warheads)	Cylindrical charge of mass C and flyer shell of mass M
Spherical charge	$$\frac{V}{\sqrt{2E}} = \left(\frac{M}{C} + \frac{3}{5}\right)^{-\frac{1}{2}}$$	Military grenades and cluster bombs	Centre-initiated spherical charge - spherical explosive charge of mass C and spherical flyer shell of mass M
Symmetrical sandwich	$$\frac{V}{\sqrt{2E}} = \left(2\frac{M}{C} + \frac{1}{3}\right)^{-\frac{1}{2}}$$	Reactive armour on heavily armoured vehicles	Symmetrical sandwich - flat explosives layer of mass C and two flyer plates of mass M each
Open faced sandwich	$$\frac{V}{\sqrt{2E}} = \left[\frac{1 + \left(1 + 2\frac{M}{C}\right)^3}{6\left(1 + \frac{M}{C}\right)} + \frac{M}{C}\right]^{-\frac{1}{2}}$$	Explosion Welding, Reactive armour	Open-faced sandwich - flat explosives layer of mass C and single flyer plate of mass M

Equation 303: Optical Transfer Function

$$H\left(f_x, f_y\right) = \frac{G_{image}\left(f_x, f_y\right)}{G_{object}\left(f_x, f_y\right)}$$

The optical transfer function or OTF describes the ability of an optical system to image the details of an object, assuming incoherent or "light bulb" type illumination. The equation is remarkably simple in that it involves linear mathematics, once we master the idea of *frequency content* for an image. The frequency representation of a light intensity pattern follows from a Fourier Transform (see *Beautiful Equations #150*) in two dimensions. In the equation above, the Fourier Transform (FT) of the object intensity pattern is G_{object}, and the FT of the image is G_{image}. The OTF is H, and is a function of the spatial frequency coordinates $\left(f_x, f_y\right)$. The figure below is a conceptualization of the imaging process from this perspective. The beautiful and amazing thing about this description is that it summarizes what may be a very complicated optical system into a black box represented by a single function H.

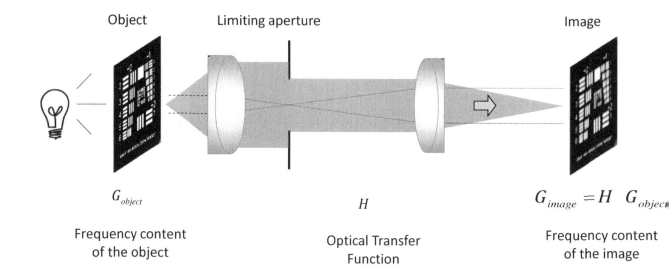

Object   Limiting aperture   Image

G_{object} H $G_{image} = H \; G_{object}$

Frequency content of the object   Optical Transfer Function   Frequency content of the image

In a diffraction-limited optical system, the OTF can be determined either by the Fourier Transform of the entrance or of the exit pupils, both of which are images of the limiting aperture in the system. If we assume that the illumination is perfectly uniform and "fills" this aperture evenly, we can calculate the OTF from the autocorrelation of the pupil, in a mathematical process that has a simple physical interpretation—the various diffracted rays coming from the object must fit through the limiting aperture of the system for them to take part in the reconstruction of the image. This is the theory of imaging proposed by Ernst Abbe over a century ago. For the very common case of a circular aperture, the modulus of the OTF, also known as the MTF, looks like this:

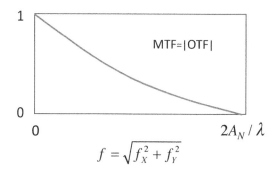

$$f = \sqrt{f_X^2 + f_Y^2}$$

The graph shows that typical incoherent optical systems have a monotonically decreasing sensitivity to the frequency content of an intensity pattern, finally leading to a hard cutoff at $2A_N/\lambda$, where A_N is the numerical aperture of the entrance pupil and λ is the mean wavelength. Aberrations distort the MTF, leading to inferior transfer characteristics except perhaps at specific frequencies where there may be an enhancement at the expense of other frequencies.

The OTF is part of an ensemble of important mathematical tools in Fourier optics—an approach to optical systems that became increasingly popular in the 1970's and afterwards. The power of this approach is that it recognizes that optical instruments, electronic circuits and communication systems often share basic properties such as linearity and invariance. This approach enables the straightforward analysis of highly complex optical systems using frequency analysis. For most of us, the Bible of frequency analysis of optical systems was authored by Joseph W. Goodman, who pioneered the introduction of Fourier Optics.

PdG

Equation 304: Euler's Number

Euler's number (after the Swiss mathematician Leonhard Euler), e is a hugely significant mathematical constant, given by today's equation and approximately equal to 2.71828. In the equation, e is given as a limit and also as an infinite sum. Like π, e is an irrational number, meaning that it is a real number that goes on forever and cannot be written as the fraction of two numbers. There are many other ways to define e, in terms of sums, recursive relations and trigonometric identities (see BE 38) and it has countless uses in mathematics. e can also be derived from a Binomial expansion (see BE 238).

Disciplines where e is used include, for example: calculation of compound interest (see BE 65), in probability theory, in the definition of the normal distribution (see B 177) and in calculus (here an interesting result is that the derivative of e is e, therefore, the integral of e is e). e also enters mathematics in the definition of the natural logarithm (of x) or logarithm to the base e (often written ln x).

e first came into mathematics in an appendix to some work on logarithms in 1610 by John Napier, but its first formal definition, as in today's BE, is credited to Jacob Bernoulli (1654-1705). The first use of e proper was in a letter by Gottfried Leibniz to Christiaan Huygens in 1690. Finally, Euler used e in its logarithmic guise to calculate the explosive force of cannons (around 1728).

$$e = \lim_{n \to \infty} \left(1 + \frac{1}{n}\right)^n = 1 + \frac{1}{1} + \frac{1}{1 \times 2} + \frac{1}{1 \times 2 \times 3} + \cdots$$

RL

Equation 305: Aspheric Lens

An aspheric lens is a type a lens profile that allows a specific optical function. Aspheric lenses are possibly the most important optical development since the telescope. Aspheric lenses can greatly reduce the size and complexity of an optical system and can achieve far superior performance than a system built of standard spherical optics.

The Aspheric lens equation is as follows

$$z(r) = \frac{r^2}{R\left(1 + \sqrt{1 - (1+k)\frac{r^2}{R^2}}\right)} + \alpha_1 \, r^2 + \alpha_2 \, r^4 + \alpha_3 \, r^6 + \cdots$$

The parameters for the aspheric equation are specified as a height z from an axis center point r=0 out to the periphery of the lens profile. The R value is typically considered as a radius or a master lens radius. The K parameter is the conic constant and defines whether the fundamental aspheric profile is a hyperbola, parabola, ellipse or a plain sphere ($K = 0$) The function is usually, but not always axisymmetric.

The additional terms are the coefficients that describe the non axisymmetric components of the optic. They are not required to describe an asphere and are specific to individual optical requirements. Very complex optical functions can be produced by just two aspheric lenses with multiple non axisymmetric coefficients.

Aspheres were first manufactured in 1620 by Rene Descartes. There were several other attempts throughout the next century. The first commercially available aspheric lense was made in 1956 by Elgeet.

Aspheric lenses are used in a wide variety of applications such as cameras, ophthalmics and many high end optical systems. It is an interesting and somewhat paradoxical observation that as computing has advanced, so has the design of complex optical systems. As complex optical systems have advanced so has computing technology. The two developments are closely aligned and are complementary in each other's development.

TS

Equation 306: Thomson Cross Section

$$\sigma_t = \frac{8\pi}{3} r_0{}^2$$

$$r_0 = \frac{q^2}{4\pi\varepsilon_0 mc^2}$$

Thomson scattering is the elastic, frequency-independent, non-relativistic scattering of electromagnetic radiation by a free charged particle, analysable using classical electromagnetism. It is just the low-energy limit of Compton scattering, which is otherwise inelastic.

In these situations, the energy hf of a photon interacting with a charged particle is much smaller than the particle's rest energy mc^2 (or equivalently, the wavelength of the radiation is much longer than the Compton wavelength of the particle).

The likelihood of this photon being scattered is represented by the Thomson cross section σ_t. In the formula above, m and q are the particle's rest mass and charge, ε_0 is the permittivity of free space, and c is the speed of light in vacuo. r_0 is called the 'classical radius' of the particle. An electron or proton interacting with electromagnetic radiation effectively behaves as if it has a spatial extent r_0. Thus even point particles have a finite scattering cross-section.

The formula tells us that, because m is much less for an electron than for a proton, the values of r_0 and hence σ_t are much greater for the electron. Thus electrons are far better scatterers, and are the particles normally involved when discussing Thomson scattering. For an electron, the classical radius and Thomson cross section work out to be 2.82×10^{-15} m and 6.65×10^{-29} m² respectively. (Note that these are fixed values: they are completely independent of the frequency of the radiation.) The Thomson formula works with X-rays incident upon electrons and gamma-rays upon protons.

The Thompson cross section is sometimes expressed in terms of the fine structure constant α:

$$\sigma_t = \frac{8\pi}{3}\left(\frac{\alpha\hbar}{mc}\right)^2$$

384

Thomson cross section

$$\sigma_T = \frac{8\pi}{3} r_0{}^2$$

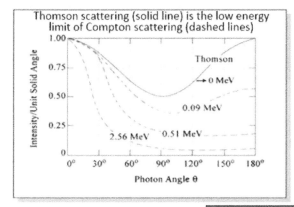

Thomson scattering (solid line) is the low energy limit of Compton scattering (dashed lines)

where 'classical radius' $\quad r_0 = \dfrac{q^2}{4\pi\epsilon_0 mc^2}$

J. J. Thomson 1856-1940

$q = charge$

$m = mass$

$\epsilon_0 = pemittivity$
$\quad\quad of\ free\ space$

$c = speed\ of$
$\quad\quad light\ in\ vacuo$

Effects of interaction

Kinds of interaction		Elastic scattering	Inelastic scattering	Total absorption
	Photon with all particles 1. $\pi D \ll \lambda$ 2. $\pi D \approx \lambda$ 3. $\pi D \gg \lambda$	1. **Rayleigh scattering** 2. Mie scattering 3. Geometric scattering	1. **Raman scattering**	

	Photon with charged particles
	1. Electrons
	2. Nucleons
		1. **Thomson scattering**
		2. Nuclear
		   Thomson scattering
			1. **Compton Effect**
			2. Nuclear
			   Resonance scattering
				1. **Photoemission**
				2. Nuclear
				   photodisintegration

	Photon with fields 1. Electric field surrounding electrons or nuclei 2. Nuclear field	1. Delbrück scattering		1. **Pair Production** 2. Meson Pair Production

MECHANISM OF THOMSON SCATTERING

A free electron will oscillate at the driving frequency of the incident electromagnetic wave. As a result, the electron will radiate at that frequency. Viewed from the 'outside', this looks like the photon hits the electron and then scatters off in a new direction. Although a scattering cross-section of $\sim 10^{-28}$ m^2 seems very small, Thompson scattering is actually one of the most important types of scattering in the Universe. Solar photons, which have a mean free path of about one cm, are very strongly Thomson-scattered by free electrons. This means it takes a photon emitted in the solar core several thousand years to struggle to the surface. Scattering from atoms involves the cooperative effect of all the electrons.

PRIMORDIAL THOMSON SCATTERING

Immediately after the Big Bang, the universe consisted primarily of antimatter plus a hot, dense electron, proton and photon plasma. It was Thomson scattering of the electromagnetic radiation by the free electrons that meant that this plasma was effectively opaque, due to the very short mean free path of the photons. As the universe expanded, it cooled to the point, just over a third of a million years after the Big Bang (the 'recombination era'), at a temperature of about 3000 K, that nearly all of the electrons and protons combined to form neutral hydrogen, which is much less effective at scattering radiation. Very quickly, photons decoupled from matter and the universe became transparent. The decoupled photons constitute what we observe today as the cosmic microwave background.

FR

Equation 307: Holography

Holography is the recording of a light field rather than an image. This can be produced by creating an interference pattern with two beams (one of which is scattered from an object to be encoded) and then using a developed film of the recorded interference pattern (see BE30) to diffract a similar beam of light to reconstruct an image of the object when it is no longer present. In its most pure form it will require a coherent laser beam for illumination and image reproduction. The process was discovered by the Hungarian-British physicist Dennis Gabor in the 1940s, based on previous work by himself and others in electron optics. The first practical holograms were produced in the 1960s after the advent of the laser, and Gabor received the Nobel prize in physics in 1971.

Gabor's original method was in-line, and suffered from the virtual and conjugate images and the reference beam all being on the same axis and affecting each other due to overlap. This was greatly improved by off-axis holography in the 1960s to separate the reference, virtual and conjugate beams.

The diagram below shows a schematic representation of the recording and image reconstruction of an off-axis hologram, in which the plate is rotated away from the reference beam axis at angle θ. The object is illuminated by the same source (laser) as the reference beam, using a beamsplitter (not shown).

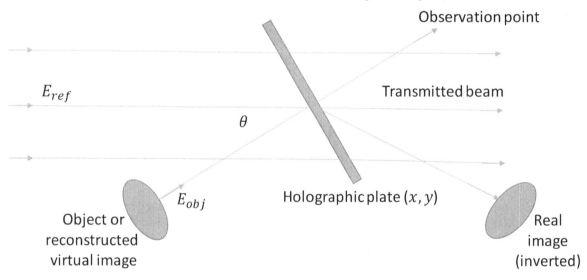

In reconstruction with the reference beam only, a virtual image is seen from the opposite side of the plate to the original object position, even when it is no longer present. This is due to the encoding of the interference pattern in the plane of the plate after exposure of the photographic emulsion on the plate (often a high resolution suspension of silver halide crystals in gelatine), the result being that part of the reference beam is diffracted to form a virtual image of the object in its original position. This is a 3D reconstruction that retains perspective (the image will appear to change according to the relative position of the observer, reproducing parallax).

In the following, the electric field strength representations are simplified in order to show the separation of the significant beams in reconstruction, without inclusion of spatial or temporal phase, and the constant $c\varepsilon_0/2$ is left out (the proportionality of field strengths is what is important). With a coherent source, the time-related part of the phase becomes unimportant but the wavelength is relevant and so is the angle θ, in terms of the spatial frequency of the interference fringes recorded on the photosensitive plate. (To clarify, for the simplest example of the reference beam, one should assume something of the form $E_{ref}(x,y) = E_{ref}e^{i2\pi\xi x}$ where x is the direction along the axis of the holographic plate shown and y is the axis coming out of the page. This phase term is equivalent to the more familiar form $e^{ikx'}$ (see BE48 for wave representation) if x' is the spatial coordinate of the beam axis where $x' = x \sin \theta$. The parameter $\xi = \sin \theta / \lambda$ is important for quantifying the spatial resolution of the interference fringes in the plane of the plate.)

The electric field at the film is the sum of the two beams

$$E_{film} = E_{obj} + E_{ref}$$

The intensity at the film is, including interference terms

$$I_{film} \propto |E_{obj} + E_{ref}|^2 = |E_{obj}|^2 + |E_{ref}|^2 + E_{ref}{}^* E_{obj} + E_{ref} E_{obj}{}^*$$

The film on the plate can be made so that its transmittance, once developed, is proportional to the square of the intensity in its plane (special types of emulsion are made for holography and low exposure levels)

$$T_{film} \propto I_{film}{}^2$$

The transmitted intensity of the developed film is

$$I_{trans} = T_{film} I_{ref}$$

so the transmitted field in terms of the original interference pattern and the reference beam, now the only beam illuminating the plate, is

$$E_{trans} = \sqrt{T_{film}} E_{ref} \propto I_{film} E_{ref}$$

Using the expression for the interference pattern above results in

$$E_{trans} = \left(|E_{obj}|^2 + |E_{ref}|^2 \right) E_{ref} + |E_{ref}|^2 E_{obj} + E_{ref}{}^2 E_{obj}{}^*$$

The first term is the beam transmitted straight through the plate (reference terms) together a "halo" surrounding it generated by the modulation of the square of the object beam with the reference beam. The second term is what produces the hologram effect – a virtual image appearing to be in the same location as the original object if observed from the other side of the plate. The third term is a real image that is often quite difficult to observe – an "inside-out", but real, conjugate image – this is generally considered less important but has been studied in holography.

At the time of Gabor's discovery, holography was very difficult and very expensive to produce. Gabor used an arc lamp as the best way of generating a coherent source. Once practical holography had become possible in the 1960s with lasers it developed rapidly and many techniques were investigated – it was often used by artists for its aesthetic results. The subject of my final year project for a BSc in applied physics was the construction of an image-plane hologram which could be observed in white light (without the need of a laser for reconstruction) – this requires arranging the equipment such that an image is brought near to the plane of the final recording medium. Nowadays holograms can be produced digitally to generate 3 dimensional "images" of objects that never existed. There are also many scientific applications of holography that have developed, such as producing certain types of gratings in spectroscopy, and holographic interferometry (used for instance to monitor displacements of rough surfaces very precisely, and analysis of fluids and vibrations on a small scale).

PB

Equation 308: Radiation Pressure

Radiation pressure is a pressure exerted on the surface of a body by electromagnetic radiation. The concept was first proposed by Johnnes Kepler in 1619. He believed this was what caused the tail of a comet to always point away from the Sun. Later in the same century (1687), Newton, believing light was made of particles or "corpuscles", calculated the pressure due to radiation from sunlight. Euler, who strongly believed in the wave nature of light also proved the existence of radiation pressure. Lebedev (1900) and Nichols and Hull (1901) verified the existence of radiation pressure using torsion balance experiments and Ashkin used radiation pressure to manipulate particles (optical tweezers as they became known) in 1986.

Three forms of radiation pressure exist. In 1862, Maxwell, who favoured the wave approach, stated that light has momentum and therefore must exert a pressure when incident on a surface. The Poynting Vector S gives the energy flux of a wave and so the radiation pressure is given as the energy flux divided by the speed of light.

$$P = \frac{\langle S \rangle}{c} = \frac{E_f}{c}$$

When light is treated as a particle, radiation pressure is the result of the transfer of momentum when striking a surface. Although photons have zero rest mass, they exhibit the property of mass as they have energy and momentum when travelling at the speed of light. The momentum is simply the energy divided by the speed of light and the radiation pressure is twice the energy flux divided by the speed of light.

$$P = \frac{2E_f}{c}$$

The thermal energy radiated from a body also has properties of energy and momentum. Energy leaving the body reduces the temperature and the change in momentum causes a reactive force i.e. a pressure on the surface. The power radiated from a black body is described by the Stefan-Boltzmann law and states that the total energy radiated per surface area across all wavelengths per unit time is directly proportional to the fourth power of the body's absolute temperature.

$$P = \frac{E_f}{c} = \frac{\varepsilon \sigma}{c} T^4$$

Our Sun exerts a pressure on all objects within the solar system. While all objects are affected, the smaller objects such as satellites, spacecraft, etc. are affected more. The pressure exerted on both an absorbing and reflecting body are given in the equations.

$$P_{absorbed} = \frac{W}{cR^2} cos^2 \alpha$$

$$P_{reflected} = \frac{2W}{cR^2} cos^2 \alpha$$

where P_x is the radiation pressure due to absorption or reflection, W is the solar constant and has a value of 1361 Wm^{-2}, c is the speed of light, R is the solar distance in AU and α is the angle between the surface normal and the incident radiation.

If we consider the Earth, at a distance of 1.0 AU from the Sun, the radiation pressure exerted on the Earth's surface is 9.08 μPa

References:

E. F. Nichols and G. F. Hull, "The Pressure due to Radiation," ApJ, vol. 17, no. 5, pp. 315-351, 1903.
Ashkin, J. M. Dziedzic, J. E. Bjorkholm and S. Chu, "Observation of a single-beam gradient force optical trap for dielectric particles," Opt. Lett. , vol. 11, no. 5, pp. 288-290, 1986.

CM

Equation 309: Conventional Mass

In equation 170 (equal arm balance) and 292 (Kohlrausch series) it was shown has masses can be compared and derived from a standard mass. There one issue was overlooked: the effect of different densities and the density of air, and the resulting difference in buoyancy. When a balance has zero indications, it does not mean that the masses on both side are the same, but the force on both pans is the same, i.e.:

$$m_1 - V_1\rho_a = m_2 - V_2\rho_a$$

where $m_{1,2}$ is the respective mass, $V_{1,2}$ the respective volume and ρ_a is the density of air. Neglecting this effect when for example weighing an aluminum piece against a steel mass standard, gives a deviation of some 0.6%. This is a tremendous error source compared to the accuracy that can be achieved by a balance. To make this correction for any weighing is simply not feasible; when weighing things for sale in a shop nobody wants to consider what is the (effective) volume of what he is buying.

To overcome this in practical and commercial weightings, the concept of conventional mass is defined. The definition of the conventional mass m_c is:

$$m_c = m.\left(\frac{1 - \frac{\rho_{ac}}{\rho}}{1 - \frac{\rho_{ac}}{\rho_c}}\right)$$

where m is the 'true' mass of the object, ρ_c is the conventional density of 8000 kg/m³, ρ_{ac} is the conventional density of air of 1.2 kg/m³ and ρ is the density of the object. The conventional values are choosen from purely practical considerations: steel objects, also mass pieces, have a density close to 8000 kg/m³ and under normal circumstances the density of air is close to 1.2 kg/m³. In words the official definition is:

The conventional value of the result of weighing of a body in air is equal to the mass of a standard of density 8000 kg/m³ at 20 °C which balances this weight at this temperature in air of density 1.2 kg/m³.

The practical consequence is that for the conventional mass not the absolute densities are important, but the relative deviations. Moreover one can neglect the density of an object when speaking of its conventional mass, if it is weighed on a balance against steel references. This means that world-wide balances are calibrated using steel mass pieces with a nominal density of 8000 kg/m³, and after that they are used to weigh anything with the restriction that the weighing result gives the conventional mass of an object. This does not take away the problem of measuring steel mass pieces against the Platinum-Iridium mass standards that act as national standards. For that comparison the limiting factors are in the determination of the air density that is related to temperature, pressure, and the density of the mass piece and standard. By the definition of conventional mass this painstaking task is given to the national laboratories and kept outside daily life.

HH

Equation 310: Potential Energy due to Gravity

In classical mechanics, the energy of an object consists of kinetic energy and potential energy. Kinetic energy is the energy that an object is exhibiting due to its motion. Potential energy, on the other hand, is energy 'stored' in an object that can be converted to kinetic energy. The simplest example of potential energy is when an object is brought to a given height h from the ground. Gravity exerts a force on the object equal to mg, where m is the object's mass and g is the acceleration of gravity.

The potential energy of the object is given by

$$P.E. = mgh$$

As the object is dropped and approaches the ground, the potential energy is converted to kinetic energy (increases in speed).

The term 'potential energy' was coined by William Rankine, a nineteenth-century Scottish engineer, physicist, and shortbread connoisseur.

MF

Equation 311: The Witch of Agnesi

Consider a point A on a circle of radius a. Let D be the point diametrically opposite to A. Draw the tangents to the circle through the points A and D. Now, pick a point B on the circle and draw a line joining A and B. The extension of this line intersects the tangent to the circle at D at C. Drop a line through C and perpendicular to the tangent at A. Finally, draw a line through B and parallel to the tangent to the circle at A. Let the intersection of the last two lines be P. The locus of this point P as B moves on the circle is known as The Witch of Agnesi.

The Witch of Agnesi is the red curve shown in Figure 1 for the case when the point A is at (0,0) and $a = 1$.

Let x and y be the horizontal and vertical axes with the origin being at A. Then, in terms of x and y the curve is governed by the equation

$$y = \frac{8a^3}{x^2 + 4a^2}$$

In parametric form, the curve is expressed as

$$x = 2at$$

and

$$y = \frac{2a}{1 + t^2}$$

where $t \in (-\infty, \infty)$.

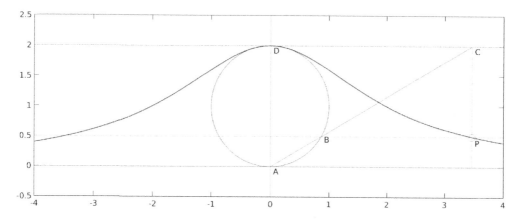

Figure 1: The Witch of Agnesi is the locus of the point P as the point B moves along the circle.

The curve was originally named *versiera* by the Italian mathematician and philosopher Maria Gaetana Agnesi who studied it in her book *Instituzioni analitiche ad uso della gioventu italiana*. The name *versiera* (meaning to ``to turn") was originally suggested by another Italian mathematician Luigi Guido Grandi in 1703. However, the first person to study the curve may have been by the French lawyer-mathematician Pierre de Fermat in 1630. The name "The Witch of Agnesi" came about during what appears to be a mistake by the English mathematician John Colson in 1801 while translating the book by Agnesi. The word *versiera* was mistaken for another Italian word *avversiera* which means "wife of the devil" or "she-devil".

The Witch of Agnesi is asymptotic to the tangent to the circle at A. The area between the curve and this tangent (asymptote) is four times the area of the circle. Suppose the curve and the asymptote through A are rotated about the line AD. The volume enclosed by the resulting surface (from the rotation of the curve) and the plane (from the rotation of the asymptote) is $4\pi^2 a^3$.

When $a = 0.5$, the equation describing The Witch of Agnesi becomes

$$y = \frac{1}{1 + x^2}$$

In turn, this equation when normalized over $(-\infty, \infty)$ becomes the standard Cauchy distribution, also known as the Cauchy-Lorentz distribution. It has the interesting property that both the mean and variance are undefined. The Cauchy distribution finds applications in many areas such as fluid mechanics, optics and quantum physics. However, the most common application is the use of the distribution as a counterexample in the probability distribution theories.

HC

Equation 312: Rumour Spreading

The spreading of rumours is an example of the way in which simple mathematical tools, along with plausible assumptions, can lead to quantitative predictions of the outcomes of seemingly intractably vague processes.

Assume that the National Physical Laboratory (NPL) has 1000 employees. The fifty senior managers attend a meeting at 09:00 on Monday, at which they are informed that NPL is to be renationalised and all its staff will become civil servants, but they must tell no-one.

However, 24 hours later, 100 employees have heard the rumour. How long will it take until 800 employees are in the know?

391

It seems reasonable to assume that the rate of rumour spreading is proportional to the number of conversations between people who already know it (call this y) and people who *don't* know it (which must be 1000 - y). As an equation:

$$\frac{dy}{dt} = ky(1000 - y)$$

where k is a constant. Solving this differential equation for $y(t)$ (the number of people in the know after t days) gives

$$y = 1000 \frac{Ae^{1000kt}}{1 + Ae^{1000kt}}$$

where A is a constant.

We know that $y(0) = 50$, so A must be 1/19. Hence

$$y = 1000 \frac{e^{1000kt}}{19 + e^{1000kt}}$$

We also know that $y(1) = 100$, so k must be

$$k = \frac{1}{1000} ln\left(\frac{19}{9}\right)$$

So, now we know the values of both A and k, we can calculate how long it will take until 800 employees have heard the rumour.

$$800 = 1000 \frac{e^{t \, ln\left(\frac{19}{9}\right)}}{19 + e^{t \, ln\left(\frac{19}{9}\right)}}$$

which is approximately 5.8 days; around 04:00 on Sunday morning.

MJG

Equation 313: Spherical Coordinate System and Laser Tracker

Coordinate systems represent a general mathematical framework in which the location and orientation of geometrical elements are quantitatively represented by numbers representing measures in each coordinate. The Cartesian (named after the French natural philosopher René Decartes) coordinate system is the most intuitive frame in daily life, where the position of an object is represented by three independent quantities: length (x), width (y) and height (z) with respect to a predetermined point called the 'origin'. Another widely used coordinate system in physics and engineering is the spherical coordinate system where the same position of the object is given by three other independent quantities; radial distance (r), elevation angle (ϕ) and azimuthal angle (θ).

The mathematical relationship between the Cartesian coordinate system and the spherical coordinate system given by the equations in the figure on the left side. These equations are often referred to as coordinate transformations and enable the user to represent a location in space using either coordinate frame.

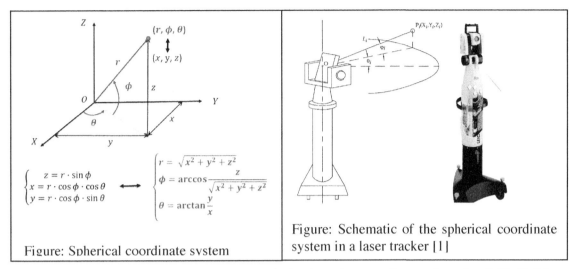

$$\begin{cases} z = r \cdot \sin\phi \\ x = r \cdot \cos\phi \cdot \cos\theta \\ y = r \cdot \cos\phi \cdot \sin\theta \end{cases} \longleftrightarrow \begin{cases} r = \sqrt{x^2 + y^2 + z^2} \\ \phi = \arccos\dfrac{z}{\sqrt{x^2 + y^2 + z^2}} \\ \theta = \arctan\dfrac{y}{x} \end{cases}$$

Figure: Spherical coordinate system

Figure: Schematic of the spherical coordinate system in a laser tracker [1]

Using the transformation equations in *Figure 1*, the Cartesian coordinate of a point $P_i(X_i, Y_i, Z_i)$ can be expressed by the corresponding spherical coordinate $P_i(\theta_i, \phi_i, L_i)$ from the angular and distance sensor respectively. Figure 2 shows a line diagram of a laser tracker. This measuring machine is similar to a modern Theodolite used by surveyors. The laser tracker consists of a laser for measuring distance, L_i, to a target mirror (not shown) located at a point of interest $P_i(\theta_i, \phi_i, L_i)$. The gimbal on which the laser is mounted has two rotary bearings that are control to follow (or 'track') the mirror as it moves around in space. Encoders on these two axes provide a measure of the spherical coordinate values θ_i, ϕ_i.

In the 1980s at the United States National Bureau of Standards, a spherical coordinate measuring system was first developed by Hocken and Lau [1] who gave 'laser tracker' its name. The laser tracker is the first instrument that physically realized the definition of each coordinate $(r, \phi$ and $\theta)$ by using two angular encoders and one distance sensor (typically a laser sensor).

The most significant advantage of laser tracker is its large measurement range (tens of meters) compared to conventional coordinate measuring machine in which a Cartesian coordinate system is used with common machines having a range of one meter or so and larger range machines tending to become successively more expensive.

Today's commercial laser trackers are prevalent in industries such as aerospace and ship building where both scale and accuracy play important roles for optimizing functionality. A picture of laser tracker manufactured by Leica is given on the right side of Figure 2. Commercial laser trackers are claimed to have an "accuracy" around the twentieth of a millimetre. Investigation of the measurement uncertainty when a laser tracker is used in different conditions are carried out by major national institutes as well as specialist contract consultants.

References:
[1]. R. J. Hocken and P. H. Pereira, Coordinate Measuring Machines and Systems 2nd, 2011, CRC Press.

KN

Equation 314: Golden Ratio

$$\varphi = \frac{1 + \sqrt{5}}{2} = 1 + \cfrac{1}{1 + \cfrac{1}{1 + \cfrac{1}{1 + \cdots}}} = 1.61803\,39887\ldots$$

a is to b as $a+b$ is to a

The golden ratio (also called the golden mean or golden section from the Latin: *sectio aurea*) is an irrational number, that is defined as the ratio of two quantities, where the ratio of their sum to the largest quantity is equal to the ratio of the larger to the smaller quantity.

The number was know to ancient Greek mathematicians because it appears often in geometry. For example in a regular pentagon the sides are in inverse golden ratio to the line connecting two non adjcent vertices. The golden ratio is also associated with the proportion of shape playing a role in human perception of beauty. For example the facade of the Greek Parthenon is build with golden ratio proportion.

Whether the golden ration was known to Egyptian mathematician too is not known. In fact, Egyptian piramids proportions (in particular the Great Pyramid of Giza) are close to a golden triangle. The golden triangle is defined as the square triangle where the height to base ratio is equal to the square root of the golden number, and the hypotenuse to base ratio is equal to the golden number (for example 3,4,5 being the size of base, height and hypotenuse).

The golden ratio can also be obtained as the ratio of successive Fibonacci numbers, and it is found in the proportion of animal, plants and humans. For example spiral growth of sea shell of plant cells follow a golden spiral shape [1]. All the key facial features of the tiger fall at golden sections of the lines defining the length and width of its face [1]. In 2010, the journal *Science* reported that the golden ratio is present at the atomic scale in the magnetic resonance of spins in cobalt niobate crystals [2].

References:
[1] http://www.goldennumber.net/nature/. Access on 06/11/2015.
[2] Quantum Criticality in an Ising Chain: Experimental Evidence for Emergent E_8 Symmetry, R. Coldea et al., Science 8 January 2010: 327 (5962), 177- 180. [DOI:10.1126/science.1180085].

GM

Equation 315: Specific Steam Consumption

In the UK we have a passion about steam locomotives. There are many preserved lines and steam engines still running that contribute greatly to the UKs tourist revenue.

Indeed, 2009 witnessed widespread public interest and enthusiasm when Tornado, the first main line steam locomotive to be built in the UK since Evening Star, made its public debut

However, Tornado was built from existing designs using 1960s technology. Another group of enthusiasts asked the question; 'What sort of steam engine could we build making use of current technology?'

The 5AT Project as it was called aimed to build the first new steam locomotive incorporating technical advances that have been developed since 1960 in order to generate the power, speed, range and reliability that may one day be needed to maintain an ongoing presence for steam traction on the main line rail network of the future. Sadly, the 5AT Project was suspended in 2012 due to lack of financial support but not before the Project Feasibility Study was completed in 2010 and that included a complete set of Fundamental Design Calculations.

One of the fundamental design calculations involves the specific steam consumption. The specific steam (or fluid) consumption is stated as the steam consumed in a power plant to generate one unit power (kW). Specific Steam Consumption is defined as the steam consumed by a locomotive's cylinders per unit output of power. It is typically measured in kg/kWh or kg/KJ.

A locomotive's Specific Steam Consumption carries important implications as may be deduced from the following equation

Power(kW) = Steam Production (kg/h) /Specific Steam Consumption (kg/kWh)

Thus for any given boiler output, a locomotive's power can be increased by reducing its specific steam consumption - in particular, by increasing its cylinder efficiency and reducing steam leakage. Or as Porta (see note) put it, "the power is limited by [the amount of steam supplied by] the boiler, while the function of the cylinders is to extract the maximum work from the steam supplied".

Note: A large number of papers covering a whole spectrum of topics about Modern Steam were written by Ing L.D. Porta, however most remain in hand-written manuscript form. The few that have been published include "Porta L.D., Advanced Steam Locomotive Development - Three Technical Papers" published by Camden Miniature Steam.

You may ask what happened to the 5AT project. The Advanced Steam Traction Group (AST Group) was established in 2012 following the closure of the 5AT Project. The Group seeks to promote the ongoing development of steam traction with the aim of prolonging steam operation of both main line and heritage line workings for the benefit of present and future generations. See http://www.advanced-steam.org/ accessed 12 November 2015

DF

Equation 316: Johnson-Nyquist Noise

$$\langle I^2 \rangle = kt \left(\frac{2}{\pi}\right) \int\limits_{0}^{\infty} R(\omega)|Y(\omega)|^2 \, d\omega$$

where k is Boltzmann's constant = 1.380 648 52(79) x 10^{-23} [J·K^{-1}], T is absolute temperature [K], I is current through a circuit element [A], $R(\omega)$ is the resistive component of a circuit element to be measured [Ω], $Y(\omega)$ is the ratio of the output current from an amplifier given the input voltage originating at the resistive component of the circuit element under test [A·V^{-1}] and $\omega = 2\pi v$ represents the frequency of the disturbance in the circuit element [rad·s^{-1}] with v being the linear frequency in Hz.

The triangular braces ◇ indicate that we are considering the expected value for the square of the current. In practice this will be varying randomly (for all intents and purposes 'random') with a very broad range of frequencies.

Listen very carefully to any audio system and, depending on the acuity of your hearing, you will always detect a faint hiss. This has puzzled, and annoyed, scientists from the first days of electrical measurements.

In the mid-1920s two researchers at Bell Laboratories finally gave a physical and mathematical foundation explaining the origin of this noise and enabling the calculation of its effects. John Bertrand Johnson (1887–1970) performed careful experiments to determine the values for many different types of resistive element (it turns out the type of resistor does not matter, they all have the same thermal noise) while Harry Nyquist (who also shows up in BE 44 and 299) was able to provide a rigorous mathematical foundation based upon thermodynamic principles and the new statistical mechanics.

To determine the noise that would be measured in a simple resistor, Nyquist imagined an experiment where the measurement process itself utilized a resistor of similar value. To determine the noise, it was simply a matter of examining all possible voltage fluctuations in the parallel resistor pair. In fact Einstein had stated that, for any system containing independent vibrational modes that can store energy, the total thermal energy is distributed evenly among modes, called the equipartition theorem (or, sometimes, law). Given this assumption, the average energy $\bar{E}$ of each mode is given by

$$\bar{E} = \frac{hv}{e^{\frac{hv}{kT}} - 1}$$

where h is Planck's constant (see BE 73).

Nyquist's task was to find an expression for the modes of this system. To simplify this problem, he took an unusual step and imagined that the two resistors were connected by a perfect and very long transmission line. In this thought experiment, after the system had maintained thermal equilibrium, the resistors were suddenly removed leaving the energy trapped in the transmission lines. The modes and their frequency distribution can then be deduced using simple one dimensional wave equation solutions. From this, it is possible to determine the (one sided) power spectral density, $S_{vv}(v)$ (see BE 132), of the measured noise. For the two equal resistors, this becomes

$$S_{vv}(v)dv = \frac{4Rh}{e^{\frac{hv}{kT}} - 1} dv \qquad [v^2]$$

For typical practical measurement frequencies the argument of the exponent is small and the above can be reduced to the simple equation

$$S_{vv}(v)dv \approx 4RKTdv \qquad [v^2]$$

The factor of 4 appears due to there being two field components (electric and magnetic) plus the fact that the total current is split between the two resistors of this particular circuit. Johnson, who originated this study, was left with the task of determining the noise produced by various circuit elements. In practice, these voltages and currents are too small for practical meters and it was necessary to use an amplifier. In these days, amplifiers were constructed from vacuum tubes (these have an operating characteristic that, although inverted, is similar to the modern Field Effect Transistor or FET) and it was necessary to derive an expression that would include the vacuum tube characteristic given by the beautiful equation at the top of this article. Since this early work, there have been substantial studies to corroborate these equations and determine other sources of noise. When currents are induced into circuit elements additional noise is generated by fluctuations in current flow (called 'shot' noise because it sounds like shot falling on a drum skin) and resistance (producing a low frequency fluctuation with a $\frac{1}{v}$ characteristic often called 'flicker' noise that occurs, for differing reasons, in a remarkably broad range of measurements).

The importance of this theory is that it provides a framework for understanding the limits of every measurement for determining signals that might be buried in noise. Extending this work it is possible to predict the limits of optical detectors. An immediate consequence of this equation is that, to this day, to minimize noise, it is necessary to measure for a long time (low bandwidth) and, wherever practicable, cool the detectors to as close to absolute zero as possible. Our current efforts to detect habitable planets in the universe are an excellent example of the desire of astronomers to not only 'see' these habitable planets (that produce very little light compared to the stars that are in the immediate neighbourhood) but also to measure the optical emission characteristics in an effort to determine molecular signatures indicating the presence of life.

References:

Johnson J.B., 1926, Proceedings of the American Physical Society: Minutes of the Philadelphia Meeting December 28th – 30th, 1926, published in *Phys. Rev.*, **29**, 367-368 (1927) – a February 1927 publication of an abstract for article #57 appearing on page 367- entitled 'Thermal agitation of electricity in conductors' - presented by J. B. Johnson during the December 1926 APS Annual Meeting.

Johnson J.B., 1928, Thermal agitation of electricity in conductors, *Phys. Rev.*,**32**, 97 – 109.

Nyquist H., 1928, Thermal agitation of electric charge in conductors, *Phys. Rev.*,**32**, 110 – 113.

SS

Equation 317: Fourth Order Runge Kutta

In equation 312 we introduced the concept of "numerical integration" a means of estimating the value of an integral without having to analytically calculate its function. This is particularly useful if the equation that expresses the function being integrated is extremely complex or if the function isn't expressed as an equation at all and can only be measured.

Equation 312 also introduced a toy example, that of a rocket, travelling at velocity v, where $v = t^2$. This allows us to analytically calculate the integral of the velocity, which gives the displacement d of the rocket as a function of time, $d = (t^3)/3$ and compare our estimate with the real answer.

Euler's method for numerical integration can have problems with accuracy. The reason for the inaccuracy is that Euler's method estimates the area under a curve as a series of rectangular strips, and when the curve bends, the flat top of the strip either over or under estimates the area. An upwards bend results in overestimation and a downwards bend results in underestimation.

The Runge-Kutta (pronounced "roo ng-uh koo t-ah") method attempts to compensate for this by finding a new strip height that is, in effect, a weighted arithmetic average of previous samples, thus being less sensitive to bends in the function. As the name suggests, this form requires four samples to provide an estimate, and whilst other orders of the equations exist (1st order being equivalent to the Euler method) the fourth order is by far the most common.

$$\widehat{f_{t=T}} = \widehat{f_{t=T-\Delta t}} + (1/8) * (k_1 + 3k_2 + 3k_3 + k_4)\Delta t$$

$$k_1 = g(t_T, f_T)$$

$$k_2 = g(t_T + \Delta t/3, f_T + \Delta t k_1/3)$$

$$k_3 = g(t_T + 2\Delta t/3, f_T - (\Delta t k_1/3) + \Delta t k_2)$$

$$k_4 = g(t_T + \Delta t, f_T + \Delta t k_1 - \Delta t k_2 + \Delta t k_3)$$

EQUATION SET 1: GENERALISED 4TH ORDER RUNGE-KUTTA
The general 4th order Runge-Kutta equations given above can solve situations where the function being integrated is a closed loop (i.e. it depends on both time and the last value of the function). This technique relies on not only iterating through the function in time, but also evaluating the gradient of the function at various heights to find an appropriate predictor for the direction of the curve. However, for our rocket example, things are simpler as the velocity function we're integrating is only a function of time. We will focus on this, simplified situation where the equations can be simplified into Equation 2.

$$\widehat{f_{t=T}} = \widehat{f_{t=T-\Delta t}} + (1/8) * (x_n + 3x_{n-1} + 3x_{n-2} + x_{n-3})\Delta t$$

EQUATION 2: 4TH ORDER RUNGE-KUTTA FOR A PURELY SAMPLE-BASED INTEGRATOR
Note that for Euler's method, Δt was both the length of time between estimates AND the length of time between samples. Now Δt is only the length of time between estimates, and we are mandated to sample the function at equal time intervals, four times between estimates.

For example, if the oldest sample we were using (x_{n-3}) was at $t = 0$, then x_{n-2} should be taken at $t = \frac{\Delta t}{3}$, x_{n-1} should be taken at $t = 2\Delta t/3$ and x_n should be taken at $t = \Delta t$.

Figure 1 is a series of samples we've taken by reading a speedometer on our rocket. A new strip is produced that is a weighted sum of those previous samples. In this form of the equation, samples in the middle of the range get three times more weighting than those at the edges of the sample window. Variants exist of the equation that only double the weighting of the two middle samples (requiring the 1/8 to be adjusted to 1/6 as per the weighted average equation).

The disadvantage of Runge-Kutta is that for every estimate, you need four measurements/evaluations of the function being integrated. However, because of the vast improvement in accuracy, using Runge-Kutta allows you to massively reduce the frequency with which you need to make estimates in order to maintain an accurate model of the value of the integral.

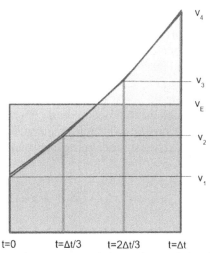

Figure 1: The four samples v_1, v_2, v_3, and v_4 are passed through a weighted average to produce an equivalent height v_E. The blue square is the new estimating strip for the area shaded in green. Note how in this example, the overestimation of the rectangle is cancelled out by an underestimation for the rest of the curve.

The table below illustrates the results of evaluating the function for 9 seconds (a multiple of 3 was used to make producing the data easier!)

Table 1: A comparison of the two numerical integration techniques for different sizes of time window

	Δt=0.15		Δt=0.9		Δt=1.8	
	R-K	Euler	R-K	Euler	R-K	Euler
Estimate after 9 seconds	243	249.11	243	280.67	243	320.76
Error	0	6.11	0	37.67	0	77.76

In fact even if using a single time step from 0 to 9, Euler's method produces an estimate of 729m (486m out) and Runge-Kutta still predicts exactly 243m (0m out), though this is obviously an extremely simple function to predict. This means that, in this case, Runge-Kutta can estimate more accurately making 4 measurements and one calculation in a 9 second window than Euler can making 60 measurements and 60 calculations in the same time. Whilst this is a drastic example of the effects, this can mean that employing Runge-Kutta in simulations and computer games tends to yield the same accuracy, if not more so, than Euler, for less computational power, allowing more complex systems to be analysed and more realistic virtual worlds to be rendered.

RFO

Equation 318: 10! Seconds = 6 Weeks

If not beautiful then it has a certain amount of cheeky charm.

Factorial Numbers

In mathematics 10! means ten *factorial*. It is possible to calculate the *factorial* of any positive integer

n by computing the product of all the positive integers less than or equal to n. For example $5! = 5 x 4 x 3 x 2 x 1 = 120$. n factorial also describes how many ways a set of n objects can be arranged into a sequence. In fact 0! can also be expressed and by convention $0! = 1$. This is because it is considered an empty product and therefore equal to the multiplicative identity 1. Or put another way, how many ways can zero objects be arranged? Answer: One way, no way..... YES, WAY! It is even possible to calculate factorials of fractions, with some very interesting results. This will be covered in a future beautiful equation.

So what is 10! Seconds?

$10! = 10 x 9 x 8 x 7 x 6 x 5 x 4 x 3 x 2 x 1 = 3,628,800$ (seconds)

3,628,800 seconds = 60,480 minutes = 1,008 hours = 42 days = 6 weeks

Or put more mathematically the factors of 10! are the same as the factors of 60 x 60 x 24 x 7 x 6

10 x 9 x 8 x 7 x 6 x 5 x 4 x 3 x 2 x 1 = 60 x 60 x 24 x 7 x 6

10 x 9 x 8 x 7 x 6 x 5 x 4 x 3 x 2 x 1 = **10** x 6 x 60 x 24 x 7 x 6

10 x 9 x 8 x 7 x 6 x 5 x 4 x 3 x 2 x 1 = **10** x 6 x 6 x 60 x 24 x 7

$10 \times 9 \times 8 \times 7 \times 6 \times 5 \times 4 \times 3 \times 2 \times 1 = \mathbf{10} \times 36 \times 60 \times 24 \times 7$

$\mathbf{10} \times \mathbf{9} \times 8 \times 7 \times 6 \times 5 \times 4 \times 3 \times 2 \times 1 = \mathbf{10} \times \mathbf{9} \times 4 \times 60 \times 24 \times 7$

$\mathbf{10} \times \mathbf{9} \times 8 \times 7 \times 6 \times 5 \times 4 \times 3 \times 2 \times 1 = \mathbf{10} \times \mathbf{9} \times 24 \times 4 \times 60 \times 7$

$\mathbf{10} \times \mathbf{9} \times \mathbf{8} \times 7 \times 6 \times 5 \times 4 \times 3 \times 2 \times 1 = \mathbf{10} \times \mathbf{9} \times \mathbf{8} \times 3 \times 4 \times 60 \times 7$

$\mathbf{10} \times \mathbf{9} \times \mathbf{8} \times \mathbf{7} \times 6 \times 5 \times 4 \times 3 \times 2 \times 1 = \mathbf{10} \times \mathbf{9} \times \mathbf{8} \times \mathbf{7} \times 3 \times 4 \times 60$

$\mathbf{10} \times \mathbf{9} \times \mathbf{8} \times \mathbf{7} \times 6 \times 5 \times 4 \times 3 \times 2 \times 1 = \mathbf{10} \times \mathbf{9} \times \mathbf{8} \times \mathbf{7} \times 60 \times 3 \times 4$

$\mathbf{10} \times \mathbf{9} \times \mathbf{8} \times \mathbf{7} \times \mathbf{6} \times 5 \times 4 \times 3 \times 2 \times 1 = \mathbf{10} \times \mathbf{9} \times \mathbf{8} \times \mathbf{7} \times \mathbf{6} \times 10 \times 3 \times 4$

$\mathbf{10} \times \mathbf{9} \times \mathbf{8} \times \mathbf{7} \times \mathbf{6} \times \mathbf{5} \times 4 \times 3 \times 2 \times 1 = \mathbf{10} \times \mathbf{9} \times \mathbf{8} \times \mathbf{7} \times \mathbf{6} \times \mathbf{5} \times 2 \times 3 \times 4$

$10 \times 9 \times 8 \times 7 \times 6 \times 5 \times 4 \times 3 \times 2 \times 1 = 10 \times 9 \times 8 \times 7 \times 6 \times 5 \times 4 \times 3 \times 2 \times 1$

MORE THAN A COINCIDENCE?

Perhaps slightly more, this is because the number systems relating to the passing of time have themselves been adopted because of the fact they have many shared factors. Both hours in a day and minutes in an hour being divisible by 12 made clock development a lot easier than it could have been. This would certainly increase the chances of finding a specific whole number of weeks that was a factorial number.

THE TIME OF DAY

So why are there 60 seconds in a minute, 60 minutes in an hour and 24 hours in a day? Well, even though today we think that counting in 10s (base 10) and recording numbers in decimal format is by far the most natural, probably because we have 10 fingers. It is actually not the most practical in terms of dividing numbers and counting numbers larger than 10 on your fingers. Ask any child who is trying to add 2 numbers larger than 10 for the first time. In fact ancient civilisations all seemed to use the more 'complicated' number systems of base 12 (duodecimal) or base 60 (sexagesimal). One explanation for this is that it is possible to count to 12 on one hand by just using your thumb to point to one of the three sections of each finger. The digits of the other hand can then be used to tally how many 12s you have and 5 x 12 gives 60 and perhaps 60 is big enough to count most things you would want to stand and count. This number system is also extremely convenient for expressing fractions, being divisibly by 2, 3, 4, 5, 6, 10, 12, 15, 20 and 30.

THAT'S LIFE...

The Egyptians are credited with dividing the day into 2 twelve hour sections; night and day. During the day sundials were used to track time and by using t-shaped bars to cast the shadow they could divide the day into 12 distinct parts. Egyptian life also required the measurement of time after dark. So early astronomers identified sets of stars to use during twilight and darkness hours. From around 1500 BC they also used water clocks and with day and night divided into 12 parts the 24 hour day was born. It wasn't until 150 BC that Greek astronomer Hipparchus suggested that these hours should all be the same length and so proposed using the 24 hours on the equinox days to be the standard hours. However, it wasn't until around the 14th century, when mechanical clocks were commonplace, that a fixed length for an hour became widely accepted.

...the universe

Well, at least our solar system has a part to play in this story. The Babylonians were one of the earliest civilisations known to use a 7 day week. It is likely they chose this because it is convenient to divide the lunar month into 4 periods. These periods relate to the moon phases of new moon, waxing half moon, full moon and waning half moon, each indicating a passage of roughly 7 days. And although there have been many different week lengths over the centuries, and even in recent times some countries have abandoned the 7 day week, like USSR in 1929 switching to a 5 day week, 7 days has now become the de facto standard. The 7 days of the week have also been linked to the 7 most visible celestial bodies and in some languages their names are still synonymous. Monday – Moon's day, Tuesday – Mars (French & Spanish), Wednesday – Mercury (French & Spanish), Thursday – Jupiter (French & Spanish), Friday – Venus (French & Spanish), Saturday – Saturn's day and Sunday – Sun's day.

So if you're looking for an answer to life, the universe, and possible everything else you could do worse than choose 10! seconds, which is 6 weeks, or if you prefer 42 Days.

IL-B

Equation 319: Parallel Plate Capacitor

Capacitors are important components in many electrical circuits. They are used in circuits for applications such as energy storage, pulsed power supply, power conditioning, motor starters, signal processing and sensing. The main application is to suppress noise in electrical components (they are referred to as decoupling capacitors).

Parallel plate capacitors are the easiest format used in teaching the basic theory as well as having many practical applications. Two parallel electrically conducting (usually metal) plates are connected in series via a power supply (i.e. a potential difference is applied between them). However, between the two plates is a dielectric material, so it would seem that no charge can flow and the electric field will be a constant value. The reason why the electric field is a constant is the same reason why an infinitely large charged plate's field is a constant. Imagine yourself as a point charge looking at the positively charge plate. Your field-of-view will enclose a fixed density of field lines. As you move away from the plate, your field-of-view increases in size and simultaneously there is also an increase in the number of field lines such that the density of field lines remains constant, that is, the electric field remains constant. However, as you continue moving away your field-of-view will be larger than the finite size of the plates. That is, the density of field lines decreases and therefore, the electric field decreases as well as the potential field. When you increase the distance between plates - capacitance drops, but stored charge remains the same, as electrons have nowhere to go. In the equation, C is the capacitance value (or capacity to store charge given in coulombs per volt, or farads), A is the plate area, d is the plate separation and ε is the permittivity of the material. Also shown, is a version of the equation where k is the relative permittivity of the material and ε_0 is the permittivity of free space (roughly 8.85×10^{-12} = F m-1).

If the electric field gets high enough, the dielectric material will "break-down" and current will flow across it. At this point the capacitor will basically be destroyed, usually accompanied by a pop and some smoke (I clearly remember this happening with polarised decoupling capacitors when I had wired them up around the wrong way).

In 1745, Ewald Georg von Kleist (Germany) found that charge could be stored by connecting an electrostatic generator by a wire to a volume of water in a hand-held glass jar. Von Kleist's hand and the water acted as conductors, and the jar as a dielectric. He found that touching the wire gave off a painful spark. Basically, his hands and the water were acting as the conductors and the jar as the dielectric. Later, the same principles were used to develop the so-called Linden jar, several of which were connected to produce batteries (Franklin played with these).

$$C = \frac{\varepsilon A}{d} = \frac{k\varepsilon_0 A}{d}$$

401

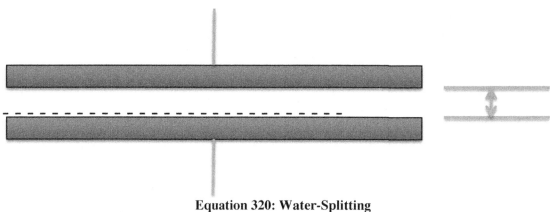

Equation 320: Water-Splitting

$$2H_2O \leftrightarrow 2H_2 + O_2$$

'Water splitting' is a generic term used to describe chemical reactions in which water is separated into hydrogen and oxygen. The stoichiometric ratio of its products is 2:1. Several water-splitting methods have been invented - often patented.

Since not much freely available hydrogen gas is to found, it must be extracted from various compounds. Generating hydrogen using water-splitting requires a lot of input energy, so most of the hydrogen now produced is derived from increasingly depleted non-renewable resources). Were economical and efficient water-splitting technology to be available, this would constitute an important component of a hydrogen economy. A hydrogen economy is a would-be hydrogen-based system for storing and delivering energy. Hydrogen can be used in fuel cells to generate electricity, burned to produce heat, and used in internal-combustion engines to power vehicles.

Water-splitting

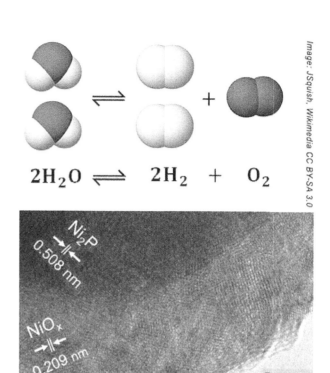

$$2H_2O \rightleftharpoons 2H_2 + O_2$$

Electrolysis method. This device uses low-cost nickel-iron oxide as a catalyst for both electrodes, generating bubbles of hydrogen on one electrode and oxygen on the other. The catalyst can split water continuously for more than a week with a 1.5 V input, with an efficiency of 82 % at room temperature. *Image: Standford University*

Ni$_2$P 0.508 nm

NiO$_x$ 0.209 nm

5 nm

Electrolysis method.
Nickel phosphide nanoparticles loaded onto a carbon electrode. A core-shell structure is adopted, with the phosphide core (dark grey) being encased in an active nickel oxide species (the shell - light grey).
Image: adapted from Xile Hu, Swiss Federal Institute of Technology

As a result of all sorts of problems that have come up during research into efficacious hydrogen production, the notion that hydrogen would be the long-term energy storage solution for the planet came to be seen by many as hype. Thus much of the research focus has shifted to batteries. Nonetheless, some researches feel that through greater understanding of reaction mechanisms the promise of a hydrogen economy may be realised.

Water-splitting techniques include the following.

- Electrolysis of the water, which is decomposed by means of an electric current at high or low pressure, perhaps with the addition of heat. Recently, low-cost and stable electro-catalysts have been developed. Electro-catalysts are a specific type of catalyst that operate at electrode surfaces, participating in electrochemical reactions. (They may be in fact be the electrode surface itself.) An example is the improvement of catalytic activity obtained through the use of ultra-small transition metal oxide nanoparticles.
- Photoelectrochemical water-splitting, where the electrical current is produced by photovoltaic systems.
- Thermochemical water-splitting, which uses high temperatures via concentrated solar power or nuclear reactor waste heat, and chemical reactions.
- Photocatalytic water splitting is an artificial photosynthesis process whereby solar photons, water, and a catalyst are used for the dissociation of the water.

Historically, subsequent to Martinus van Marum's use of electrolysis in 1785, William Nicholson and Anthony Carlisle were the first known individuals to decompose water into hydrogen and oxygen in 1800. Michael Faraday's Laws of Electrolysis followed in 1833.

FR

Equation 321: Earth's Magnetic Field

The magnetic field of the earth is thought to be generated by the motion of convective fluids in its core. The average radius of the earth is 6370 km. The core consists of a solid inner core (mostly iron-nickel) of radius 1200 km, surrounded by a molten outer core (similar material) which stretches out to about 3400 km where it reaches the solid mantle. The temperature of the inner core surface is estimated at about 6000 K, and there is a temperature difference of about 1200 K between the inner and outer surface of the molten part of the core. The movement of the outer core, resulting from the earth's rotation, heat convection and possibly also contributed to by radioactivity, generates circulating currents which in turn generate a magnetic field (a self-sustaining dynamo effect). This can be visualised schematically (see below) as a large bar magnet inside the earth, with current (yellow) circulating around its axis and magnetic field lines (red) emerging from/arriving at the magnetic poles. Magnetic north is by convention near geographical north. Note that the geomagnetic north pole is a south-seeking pole and the core's south therefore points towards north.

The current direction of the magnetic axis is just under 10° from the axis of rotation and is continually moving, currently in a north-westerly direction, due to changes in the earth's core. Many aspects of the earth's magnetic field are still not well understood, such as how it originally came into being (its age has been estimated from paleomagnetic data to be at least 3450 million years), its direction, and geomagnetic pole reversals which are known to have happened many times in the past from measuring the alignment of ferromagnetic crystals trapped in solidified lava flows. It is now understood that pole reversals can happen relatively quickly, possibly within a 100-year period. The magnetic axis is not linear all the way to the surface, and the geomagnetic pole is an approximation. The location where the field is actually vertical to the surface is called a "dip" pole, and does not coincide with geomagnetic north: it wanders faster and is currently about 3° from the physical north pole.

Variations in the earth's magnetic field can be measured with magnetometers, on the ground or in space. On the earth's surface, the strength ranges from 25 to 65 microtesla (μT). The tesla is the SI unit of the magnetic flux density $\mathbf{B}$). The field can be calculated approximately at any point using a magnetic dipole model, which is the subject of this beautiful equation. The magnetic potential can be described as

$$V(\mathbf{r}) = -\frac{\mu_0}{4\pi r^3} \mathbf{M}.\mathbf{r}$$

where $\mathbf{M}$ is the magnetic dipole moment (a vector), $\mathbf{r}$ is a vector indicating the direction of the dipole and μ_0 is the permeability of free space, the constant $4\pi \times 10^{-7}$ kg m A^{-2} s^{-2}. The magnetic dipole moment is the aligning torque generated by a magnetic field to the field strength, and has units Nm/T or alternatively Am^2. The estimated value for the earth's dipole is 7.94×10^{22} Am^2.

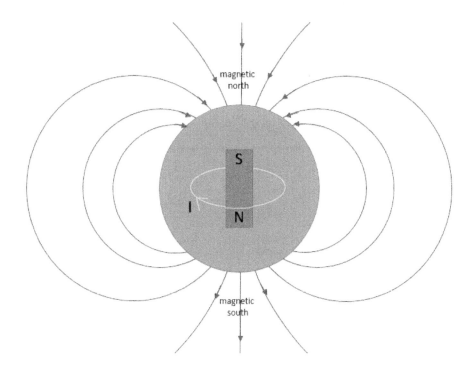

Due to the symmetry involved, the problem is helped by using spherical polar coordinates.

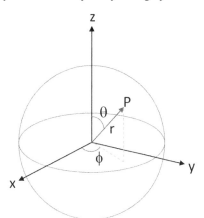

If the z-axis is aligned with the magnetic dipole, the field can be assumed to be symmetric in the longitudinal direction ϕ and hence only varies in r and θ. The dot product $\boldsymbol{M}.\boldsymbol{r} = Mr\cos\theta$. To determine the magnetic field $\boldsymbol{B}$ we need the relevant partial derivatives of the potential V

$$\mathbf{B}\,(r,\theta,\phi) = \nabla V = \frac{\partial V}{\partial r}\hat{\boldsymbol{r}} + \frac{1}{r}\frac{\partial V}{\partial \theta}\hat{\boldsymbol{\theta}} + \frac{1}{r\sin\theta}\frac{\partial V}{\partial \emptyset}\hat{\boldsymbol{\phi}}$$

where the bold items are unit vectors and the last term is zero. $\boldsymbol{B}$ can be separated into two orthogonal components with varying magnitude, a vertical component that is parallel to the direction of the magnetic pole z and a horizontal one perpendicular to this (the direction of the magnetic equator from the centre). Calculating first derivatives of V and using the dot product shown above,

$$B_r = \frac{\partial V}{\partial r} = \frac{2\mu_0 M}{4\pi r^3}\cos\theta$$
$$B_\theta = \frac{1}{r}\frac{\partial V}{\partial \theta} = \frac{\mu_0 M}{4\pi r^3}\sin\theta$$

405

from which it is immediately apparent that the field at the magnetic pole is twice as strong as it is at the equator, and the strength of the magnetic field (here represented as the magnitude of the flux density) at any point, using the identity $\sin^2\theta + \cos^2\theta = 1$, becomes

$$|B| = \sqrt{B_r{}^2 + B_\theta{}^2} = \frac{\mu_0 M}{4\pi r^3}\sqrt{1 + 3\cos^2\theta}$$

Another useful parameter, the relative strengths of the components in terms of latitude, is the magnetic inclination: $\tan I = {}^{B_r}\!/_{B_\theta} = 2\cot\theta$.

The actual observed values of the magnetic field differ somewhat from those predicted by the simplistic dipole model (although the model can be used as a large scale approximation, for example to predict past plate motions from paleomagnetic data). The field is not consistent in a simple way across the earth's surface and can be seen not to be spherically symmetric from detailed magnetic mapping. For a more accurate analysis, magnetic data can be modelled using spherical harmonics (solutions to a Laplace equation in the magnetic field), in which there are some "Gaussian coefficients" to be determined. The accuracy depends on the extent of the summation of terms in the associated Legendre polynomials in the solution. The subject of spherical harmonics was covered in BE 198.

An understanding of the earth's magnetic field is linked to navigation and dates back more than 1000 years. It was discovered in the 15th century that magnetic and geographic north did not coincide, and that the difference (the "declination") was different depending on the location on the surface. Measurements of this were made in 1596 by Willem Barentz in Norway, and soon afterwards the first scientific exploration of the field started with the English astronomer/physicist William Gilbert, published in 1600 (a magnetic unit, the vector potential, is named after him). Hans Christian Ørsted made the first discovery of the connection with electric current. Accurate measurements began in 1840 and Gauss published the first realistic model of the geomagnetic field (the cgs unit of the field B is the gauss, equivalent to 10^{-4} T). Detailed investigations of magnetic storms and the effect of solar activity on the northern lights were undertaken by the Norwegian scientist Kristian Birkeland around the turn of the 20th century.

PB

Equation 322: Number of Point Defects in Crystals

The equation giving the number of intrinsic point defects in an ionic crystal is based on the assumption that the defects are truly isolated from each other, so that the enthalpy associated to the formation of defects can be proportional to the number of defects and the enthalpy of formation of a defect pair (Equation 2).

For a Schottky defect, the equation is

$$n_S = N \cdot e^{-\frac{\Delta H_S}{2kT}} \ (1)$$

For a Frenkel defect, the equation is

$$n_F = \sqrt{N \cdot N_i} \cdot e^{-\frac{\Delta H_F}{2kT}} \ (2)$$

Where, ΔH_S is the enthalpy for the formation of a pair of Schottky point defects; ΔH_F is the enthalpy for the formation of an interstitial point defect; n_S, n_F are the number of Schottky and Frenkel defects, respectively; N and N_i are the number of lattice and interstitial sites, respectively.

The enthalpy of formation of defects in an ionic crystal varies between 60 and 600 kJ/mol. The concentration of defects is the ratio between the number of defects and the number of lattice sites in the crystal. Typical concentrations of defects at room temperatures hugely vary between 10^{-27} to and 10^{-6}, for enthalpies of 300 and 60 kJ/mol, respectively. However, at a temperature of 1000 K, the concentration of defects varies between 10^{-8} and 10^{-2}, for enthalpies of 300 and 60 kJ/mol respectively.

These equations are important, as they allow the total enthalpy contribution of defects in the crystal as

$$\Delta H = n_S \cdot \Delta H_S + n_F \cdot \Delta H_F \ (3)$$

THERMODYNAMICS OF CRYSTAL DEFECTS

From a thermodynamic point of view, solids containing point defects can be seen as solid solutions, in which the defects represent the solute and the solid crystal represents the solvent. Similarly, the defect equilibria can be treated in terms of thermodynamics of chemical solutions. Two fundamental conditions of intrinsic point defects in solids are:

1) Conservation of mass
2) Conservation of charge neutrality

This implies that, vacancies (missing ions) in an ionic crystal must form in pairs – that is to say, a cationic and an anionic vacancy must generate at the same time. This condition is also known as Schottky disorder. For a metal oxide compound, this means that vacancies of metal cations must be in the same number of oxygen vacancies.

A crystal is thermodynamically more stable with a finite number of defects in it than without defects at all. This fact is due to an entropic contribution – that is to say, the presence of defects in a crystal increases the entropy of the system – which lowers the Gibbs free energy of the crystal (Figure 1). However, a very high number of defects is no longer thermodynamically favourable, as it corresponds to an excessive increase in the enthalpy of the system.

TYPES OF CRYSTAL DEFECTS

Defects in a crystal can be overall categorised as:

1. Point defects, such as vacancies, interstitial defects and F-centres
2. Linear defects, such as dislocations
3. Bi-dimensional defects, such as grain boundaries
4. Tri-dimension defects, such as voids.

Point defects are dominant in ionic crystals and determine the electronic and catalytic properties of these crystals. Dislocations and grain boundaries are very common in metals and alloys, and play a fundamental role in the mechanical behaviour of these materials.

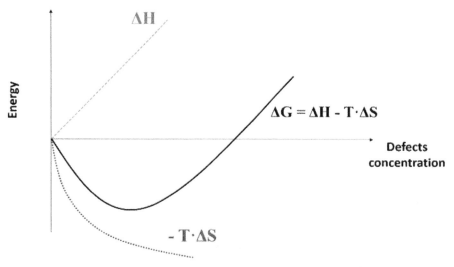

Figure 1. Energy contribution and Gibbs free energy (ΔG) of defects formation in an ionic crystal.

POINT DEFECTS

Two main types of point defects in an ionic crystal are (Figure 2):

1. Schottky defects: a pair of vacancies, respectively of a cation and an anion, which migrated to the crystal surface (Figure 2a).
2. Frenkel defects: an ion displacement from its lattice site into an interstitial site (Figure 2b).

These defects play an important role in the conduction of inorganic materials and in semiconductor technology.

Another typical point defect in ionic crystals such as zinc oxide, lithium chloride and potassium chloride, is the *F*-centre (Figure 2c). This defect consists in the presence of an electron within an anionic vacancy. The electron is energetically stabilised by the electrostatic potential of the surrounding cations. This defect takes

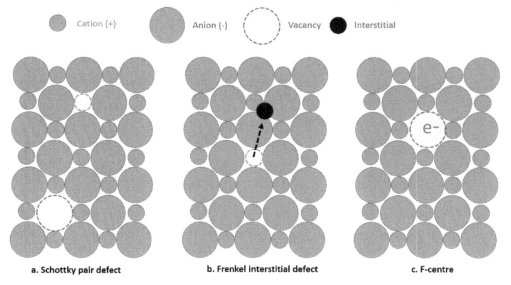

Figure 2. Common types of point defects in ionic crystals.

its name not by Frenkel, but by the German word "Farben", which means colour. This name is due to the property of these centres of easily absorbing visible lights, therefore allowing the otherwise transparent crystal to become coloured.

F-centres can be easily generated by heating ionic crystals, as an increase in temperature increases lattice vibrations, with consequent displacement of some anions, which leave some electrons behind to fill the vacancies in the crystal lattice. As a result, *F*-centres are always electrically neutral. The greater is the presence of these defects in an ionic crystal, the more intense is its colour. These electrons are often paramagnetic centres and can be studied through Electron Spin Resonance (ESR) spectroscopy.

SCHOTTKY AND FRENKEL DEFECTS

Walter H. Schootky was a German physicist, who lived between 1886 and 1976. He developed the concept of "Schottky's barrier" in 1914 within the theory of electron and ion emission phenomena. This barrier is the interaction energy between a point charge and a flat metallic surface. Schottky barriers are fundamental in semiconductor technologies, including metal-semiconductor junctions. They also play an important role in the theory of thermionic emission and field electron emission.

Yakov (or Jacov) Frenkel, was a Russian physicist who lived from 1894 to 1952 and worked at the Physico-Technical Institute of St. Petersburg. After working on Earth's magnetic field and atmosphere physics, he fully focused on condensed state matter and solid state physics. He introduced the concept of "holes" in electronics, which play a fundamental role in the physics of semiconductors and transistors, and gave a very important contribution in the theory of dislocations and plastic deformation of materials.

AT

Equation 323: The Lorentz Transform

The Lorentz transform describes how two observers that have a constant speed relative to each other, experience the laws of nature that take place in their own and each other's time- and space frame. The 'stationary' observer has coordinates x, y, z and time t; the 'moving' observer has coordinates x', y', z' and time t'. For $t = t' = 0$ we have $x = x' = 0$, $y = y' = 0$ and $z = z' = 0$. When aligning the x-axes along the direction of the movement, and the movement takes place with a constant speed v, the Lorentz transformations are

$$t' = \gamma \left(t - \frac{vx}{c^2} \right)$$
$$x' = \gamma(x - vt)$$
$$y' = y$$
$$z' = z$$

where

$$\gamma = \frac{1}{\sqrt{1 - \left(\frac{v}{c}\right)^2}}$$

and c is the speed of light. This is all, so what? In the first place there is no way an object can have a speed larger than c. The speed c seems to be the same for both observers. The equation for t' gives the time dilation that was already described in BE 8 and the equation for x' describes the Lorentz contraction that makes an observer see objects shorter if they pass at a speed close to the speed of light. This is the popular interpretation, in fact an observer sees an object somewhat rotated.

The equations were derived by Lorentz and some others in order to explain the outcome of the Michelson-Morley experiments in 1887. At that time it was unthinkable that a 'vacuum' could exist. Physics could only be understood if there was an 'aether', that enabled light to propagate. The Michelson-Morley experiment consisted of an equal arm interferometer that would show shifted fringes when one or the other arm of the interferometer would be parallel to the movement of the earth through the aether. The Michelson-Morley experiment measured no such effect. The Lorentz transforms were derived in order to describe that the length of one arm of the interferometer was contracted by the 'ether wind' by just the amount to compensate for the expected change of light velocity. The time-transformation was added to make the Maxwell equations invariant for the transformation to a different inertial frame. Lorentz considered this as an unsatisfactory mathematical trick and called this 'local time'.

Lorentz kept this interpretation for quite some time, even after Albert Einstein had published his special theory of relativity. In due course he was willing to admit that the physical laws are independent of the speed of the aether, but the logical conclusion that the whole aether does not exist was for him difficult to accept. The Lorentz transform plays a crucial role in Einstein's theory of relativity. In Einstein's interpretation, the Lorentz transform do not only keep the Maxwell equations valid in different inertial frames, but it describes how all physics is transformed, including the notion of place and time, that become relative in the literal sense.

Hendrik Antoon Lorentz (1853 ~ 1928) was a Dutch physicist who was a nestor in physics in the times when the relativity theory and later the quantum physics was developed. He made important contributions to these fields (Lorentz force, Lorentz-Lorenz equation, etc).

HH

Equation 324: The Stereographic Projection

Consider a sphere (assumed to be of radius 1) with the north pole $N = (0,0,1)$ as shown in the figure below. Let us define a coordinate frame (x, y, z) such that the origin of the coordinate system is at the centre of the sphere and the z-axis passes through the north pole. Furthermore, let $z = -1$ be a tangent plane to the sphere passing through the south pole S=(0,0,-1) and P an arbitrary point on the sphere with the coordinates (x, y, z). Then, a stereographic projection of P is the intersection P' of the line connecting N and P with the plane $z = -1$. The point N is also known as the projection point.

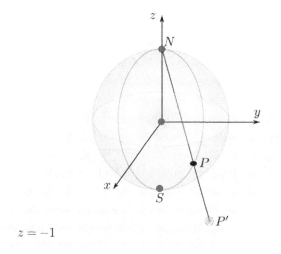

Figure: Stereographic projection with $z = -1$ as the projection plane.

410

Suppose that we denote the coordinates of P' by (s,t). Then, (s,t) are related to (x,y,z) through the following equations

$$s = \frac{2x}{1-z}$$

and

$$t = \frac{2y}{1-z}$$

Each point on the sphere, other than N, maps to a unique point on the projection plane. The point N maps to infinity on the projection plane. Great circles passing through N (lines of longitude) map to straight lines radiating from the south pole. The lines of latitude map to concentric circles with the south pole as the centre. In particular, the equator (a circle of radius 1 at $z = 0$) maps to a circle of radius 2 on the projection plane. The lines of latitudes below the equator appear as concentric circles within this circle (the projection of the equator) whereas the lines of latitude above the equator are projected as concentric circles exterior to this circle.

Another beautiful property of the stereographic projection is that any circle on the sphere and not passing through N is mapped to a circle on the projection plane. And, a circle on the sphere passing through N is mapped to a straight-line.

The plane $z = -1$ is by no means the only choice for stereographic projection. $z = 1$ and $z = 0$ are also commonly used as projection planes. When the former is chosen as the projection plane, the south pole is used as the projection point. When $z = 0$ is the projection plane, either the north pole or the south pole can be used for the projection point.

The mapping is conformal i.e., the angle between any two curves on the sphere is preserved by the projection. However, distances and areas are not preserved.

Stereographic projection was first introduced by the Greek astronomer Hipparchus (190 BCE - 120 BCE) considered to be the greatest astronomer of the ancient times. That the stereographic projection is conformal was also first shown by Hipparchus. Stereographic projection was used by Hipparchus in developing a first version of the astrolabe to measure the time of day, the times of sunrise and sunset etc. Astrolabes went onto become the tools of choice in astronomy and for navigation in many parts of the world until as recently as the mid-19th century.

In addition to astronomy and navigation, stereographic projections find uses in cartography, mineralogy, materials science and photography. The first ever use of stereographic projection in cartography appears to have been by Gaultier Lud in 1507. Since angles between lines are preserved during stereographic projection, the maps of the earth using stereographic projection are preferred in navigation. However, it should be kept in mind that the areas are not preserved during the projection. Since the great circles passing through N are projected as straight lines emanating from S, true distances from a place A to another place B can be readily calculated by taking S to be coincident with A. Stereographic projection is still one of the most commonly used azimuthal projection methods in cartography. In materials science, crystallographic texture is the preferred orientation of the grains constituting a polycrystalline material. The texture of a given polycrystalline material is often represented using pole figures which are the stereographic projections of various crystallographic directions.

Reference
Howarth, R. J. (1996). History of the stereographic projection and its early use in geology. Terra Nova, 8(6), 499-513.

HC

Equation 325: Paschal Full Moon Date Equation

The celebration of Easter differs for Christian denominations depending on which calendar is used. While most Christian denominations follow the more recent Gregorian calendar, some Eastern Orthodox communities still use the older Julian calendar. For this reason, Easter celebrations in the two calendars don't always coincide. While the tabulated Easter dates were first discussed in the 4th century AD, the method to calculate the dates arithmetically was introduced in the year 1582.

Easter Sunday is the Sunday following the Paschal Full Moon date, which is determined using the equation below.

$$PFMd = 45 - (Y \bmod 19 \times 11) \bmod 30$$

where Y is the year for which the Paschal Full Moon date is being determined. If the resulting date is below 31, the Paschal Full Moon is in March. If the date is larger than 31, then the Paschal Full Moon occurs on April and the date is $PFMd - 31$. The following conditions apply for the PFMd equation:

If $Y \bmod 19 = 5$ or 16, then 29 must be added to the $PFMd$.

If $Y \bmod 19 = 8$, then 30 must be added to the $PFMd$.

MF

Equation 326: Stature Equations

One of the primary functions of applied mathematics is to wrest meaning from measurement data. Where there is a physical law behind such data, a precise mathematical description can be formulated, as in the cases of Galileo's, Kepler's, Newton's and Einstein's laws.

In biology however, many phenomena occur, not as the expression of a law of nature but through the complex interaction of many causes. An example of this is the relation between the stature of a person and the dimensions of their body parts. While it is obvious that taller people tend to have larger body parts, there is no precise relationship to be discovered. (Despite the conclusion of the Roman engineer Vitruvius that the ideal human form has a precise set of proportions, leading to the proliferation of copies of the image of the Vitruvian man):

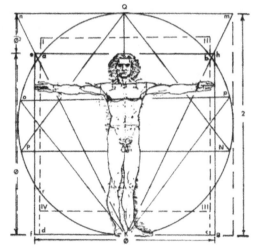

However, there are two areas in which it is very useful to derive stature from (in particular) bone length: archaeology and forensic science, in both of which a few bones may be the only source of information about their former owner.

Such topics are the province of regression analysis, which helps pick out relationships from (ideally) large datasets. The simplest kind of relationship is one in which the value of one parameter is directly proportional to the value of another, in which case one would expect that plotting the values of one variable against the other would yield a straight line. Where data are inaccurate or there are other factors at pay which modify the relationship from a perfectly linear one, one would expect the data points to be scattered around a straight line. To infer what linear relationship "lies behind" such data, regression analysis can be used. The simplest and commonest way to do this is to find the line which is the closest possible to all the points, "closest possible" here meaning that the sum of the squares of the distances of the points from the line must be as small as possible. The following equation gives this as a formula for n data points, in which r_i is the distance of the rth data point from the line.

$$S = \sum_{i=1}^{n} r_i^2$$

In applying regression analysis to the prediction of stature from bone lengths, account must be taken of the fact that Vitruvian men are quite rare. Not only do women and infants have different proportions, there are also great variations with race. However, by collating data on bone lengths of different ages, races and genders, a standardised set of stature equations has been derived by regression analysis:

BONE	RACE	MALE EQUATION	FEMALE EQUATION
Femur	Caucasoid	2.32 * femur + 65.53 ± 3.94 cm	2.47 * femur + 54.10 ± 3.72 cm
Femur	Negroid	2.10 * femur + 72.22 ± 3.91 cm	2.28 * femur + 59.76 ± 3.41 cm
Femur	Mongoloid	2.15 * femur + 72.57 ± 3.80 cm	
Tibia	Caucasoid	2.42 * tibia + 81.93 ± 4.00 cm	2.90 * tibia + 61.53 ± 3.66 cm
Tibia	Negroid	2.19 * tibia + 85.36 ± 3.96 cm	2.45 * tibia + 72.56 ± 3.70 cm
Tibia	Mongoloid	2.39 * tibia + 81.45 ± 3.24 cm	
Fibula	Caucasoid	2.60 * fibula + 75.50 ± 3.86 cm	2.93 * fibula + 59.61 ± 3.57 cm
Fibula	Negroid	2.34 * fibula + 80.07 ± 4.02 cm	2.49 * fibula + 70.90 ± 3.80 cm
Fibula	Mongoloid	2.40 * fibula + 80.56 ± 3.24 cm	
Humerus	Caucasoid	2.89 * humerus + 78.10 ± 4.57 cm	3.36 * humerus + 57.97 ± 4.45 cm
Humerus	Negroid	2.88 * humerus + 75.48 ± 4.23 cm	3.08 * humerus + 64.67 ± 4.25 cm
Humerus	Mongoloid	2.68 * humerus + 83.19 ± 4.16 cm	
Ulna	Caucasoid	3.76 * ulna + 75.55 ± 4.72 cm	4.27 * ulna + 57.76 ± 4.30 cm
Ulna	Negroid	3.20 * ulna + 82.77 ± 4.74 cm	3.31 * ulna + 75.38 ± 4.83 cm
Ulna	Mongoloid	3.48 * ulna + 77.45 ± 4.66 cm	
Radius	Caucasoid	3.79 * radius + 79.42 ± 4.66 cm	4.74 * radius + 54.93 ± 4.24 cm
Radius	Negroid	3.32 * radius + 85.43 ± 4.57 cm	3.67 * radius + 71.79 ± 4.59 cm
Radius	Mongoloid	3.54 * radius + 82.00 ± 4.60 cm	

In using such equations the spread of the original data is as important as the best-fit formula, since the amount of that spread measures the precision with which the actual values of height are related to bone length. Hence, what can be inferred from the length of a bone is actually a range of values, rather than a specific one; hence the +/- values in the table.

MJG

Equation 327: Triangulation

Triangulation is a general geometric method for determining the location or the dimension of an object by the measurements of angular quantities. It has been widely used in everyday life, from surveying and navigation to astrometry and metrology. The use of triangles and their trigonometric properties rooted in the wisdom of ancient mathematicians. Thales (c. 624 – c. 546 BC), measured the height of a pyramid by the length of its shadow and knowing his body height and his shadow [1]. Thales' method is based on the theorem of similar triangles, which is the same as the measurement of the height of an island by the Chinese mathematician Hui Liu (c. 225 – c. 295) [2].

Modern sensor technology takes advantages of this idea to measure geometric quantities in industrial applications. Figure 2 shows the schematic of using a one dimensional laser triangulation sensor to measure displacement of a target. When a beam of light coming out the source hits the surface of a target, the light will be reflected. The imaging lens shown in the figure collects the reflected light spot to the detector, where the position of the target is coded to the location of the reflected spot on the detector. The equation shows how the displacement of Δx is determined by the location of the image Δh as a function of the angle θ formed by the axis of imaging lens and the direction of incident beam as well as the optical magnification of the imaging lens M.

[1] http://www.anselm.edu/homepage/dbanach/thales.htm
[2] Frank J. Swetz *The sea island mathematical manual: surveying and mathematics in ancient China* 1992.

KN

Equation 328: Srinivasa Ramanujan's Infinite Series For $1/\pi$

$$\frac{1}{\pi} = \frac{2\sqrt{2}}{9801} \sum_{k=0}^{\infty} \frac{(4k)!\,(1103 + 26390k)}{(k!)^4 316^{4k}}$$

Srinivasa Ramanujan Lyengar was an Indian mathematician (1887-1920), who, with almost no formal training, made important contributions to various fields of mathematics. He started to study and develop mathematics in isolation, but was quickly recognised by an Indian mathematician. He ultimately moved to England and started a collaboration with the English mathematician, G.H. Hardy.

One of his famous results is his formula to express the inverse of pi as an infinite series. The Ramanujan's series converges to π

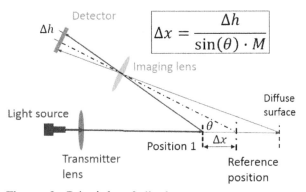

$$\Delta x = \frac{\Delta h}{\sin(\theta) \cdot M}$$

Figure 2: Principle of displacement measurement by laser triangulation.

exponentially and forms the basis of the fastest algorithm used to calculate the constant. Truncating the sum to the first term gives an approximation corrected to six decimal places. Despite its usefulness and importance, the equation was constantly voted as the most ugly equation in a study published in Frontiers of Humam Neuroscience in February 2014 [1]. If the reader is a trained expert mathematician, he/she can appreciate its deep and not immediately obvious mathematical meaning [2], otherwise the equation might help him/her to better appreciate the other 364 beautiful equations.

References

Zeki S, Romaya JP, Benincasa DMT and Atiyah MF (2014) The experience of mathematical beauty and its neural correlates. *Front. Hum. Neurosci.* **8**:68. doi: 10.3389/fnhum.2014.00068

Ramanujan's series for 1/π: A survey (with N. D. Baruah and H. H. Chan), Math. Student (Special Centenary Volume 2007), 1-24; Amer. Math. Monthly 116 (2009), 56787

http://www.math.uiuc.edu/~berndt/articles/monthly567-587.pdf

GM

Equation 329: Electrical Resistivity

Electrical resistivity is an intrinsic property that quantifies how strongly a given material opposes the flow of electric current. A low resistivity indicates a material that readily allows the movement of electric charge.

Resistivity is commonly represented by the Greek letter rho (ρ). The SI unit of electrical resistivity is the ohm·metre.

Electrical conductivity or specific conductance is the reciprocal of electrical resistivity, and measures a material's ability to conduct an electric current. It is commonly represented by the Greek letter sigma. Its SI unit is siemens per metre (S/m).

A conductor such as a metal has high conductivity and a low resistivity while an insulator like glass has low conductivity and a high resistivity.

Many resistors and conductors have a uniform cross section with a uniform flow of electric current, and are made of one material. In this case, the electrical resistivity ρ is defined as

$$\rho = R\frac{A}{l}$$

where R is the electrical resistance of a uniform specimen of the material (measured in ohms, Ω), l is the length of the piece of material (measured in metres, m) and A is the cross-sectional area of the specimen (measured in square metres, m^2).

An equation also exists in the case of a non-uniform cross-sectional area but we'll save that for another day.

Resistivity is an intrinsic property, unlike resistance. So if we take the example of copper all copper wires, irrespective of their shape and size, have the same resistivity, but a long, thin copper wire has a much larger resistance than a thick, short copper wire.

In some applications where the weight of an item is very important resistivity density products are more important than absolute low resistivity.

We discussed how the length of a material changes with temperature, well you may be surprised to learn that the same equation applies to resistivity.

The electrical resistivity of most materials changes with temperature. If the temperature T does not change too much, a linear approximation can be used

$$\rho_T = \rho_0[1 + \alpha(T - T_0)]$$

where α is called the temperature coefficient of resistivity, T_0 is a fixed reference temperature (usually room temperature), and ρ_0 is the resistivity at temperature T_0. The parameter alpha is an empirical parameter fitted from measurement data.

DF

Equation 330: Brownian Motion

$$\frac{p_i^2}{2m} = \frac{1}{2}mq_i^2 = \frac{kT}{2} = \frac{\kappa q_i^2}{2}$$

where p_i is momentum associated with a coordinate (or 'state') q_i, $k = R/N_A$ is Boltzmann's constant that is, in turn, given by the ratio of the gas constant R (8.3144598(48) J·K^{-1}·mol^{-1}) and Avogadro's number 6.022140857(74)×10^{23} mol^{-1}. T is the absolute temperature (K), m is a mass (kg) associated with the coordinate and κ is the stiffness (N·m^{-1}) associated with the coordinate.

You have finished playing guitar for the evening, put it in its case, placed it in the cupboard and gone to bed. You might have finished, but your guitar is still humming along (not very harmoniously). So why are the strings still moving? Answering this question was a puzzle fit for Albert Einstein to 'cut his teeth on' and is related to thermal agitation (Johnson noise) causing electrical noise (BE 316).

As with any topic addressed by Einstein, this is both complex to understand fully and profound in its implications. In this short essay, only a small aspect of our understanding of this phenomenon will be possible. Our starting point is to use statistical mechanics to evaluate the mean energy of a particle within a system of many particles (in our case this will be a solid material made up, of course, from atoms held together by their mutual cohesive forces). Although this analysis can include energy states due to atomic level interactions, these will be ignored and we will think of the system as being made up of particles capable of containing energy of motion (kinetic energy) held together by springs capable of storing potential energy. In general, for a system of masses and springs the total energy is called the Hamiltonian, $H(\dot{q}_i, q_i)$, of a system and, for our simple example, conveniently splits into potential and kinetic terms only. The mean energy of any particular degree of freedom can be found from

$$\frac{1}{2}m_i\langle\dot{q}_i^2\rangle = \frac{\frac{1}{2}m_i\int_{-\infty}^{\infty}\dot{q}_i^2 e^{-\frac{H(\dot{q}_i)}{kT}}d\dot{q}_i}{\int_{-\infty}^{\infty}e^{-\frac{H(\dot{q}_i)}{kT}}d\dot{q}_i} = \frac{1}{2}kT$$

Similar considerations for the potential energy of the system provide the beautiful equations at the top of this article. This provides a first insight in that for any coordinate (or 'state') of ANY system each contains, on average, an equal amount of energy. This is called the equipartition theorem. As an aside, if energy is quantized at $h\upsilon$ as assumed by Planck and Einstein, the energy, E, of the particle is given by

$$E = \frac{\sum_{n=0}^{\infty} nh\upsilon e^{-\frac{h\upsilon}{kT}n}}{\sum_{n=0}^{\infty} e^{-\frac{h\upsilon}{kT}n}} = \frac{h\upsilon}{e^{\frac{h\upsilon}{kT}} - 1}$$

which is the foundational equation used for BE 316 discussing Johnson noise in electronics.

To appreciate the implications of the above, let us return to the guitar string. Skipping a large number of steps, the solution for motion of a guitar string of mass per unit length ρ (kg m^{-1}) and subject to a tension force F_0 (N) in a fixed plane (i.e. with the string moving perpendicular to the face of the guitar) is

$$y(x,t) = \sum_{s=1}^{\infty} \left[\sin\left(\frac{s\pi x}{l}\right) (a_s \cos(\omega_s t) + b_s \sin(\omega_s t)) \right] = \sum_{s=1}^{\infty} \phi_s(x) q_s(t)$$

where x (m) is the distance along the string and

$$\omega_s = \frac{s\pi}{l} \sqrt{\frac{F_0}{\rho}}$$

are the resonances (rad·s^{-1}).

Within this summation, the term on the right is the dimensionless mode shape, φ, and represents half of an odd Fourier series representation of an arbitrary shape of period $2l$. The term on the right hand side has units of length and can be used to represent what are called the normal (or modal) coordinates, q_i, of this system. It is noted that this representation of components of a solution as being coordinates is often a barrier to understanding that, upon first exposure to the idea, many find difficult.

In this case the potential energy V and kinetic energy T_{KE} can be expressed in terms of the generalized coordinates by

$$V = \frac{F_0}{2} \int_0^1 \left(\sum_{i=1}^{\infty} \left(\frac{i\pi}{l}\right) \cos\left(\frac{i\pi x}{l}\right) \right)^2 dx = \frac{F_0 l}{4} \sum_{i=1}^{\infty} \left(\frac{s\pi}{l}\right)^2 q_i^2$$

$$T_{KE} = \frac{\rho l}{4} \sum_{i=1}^{\infty} \dot{q}_i^2$$

From this it is apparent that these are in the requisite simple quadratic form for equipartition to apply. Consequently, for each individual normalized coordinate, the mean square motion due to thermal excitation will be

$$\frac{F_0 l}{4} \left(\frac{s\pi}{l}\right)^2 \overline{q_i^2} = \frac{kT}{2}$$

$$\frac{\rho l}{4} \overline{\dot{q}_i^2} = \frac{kT}{2}$$

These immediately provide a value for the effective stiffness and mass corresponding to each coordinate. The mean square of the displacement can be obtained from the sum

$$\overline{y(x,t)^2} = \left\langle \left(\sum_{s=1}^{\infty} \varphi_s(x) q_s(t) \right)^2 \right\rangle = \sum_{s=1}^{\infty} \varphi_s^2(x) \overline{q_s^2(t)} = \sum_{s=1}^{\infty} \sin^2\left(\frac{s\pi x}{l}\right) \overline{q_s^2(t)}$$

$$= \left(\frac{2kTl}{F_0 \pi^2}\right) \sum_{s=1}^{\infty} \frac{1}{s^2} \sin^2\left(\frac{s\pi x}{l}\right)$$

417

Observing the motion of the midpoint of the string, the sine term disappears for even values of s and is equal to 1 for odd values. For the middle point of the string

$$\overline{y(x,t)^2} = \left(\frac{2kTl}{F_0\pi^2}\right) \sum_{s=1,3,5}^{\infty} \frac{1}{s^2} = \left(\frac{2kTl}{F_0\pi^2}\right)\frac{\pi^2}{8} = \frac{l}{F_0 4}kT$$

This reveals the amazing fact that through the analysis of Brownian motion we can derive expressions for the effective mass and stiffness of even macroscopic processes such as strings, beams, plates and any other structures. This extends from large scale structures down to subatomic mechanics and beyond. All that is necessary to measure stiffness at a point and in a given direction is to measure the Brownian motion at that location! For our guitar string we can determine the motion at any location x along the string by knowing its stiffness (i.e. resistance to plucking) at that location. Finally, we can write down an expression for the average (root mean square) amplitude of oscillation at any point along the string as

$$\overline{y(x,t)^2} = \frac{x(l-x)}{F_0 l}kT$$

Although this is a parabola, the instantaneous shape of the string is arbitrary or, as was the case for Johnson noise, random as if the string were being plucked by an infinite number of uncoordinated fingers. Another incredible point is that adding damping (putting your guitar in a bucket of oil) doesn't change the amplitude of these oscillations and is necessary because there must always be some external interaction necessary for the system to maintain its equilibrium temperature. Another misconception is that this jiggling motion can be modelled as collisions of particles or any other definitive source. This is not so, it is meaningless to ascribe any source other than the rather abstract concept of the states of the system itself.

Space does not permit the discussion of many other amazing aspects of this theory that started with the 1827 publication of an article by the Scottish Naturalist Robert Brown (1773 – 1858) writing on his observations of the jiggling motion of pollen on the surface of water. This led to many (doomed) speculations of the origins of this phenomena until it caught Einstein's attention nearly a hundred years later.

It must be said that this essay has become rather long and mathematically involved. Notwithstanding the complexity, it is reassuring to known that while we are sleeping all of the guitars in the world are still being played by an unimaginably large number of invisible fingers.

References

Bowling Barnes R., and Silverman S., 1934, Brownian motion as a natural limit to all measuring processes, *Rev. Mod. Phys.*, **6**, 162 – 192.

Einstein A., 1906, On the theory of the Brownian movement, *Annalen der Phys.*, **17**, 371 – 381, see also Einstein A., 1956, Investigations on the theory of the Brownian movement, Dover Publications Inc., ISBN 486-60304-0.

McCombie C.W., 1953, Fluctuation theory in physical measurements **16**, 266 – 320.

SS

Equation 331: Dynamic (Shear) Viscosity

The viscosity of a fluid can be determined as its ability or quantity of resistance to a plane plate moving (or shearing) through it with respect to the distance from a fixed plane plate. The dynamic viscosity will determine how easily

The force presented by the dynamic shear viscosity can be found with the following equation

418

$$F = \mu A \left(\frac{u}{y}\right)$$

where u/y is the local shear velocity, F is force, u is velocity, μ is dynamic viscosity (proportionality factor), y is the vertical separation dimension and A is the area of each plate.

There are some assumptions and requirements for this equation to hold true, they are:

- Flow is moving parallel to the fixed and moving plates
- The y axis is perpendicular to the direction of flow
- The fluid has to be homogeneous

Isaac Newton expressed the viscous forces as a differential equation which still uses the local shear viscosity as the proportionality factor but is expressed in a different way.

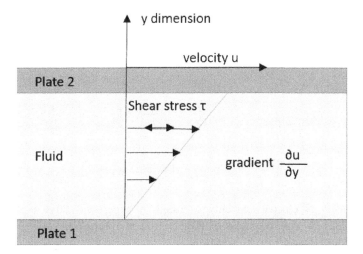

$$T = \mu \left(\frac{\partial u}{\partial y}\right)$$

The equation can be used even in nonlinear systems where the shear force does not vary linearly with the y dimension, this is due to the proportionality factor given by the local shear velocity quantity or gradient $\partial u / \partial y$.

Knowledge of dynamic viscosity forces and related fluid dynamics parameters have a wide variety of uses. Many fluid pumping applications and mechanical systems utilize fluid properties to make systems perform to prescribed criteria.

Viscosity measurement is typically done by a capillary based system that measures the amount of time it takes a fixed quantity of fluid to run through a small diameter tube at a given temperature. The measurement apparatus is chosen to give a sensitive measurement and the viscosity of a fluid can be easily and accurately obtained.

TS

Equation 332: The London Equations

The London equations give a phenomenological explanation of two fundamental properties of superconductors.

1. Zero resistance conductivity – a persistent current exists even in absence of an external electric field (when an electric field is removed).
2. Perfect diamagnetism – a superconductor expels an external magnetic field, no matter if it was switched on before or after the time the material underwent the transition to superconductor

The second property is also known as Meissner- Ochsenfeld effect, which was discovered in 1933 by the two German physicists Walther Meissner and Robert Ochsenfeld, 22 years after the discovery of superconductivity by the Dutch scientist Kamerlingh Onnes, who won the Nobel Prize for the Physics in 1913 for his discovery.

German brothers and physicists, Fritz and Heinz London in 1935, developed the London equations, when working at the Clarendon Laboratory at the University of Oxford in England, where they moved following to escape from the Nazi's racial laws.

The London equations are the following:

First London Equation,

$$\frac{dj_s}{dt} = \frac{n_s e^2}{m} E \ [1]$$

Second London Equation,

$$\boldsymbol{\nabla} \times \boldsymbol{j_s} = -\frac{n_s e^2}{mc} \boldsymbol{B} \ [2]$$

where $\boldsymbol{j_s}$ is the density of electric current generated by superconductive electrons. n_s is the density of superconductive electrons, m is the mass of the electron, c is the speed of light, $\boldsymbol{E}$ is the external electric field and $\boldsymbol{B}$ is the external magnetic field.

CLASSICAL FOUNDATION
The London equations rely on the classical theory of the electromagnetism and are built on the assumption that in a superconductor there is a specific density of electrons that behave as superconductive electrons – i.e. move across the crystal lattice without being scattered by the lattice vibrations. This density can be equal or less than the total density of electrons in the system.

The first London equation, which is a consequence of the Ohm's law for a normal conductor, states that the temporal variation of a density current of superconductive particles is proportional to the external electric field. This implies that a constant current can persist in the superconductor even when the external electric field is zero.

The second London equation can be obtained by applying the curl operator to both terms of the first equation, and combining the result with Faraday's law:

$$\boldsymbol{\nabla} \times \boldsymbol{E} = -\frac{1}{c}\frac{dB}{dt} \ [3]$$

As a result, London obtained a differential equation:

$$\frac{d}{dt}\left(\boldsymbol{\nabla} \times \boldsymbol{j_s} + \frac{n_s e^2}{mc}\boldsymbol{B}\right) = 0 \ [4]$$

This equation has two solutions:

1. A time-independent solution, which leads to a constant magnetic field different from zero that exists inside the material, due to the combination of an external magnetic field B and a magnetic field generated by the superconductive density current. This would be the case of a perfect conductor, when an external magnetic field exists before the transition of the material to the perfect conductive state.

2. Another solution, which assumes that there cannot be a residual magnetic field within the superconductive material (no magnetisation). London considered this the only possible solution for a superconductor, leading to the second London equation, Equation 2.

MEISSNER-OCHSENFELD EXPLAINED

The second London equation explains the Meissner-Ochsenfeld effect when considering the Ampere's circuital law for a superconductor:

$$\mathbf{\nabla} \times \mathbf{B} = \mu_0 \, \mathbf{j}_s \quad [5]$$

where μ_0 is the magnetic permittivity.

In fact, by using vectorial calculus properties and by re-arranging the equations according to the Ampere's law, a second order differential equation can be obtained from the second London equation:

$$\mathbf{\nabla}^2 \mathbf{B} = \frac{1}{\lambda_l^2} \mathbf{B} \quad [6]$$

The resolution of this equation in one dimension corresponds to the exponential decay of a magnetic field with the distance (x) from the surface of the superconductor:

$$\mathbf{B} = \mathbf{B}(0) e^{-\frac{x}{\lambda_l}} \quad [7]$$

This equation is the theoretical explanation of the Meissner-Ochsenfeld effect, as it predicts that an external magnetic field will penetrate the superconductor only in the region very closed to the surface and will rapidly decrease to zero when getting further inside the superconductor. The key parameter of this equation is the London's *penetration depth* (lambda), expressed as:

$$\lambda_l = \sqrt{\frac{mc}{\mu_0 n_s e^2}} \quad [8]$$

whose numerical value for an YBCO high-temperature superconductor is approximately 10-100 nm at a temperature approaching 0 K. The corresponding density of superconductive carriers is about 10 to the 28 / cubic metre, which amounts to the total electron density of the material.

As temperature increases, also London's penetration depth increases and the density of superconductive electrons reduces to zero, which will occur at the critical temperature for the superconductive transition. At this temperature, superconductivity is broken and the material is back in its state of normal conductor.

DIFFERENCE BETWEEN PERFECT CONDUCTOR AND SUPERCONDUCTOR

A perfect conductor behaves differently from a superconductor from the point of view of their magnetic properties. If an external magnetic field is switched on, after a perfect conductive metal and a superconductive metal have been cooled down to a temperature below the critical transition, the two materials will show similar magnetic properties. In particular, there will be currents on the surface of the material, generating a magnetic field that opposes the external field, so that the total internal magnetic field will be zero. For a perfect conductor this is in agreement with Lenz's law, stating that a variation in the magnetic field generates a current with a direction that counteracts the change in flux of the magnetic field.

However, a perfect conductor changes its behaviour in the case of an external magnetic field that is switched on before the metal undergoes the transition. In this case, once the metal underwent the transition to perfect conductor, the induced current will generate a field that freezes the value of magnetic field inside the material, in order to oppose a variation in the change of magnetic flux through the material (in agreement with Lenz's law). If the external field is subsequently removed, the current inside the perfect conductor will persist and a uniform magnetic field will be present within the region around the perfect conductor. On the contrary, a superconductor will have superficial currents which generate a magnetic field that will act against the external magnetic field so that the internal magnetic field will be zero (Figure 1).

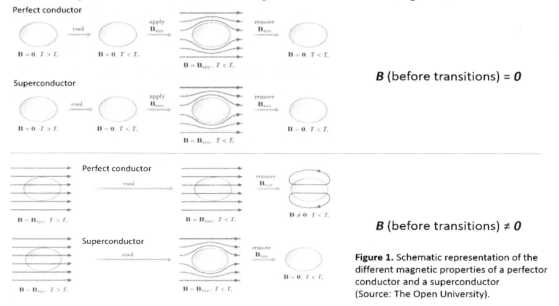

Figure 1. Schematic representation of the different magnetic properties of a perfector conductor and a superconductor (Source: The Open University).

Figure 1. Schematic representation of the different magnetic properties of a perfect conductor and a superconductor (Source: The Open University).

MEISSNER EFFECT AND MAGNETIC LEVITATION
One of the consequences of the Meissner effect is the levitation of a superconductor over a magnet (or vice versa), resulting from the expulsion of that the external magnetic force field lines by the superconductive material. However, this type of levitation would not be particularly stable and the superconductor would tend to fall because of the repulsive forces from the magnet (or vice versa). In several practical demonstrations showing the levitation of a thin superconductor over a magnet or a magnetic track, a slightly different phenomenon takes place, which is the result of purely quantistic effects in superconductors. Scientists also call this phenomenon "quantum levitation or quantum locking".

In quantum levitation, the superconductor is locked into a stable position over the magnet. If we tilt the superconductor at an angle with respect to the magnet (or a magnetic track), it will retain this position while levitating. Moreover, if the magnet and the superconductor are flipped upside down, the superconductor will remain locked in its position under the magnet, without falling on the floor as we would expect from the action of the magnetic force and the gravity force. The phenomenon is described sometimes as if there was an imaginary magnet on the other side of the superconductor, which balances the forces exerted by the magnet. As illustrative case, see this video, https://www.youtube.com/watch?v=Ws6AAhTw7RA.

TYPE-I AND II SUPERCONDUCTORS AND QUANTUM LEVITATIONS
An external magnetic field can break the superconducting behaviour of a material, restoring its normal conductivity when the magnetic field is equal or greater than critical value of the magnetic field. According

to their magnetic properties, superconductors are classified as type I and type II. Type I superconductors undergo a sudden transition in the internal magnetic field, which leaps from zero to a linear behaviour for external fields greater than critical value. Type II superconductors have a lower and a higher critical magnetic field. When an external magnetic field is in between the lower and higher critical values, it partially begins to penetrate the interior of the superconductor, while the material retains its superconductivity (Figure 2).

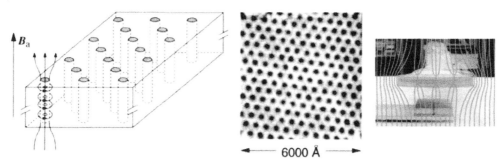

Figure 2. Schematic representation of "fluxons" (or "flux tubes") in a type-II superconductor (Source: Wikipedia; Joe Khachan and Stephen Bosi).

This type of behaviour allows quantum levitation to occur, as the partially penetrating magnetic field acts in such a way that it locks the superconductor on the magnet. In particular, the flux of the external magnetic field through the material is quantised – the term used by physicists is "fluxons". Fluxons, also called flux tubes or vortex states, are essentially regions of normal conductivity within the superconductive material (Figure 2), which allow the penetration of external magnetic field, when its intensity is in between the lower and higher critical values. The movement of the magnet producing the external magnetic field would make the vortex filaments move; however, crystal defects and grain boundaries in the material stop the motion of the vortices, determining a "flux pinning". On a 76 mm diameter and 1-micrometer thick superconductive disk, within a magnetic field of 350 Oersted, there are about 100 billion flux tubes that can hold 70,000 times the superconductor's weight. This phenomenon provides the "locking" effect, hence the stability of the superconductor over the magnet. The thinner the superconductor, the stronger the pinning effect when exposed to magnetic fields. Several companies in their effort to develop faster and more efficient transportation, such as hovercraft-like transportation systems and trains, have investigated quantum levitation.

THE BCS THEORY
The flux quantisation of an external magnetic field through a type II, thin superconductor is a quantistic effect and it originates from a generalisation of the second London equation, within the Bardeen-Cooper-Schiffer theory (BCS), which explains the phenomenon of superconductivity at low temperatures. The three scientists were awarded with the Novel prize for physics in 1972 for this theory. The basic concept of the BCS theory states that at temperature below the critical temperature of superconductivity two electrons can undergo a long-range attractive interaction, by coupling with phonon lattice vibrations in the crystal. In a normal conductor, the phonon vibrations, which cause the scattering of electrons and therefore the origin of resistivity; however, in a superconductor at very low temperatures, the vibrations are extremely slow and can actually favour an attractive interaction of two electrons with opposite momentum and spin. This interaction results in the formation of several electron-electron couples, also known as Cooper pairs, which can be modelled as a new particle with twice the mass and charge of an electron. A more important consequence is that the total spin of these new particles is 1 and no longer ½, which means that they are now Bosons and no longer Fermions. At temperatures below the critical temperature for superconductivity, all the bosons collapse in the same energy state, so that a macro wave function can be associated to this state. The application of the London and Ampere's laws to this equation leads to the theorisation of fluxons, which fully explain the phenomenon of quantum locking of superconductors in an external magnetic field.
AT

Equation 333: Ultrasound Thermal and Mechanical Index

Ultrasound has been used as a diagnostic tool in medicine since 1942. While used on a variety of patients for many reasons, ultrasound has become the main tool for visualising the foetus in the womb of pregnant women. This is known as obstetric ultrasound. Like all tools used in hospital and medical research centres, the question of whether ultrasound is safe for the baby (or babies) is one that many expecting mothers (and fathers) will be asking.

Medical ultrasound works in the same principle as ultrasound used in NDT environments. High frequency sound waves propagate into the tissue from a probe (the probe acts as both transmitter and receiver) and is reflected from interfaces to generate an image. It also has the advantage that it does not use ionising radiation.

However, ultrasound does have potential harmful effects on the unborn baby (or babies). There are two potential causes of foetal damage using ultrasound:

1. Thermal
2. Mechanical

THE THERMAL INDEX (TI)
The propagating ultrasonic beam into the tissue is partially reflected and partially absorbed. The absorbed energy (intensity dependant) into the tissue causes heating, raising the tissue temperature. The thermal index is given below and is defined as the ratio of the power used to visualise the foetus to the power required to produce a temperature rise of 1°C.

$$TI = \frac{W_p}{W_{deg}}$$

where W_p is the acoustic power at the depth of interest and W_{deg} is the power required to raise the temperature of the tissue by 1°C.

The thermal index is a linear scale and a value of $TI = 1$ will result in a temperature rise of 1°C. Current guidelines suggest this is the maximum value to be used in obstetric ultrasound.

The thermal index is further split into three categories:

1. Thermal index in soft tissue (TIs)
2. Thermal index in bone (TIb)
3. Thermal index in the cranium (TIc)

In liquids such as embryonic fluid, the scattered ultrasonic energy dissipates in all directions longitudinally. In bone however, transverse shear waves are generated on the surface, spreading to surrounding soft tissue. This becomes a major concern to the developing brain and spinal cord especially in the third trimester when mineralisation of the bones occurs.

THE MECHANICAL INDEX (MI)
The mechanical index gives an estimate of the risk of non-thermal effects on the foetus, such as cavitation. It is defined as the ratio of the maximum negative pressure of the ultrasound wave to the square root of the frequency and is given in the second equation.

Ultrasound is a mechanical wave and the negative pressure component of the wave can form bubbles which in turn collapse. In extreme cases, the collapsing bubbles can cause large shockwaves. It has also been estimated that while an MI value is given at the time of a scan, the true value can be as much as 90% lower, taking into account system accuracy and steering angles.

The thermal and mechanical indices do not take exposure time into consideration. According to the British Medical Ultrasound Society (BMUS), a foetus under 10 weeks must have TIs and MI values monitored and given an index of 1.0, the exposure time cannot exceed 30 minutes. After the foetus is 10 weeks, TIb and MI values must be monitored, and again given an index of 1.0, the exposure time cannot exceed 30 minutes. The mechanical index is given as

$$MI = \frac{P_N}{\sqrt{F_c}}$$

where P_N is the maximum negative pressure (in MPa) and $\sqrt{F_c}$ is the centre frequency of the ultrasonic wave (in MHz).

The image shows a typical ultrasound scan at nine weeks, in this case showing twins, Killian and Grayson McKee. Values of TIs and MI are given as 0.1 and 1.0 respectively.

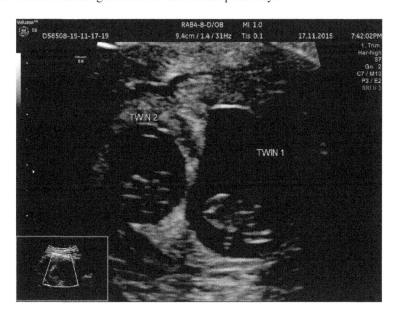

T. A. Bigelow, C. C. Church, K. Sandstorm, J. G. Abbott, M. C. Ziskin, P. D. Edmonds, B. Herman, K. E. Thomenius and T. J. Teo, "The Thermal Index: Its Strengths, Weaknesses, and Proposed Improvements," J. Ultrasound Med., vol. 30, p. 714, 2011.
N. de Jong, "Mechanical Index," Eur. J. Echocardiography, vol. 3, pp. 73-74, 2002.
"British Medical Ultrasound Society," [Online]. Available: https://www.bmus.org/.

CM

Equation 334: Young's Double Slit Experiment

The double-slit experiment is a foundational standard of physical optics. The apparatus is a classical division of wavefront interferometer, in which the interference fringes appearing on the screen at the right-hand side of the figure above result from the difference in path length of the two rays A and B as a function of inclination angle. This very simple experiment illustrates many profound ideas, one of which is expressed by the beautiful equation $\lambda = \Delta x\, z/D$, which says that we can measure the wavelength λ of light simply by measuring the spacing Δx of interference maxima on the screen placed a distance z away from two slits spaced by D. Another bit of physics is the role of spatial coherence, which in the figure above is controlled in part by the single slit to the left.

The double slit experiment is associated with Thomas Young (1773 – 1829) and the debate about the nature of light that took place around 1800 regarding the true nature of light. The prevailing belief at that time was that light behaved ballistically, like a swarm of tiny cannon balls. Isaac Newton was responsible for the stubborn dominance of this "corpuscular" idea, since he seemed to have been right about almost everything, and wrote a nice book about optics in 1704, famous today for its discussion of prism rainbows and for the quaint spelling of "Opticks." The odd behavior of light in diffraction and interference experiments was attributed to "fits" or to the effect of particles bouncing off the edges of slits. So ponderous was Newton's scientific opinion that the particle theory suppressed Huygens wave theory, already developed in 1680 but out of favour thanks to Isaac.

Young devised and performed an experiment to measure the wavelength of light to settle the matter. In his 1804 paper entitled "Experiments and Calculations Relative to Physical Optics," he describes passing sunlight through a pinhole and directing the pinhole beam to a paper card, splitting the beam into two beams, in what amounts to a double-slit geometry. These two beams interfere when projected onto a screen, creating one of those patterns that are difficult to explain without relying on Huygens' wave theory. From the spacing of the interference maxima, Young was able to measure the wavelength of light—a truly remarkable achievement with such a simple apparatus. A modern version of this experiment measures the width of a hair using a laser pointer.

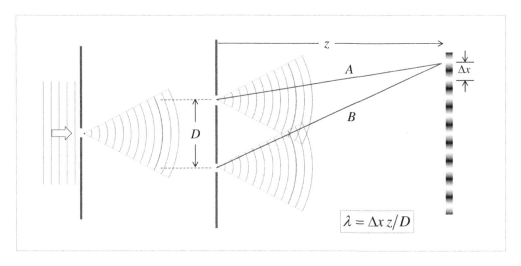

$$\lambda = \Delta x\, z / D$$

In more recent times, experiments at the limit of extremely low light levels have shown that the light distribution remains the same, even though only one photon is present in the apparatus at a time. The quantum-mechanical explanation is that the *probability* distribution for the location of a detected photon at the screen is governed by the classical wave results described by today's beautiful equation, even though the light energy is carried by a single photon. This insight explains why it has been possible to repeat the double-slit experiment with electrons, atoms and even some molecules. Speaking about the experiment, Richard Feynman wrote in Chapter 37 of his famous lectures: "We choose to examine a phenomenon which is impossible, absolutely impossible, to explain in any classical way, and which has in it the heart of quantum mechanics. In reality, it contains the only mystery."

PdG

Equation 335: Lovelock's Biological Homeostasis

The daisy population change is

426

$$\frac{da}{dt} = a(x\beta - \gamma)$$

where a is the population being assessed and γ is the death rate.

β, the fitness function is

$$\beta = 1 - 0.003265(22.5 - T_l)^2$$

where T_l is the local temperature of the daisy and can be replaced by w or b to denote white daisy temperature or black daisy temperature

x the total space left on the planet

$$x = p - (a_w + a_b)$$

where the subscripts w and b refer to white daisies and black daisies respectively

Local temperature is

$$T_l = q'(A - A_l) + T_e$$

where q' is a positive constant that defines the rate of conductivity of temperature across the planet, A is the effective albedo of the planet taking into account daisy population, A_l is the albedo of the daisy species of interest and T_e is the planet's temperature determined by

$$\sigma(T_e + 273)^4 = SL(1 - A)$$

where σ is Stefan's constant, S is the solar output of the sun in flux and L is solar luminosity.

Equations from "Biological Homeostasis of the Global Environment: The Parable of Daisyworld", by Andrew J. Watson and James E. Lovelock, Tellus 1983

Gaia theory is based on the premise that you can model a planet as a single, living organism, in much the same way as we model a collection of cells as a single macroscopic lifeform. This encourages analysis at the system-level, asking that actions be considered not just from the perspective of a single species or biome, but from the perspective of the health of the planet.

This model has been championed by environmentalists and scientists alike as a way of viewing Earth. Gaia Theory's creator, James Lovelock, proposed that one way that a planet is akin to a single creature, is the presence of homeostatic systems. All macroscopic creatures have homeostatic systems i.e. regulatory systems that maintain their internal environment to ensure that the cells that they are formed from can survive. One example of a homeostatic system is the way warm blooded animals maintain their body temperature, making their internal environment suitable for the various cellular fauna that live within it.

Lovelock drew parallels between homeostasis in animals and the atmospheric conditions of planets. On "dead" planets the atmosphere is ruled by chemical equilibrium, with little variation in overall atmospheric conditions without some external event, such as the ever increasing thermal output of the sun, disturbing that equilibrium. On Earth different species of bacteria and algae have a huge impact on the concentrations of oxygen, nitrogen and carbon dioxide in the atmosphere, resulting in complex, dynamic cycles. Lovelock suggested that an emergent property of the interaction between atmospheric gases and life on the planet results in a form of regulatory system: homeostasis on a planetary scale. This model has come under a lot of criticism, not least because it is difficult to generalise about the effect of life on planets when we only have a sample size of one!

To illustrate how life could have an impact on environmental conditions, Lovelock and the mathematician Andrew Watson built a simplified model that could be explored and analysed. Not only does this illustrate the concepts that Lovelock was trying to communicate, but it also highlights the power of mathematics. With these few equations, it is possible to tell a story about life on a planet. The story is not detailed, but a caricature of reality that can explore themes and concepts that have a real impact on our world.

The equations describe "Daisyworld". On Daisyworld the soil is perfectly grey, reflecting and absorbing heat from the sun in equal measure. The soil is permeated with the seeds of two species of daisy, black daisies and white daisies. In this model, the only thing that a daisy needs to survive is space on the planet and the right temperature. Like all stars, Daisyworld's sun gets hotter as it ages, providing external change to this environment.

By incorporating these equations into a simulator we can use numerical integration to evaluate the differential equations for population (see the BE about Euler Integration) and play out this scenario. Figure 1 shows the populations of black and white daisies over time.

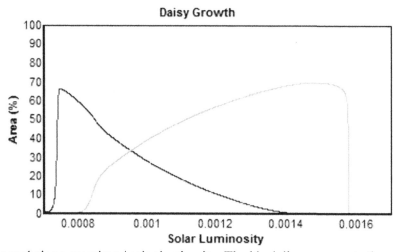

Figure 1. Daisy populations over time / solar luminosity. The black line represents the population of black daisies and the grey line represents the population of white daisies. Both are expressed as a percentage of the planet's total area.

From Figure 1 we can see that eventually the solar output from the sun makes the planetary temperature suitable for life. At this point, when a white daisy grows, it reflects solar energy, lowering the temperature of the world around it. However, as the planet is precariously sitting on the threshold of being too cold for life, the cooling effect of the white daisy kills it. Conversely, black daisies absorb heat, making the world around them warmer and pushing the local temperature from being "just warm enough" towards "perfectly warm" as far as daisies are concerned! This allows the black daisies to thrive. As the centuries move forward, the combination of increased solar output and the warming effect of black daisies, means that some white daisies can grow and survive, although they must compete with black daisies for the limited space on the planet. If too many white daisies grow, their local temperature falls too far and they die, but small numbers can survive. Fast forward many centuries and now black daisies are struggling to survive in the excessive heat of the sun but the reflective properties of white daisies allows them to thrive. Eventually the solar output of the sun means that not even a planet full of highly reflective white daisies is enough to cool the global atmosphere, and all life on the planet dies.

The important result from this experiment is to compare the average temperature on the planet with another, identical planet with no life. In Figure 2 the temperatures of the two planets are compared.

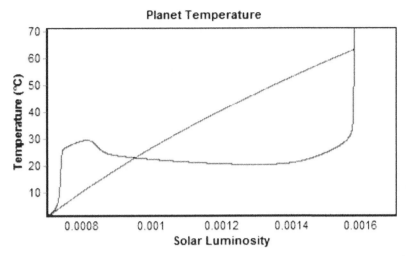

Figure 2. The average temperature of Daisyworld (in red) and a dead planet of the same size and colouring (in green)

The dead planet's temperature is entirely dependent on solar output. As a result it spends relatively little time in the temperature region where life could survive. However, Daisyworld's temperature is partially regulated by the reflectivity of the daisies and has a relatively stable temperature, staying in the optimal region for life for many centuries.

It's important not to take Daisyworld too seriously as a model for Earth. The factors for competition, the number of variables that need to be controlled and the complexity of the physics make for a much more difficult analytical problem. However, as a parable, Daisyworld illustrates that without any centralised information or conscious effort, it is possible for a form of control to emerge from the interaction between life on a planet.

Many thanks to John Wesley for permission to use his Daisyworld Simulation tool for the graphs in this article.

RFO

Equation 336: The Klein-Nishina Formula

$$D(\theta) = \tfrac{1}{2}\, \alpha^2 r_c^{\,2} \mathrm{p}^2 (\mathrm{p} + p^{-1} - \sin^2\theta)$$

An artist or musician may not be noted for their physical attractiveness, yet may produce very appealing art or music. In this sense, one may be justified in speaking of a 'beautiful' artist. By the same token, for someone glancing at the Klein-Nishina formula's algebraic representation, the phrase 'beautiful equation' might not spring immediately to mind. However, when espying the equation's graphical representation, they may well concur that the image is aesthetically pleasing as well as being informative.

The Klein-Nishina formula gives what is perhaps confusingly called the 'differential cross section' for Compton scattering - denoted as $D(\theta)$ above - of photons scattered at an angle θ from a single free electron. $D(\theta)$ is calculated using lowest order quantum electrodynamics. At low frequencies (e.g., visible light or UV), this formula converges to the Thomson formula for differential cross section.

The formula was derived by Swedish Oskar Klein and Japanese Yoshio Nishina. For low-energy photons (e.g. 2.75 eV in the diagram), it demonstrates almost symmetrical scattering on either side of any direction at 90° to the beam direction. In the diagram, 0° and 180° represent the directions of unscattered and totally back-scattered photons respectively.

Klein-Nishina formula

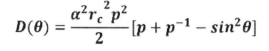

$$D(\theta) = \frac{\alpha^2 r_c^{\,2} p^2}{2}[p + p^{-1} - sin^2\theta]$$

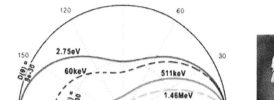

Image: Adapted from DScraggs, Wikimedia Commons

Oskar Klein
1894-1977

Yoshio Nishina
(仁科 芳雄)
1890-1951

where

$D(\theta)$ is the so-called 'differential cross section'

θ is the scattering angle

α is the fine structure constant

r_c is the reduced Compton wavelength of the electron $\frac{\hbar}{m_e c}$, $\hbar$ being (the Planck constant)/2π

$$p = P(E_\gamma, \theta) = \{1 + (\frac{E_\gamma}{m_e c^2})(1 - cos\theta)\}^{-1}$$

is the ratio of the energy of the photon, scattered at angle θ, to its incident energy

E_γ is the incident photon energy

m_e is the mass of an electron

c is the speed of light *in vacuo*

With increasing photon energy (e.g. 60 keV), forward scattering predominates; for energies more than five times the rest energy 511 keV of the electron, the scattering is almost entirely confined within a forward angle of 30°.

The differential cross section, here denoted D(θ), is much more often represented by the symbol $d\sigma/d\Omega$, where σ and Ω indicate cross-section and solid angle respectively. Some authors have pointed out that as dσ/dΩ in the formula does not actually correspond to a derivative in the context of calculus, its continued use is misleading.

In the formula above, α is the fine structure constant, r_c is the reduced Compton wavelength of the electron, and p is the ratio of the energy of the photon, scattered at angle θ, to its incident energy.

FR

Equation 337: Einstein's Field Equations of General Relativity

$$G_{\mu\nu} = R_{\mu\nu} - \frac{1}{2}g_{\mu\nu}R = \frac{8\pi G}{c^4}T_{\mu\nu}$$

This beautiful equation (or set of equations) marks a centenary at the time of writing this article. Einstein completed his tensor field equations for general relativity in a series of lectures presented at the Prussian Academy of Sciences in November 1915. In the last of these lectures on 25 November 1915, he presented the final corrected version of the equations, and these were published on 2 December 1915 in the Proceedings of the Academy, entitled "Zur allgemeinen Relativatstheorie". It had taken him eight years of hard work which arose from an idea that occurred to him in 1907 called the "principle of equivalence", a couple of years after he had presented his special theory of relativity.

Two main principles are required for general relativity which is a relativistic theory that takes gravity, and hence accelerating frames of reference, into account. The first is special relativity (this relates to inertial reference frames only) which results in changes in space and time (contraction and dilation) depending on the relative speed of motion of observers, and its overall consequence of mass-energy equivalence, which we will take as read (refer to BE 8 and BE 74). The second is the principle of equivalence. This principle implies that, in a local reference frame, often depicted in a thought experiment as an elevator or laboratory on the ground or accelerating in free space, the effect of a gravitational force (e.g. from the earth, or more specifically the reaction force by the ground against it, equivalent from Newton's third law of motion) is indistinguishable from the effect of a force that would be experienced by the observer in a non-inertial frame of reference with an equivalent acceleration (g in the earth's case). Note that in freefall, the effect of gravity is similar to that experienced by an observer in an inertial frame, resulting in equivalence between gravitational and inertial mass. This is a profound difference from the previous picture: from Newton's point of view, a body at rest on the ground was inertial and freefall was not. For the principle of equivalence to work, there must be no effective tidal forces: the frame must be small compared with any variation in gravity (gravitational field of the massive object such as the earth). This allows for calculations to be made in which the effect of gravity can be assumed to be equivalent to a unidirectional acceleration. An immediate consequence of this principle (or more specifically the "strong" equivalence principle which applies to all the laws of physics) is that if these two scenarios are equivalent locally, and a passing photon emitted within the accelerating reference frame is seen by its observer to have a slightly curved path, the same must be true for the reference frame on the ground experiencing gravity. (Our view from the outside, of course, is a straight path for the photon and a movement of the observer and reference frame). Since photons, massless particles, are expected to follow the straightest path possible between two points, Einstein's conclusion was that their path will be curved in a gravitational field. This curvature is very small in the presence of a relatively weak gravitational field such as that of the earth or the sun, but significant in a very intense gravitational field. In these extreme cases there are other consequences such as gravitational time dilation and the frequency shifting of light between observers at different locations in a gravitational field. There are very small changes that can be observed on earth compared with in orbit, for instance. Many interesting consequences such as differences in ageing relative to motion, a restricted type of time travel, are often speculated on, as they have been in science fiction for decades, but the subject of this article is about the field equations themselves and how they work. Physically important consequences of the theory, such as gravitational waves, black holes, the big bang and dark energy, have been the subject of intense study.

The principle of equivalence is valid for a small mass in a local region and is not valid if compared with someone, for example, on the other side of the earth within the same frame – there is no such thing as a global inertial frame - clearly a more general theory is required for universal gravitation, however the equivalence principle is a good enough approximation for locally observed effects to infer that the spacetime continuum is in some way curved by the presence of matter. This spacetime curvature is the groundwork for Einstein's field equations, which had to be formulated in such a way as to be able to perform important calculations such as vector calculus in a curved geometry: they provide a direct relationship between matter and the geometry of spacetime. The 2-way process is encapsulated in a well-known comment by the American theoretical physicist John Archibald Wheeler: "spacetime tells matter how to move, and matter tells spacetime how to curve" – effectively, gravity and spacetime curvature are the same thing. Newton's gravitational theory assumed a flat, unaffected environment of space and time, an independent backdrop along which all motions and forces could be calculated. This is applicable to a large degree in many practical calculations, even involving spacecraft missions, but becomes inaccurate in regions of high density and hence strong gravity.

The field equations in their simplest form are

$$G_{\mu\nu} = \frac{8\pi G}{c^4} T_{\mu\nu}$$

431

This is a mathematically compressed description, and actually represents 10 second order partial differential equations (quasi-linear) in a 4-dimensional spacetime. The terms with double subscripts on the left and right are tensors which are described below. Tensors, here written in abstract index notation, are multidimensional mathematical objects that describe linear relationships between vectors, scalars and other tensors: linear operators that can be represented in a multidimensional array of numerical values. The number of dimensions is generally described as the order or rank: in this general concept, a scalar is itself a 0 order tensor (a single number), and a vector (a one-dimensional row or column of numbers) is an order 1 tensor. A 2-dimensional matrix or linear map is order 2. The tensors used here are constructed in such a way that general covariance is followed: it is important that the laws of physics remain equivalent in all reference frames. The tensor is defined at a point in spacetime – in the case of curved geometry this is associated with a "tangent" space at the point on the curved surface.

The equations above relate to a concept in differential geometry: a manifold. A manifold is a (typically) smooth multidimensional space that is equipped with a "metric". The metric is what the equations are to be solved for, this is the "field" quantity that is analogous to the Newtonian gravitational potential, and fully describes the curved space. Tensor fields on a manifold map a tensor to each point. The geometry of curved surfaces (non-Euclidean geometry) had been discovered by Gauss in the early 19th century, but it was Bernhard Riemann, a student of Gauss at Göttingen, who extended the process to higher dimensional spaces in the 1850s. In 1912, in his search for a set of equations that would describe how matter influences spacetime curvature and vice versa, Einstein began to make use of Riemannian geometry to replace the flat Minkowski spacetime, which he had previously used for special relativity in the absence of gravity, with a curved spacetime. The straight "world lines" in Minkowski space are generalised as "geodesics" – the paths of shortest distance between two events in curved spacetime. These represent a kind of inertial motion in general relativity: in the absence of any other external forces, natural motion is along these geodesic lines. In special relativity, parallel geodesics never cross paths; in general relativity they typically do: they can be likened to lines of longitude on the earth which are parallel at the equator but meet at the poles, although they still represent the shortest distance between two points on the earth's surface. In deriving his general relativity Einstein used the concept of a "test particle" (of negligible mass) that moves along a geodesic. A geodesic equation, together with the field equations which define the curvature, is used to predict the motion of a test particle in a gravitational field.

The tensor on the right is usually called the stress-energy tensor (more accurately the energy-momentum tensor, which contains components of stress). This is a symmetric order 2 tensor which contains the flux of the components of the momentum density vector across a surface. It represents the source of the gravitational field, and from special relativity is not due to matter alone (as in Newtonian mechanics). It can be seen as being analogous to Maxwell's charge-current vector J in electromagnetism. The first term (time-time index) is equivalent to the density of relativistic mass ($E_{density}/c^2$): in general relativity, the "4-momentum" vector is used – it includes the energy density in the first term. The momentum density is equal to the energy flux/c^2: beyond the first row/column, the diagonal spatial coordinate terms are the normal stress, or pressure (the density of the spatial components of linear momentum), and off-diagonal components are shear stress (like a typical stress tensor, eg as encountered in fluid dynamics) in spacetime. The tensor basically does a mapping of one type of vector to another: it gives us the momentum passing through a surface that is defined by the position/time vector.

The tensor on the left is called the Einstein tensor - this represents the spacetime curvature using a pseudo-Riemannian manifold (the "pseudo" description means that the metric tensor does not have to be positive definite). Einstein had to try out several versions before he achieved the consistency and conservation of energy and momentum that was required. The values of $G_{\mu\nu}$ and $T_{\mu\nu}$ need to be made equivalent (apart from the constant) for the same event in spacetime. They are only possible in differential form: there is no integral conservation law for the energy-momentum tensor. The Einstein tensor is an abbreviated notation, a function of the metric and its derivatives, and can be separated out on the highest level as

$$G_{\mu\nu} = R_{\mu\nu} - \frac{1}{2}g_{\mu\nu}R$$

where $R_{\mu\nu}$ is called the Ricci tensor and $g_{\mu\nu}$ the metric tensor. The ½ term arises from differential calculus. The metric tensor is the result of solving the field equations and can be written as a 4x4 symmetric matrix which has 10 independent components. The metric describes the geometric properties of the spacetime, allowing for local concepts such as distances and angles to be represented on the manifold. For example, in the more familiar flat surface-based Cartesian coordinate system, the distance (metric) between two points is given by

$$s^2 = \Delta x^2 + \Delta y^2$$

which is basically a re-statement of the Pythagorean theorem (BE 11). In spherical polar coordinates (see for example BE 198) or any other curved surface, this is not the case and we need a generalised metric which takes into account the transformations required, so that for example between two vectors

$$s^2 = g_{\mu\nu}v_\mu v_\nu$$

In a 3-D Cartesian coordinate system this is simply

$$g_{\mu\nu} = \begin{pmatrix} 1 & 0 & 0 \\ 0 & 1 & 0 \\ 0 & 0 & 1 \end{pmatrix}$$

which is equivalent to the Kronecker delta function $\delta_j^i = \begin{cases} 1, & i = j \\ 0, & i \neq j \end{cases}$.

The Ricci tensor is a reduced version of the more complex Riemann tensor, a rank 4 tensor with 256 components which can be investigated if more detail on the curvature of the manifold is required, and this is what Einstein started with in his derivations. More specifically, the Riemann tensor is "contracted" by cancelling the first and last indices, which is equivalent to taking the "trace" of a rank 4 tensor (elements along the central diagonals). The Ricci curvature can be described as the amount the volume of a geodesic sphere in the curved manifold deviates from that in the flat case – in general it reduces volume with the strength of the field. It is itself a complex function of the metric and its derivatives, including components that ensure that the derivatives are covariant. Covariant derivatives satisfy the transport of parallel lines between spaces for differentiation of vector fields. The last term, R, is called the scalar curvature – a further contraction of the Ricci tensor - relating to a local radius of curvature for the manifold.

The Einstein field equations by themselves do not predict motion, but the equations of motion can be derived using them. These are called the "geodesic equations". Without going into detail, the process is as follows: the field equations are solved for the metric; functions called "Christoffel symbols" appear as functions of the metric; finally, the geodesic equations are solved to obtain the path of accelerated motion of a test particle, which depends on its velocity. The path of time-like curves (bodies travelling at less than the speed of light) can be calculated – in fact the near-circular orbits of the planets around the sun in space can be reproduced from the geodesics being followed in curved spacetime projected onto 3-dimensional space.

There are a few important points to conclude with here:

1. The maths Einstein had to use to formulate his general relativity was very complex, particularly the Riemannian geometry, and was therefore unlikely to have been achieved without the help of others, in particular Marcel Grossmann and David Hilbert.
2. In 1916, the first exact solution to the field equations (other than the trivial flat one) was found by Karl Schwarzschild (see BE 4) for a spherical massive object in free space. This resulted in a limiting radius describing the event horizon of a black hole. It can also be used as an approximation for

many slowly rotating objects such as planets and stars like the earth and sun. In a region of empty space in which there is no mass, $T_{\mu\nu} = 0$ and it follows that $R_{\mu\nu} = 0$, i.e. there is no Ricci curvature. This is the "vacuum" equation, or "Ricci-flat" condition. The Ricci curvature is closely identified with the source as we know, but this does not mean the entire Riemann space is flat. The "Ricci part" is associated with volume change (although in some special cases, freely moving massive particles, it can have a distorting effect on time-like geodesics). If we remove the Ricci component from the Riemann we extract something called the Weyl or conformal tensor. Having removed the source, this provides the background degrees of freedom of the gravitational field itself, and leads to the prediction of gravitational waves, which are still being searched for in modern astronomical experiments.

3. In 1917, Einstein introduced a term in his equations called the cosmological constant, expanding to the following form

$$R_{\mu\nu} - \frac{1}{2} g_{\mu\nu} R + \Lambda g_{\mu\nu} = \frac{8\pi G}{c^4} T_{\mu\nu}$$

This was introduced as a tiny adjustment (only observable on cosmological scales) to allow for a "steady-state" (static spatially closed) universe. It became clear in 1929 from Edwin Hubble's observations that the universe was expanding (BE 166) and Einstein withdrew it, referring to it as his "greatest mistake". However, recent observations of supernovae suggested that the expansion of the universe is accelerating and have led to physicists re-introducing a constant similar to Λ to account for the suspected "dark energy" repulsive component of expansion on a cosmological scale.

4. Two tests of Einstein's theory were conducted in 1919: the slight predicted changes in the perihelion of Mercury (the part that is not related to the pull of the other planets or the non-sphericity of the sun, which had been known about since 1859) and the gravitational deflection of light by the sun. Einstein used his theory to predict the extra few seconds of arc per century of perihelion and came up with good agreement, and Arthur Eddington tested the gravitational bending of light by the sun by simultaneous observations of star positions during a solar eclipse in Brazil and Africa. Despite its effect on the popularity of general relativity the accuracy of this first test was not good, but similar tests were repeated many times since, and eventually had to be accepted when the experiments were made possible at radio frequencies. Many other tests (including those for gravitational redshift) have continued to agree with general relativity.

5. General relativity, as a classical theory, which is smooth and continuous, has issues with compatibility with quantum theory on very small scales. In quantum mechanics particles interact by the exchange of other particles. The renormalisation that works for the other three forces does not work for gravity. It has been speculated that spacetime may become discontinuous at around the Planck scale (BE 215). The fundamental particle of gravity, a spin-2 massless particle called the graviton (as yet undiscovered), which is different from all the other force-mediating particles (gauge bosons), is present in theories of quantum gravity such as superstring theory and loop quantum gravity - these require a larger number of spatial dimensions to work. In this scenario, the quantum mechanics of interacting gravitons could potentially give rise to general relativity and space curvature. Physical conditions at which general relativity and quantum mechanics are expected to overlap are regions of extreme density such as black holes and the very early stages of the universe. In these cases, there could be a singularity (an infinitesimally small condition of infinite density). In addition to the 100 years since Einstein's field equations were presented, it is also exactly 50 years since the first rigorous black hole singularity theorem using general relativity was presented by Roger Penrose while at Birkbeck College, now considered to be a milestone in the revival of general relativity theory. Shortly afterwards, Penrose and his associate Stephen Hawking developed a theory that indicates that an expanding universe has its origin in an instantaneous singularity: an infinitely dense state which we call the big bang.

PB

434

Equation 338: Coriolis Effect

The Coriolis Effect shows the deflection of a moving object relative to a rotating reference frame. Although noted by others as early as 1651, it was Gaspard-Gustave de Coriolis (21 May 1792 – 19 September 1843), a French mathematician, mechanical engineer and scientist, who developed the, mathematical expression in 1835 while working on the theory of water wheels.

In a rotating reference frame, the Coriolis Effect acts like a real force, i.e. it causes acceleration and has real effects. This is demonstrated in the given equations. The Coriolis Effect shows the deflection of a particle from its path within a rotating reference frame. The particle only appears to deviate from its path because of the motion of the coordinate system. A particle moving in the longitudinal direction, i.e. from pole to pole will appear to be deflected westward. There are two reasons for this:

1. The Earth rotates eastward.
2. The velocity of a point on the Earth is a function of latitude, i.e. the surface of the Earth moves faster at the equator, and not at all at the poles.

To illustrate this, if a particle is fired in a northerly direction from the equator, it would land to the east of its target point. This is because the particle is moving eastward faster at the equator than the target was moving at its more northerly point. Likewise, if the particle was fired from a northerly point towards the equator, the particle would land west of its target point. The higher velocity at the equator means the target would have moved east before the particle would reach it.

The Coriolis Effect influences global weather and wind patterns. Wind blowing from subtropical high pressure to low pressure in the north gets deflected to the right, giving the UK its prevailing south westerly wind. Likewise, the oceans are subjected to the accelerations from the Coriolis force. These can affect wind patterns, the rotation of storms and oceanic currents. They also play a role in astrophysics, such as the Great Red Spot on Jupiter and sunspots on our Sun.

$$a_C = -2\,\Omega{\times}v$$

$$F_C = -2\,m\,\Omega{\times}v$$

a_C is the acceleration of the particle in the rotating system, F_C is the Coriolis force, Ω is the angular velocity vector, v is the velocity of the particle with respect to the rotating system and m is the mass of the particle.

F_C is known as a fictitious force, in this case proportional to the mass m.

CM

Equation 339: Law of Titius-Bode

The law of Titius-Bode, commonly wrongly called Bode's Law, predicts the distance from the sun as a function of the number of a planet.

The equation is

$$a = 0.4 + 0.3 \cdot 2^{n-2}$$

where a is the distance of the planet from the sun in astronomical units ($a = 1$ for the earth-sun distance) and n the number of the planet.

This equation is beautiful in its simplicity, but is has two peculiar, contradictory properties

1) It is purely empirical, and cannot be derived from any basic physics law, from simulations, or whatever.
2) The 'law' was discovered before the planetoids and Uranus were discovered. The law originally was accepted despite the gap between Mars ($n = 4$) and Jupiter ($n = 6$). The search for Uranus (discovered in 1781) was inspired by this law and its discovery fitted the law well with $n = 8$, and the planetoid Ceres (discovered in 1801) well coincided with $n = 5$. This means that at that time the law had predictive properties that appeared to confirm its general and scientific validity.

Neptune (discovered in 1846) is rather off, however if Pluto (discovered in 1930) is taken instead it well conforms to $n = 9$. Mercury is quite off as well, but can be fitted better with $n = -\infty$. Today it is considered that the law is a peculiar coincidence indeed, and probably a similar coincidence can be found in any stable planetary system as planets cannot be too close together anyhow. Also it must be considered that taking the logarithm of a series of data points tends to produce a rather straight line in many cases.

The 'law' was discovered by the German astronomer Johan Daniel Titius (1729-1796) in 1765, and was published in 1768 by the German astronomer Johann Elert Bode (1747-1826).
The table below gives the planets, the distance to the sun as calculated by the law and the real distance.

N	planet	a calculated	a in reality
1	Mercury	0.55	0.387
2	Venus	0.7	0.723
3	Earth	1.0	1.000
4	Mars	1.6	1.524
5	Planetoids	2.8	about 2.8
6	Jupiter	5.2	5.203
7	Saturn	10	9.537
8	Uranus	19.6	19.19
9	Neptune (Pluto)	38.8	30.07 (39.48)

HH

Equation 340: Celsius To Fahrenheit Conversion

While most countries in the world have officially adopted degrees Celsius (°C) as the unit of temperature, some[1] still use degrees Fahrenheit (°F). Given a temperature in degrees Celsius T_C, the corresponding temperature in degrees Fahrenheit T_F is given by the following equation [1]

$$T_F = T_C \left(\frac{9}{5}\right) + 32$$

The Fahrenheit scale was first proposed by German physicist and bratwurst enthusiast Daniel Gabriel Fahrenheit in 1724, 18 years before the Celsius scale was first proposed. In this first proposal, zero degrees was defined as the temperature of a solution consisting of 1 part water, 1 part ice, and 1 part ammonium chloride. This solution was chosen as a reference due to the fact that its temperature reaches a stable equilibrium value regardless of the original temperatures of the individual components. 32 degrees Fahrenheit was designated as the temperature of an equal mixture of ice and water, while 96 degrees Fahrenheit corresponded to (roughly) the heat of the human body. The Fahrenheit scale was revised several times prior to its current definition. Fahrenheit was officially used among most English speaking countries until a trend towards accepting the Celsius scale occurred in the 1960s and 1970s.

It should be noted that absolute temperature is given by the Kelvin scale, where zero—absolute zero—corresponds to the temperature in which there is no thermal motion. The Celsius scale is offset from the Kelvin scale by +273.15 °, that is, 0 °C = 273.15 °K. The Fahrenheit scale is offset by +459.67 and Fahrenheit graduations are smaller than Kelvin and Celsius graduations by a factor of (9/5) = 1.8.

[1] Countries whose official temperature unit is degrees Fahrenheit: The United States, the Bahamas, Belize, the Cayman Islands, and Palau.

[1] Walt Boyes (2010). Instrumentation Reference Book, 4th Edition. Butterworth-Heinemann. pp. 273–274

MF

Equation 341: The Young-Laplace Equation

Why are water drops and soap bubbles spherical? Why do paint brush bristles stick together when wet but separate when completely immersed in water? Why does water rise in a glass tube when the tube is immersed in a bath of water? How are water striders able to walk on water? The explanation to all of these lies in the concept called surface tension. Surface tension is the tensile force that exists in the surface of a fluid much like the tension in stretched membranes.

Surface tension arises out of intermolecular cohesive forces. The molecules in the bulk of a fluid experience the cohesive forces from all directions whereas the molecules on the surface (or an interface) experience the forces from like molecules on only one side. This causes the top surface of the fluid to behave like a membrane in tension. Thus, if a line is drawn on the surface, these forces act perpendicular to the line and parallel to the surface. Surface tension is this force per unit length. It has the units of force per unit length. Water at room temperature has a surface tension of approximately 0.07 Newtons/meter whereas liquid mercury has a significantly higher value of approximately 0.485 Newtons/meter.

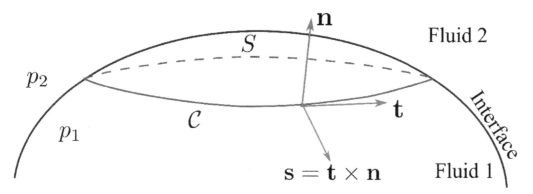

Figure 8: A schematic of two immiscible fluids in equilibrium. The net force acting on S due to the fluid pressure on either side of the interface is balanced by the net force due to surface tension on C. This figure is based on the lecture notes of Professor John W.M. Bush, Department of Mathematics, MIT. (http://web.mit.edu/1.63/www/Lec-notes/Surfacetension/Lecture2.pdf)
Surface tension plays an important role in many naturally occurring phenomena involving interfaces, particularly when these interfaces are between two immiscible fluids.

Consider two immiscible fluids, 1 and 2, in static equilibrium as shown in Figure 1. Let the pressures exerted by the two fluids on the interface be denoted by p_1 and p_2. If $p_1 \neq p_2$, the interface must be curved in order for the surface tension to balance the net non-zero force due to the pressure difference. Suppose that S (see Figure 1) is a segment of the interface containing a point of interest and C is the curve that bounds this segment. Let n be the unit normal vector to the surface (pointing from fluid 1 to fluid 2) at a point on C and t, the unit tangent vector to C at the same point. Then, the surface tension points in the direction s. By requiring that the S be in equilibrium under the action of the surface tension and the fluid pressures, the following beautiful equation that relates the pressures to surface tension is obtained

$$p_1 - p_2 = \gamma \left(\frac{1}{R_1} + \frac{1}{R_2} \right) = \frac{2\gamma}{R_m}$$

Here, γ is the surface tension measured in Newtons per meter, R_1 and R_2 are the two principal radii at the point of interest. R_m is the mean radius of curvature. This equation is known as the Young-Laplace equation. The English physician-physicist, Thomas Young introduced the concept of surface tension in 1804. Based on Young's work on surface tension, the French mathematician-physicist Pierre-Simon Laplace derived the above equation in 1806.

The Young-Laplace equation states that the pressure difference across an interface is proportional to the surface tension. When the interface is a flat surface, both the radii go to infinity and $p_1 = p_2$. Thus, for a flat interface, the pressure difference is zero. However, when the interface is curved, the right-hand side is non-zero and therefore, there will be some pressure difference across the interface.

For air bubbles in water, it is reasonable to assume that the pressures inside and outside the bubble are constant. Consequently, the pressure difference $p_1 - p_2$ is also constant along the interface which, by the above equation leads to the conclusion that the mean curvature of the interface is constant. The only surface for which the mean curvature is constant is the sphere. Thus, the shape assumed by the bubbles is spherical, the mean curvature equals the radius of the sphere and the Laplace-Young equation reduces to

$$p_1 - p_2 = \frac{2\gamma}{R}$$

Where R is the radius of the spherical bubble. From this, we conclude that the smaller bubble size, the larger the pressure inside the bubble.

The Laplace-Young equation can also be used to find the capillary rise and capillary pressure in a tube. The capillary height is given by

$$h = \frac{2\gamma \cos \theta}{\rho \, g \, r}$$

where r is the radius of the capillary tube, θ is the so-called contact angle that measures the angle at which a liquid-gas interface meets a solid surface (the inner surface of the tube), ρ is the density of the liquid, and g is the acceleration due to gravity. This equation shows that the capillary rise is inversely proportional to the tube diameter.

Surface tension is a key factor in the inflation and deflation of the alveoli during breathing. The alveoli are surrounded by a mucous liquid and a surfactant. Without the presence of the surfactant, the alveoli will have to overcome a substantial surface tension of the mucous liquid during respiration. The surfactant relieves the surface tension and therefore, decreases the pressure required for the inflation of the alveoli and enables normal respiration.

HC

Equation 342: Survival Function

However unattractive death may be for most of us, for mathematicians it has several highly satisfactory characteristics: it is unambiguous, it is universal, it occurs at a precise time (that is, it is not a gradual process) and it is irreversible. As a result, the statistical analyses of deaths within populations are relatively straightforward. The fundamental variable in such analyses is T, the time until death.

The survival function $S(t)$, is the probability of survival of an individual who is currently alive after a specified elapsed time t.

Since death is irreversible, it follows that $S(t)$ always increases as $(T - t)$ increases, and since it is inevitable, $S(t)$ must tend to 1.

The lifetime distribution function, $F(t)$, is defined simply as the complement of the Survival function: $F(t) = (1 - S(t))$. The differential of $F(t)$ is denoted $f(t)$ and is the rate of death per unit time.

In medical science, anthropology and actuarial science, $S(t)$ is of great interest, but to determine $S(t)$ from actual data is problematic, since real datasets are almost always incomplete ("censored"). The main reason for this is the difficulty in keeping track of all individuals - if a person disappears from a population, one does not know whether they have died or not. Conversely, some individuals may still be alive when the period of observation comes to an end. Finally, one may wish to add new individuals to the population after the study has begun (in the case if medical trials, for example). To deal with such populations, the Kaplan-Meier method is used.

The Kaplan-Meier method means that an individual can contribute to the survival curve for the length of time his/her status is known, but can be removed from the curve thereafter. For example, if a patient has been in the trial for three years and then disappears (or the trial comes to an end), the fact that (s)he has lived three years should contribute to the survival data for the first three years of the curve, but not to the part of the curve after that.

Given a sample with N members from a population of interest, we assume that they are numbered in the order in which they will die (i.e. member 1 dies first and member N dies last). Observed time until death of the ith member is denoted t_i, so we have

$$t_1 \leq t_2 \leq t_3 \leq \cdots \leq t_N$$

We also define n_i as the number "at risk" just before time t_i, and d_i as the number of deaths at time t_i.

If the status of all N members is known, n_i is just the number of survivors, but if x members have unknown status (that is, the data is censored), then $n_i = number\ of\ survivors - x$.

The maximum likelihood estimate of $S(t)$ is then given this equation, the Kaplan Meier estimator.

$$\hat{S}(t) = \prod_{t_i < t} \frac{n_i - d_i}{n_i}$$

The following chart shows Kaplan Meier curves representing the effectiveness of a drug intended to extend the lives of terminally ill patients. The blue curve corresponds to the test group, the black to the control group.

While most commonly applied to survival, the same formulae and approach can be applied to the distributions of the durations of residence in a city, of employment in a particular job, of intervals between births, of hospital stays and many other examples.

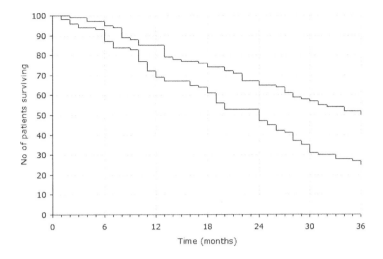

MJG

Equation 343: Abbe Principle

Known as the first principle of precision instrument design, the Abbe Principle is named after the German scientist Ernst Abbe (1840 - 1905), who laid the foundation of modern optics in the 19th century. As a co-founder of Carl Zeiss AG, Abbe stated "if errors of parallax are to be avoided, the measuring system must be placed co-axially the line in which displacement is to be measured on the work-piece." The violation of Abbe Principle leads to the Abbe error, which occurs when two mechanical imperfections happen simultaneously: first, the line of sensing is spatially separated from the line of reference; second, the positioning unit of the instrument has angular error motion in the plane of the reference and sensing line.

Figure on the left side illustrates the Abbe principle when the outer diameter of an object is being measured by a calliper. The line of reference is made by the two scale markers on the calliper's scale and the line of sensing by the contact points of the object's edge. The distance of the lateral separation between those two lines is denoted as L_A, which is known as the Abbe offset. In contrary, when the same object being measured on a micrometer, the Abbe offset is nearly zero since the reference line (i.e. thread tooth) and the sensing line is overlapped shown in the right on the right side.

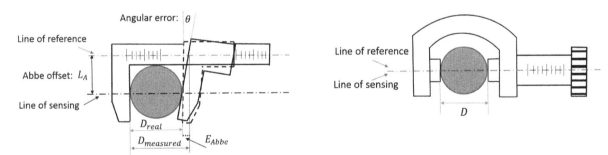

Illustration of Abbe Principle on a calliper   Illustration of Abbe Principle on a micrometer

The Abbe error can be calculated by the following equation:

$$E_{Abbe} = D_{measured} - D_{real} = L_A \times \tan \theta$$

where L_A is the Abbe is offset and θ is the angular error of the positioning unit. As a guideline for instrument designer, this equation implies the effort to reduce the Abbe offset and prevent the angular error. However,

441

a finite amount of Abbe offset is nearly evitable as well as the angular error motions in practice. So this equation also indicates possible ways to compensate the Abbe error if the prior knowledge of the Abbe offset is available and the angular error motion is measurable. Nevertheless, the uncertainty of the measurement of the Abbe offset and the angular error determine the "quality" of reducing Abbe error in measuring devices or machine tools.

Abbe principle is heavily studied and applied by pioneering precision engineers and dimensional metrologists and it should be kept in mind for generation of precision engineers yet to come. Restatement and generalization of the Abbe Principle at various application scenarios are given by Bryan [2] and Zhang [3] and a recent definition and review of Abbe Principle is given by Leach [4].

References
[1] Ernst Abbe, "Meßapparate für Physiker", *Zeitschrift Für Instrumentenkunde, 1890; 10: 446-448.*
[2] Jim Bryan, *The Abbe principle revisited: an updated interpretation.* 1979, Precision Engineering 1:129–132
[3] Guoxiong Zhang, *A study on the Abbe Principle and Abbe Error.* Annuals of CIRP 38(1): 525-528
[4] Richard Leach, *Abbe Error/Offset*, CIRP Encyclopedia of Production Engineering, 2014

KN

Equation 344: Intelligence Equation

$$F = T\nabla S_\tau$$

where F is the force of intelligence, T is the strength, ∇ is the nabla operator, S_τ is the entropy field of all states reachable in the time horizon τ.

"Intelligence should be viewed as a physical process that tries to maximize future freedom of action and avoid constrain in its own future". Cit. Alexander D. Wissner-Gross [1].

In a paper published in 2013 [2] Alexander Wissner-Gross uses a single equation to describes how an intelligent system should behave. The equation states that the intelligence is a force, with a strength T, pointing in the direction of maximum slope of the entropy field (S) of all states reachable in the time horizon τ. In other words, intelligence is a force that pushes in the direction to reach a state with the largest number of possibilities. The concept is also briefly explained in the author's TED talk [3] and examples of how this concept is applied and behaves in different systems can be viewed in a video by the same author [4].

Different system behaves intelligently without any training or instructions apart from the described equations.

The example I find more fascinating is the pendulum on a rigid string: according to the equation the force moves the system to reach the configuration where the pendulum is upright and is kept in balance. The process resembles the evolution of mankind when we moved to an up-right position from which we now have the choice to bend, crawl or lay depending on our needs.

Another fascinating example is how humans and animals act cooperatively. An intelligent direction as a group/society/humankind is the one that maximize our chance of survival whereas a non-intelligent, less entropic direction is the one that would bring us to be confined into a single fixed irreversible state: extinction.

I like to see the strength T, as our knowledge which can drive us to a state with more possibilities, as for example any scientific or technological development that can make our life better. However, if the understanding of the possible states is erroneous, having a wide scientific knowledge will point us to move faster in the wrong direction, eventually leading to self-destruction.

[1] http://www.alexwg.org/
[2] A. D. Wissner-Gross, C. E. Freer Physical Review Letters 110, 168702 (2013)
[3] https://www.ted.com/talks/alex_wissner_gross_a_new_equation_for_intelligence
[4] https://www.youtube.com/watch?v=rZB8TNaG-ik

GM

Equation 345: And a Partridge in a Pear Tree

As it's getting near Christmas I thought we would have a Christmas themed equation.

The old song goes …

On the Twelfth day of Christmas,
My true love sent to me
Twelve drummers drumming
Eleven pipers piping
Ten lords a-leaping
Nine ladies dancing
Eight maids a-milking
Seven swans a-swimming
Six geese a-laying
Five golden rings
Four calling birds
Three french hens
Two turtle doves
and a partridge in a pear tree

So, quickly, how many presents did my true love send to me? Well you could sit there and add them up but a mathematician would give you the answer in a blink of an eye. How do they do it?

Well the answer lies in the equation for the sum of the first n natural numbers.

So the sum of the first n natural numbers is given by the equation $n(n + 1)/2$ in this case $12 \times 13/2 = 78$

This equation was known to the Pythagoreans as early as the sixth century B.C.E

Although the series seems at first sight not to have any meaningful value at all, it can be manipulated to yield a number of mathematically interesting results, some of which have applications in other fields such as complex analysis, quantum field theory, and string theory.

The infinite series whose terms are the natural numbers $1 + 2 + 3 + 4 + \cdots$ is a divergent series. The nth partial sum of the series is the triangular number.

It can be shown that $1 + 2 + 3 + 4 + \ldots = -1/12$

A simple plot in Excel of the first few terms will show that the intercept of the curve is approximately -1/12.

DF

443

Equation 346: Stretching of a Guitar String (Introduction to Elliptic Integrals)

$$\frac{\delta L}{L} = \left(\frac{\sqrt{1 + \psi^2}}{2\pi} E(kx, m) - 1 \right)$$

Plucking a string so that it freely vibrates vertically with respect to the guitar it has been shown in previous equations (BE's 48, 81, and 330) that the motion is made up from a harmonic series of sinewaves each being an integer multiple of the longest shape and vibrating at a frequency that is a corresponding integer of the lowest tone. When the string is vibrating, briefly touch it at its centre point and the tone will immediately appear to double (i.e. rise an octave). This is because you contacted the string where the slowest vibration shape is moving and the next one has a point of no displacement (called a node) at the centre. Hence by touching the string in the centre, the lowest mode is damped while the second (and fourth, and sixth, etc.) mode is unaffected.

Ignoring the higher modes, if you were to take a snap-shot of the string at any instant, its shape would appear to be a sinewave of a single period between the two bridges supporting the string at each end.

Before being plucked, the stationary string is pulled to be a straight line. After, it has curvature. Knowing that the shortest distance between two points is a straight line (an interesting proof of itself) there must be a finite stretching of the string as it vibrates. How much the string extends has posed a problem to mathematicians for the last three centuries.

To understand the above equation, consider the equation for the shape of a string of length, L, oscillating at an amplitude, A, given by

$$y = A \sin(kx), \, k = \frac{2\pi}{L}$$

where y is the displacement of the string perpendicular to its axis at location x along the string measured from one end and the constant k is the wave number for the string shape. It is relatively simple to determine the equation for the length, s, of the distorted string from the integral equation

$$s = \int_0^x \sqrt{1 + \left(\frac{dy}{dx}\right)^2} \, dx = \int_0^x \sqrt{1 + (Ak)^2 \cos^2(kx)} \, dx$$

Unfortunately, the above integral turns out to be surprisingly intractable. As a first step in its solution, it is necessary to make some substitutions to yield a more general form

$$s = \frac{L}{2\pi} \int_0^{kx} \sqrt{1 + \psi^2 \cos^2(\xi)} \, d\xi = \frac{L}{2\pi} \int_0^{kx} \sqrt{1 + \psi^2 - \psi^2 \sin^2(\xi)} \, d\xi$$

$$\xi = kx, \psi = Ak$$

Further manipulation yields

$$s = \frac{L}{2\pi} \sqrt{1 + \psi^2} \int_0^{kx} \sqrt{1 - \frac{\psi^2}{1 + \psi^2} \sin^2(\xi)} \, d\xi = \frac{L}{2\pi} \sqrt{1 + \psi^2} \int_0^{kx} \sqrt{1 - m \sin^2(\xi)} \, d\xi$$

$$= \frac{L}{2\pi} \sqrt{1 + \psi^2} E(kx, m)$$

444

$$m = \frac{\psi^2}{1 + \psi^2}$$

As a final step the stretching of the string, $\delta L = s - L$, is obtained by subtracting the length between the supports from the length of the distorted string after which the top equation follows. The integral function $E(kx, m)$ is called an incomplete elliptical function of the second kind and crops up in the solution for a broad range of physical problems involving physical processes bounded by cylindrical, elliptical, and sinusoidal geometries. It is interesting to note that, for any value of ψ, m can only take on a value ranging from 0 to 1. A Matlab™ script is shown below for computing the length, s, of a sine wave using the above formula.

The history of determining the length of a sine (and, of course, cosine) curve is not clear (at least to this writer). However, Leonard Euler (1707 – 1783) and Carl Friedrich Gauss (1777 – 1855) had studied the properties of elliptic functions. Their studies helped to develop a foundation for Niels Henrik Abel (1802 – 1829) and Carl Gustav Jacob Jacobi (1804 – 1851) to recognize and establish the general foundations of elliptic integrals, for which three different kinds were identified. The confusion at this time about the origins of the theory was brought to the attention of Adrian-Marie Legendre (1752 – 1833) who also made significant contributions during this early period. Since this time, because of their importance across a broad range of physical and mathematical fields, the properties and applications of these function have been substantially researched resulting in an extensive body of mathematics on this topic.

References
Abel N.H., 1827, Recherches sur les fonctions elliptiques. *Journal für die reine und angevandte Mathematik*, 2, *101*-181.
Jacobi C.G.J., 1827, De residuis cubicis commentatio numerosa, *Astr. Nach.*, VI (#123)
Whittaker, E. T. and Watson, G. N. *A Course in Modern Analysis, 4th ed.* Cambridge, England: Cambridge University Press, 1990.

A short Matlab™ program for computing this is shown below.

```
clear all
L=2*pi; % Length of sine period
A=1; % amplitude of sinewave
k=2*pi/L; % Wave number
psi=A*k; % psi in the equation
x=L; % Set this to whatever you wish to compute
m=psi*psi/(1+psi^2); % Elliptic integral constant
k*x; % Integral limit
funcell=@(efun,mfun) sqrt(1-mfun*(sin(efun).^2)); %Integralfunction
intk=quadgk(@(x)funcell(x,m),0.000001,k*x); %Gauss-Kronrod integrator
% I like this, fast and
% precise
s=L*(sqrt(1+psi^2)/(2*pi))*intk; % Computes length
fprintf(1,'For a sine wave of amplitude %2.3f and Periodic length %2.3f\n',[A L])
fprintf(1,'the length of the curve at distance %2.3f is %2.7f\n',[x s])
```

For the program shown, Matalab produces the following output:
For a sine wave of amplitude 1.000 and Periodic length 6.283 the length of the curve at distance 6.283 is 7.6403942

SS

Equation 347: Linear Congruential Generator

A linear congruential generator is a mechanism by which to generate psuedo-random numbers. The term "random" seems initially obvious but when you look closer at the requirements for something that is random and then use a machine to generate "random" numbers the situation becomes much more complex. There are many methods for generating "random" numbers. This equation describes the Linear Congruential Generator

$$X_{n+1} = (aX_n + c) \bmod m$$

where X is the sequence number, m is the modulus, a is the multiplier, c is the increment and X_0 is the seed value.

There are some prerequisites for an efficient and effective equation. m and c are typically prime numbers with the c increment parameter being relatively small and the m modulus being relatively large prime numbers. $a - 1$ should be divisible by all prime factors of m.

The seed value is really a starting value and this is typically chosen from an indexed sequence of numbers so that the same set of random number sequence is generated at every iteration or run time of the calculation.

TRUE V PSEUDO RANDOM NUMBERS

It is relatively difficult for a machine to generate a sequence of numbers that are truly random or that appear random with no patterns and without any particular bias, distribution or recognizable pattern. True random number generators use some sampling of environmental variables such as temperature, voltage fluctuation of other parameter that is disconnected or independent of the machines function or operation. Such environmental parameters used within a random number generator algorithm ensure a more truly random number, however they become quickly limited by the data rate, frequency, sampling and variance of the environmental parameter.

If you need to very quickly generate a lot of random numbers you typically need to use a pseudo random number generator, which to all intents and purposes for most applications can be considered random. Pseudo random number generators can often suffer from a phenomenon called "clumping" or "clustering". This is where a series of numbers of similar values occur in close proximity to each other. This can give undesirable results in some applications so it is not uncommon for people to run "declustering" algorithms one a sequence of generated numbers. Linear Congruential Generators suffer from such challenges among others. Linear Congruential Generators are especially well suited to applications that have limited memory as they require very little memory and processing load to generate large sequences of random number series.

Pseudo random number generators are sometimes used to generate session keys in cryptographic data transfers. While this method is reasonably valid it is susceptible to the strength of the program that uses it. If the initial seed set is known or relatively obvious, such as a time stamp, and the algorithm can be known (as can be determined) this then makes the program more vulnerable to hacking. Hackers have developed very clever techniques for reducing the number of possible options in cryptographic and random number sequence techniques. A particular method is referred to a rainbow tables which show sequences that have a higher likelihood of being duplicated multiple times and as such are more vulnerable to attacks.

References
Brunner, D. and Uhl, A. "Optimal Multipliers for Linear Congruential Pseudo Random Number Generators with Prime Moduli: Parallel Computation and Properties." BIT. Numer. Math. 39, 193-209, 1999.
Pickover, C. A. "Computers, Randomness, Mind, and Infinity." Ch. 31 in Keys to Infinity. New York: W. H. Freeman, pp. 233-247, 1995.

IL-B

Equation 348: Treegonometry – equations of festive beauty

$$B = XT \cdot \frac{\sqrt{17}}{20}$$

$$T = XT \cdot \frac{13\,\pi}{8}$$

$$L = XT \cdot \pi$$

$$H = \frac{XT}{10}$$

SO WHAT DOES ALL THIS MEAN?

Well according to SUMS (the University of Shefield's maths society) these equations capture the perfect ratio of baubles, tinsel, lights and even the optimum size of the star / fairy on top! So once your perfect tree has been selected you can use these equations with the following definitions to perfect your festive centrepiece.

B is number of baubles, T is total length of tinsel in cm, L is total length of lights in cm, H is the height of the top decoration fairy or star in cm and XT is the total height of your Christmas Tree in cm.

For example, a 180cm (6ft) Christmas tree would need 37 baubles, around 919 cm of tinsel, 565 cm of lights and an 18cm star or angel is required to achieve the perfect look.

SAM HAYES, PRESIDENT OF SUMS…

"I'm often asked what the point of maths is; the Christmas tree formula - which was actually created by two former students and members of SUMS a couple of years ago - is something which everyone can relate to and shows how maths can be used in everyday life.

"It's a bit of fun and if gets people engaging with maths while enjoying the festive season then it's a good thing."

SO WHY DECORATE A TREE AT CHRISTMAS?

It is frequently traced to the symbolism of trees in pre-Christian winter rites, in particular through the story of Donar's Oak and the popularized story of Saint Boniface and the conversion of the German pagans, in which Saint Boniface cuts down an oak tree that the German pagans worshipped, and replaces it with an evergreen tree, telling them about how its triangular shape reminds humanity of the Trinity and how it points to heaven.

According to the Encyclopædia Britannica, "The use of evergreen trees, wreaths, and garlands to symbolize eternal life was a custom of the ancient Egyptians, Chinese, and Hebrews. Tree worship was common among the pagan Europeans and survived their conversion to Christianity in the Scandinavian customs of decorating the house and barn with evergreens at the New Year to scare away the devil and of setting up a tree for the birds during Christmas time.

Famous tree facts

- The world's tallest Christmas tree would, at 2,600ft tall, need more than 16,000 baubles, over 4,000 metres of tinsel, almost 2,500 metres of lights, and an 80 metre tall star.

- This year's Trafalgar Square tree is about 20 meters tall, meaning it would need 412 baubles, over 100 metres of tinsel, 62 metres of lights and a two metre tall star.

ILB

Equation 349: Gaussian Integral

Today's beautiful equation relates two fundamental constants of mathematics, i.e. e (see BE 304) and π (see BE 66). It is known as the Gaussian integral since it integrates the Gaussian function e^{-x^2}, which is the standard bell-shaped curve found in many mathematical and physical applications, especially in statistics, where the Gaussian or normal distribution is one of the common distributions of random data (see BE 177). An integral basically calculates the area under a curve; it is certainly not obvious at first glance that the area under the Gaussian curve is the square root of π. There are many ways to solve the integral, but the most common involve a change in coordinates from Cartesian to polar.

The integral is names after the famous German physicist Johann Carl Friedrich Gauss (1777 –1855), although it was Pierre-Simon Laplace (French) that proved it in 1778.

$$\int_{-\infty}^{\infty} e^{-x^2} dx = \sqrt{\pi}$$

RL

Equation 350: Wadell's True Sphericity

$$\Psi = \frac{s}{S}$$

There are a number of so-called 'compactness measures' of particle shape. One of these is sphericity, which measures the degree to which a particle approaches a spherical shape. Sphericity is not the same as another parameter, the roundness of the particle, which refers to the sharpness of the corners and edges. The former applies to the 3D characteristics of the object, whereas classical roundness alludes to 2D particle silhouettes (although some later definitions extend to three dimensions.) The dodecahedral structure of garnet (see image) has a high sphericity in comparison to its roundness, which is influenced by the sharp edges and corners.

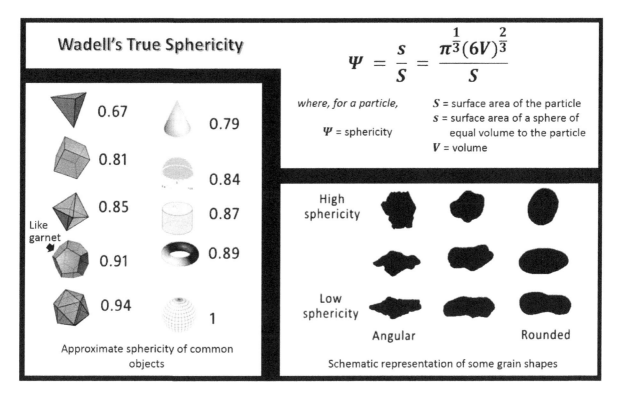

The 'true sphericity' Ψ of a rock particle, as defined in 1932 by Hakon Wadell, is the ratio of the surface area s of a sphere that has the same volume as the given particle, to the surface area S of the particle in question[1]. By definition, the sphericity of a sphere is unity. Any particle which is not a sphere will have sphericity less than 1. Thus $0 < \Psi \leq 1$.

Sphericity has applications in geology, for classifying particles according to shape. The shape of grains is a significant factor in the way sediment is deposited. The definitions of sphericity used in geology differ from those used in dimensional metrology.

The presence of surface roughness is likely to produce misleading values for Wadell's True Sphericity, especially if there are fissures present in the object. This is because of the increased surface area which is likely to distort the result. Some measures of sphericity are less sensitive to surface roughness than is Wadell's True Sphericity.

[1] Volume, Shape, and Roundness of Rock Particles, Hakon Wadell, The Journal of Geology, Vol. 40, No. 5 (Jul. - Aug., 1932), pp. 443-451

FR

Equation 351: The Standard Model Lagrangian

The Standard Model (SM) of particle physics is a "gauge invariant" quantum field theory (QFT) which was developed in the 1960s and describes the fundamental particles of what is now described as *ordinary* matter and the force-mediating particles that have been established to some degree of certainty from experiments in particle accelerators, and their interactions. Although incomplete (as it neglects gravity which is not yet represented as a quantum field), it is extremely efficient at predicting particles and their interactions and verifying the results of particle collision experiments. A QFT represents particles as excited states of an underlying field, whose properties can be represented mathematically as "adjoint operators" on a complex Hilbert space. An adjoint operator (also called a Hermitian conjugate operator) can be thought of as the

449

general matrix form of a complex conjugate, in multiple dimensions. The matrices involved have symmetric properties, and in particle physics the symmetry is described as "special unitary (SU)": unitary is the complex analogue of the property of orthogonality, and unitarity is required to preserve the important norms such as probability amplitudes and their complex conjugates – the main reason why complex numbers are inherent in quantum theory. The invariance is important to preserve the interactions which involve derivatives. The concept is similar to that required in relativity and the spacetime dependence of group transformations: the directions of spacetime are meaningless unless a connection (metric) is specified that will preserve the laws of physics regardless of the relative location and orientation of an event.

The table below summarises the fundamental particles represented by quantum fields in the standard model. The particle interactions are carried via boson gauge fields (the "force" particles), some of which also have mass, and the scalar Higgs field and the recently discovered massive boson associated with it. With gravity absent, the 3 fundamental forces are the strong nuclear force (carried by the gluon), the weak nuclear force (W^-, W^+ and Z^0), and electromagnetism (photon). The fundamental particles of ordinary fermionic matter (as distinct from yet-to-be-verified dark matter) currently consist of two main types of fermions ("spin ½" particles) and their anti-particles, expressed as fields: 1) quarks, which are the "heavy" constituents of the hadrons, not observed in groups of less than 2 due to a confinement property of the strong force, which include the familiar baryonic protons and neutrons that make up all atomic nuclei, together with some heavier less stable forms; and 2) leptons, the "lighter" particles, which include the familiar electrons that determine atomic and molecular electromagnetic phenomena, neutrinos which are observed together with electrons in weak nuclear interactions, and again some heavier less stable types. There are 6 types or "flavours" of each. Under certain conditions in weak interactions, such as for example beta decay or neutrino oscillations, there can be changes from one flavour to another due to broken symmetry – but in strong nuclear interactions flavour is conserved.

STANDARD MODEL PARTICLES/FIELDS

fermions (spin ½)		Bosons	
quarks	*Leptons*	*gauge (spin 1, "vector")*	*spin 0, "scalar"*
up (u)	electron-neutrino (ν_e)	photon (γ)	Higgs (ϕ)
down (d)	electron (e)	W^-	
strange (s)	muon-neutrino (ν_μ)	W^+	
charm (c)	muon (μ)	Z^0	
top (t)	tau-neutrino (ν_τ)	gluon (g)	
bottom (b)	tau (τ)		

The quarks have masses ranging from 2.3 MeV (up) to 173 GeV (top); the leptons from 0.5 MeV (electron) to 1.8 GeV (tau), with neutrinos of almost negligible mass (probably less than 2 eV for the electron-neutrino); the photon and gluon are massless; the W and Z particles have a mass of about 80 GeV and 90 GeV respectively; the Higgs has a mass of about 125 GeV. The SM does not predict particle masses (or coupling strengths, see below) and these can be entered externally. The masses of larger particles like protons, and charged particles in general, can be measured by their deflection in a magnetic field: the radius is proportional to the mass (Lorentz force) and the curvature of the path depends on the charge and mass. From mass-energy equivalence (BE 74) relating to fast-moving moving particles ($E^2 = m^2 c^4 + p^2 c^2$) the mass can be determined from the path taken in eg a mass spectrometer. With fundamental particles, particularly unstable ones, the mass can be inferred approximately from the total energy and momentum of its decay products, eg in a particle accelerator. The energetic resonances determined from scattering cross-sections are however of finite width which puts constraints on the accuracy of mass determination, and from quantum mechanics the uncertainty in mass (energy) measurement is inversely proportional to the life time of the particle. In the case of quarks, lattice QCD (see later for a description of QCD) has been shown to be an accurate method in estimating the quark masses by solving the QCD of quarks and gluons in the system. This typically requires the application of Monte Carlo numerical methods which require a lot of computational processing.

The SM makes use of gauge symmetry (see below for description) and unification of two of the four known fundamental forces: electromagnetism and the weak nuclear interaction. Above particle collision energies of 100 GeV, the electromagnetic and the weak nuclear interactions are unified. The strong interaction is treated as a part of the theory within the symmetry framework, but the unification of this with the other two, if achieved, would be called a grand unification theory (GUT) – but this can only happen at an extremely high energy level of about 10^{12} TeV – far beyond the capability of any particle collision experiment available now or in the foreseeable future. The SM gauge symmetry groups covering the interaction types are currently $SU(3) \times SU(2) \times U(1)$ - invariance under $SU(3)$ gauge transformations for strong interactions, and under $SU(2) \times U(1)$ gauge transformations for electroweak interactions, however, observations of the coupling constants (usually denoted as g in the SM, a number representing the strength of the interactions) suggest that at very high energies, such as those present around the time of the big bang, all these strengths would converge and become equivalent for the known forces and interactions.

The relative force/field strengths (coupling constants, normalised to strong interaction) are:

$$g_S = 1; \; g_{EM} = 10^{-2}; \; g_W = 10^{-6}; \; g_{gravity} = 10^{-38}$$

Electromagnetism was naturally the first interaction to be formulated as a relativistic QFT. This was called quantum electrodynamics (QED) and developed by Richard Feynmann and others in the 1940s, following work by Paul Dirac, as described by the Dirac equation: the Schrödinger equation (BE 17) had been generalised to include spin by Wolfgang Pauli and then further by Dirac to incorporate the effects of special relativity into the quantum theory of the electron. This is required due to the fast speed of particle interactions. The introduction of special relativity into the picture predicted the annihilation and production of particles and antiparticles which led to the requirements for the particles being represented as fields. QED can be represented using a $U(1)$ symmetry group, an Abelian group that is mathematically equivalent to the basis of a unit circle in the complex plane.

Electromagnetism was unified with the weak nuclear interaction as an "electroweak" theory by Abdus Salam, Steven Weinberg and Sheldon Glashow in the 1960s, based on earlier work in the "gauge" theory by Chen Ning Yang and Robert Mills in the 1950s – this is the core of the representation of electroweak unification and strong interaction in the standard model. The *gauge* term, somewhat outdated now as a descriptor which was originally coined by Hermann Weyl in 1918, relates to the redundant degrees of freedom in the Lagrangian (more specifically here the Lagrangian density), which is representative of the kinetic and potential energy of a quantum system. The $U(1)$ symmetry of QED was replaced by another $U(1)$ group representing the source of the unified interaction as a weak hypercharge (usually denoted as Y) instead of the more familiar electric charge (Q) – the electric charge, weak hypercharge and another property, the third component T_3 of the weak isospin (the weak analogue of the angular momentum quantum number in QED) are related by a surprisingly simple formula ($Q = T_3 + Y/2$). Yang and Mills had extended the gauge theory to a non-Abelian group (specifically a semi-simple Lie algebraic group). A proper explanation of these algebraic groups is unfortunately too detailed to describe here; basically the reason for their application is that the weak nuclear interaction in electroweak theory is not symmetric in reflection (it has "chiral" symmetry, requiring $SU(2)$ – the weak isospin of a "right-handed" fermion is zero); and the strong nuclear interaction requires another degree of freedom to account for the additional colour parameter, requiring $SU(3)$.

Initially the quanta had to be massless for the gauge invariance to work, but in 1960 Jeffrey Goldstone, Yoichiro Nambu and Giovanni Jona-Lasinio proposed the idea of symmetry breaking in the massless theory as a mechanism by which particles can acquire mass. Following this idea, the Yang-Mills theory was further developed in the formulation of electroweak theory and another quantum field theory of the strong interaction in the 1960s, quantum chromodynamics (QCD). QCD was developed in the most part by Murray Gell-Mann, who expanded the quark model and identified their flavour and colour symmetry, and the analogous quark isospin and its extension to the property of strangeness found in large quarks being observed at the time. He also modified the earlier integral charge model of Han and Nambu to allow for the non-integral charge indicated for quarks. As in the weak interaction, there are also hypercharge and isospin parameters relating to the strong interaction.

The mechanism that was employed in the standard model is a spontaneous symmetry breaking effect now known as the Higgs mechanism, which initially requires an additional scalar doublet field to interact with the massless gauge boson fields and account for the masses of the W^+, W^- and Z^0 particles observed in electroweak interactions in particle accelerators. The Higgs field can interact with itself to produce the recently-discovered Higgs boson (LHC, 2012). The Higgs field can also be used to give rise to the masses of the individual fundamental fermions, via a different mechanism, the Yukawa interaction, also used to describe the interaction between nucleons and pions (unstable quark-antiquark pairs that were known of at the time). The main part of bulk atomic mass, however, arises from hadrons such as protons and neutrons, and most of their mass is not due to the individual quark masses but due to the binding energy of virtual quarks and gluons (the contribution to atomic mass from electrons is comparatively small). For example, the proton mass is 938 MeV, and the individual up and down quark masses are a tiny fraction of this. The binding energy of baryonic matter is completely described by QCD and its chiral symmetry breaking with the gluon colour charge, hence a common conception that the Higgs field and boson accounts for all mass is rather inaccurate: it is only responsible for a small quantity of it at a fundamental level. Most of the ordinary (baryonic) mass observed in the universe is QCD-derived.

The standard model does not include gravity, largely due to "radiative correction" issues relating to virtual particles. Virtual particles arise naturally over a short time period due to the energy-time aspect of Heisenberg uncertainty ($\Delta E . \Delta t \geq \hbar/2$) that is inherent in quantum mechanics. This can violate the conservation of energy ("borrow from the vacuum") over very small timescales. In the case of the three forces described in the standard model, the effects can be cancelled out by a scaled renormalisation group, but this currently cannot be made to work for gravity in its current form as described by general relativity (BE 337).

The fundamental boson fields in the standard model are usually denoted as B, W (types 1- 3) and G. The B and W bosons "mix" at lower energies via the Higgs mechanism to generate the more familiar photon of electromagnetism and the W and Z bosons that relate to weak nuclear interactions. They are produced by different combinations of the fundamental electroweak fields and a parameter called the Weinberg angle. (For example the photon arises as $A_\mu = W_{11\mu} \sin \theta_W + B_\mu \cos \theta_W$ where θ_W is the Weinberg angle). The weak hypercharge of U(1) acts on B and the Higgs field (ϕ). The weak isospin of SU(2) acts on W and ϕ. The B field is symmetric and does not self-interact, as observed with the photon in electromagnetism, but the W and G types do. These are all *vector* bosons of "spin" 1: they transform like typical 4-vectors in a Lorentz transformation (a transformation between reference frames moving relative to each other but not accelerating). G is the gluon tensor field of the strong interaction. All of these fields are called gauge fields.

The Higgs field ϕ, of which the Higgs boson is an excitation, is scalar and has spin 0 (it has no direction component and is uniquely associated with the "mass" property of particles, as described above).

The fields representing the fermionic matter particles, the fermion fields, are different in that they have spin ½. This means that they are not true vector fields: a rotation results in a "half turn" compared with a vector under a Lorentz transformation – the fermion (Dirac) fields are called *spinors* (ψ). The original Dirac spinors contained kinetic and mass terms but they are often separated out in SM Lagrangian terms. There are 18 spinor fields for the quarks (3 quark colours x 6 flavours).

In its most compact form, the SM Lagrangian can be approximately written for all three forces and all particles as the following, in which there are four distinct parts:

$$\mathcal{L}_{SM} = -\frac{1}{4}F_{\mu\nu}F^{\mu\nu} + i\bar{\psi}\gamma^{\mu}D_{\mu}\psi + \psi_i y_{ij}\psi_j\phi + h.c. + \{|D_{\mu}\phi^2| - V(\phi)\}$$

This is similar to the form found printed on T-shirts and mugs supplied at CERN. It can also be separated out in different ways, e.g. for electroweak and QCD with kinetic and potential terms grouped together, but there still remains a separate Higgs potential term at the end. The way it is separated above is into gauge field interactions, fermion dynamics, fermion interaction with the Higgs field (mass), and the Higgs field coupling with the other fields and itself (these last two grouped in the curly bracket). The first term in the equation above represents the general gauge boson tensor fields for the three forces, without a force index shown. The term $F_{\mu\nu}$ is a tensor – the form in which the field must be written so that it behaves correctly under Lorentz transformations – the indices μ and ν are Lorentzian coordinate indices. This can be described in terms of the classical electromagnetic tensor from which the notation was derived, in terms of the electromagnetic four-potential A:

$$F^{\mu\nu} = \partial^{\mu}A^{\nu} - \partial^{\nu}A^{\mu}$$

The tensor product, as shown in the SM equation above, gives the Lagrangian density for a free electromagnetic field. The term can be expanded for all three gauge fields as

$$\mathcal{L}_{gauge} = -\frac{1}{4}B_{\mu\nu}B^{\mu\nu} - \frac{1}{8}tr(W_{\mu\nu}W^{\mu\nu}) - \frac{1}{2}tr(G_{\mu\nu}G^{\mu\nu})$$

where tr seen in the second two terms refers to the trace (only the trace of the weak isospin and strong colour tensor products can appear in the SM Lagrangian due to the requirements of gauge invariance). A summation index is left out of the description – in the case of W there are 3 components of weak isospin, in the case of G there are 8 gluon colour combinations. If the weak and strong gauge fields are expanded as in the electromagnetic case they appear similar, but there is an additional self-interaction term with a coupling constant g or g_s respectively (the non-Abelian component).

The second term in the equation is a general fermion dynamical term, and can be shown in more detail for the example of leptons, in this case the electron and its neutrino and their antiparticles:

$$\mathcal{L}_{lepton\ dynamics} = +(\bar{v}_L, \bar{e}_L)\tilde{\sigma}^{\mu}iD_{\mu}\begin{pmatrix}v_L \\ e_L\end{pmatrix} + \bar{e}_R\sigma^{\mu}iD_{\mu}e_R + \bar{v}_R\sigma^{\mu}iD_{\mu}v_R$$

The subscripts L and R represent chirality (left- and right-handedness). The entity $\begin{pmatrix}v_L \\ e_L\end{pmatrix}$ is a doublet, with each component a spinor state. Note that the dynamics are different for left-handed and right-handed particles. The $\tilde{\sigma}^{\mu}$ and σ^{μ} terms are two types of Pauli spin matrices of different dimension (these are often used interchangeably in such expressions with the 4x4 Dirac gamma matrices γ^{μ} shown in the general representation above) and the D_{μ} terms are covariant derivatives, which include contributions from the boson fields, the weak isospin and the weak hypercharge.

The third term, which involves the Higgs and fermion field Yukawa coupling and a fermion mass matrix, can be shown in the same case. The mass matrices are equivalent to the

$$\mathcal{L}_{lepton\ mass} = -\frac{\sqrt{2}}{v}\left[(\bar{v}_L, \bar{e}_L)\phi M^e e_R + \bar{e}_R\bar{M}^e\bar{\phi}\begin{pmatrix}v_L \\ e_L\end{pmatrix}\right] + h.c.$$

In some cases a neutrino mass term is added, although an accurate mass for neutrinos is still uncertain. The v term is the required non-vanishing Higgs vacuum expectation value (VEV) which is required for the SU(2) symmetry breaking and fermions to acquire mass by the Yukawa interaction, another parameter not predicted by the SM. It is the only parameter in the SM Lagrangian with a specific dimension and is equivalent to about 246 GeV (as determined from measurements of the Fermi constant in muon decay experiments and also predicted from theory). The matrix M^e can be considered as the lepton mass generation matrix equivalent to the term y_{ij} in the compact expression for the SM above; in the case of quarks this would incorporate the Gell-Mann matrices and structure constants of higher dimension. The h.c. term stands for Hermitian conjugate (this does not show up as an additional factor in the dynamical term as it is already present in the self-adjoint formulation).

The final term has two parts, the Higgs/gauge field coupling term $\left(\left|D_\mu \phi^2\right|\right)$ which allows the weak bosons their mass and the potential term which includes the self-interaction and the expected mass of the Higgs boson m_h:

$$V(\phi) = \left. m_h{}^2 \left[\bar{\phi}\phi - v^2/2\right]^2 \middle/ 2v^2 \right.$$

The theoretical Higgs field is scalar (is not affected by Lorentz transformation as the vector fields are) and is also called tachyonic (this is a misnomer and now refers to an imaginary mass field rather than faster than light speed in this context, and the symmetry breaking results in tachyon condensation generating bosons). The Higgs field has four components, two charged and two neutral. The two charged and one of the neutral components are called "Goldstone" bosons and become components of the three weak bosons; the second neutral component results in the Higgs boson by self-interaction and is also responsible for the Yukawa coupling with fermions.

PB

Equation 352: The Michaelis-Menten Equation

The biochemical reaction in which a substrate S and an enzyme E bind together to form a substrate-enzyme complex ES, which subsequently transforms the substrate into a product P is

$$E + S \underset{k_r}{\overset{k_f}{\rightleftarrows}} E \overset{k_{cat}}{\longrightarrow} E + P$$

The rate of this reaction is given by the Michael-Menten (M-M) equation

$$v = \frac{d[P]}{dt} = \frac{V_{max}[S]}{K_M + [S]}$$

where v is the reaction rate, $[P]$ is the molar concentration of the products, $[S]$ is the molar concentration of the reagents and K_M is the Michaelis constant; it is the substrate concentration at which the reaction rate is half of the maximum rate, V_{max}.

$$V_{max} = k_{cat}[E]_0$$

$$K_M = \frac{k_f + k_{cat}}{k_r}$$

$K_d = \frac{k_f}{k_r}$ is the dissociation constant of the enzyme-substrate complex ES and k_{cat} represents the turnover of the reaction, while the ratio $\frac{k_{cat}}{K_M}$ gives the catalytic efficiency, that is to say how efficiently an enzyme converts a substrate into a product.

The Michael-Merten (M-M) equation allows biochemists to calculate the reaction rates of biochemical and physiological processes catalysed by enzymes. The reaction rate in the MM equation is a function of the concentration of the "substrate".

454

In biochemistry, a "substrate" is a reactant, a molecule that is transformed into a new molecule, by the action of an enzyme. A typical example of a biochemical, enzyme-catalysed reaction is the transformation of sucrose into its constituent sugars, glucose and fructose, through the action of a family of enzymes called "sucrase". In this example, the sucrose molecule is the "substrate", glucose and fructose are the "products", and sucrase is the enzyme that allows the reaction to happen.

HISTORY

German biochemist Leonor Michaelis and Canadian physician Maud Merten proposed the equation in 1913, while studying the reaction mechanism of the enzyme invertase catalysing the hydrolysis of sucrose into glucose and fructose. The two scientists, who worked together at the University of Berlin, continued the pioneering work of French physiologist Victor Henri, who discovered in 1903 that enzyme reactions were initiated by an interaction between the enzyme and the substrate.

In 1922, Michaelis moved to the Medical School of the University of Nagoya (Japan), where started studies on the cellular membrane. He subsequently moved to the US in 1926 where he retired in 1941. He also opened the way to several applications in the field of cosmetics by discovering that thioglycolic acid can dissolve keratin. Maud Leonora Menten made also great contributions in histochemistry, inventing the azo-dye coupling reaction for alkaline phosphatase, still used in this field. She was also the first scientist to conduct electrophoretic separation of blood haemoglobin proteins in 1944.

ASSUMPTION AND IMPLICATIONS OF THE EQUATION

The M-M equation is based on the assumption that the substrate S will react with the enzyme E to form a complex ES, through an equilibrium reaction; the complex will subsequently transform into a new product P, leaving the enzyme behind, so that it is able to react again with other substrate molecules.

An important assumption of the M-M equation is that the concentration of the enzyme is much lower than that of the substrate ($[E] \ll [S]$), which is generally the case for the majority of the enzyme-catalysed biochemical reactions.

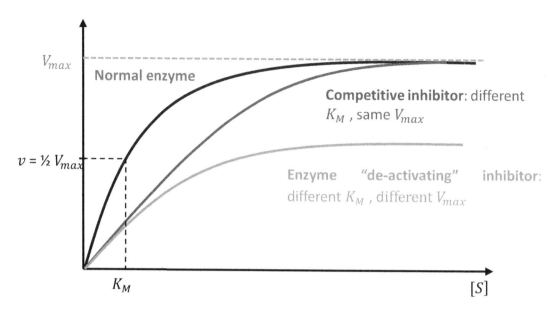

Figure 1. The trend of the reaction rate as a function of the substrate concentration. Effects of a competitive inhibitor and an enzyme de-activating inhibitor on the reaction rate. (Source: Wikipedia)

Under the M-M regime, the reaction rate initially increases linearly with an increase of the substrate concentration. Then, it asymptotically approaches the maximum rate V_{max} at higher substrate concentrations. This situation corresponds to have all enzyme's active sites bonded with the substrates. The saturation of the enzyme takes place because there is a fixed number of enzymes, each of which has a fix number of binding sites available to the substrate. By increasing the concentration of substrates, they will start bind to more active sites of the enzymes present in the reaction, up to a point where all binding sites of the enzyme are "occupied" by a substrate and each enzyme is "working" as fast as its intrinsic, maximum, rate to catalyse the reaction (Figure 1).

This situation is very similar to what predicted by the Langmuir's equation for adsorption phenomena in monolayer regime. Moreover, the M-M equation assumes that the concentration of the enzyme and the transition states are the same, or better, that the rate of loss of enzymes E is equal to the rate of formation of a transition states ES. This is called a steady-state approximation.

There are other important assumptions in the M-M equation:

1. Free diffusion of substrates and enzymes: in the environment of a living cell, where there is high concentration of proteins, this assumption is not ideal, as the cytoplasm often behaves more like a gel than a liquid, limiting the diffusion and consequently altering the reaction rate.
2. Irreversibility of the process that leads to the formation of the products from the transition state ES: in general, however, the product formation can be reversible, but irreversibility is a valid approximation if the concentration of the substrate is much higher than that of the products. This is true in In Vitro assays and many In Vivo biological reactions. Irreversibility is also a valid approximation when the energy released in the reaction is very large.

STUDYING BIOCHEMICAL REACTIONS

The M-M equation is often use to characterise a generic biochemical reaction, in the same way the Langmuir isotherm is used to study a generic adsorption phenomenon. Biochemists run a series of enzyme assays, varying substrate concentrations and measuring the initial reaction rates. By plotting reaction rates against

[S], the parameters V_{max} and K_M (the Michaelis constant) can be obtained with a fit of the data to the M-M equation. The plot is also known as "Linewever-Burk plot".

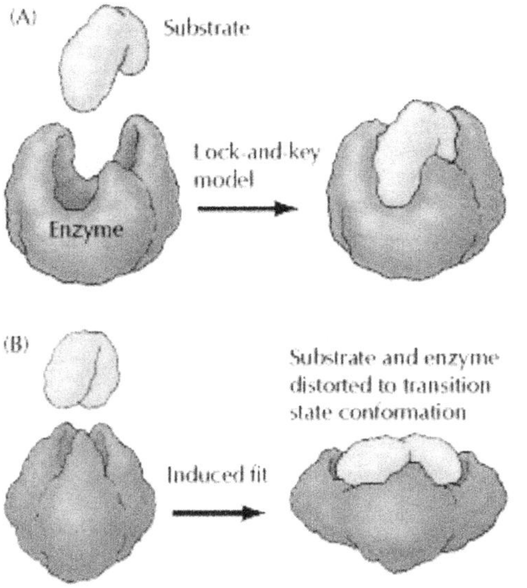

ENZYMATIC CATALYSIS

Enzymes increase the rate of a reaction up to 10 billion fold and are effective in tiny amounts. One enzyme molecule usually converts up to 1000 molecules of substrate in a minute, but some enzymes are known to convert up to 3 million molecules per minute. Enzyme's catalytic activity is very specific. While a substrate may undergo many reactions, enzymes can carry out only one specific reaction for that substrate, as a very specific interaction is needed to bind a substrate into the enzyme active site. This interaction can take place in two ways (Figure 2):

1. "Lock-and-key model": the substrate fits precisely into the active site.
2. "Induced fit model": the geometrical conformation of the substrate is changed so that it can more easily access the active site on the enzyme and resemble more closely the configuration of the transition state. This modification of geometrical configuration, in turn, can weaken critical bonds within the substrate and promote a tighter binding to the enzyme, with consequent decrease of the activation energy for the formation of the products.

Figure 2. Schematic illustration of the two mechanisms of the enzyme's action on the substrate. (Source: Geoffrey M Cooper, "The Cell: A Molecular Approach". 2nd edition, Sinauer Associates).

Substrates initially bond to the active sites of the enzyme through non-covalent interactions, including hydrogen bonds, ionic bonds and hydrophobic bones. Once bonded to the enzyme, substrates can be transformed into products by a series of multiple mechanisms. In some cases, the bonding of two or more substrates to the enzyme is necessary in order to promote the formation for a specific product (Figure 3), like in the case of the formation of a peptide bond between two amino acids.

ENZYME INHIBITION

However, the enzyme's activity can be inhibited in several was. One way involves a substance binding to a site on the enzyme in such a way to "turn off" the enzyme's function. In this case, the enzyme is "de-acticated" so that the Vmax is lower than that predicted by the MM equation; no matter how much additional substrate is added, the enzyme is "turned off" and the reaction no longer produces products. A second way to inhibit the enzyme is by introducing an inhibitor molecule that similarly binds to the active sites, preventing the actual reactants (the substrates) to access the enzyme workbench. In this case, however, the effect is not permanent and disappears when increasing the concentration of the substrate.

457

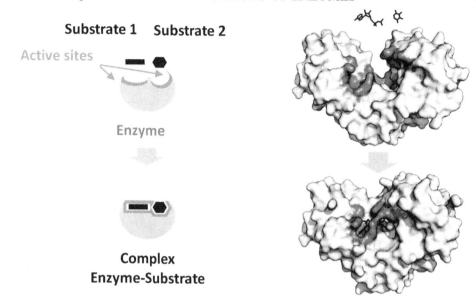

Figure 3. Schematic illustration of the enzyme's action on the substrates. Enzyme changes shape by induced fit upon substrate binding to form enzyme-substrate complex. Hexokinase has a large induced fit motion that closes over the substrates adenosine triphosphate and xylose. (Source: Wikipedia)

The M-M equation is not only used for enzyme-driven biochemical reactions, but also for antigen-antibody binding, DNA-DNA hybridization and protein-protein interaction.

The use of enzymes has been also very important in industry, for example in leather tanning, to soften and remove hair through proteases, and in brewing, to convert the starch stored in the grain to sugars and proteins to amino acids; sugars and amino acids are used by yeast to grow and alcohol production.

In cheese manufacture, proteins in milk are coagulated through the stomach enzyme rennin, obtained from veal calves.

Orange juice production relies on the use of enzymes to break down the cells of the fruits, which made of extremely tough cellulose fibres held together by pectin and hemicellulose. If the break down is carried out by heating at high temperature, it would require much more energy and it would alter the colour and flavour of the juice.

In medicine, the enzyme protease is very important to counteract thromboses, as it dissolves fibrin in blood clots, which can sometimes build up in damaged blood vessels and can accumulate in the heart or brain arteries, with potential risk of heart attack or stroke.

Figure 4. Fermentation plant at the Budweiser Brewery in Fort Collins, Colorado (US). (Source: Wikipedia)

Enzymes can be extracted from plants and animals, but the vast majority of them is produced from microorganisms such as bacteria and fungi, which are grown in bulk fermenters.

AT

Equation 353: Minimum Deviation Angle of a Prism

As we have seen in BE 3 (Snell's law), a ray of light entering a transparent material is deflected, and so it is when exiting. The sum of these two deflections is called the deviation angle. For a prism with refractive index $n(\lambda)$, with an internal angle α, in air with refractive index $n_a(\lambda)$ (see also BE 68, Edlen equation), where light enters at the minimum deviation angle $\delta(\lambda)$, it can be written that

$$n(\lambda) = n_a(\lambda) \cdot \frac{\sin\left(\frac{\alpha + \delta(\lambda)}{2}\right)}{\sin\left(\frac{\alpha}{2}\right)}$$

The equation can be derived from the Snell's law and symmetry considerations. At the minimum deviation angle the situation is as sketched in the figure

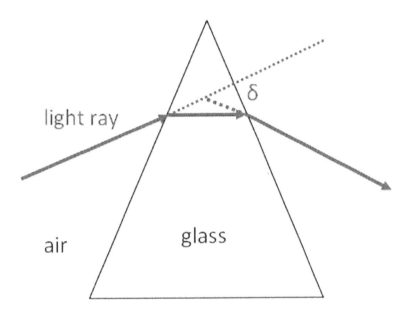

The minimum deviation angle can be used to make an accurate measurement of the refrective index of an optical material such as quartz or various kinds of glass. From a certain melt of glass a prism is prepared and the minimum angle of deviation can be measured by a goniometer (a rotating table) set-up.

Apart from being the minimum, it is also the angle at which most rays are deflected. Therefore, the angle of minimum deviation explains some meteorological phenomena, like halos and sundogs, caused by the refraction of sunlight in the hexagonal prisms of ice crystals in the air. The reflection from raindrops shows a minimum deviation angle and causes the rainbow!

The concept of minimum deflection was known to Newton, and Joseph Fraunhofer used the relationship to determine refractive indices of different glasses for spectral lines of several wavelengths.

HH

Equation 354: Forward Projection on Aligned Image Plane

When a 3D object is placed between a source of light and a 2D image plane, a projection will be formed. In mathematical terms, the position of the 3D object's projection on the 2D image plane can be mapped using forward projection equations. The term 'forward projection' is used to denote the transformation from 3D to 2D. 'Back projection' is used for transforming from the 2D projection image back to the 3D object space—a critical operation in computed tomography.

Two separate coordinate frames are introduced: a three-dimensional 'global' X-Y-Z coordinate frame and a two-dimensional U-V 'image' coordinate frame. The figure below illustrates the aligned projection system, in which the X and U axes are parallel, the Y and V axes are antiparallel, and the point-like light source is on the Z axis at a distance Z_s. The origin of the U-V frame is located at the top left corner of the image plane, while the origin of the X-Y-Z coordinate frame is centred on the image plane. The coordinates of the X-Y-Z frame are denoted (U_o, V_o) in the U-V coordinate frame.

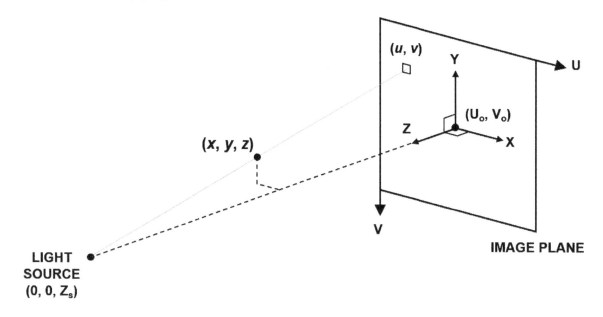

The equations below provide the transformation from a point-like object at global coordinates (x, y, z) to its projection coordinates (u, v) on the aligned image plane.

$$u = \left(\frac{Z_s}{Z_s - z}\right) x + U_o$$

$$v = V_o - \left(\frac{Z_s}{Z_s - z}\right) y$$

In X-ray computed tomography, the relationship provided by the above equations is critical for performing back projection of 2D X-ray intensity values to reconstruct the volumetric distribution of X-ray attenuation.

Concepts of forward projection are derived from perspective geometry, most commonly related to photogrammetry. The earliest mention of perspective geometry is by Leonard Da Vinci in 1480, when he wrote

"Drawing is based upon perspective, which is nothing else than a thorough knowledge of the function of the eye. And this function simply consists in receiving in a pyramid the forms and colours of all the objects placed before it. I say in a pyramid, because there is no object so small that it will not be larger than the spot

460

where these pyramids are received into the eye. Therefore, if you extend the lines from the edges of each body as they converge you will bring them to a single point, and necessarily the said lines must form a pyramid." [1]

[1] The Notebooks of Leonard Da Vinci, Complete (2004) The Gutenberg Project http://www.gutenberg.org/cache/epub/5000/pg5000-images.html

MF

Equation 355: Contact Angles and Work of Adhesion

Suppose that a drop of liquid is placed on a solid surface. Whether the drop spreads on the surface or not (i.e., wets or not) is determined by the various surface tensions present at the triple interface formed by the solid surface, the surrounding air and the liquid. A schematic of the interface is shown in the figure below. Let γ_{sg}, γ_{sl} and γ_{lg} be the surface tensions at the solid-air interface, the solid-liquid interface, and the liquid-air interface respectively.

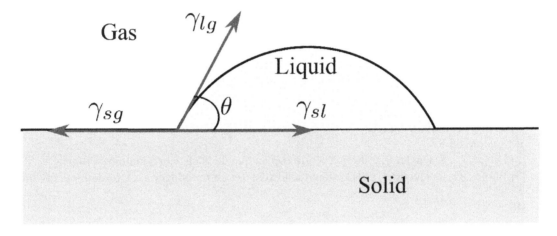

Figure 9. Various surface tensions acting at the interfaces between a liquid drop on a solid surface and the surrounding gas.

Initially, when the drop just makes contact with the solid surface, f, the net force per unit length along the triple-interface, equals

$$f = \gamma_{sg} - \gamma_{sl} - \gamma_{lg} \cos \bar{\theta}$$

Here, $\bar{\theta}$ is the angle between the liquid surface and the solid at the triple interface. This angle changes until equilibrium is established. The angle at equilibrium is called the contact angle and denoted by θ. At equilibrium, the force f is zero and we have the Young relation, first derived by Thomas Young in 1805,

$$\gamma_{sg} = \gamma_{sl} + \gamma_{lg} \cos \theta$$

Dupré in 1869 derived an equation for the work of adhesion (the work per unit area required to separate two phases from an interface). This is given by

$$W_{sl} = \gamma_{lg} + \gamma_{sg} - \gamma_{sl}$$

Upon combining the above two equations, we obtain the Young-Dupré equation

461

$$W_{sl} = \gamma_{lg}(1 + \cos\theta)$$

All the equations described above have some important implications in the study of interfaces.

By the Young relation, the contact angle between the drop and the solid surface is determined by the surface tensions between the three phases.

1. Wetting, the ability of a liquid drop to spread out on a solid, can be defined in terms of the contact angle θ.
2. A liquid is said to be wetting a surface when $\theta \leq \frac{\pi}{2}$ and non-wetting when $\theta > \frac{\pi}{2}$. The case $\theta = 0$ corresponds to total wetting, i.e., when the liquid spreads completely on the surface.
3. From the Young relation, we note that partial wetting occurs when $\gamma_{sg} > \gamma_{sl}$. On the other hand, $\gamma_{sg} < \gamma_{sl}$ corresponds to the non-wetting case since θ must be bigger than $\pi/2$.
4. Let the spreading coefficient S be defined as $S = \gamma_{sg} - \gamma_{sl} - \gamma_{lg}$. It can be thought of as a quantifying parameter for the spread of the liquid. The case of total wetting, i.e., $\theta = 0$ corresponds to $S > 0$. The case of partial wetting corresponds to $S < 0$.
5. When the drop is placed on the surface, as the equilibrium is being established, the direction in which the triple interface moves is determined by the direction of f.
6. If the contact angle and the surface tension of a liquid are known, the Young-Dupré equation allows for the calculation of the work of adhesion which is important for characterizing the strengths of contacts between a solid and a liquid such as in coatings.

References
[1] Young, T. (1805). An essay on the cohesion of fluids. Philosophical Transactions of the Royal Society of London, 65-87.
[2] Dupré, A., & Dupré, P. (1869). Théorie mécanique de la chaleur. Gauthier-Villars.
[3] Barnes, G., & Gentle, I. (2011). Interfacial science: an introduction. Oxford university press.

HC

Equation 356: Happiness

Though it has hardly received the respect, column-inches or length of library shelves accorded to goodness, power or wisdom, happiness undoubtedly ranks high on the check list of human goals. However, as anyone who has read this far will have discovered, the scope of mathematics is as wide as its reach is deep, and happiness can readily be analysed by its tools.

In 2012, Dr Robert Rutledge (University College London) gathered over 18,000 volunteers and asked them to play a simple computer game. In each round of the game, the player is asked to choose between accepting a certain sum of money, and gambling on receiving a larger sum. The chance of winning the gamble is always 50:50. After every 3 or 4 rounds, the player is asked how happy (s)he is, on a sliding scale. A series of screens from the game is shown in Figure 1, in which the choice is between accepting a certain sum (0p) or taking a chance on winning 65p / losing 36p. In this example, the player takes the chance and wins the 65p.

Three variables are defined as:

Certain Reward (CR): the amount that can be chosen without risk (0p in the example).

Expected Value (EV): the average amount you should expect if you chose to gamble (14.5p here, though clearly for a single round this value can't actually be achieved)

Reward Prediction Error (RPE), the difference between what you expect and your actual reward (65p - 14.5p = 50.5p)

When Rutledge analysed the results, he found that happiness increased with all of these three variables. So, as one might expect, one's happiness in response to good fortune increases with both senses of the "fortune": winning a fortune makes one happier if that outcome was extremely fortunate (i.e., unlikely). In short, if things are going well, we are happy, and if they are going better than expected, we are happier still.

The equation is

$$\text{Happiness}(t) = w_0 + w_1 \sum_{j=1}^{t} \gamma^{t-j} CR_j + w_2 \sum_{j=1}^{t} \gamma^{t-j} EV_j + w_3 \sum_{j=1}^{t} \gamma^{t-j} RPE_j$$

where t is the trial number, w_0 is a constant term, other weights w capture the influence of different event types, $0 \leq \gamma \leq 1$ is a forgetting factor that makes events in more recent trials more influential than those is earlier trials, CR_j is the CR if chosen instead of a gamble on trial j, EV_j is the EV of a gamble (average reward for the gamble) if chosen on trial j and RPE_j is the RPE on trial j contingent on the choice of the gamble.

Since RPE = (Outcome - EV), EV figures appears twice in the equation, first on its own, where it increases happiness, and then as part of RPE, where it decreases it again, quantifying the fact that the more likely it is that good things are coming up, the happier ones expectations are - but also that, if the highly likely good thing actually turns out to be a damp squib/slap in the face with a wet fish, the disappointment is all the keener.

The presence of a "forgetting factor" shows (one might say) that people soon get over their happy/unhappy reactions to small events; after about 10 rounds (about 2 minutes), the impact of a win or loss was zero.

Scientists reading this might be forgiven for a sense of dissatisfaction by now: is asking people "How happy are you now? And now? Ok, so what about now?" and getting them to move a cursor in response really a measurement? Is there just one kind of happiness? What if happiness is multidimensional? Should the scale be logarithmic (all other human reactions are logarithmic after all)? Won't people react to the question itself? Will their cursor-sliding behaviour not change simply through repetition? Won't boredom set in, and affect the results? And so on.

But Rutledge did make objective measurements too, by scanning the brains of all the volunteers using an MRI machine at UCL's Wellcome Trust Centre for Neuroimaging. He found that the values of CR, EV and RPE all correlated with activity in a part of the brain called the ventral striatum, and that the output of the whole equation predicts the activity level of the anterior insula. Damage to these two areas was already known to be related to addiction and depression respectively.

Isn't all this obvious? Yes, but what is impressive about the equation is that it is the first that can quantitatively predict exactly how happy a person will be from moment to moment as (s)he makes decisions and receives outcomes resulting from those decisions. And the correlations with brain activities are important steps forward into mapping where and how happiness develops in the brain. In practical terms though, Rutledge's work may be of most value in benchmarking and assessing the states of people suffering from depression and pinpointing the physiological correlates of those states; what some have called the machineries of joy.

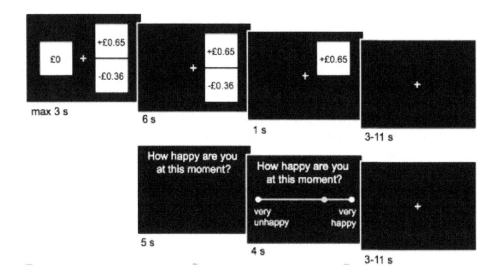

MJG

Equation 357: Structure Function

Surface characterization started as a qualitative subject with the goal of understanding tribology (the science of friction, lubrication, and wear) in the era before industrial revolution. Pioneers in the 17th century, like Leonardo da Vinci, Guillaume Amontons and Charles-Augustin de Coulomb studied the problem of friction, which is one of the fundamental functional influences in practically all instruments and machines.

In the early 1930s, Harrison developed the first analogue surface instrument which used a stylus to slide across the surface of a part and recorded its change of height as a measure of surface irregularities arising from manufacturing processes [1]. E.J. Abbot and F.A Firestone later on described their measured surface profile (a two dimensional plot) by the so called bearing area curve (also known as Abbot-Firestone curve) [2]. The Abbot-Firestone curve characterizes the material-to-air ratio as a parameter to help manufacturers understand the properties of sealing and bearing surfaces. Since then, a variety of parameters (e.g. the BE 151) have been derived to characterize the surface measurement results.

Thanks to the advancement of digital signal acquisition and processing and non-contact measurement technology, today's engineering surfaces are measured by optical instruments which can rapidly acquire a map of surface deviations over a two dimensional area. In the 1970s, characterization of surface texture based on advanced mathematical methods were pioneered by David J. Whitehouse, who used the theories of digital signal processing to study both profiles and maps of surface measurement results [3]. In particular, the linear Structure Function (SF) was first used in surface metrology in his PhD thesis in 1971 and is characterized by the equation

$$SF(\tau) = E\{[z(x) - z(x + \tau)]^2\}$$

where, E signifies the statistical expectation, $z(x)$ is the surface profile height as a function of location x and τ is a spatial separation length. The linear SF connects the height features at one point on the profile with the height feature at other point as the separation length changes across the range the measurement length. SF is tool which resolves the surface irregularity and periodicity therefore it has been used in large telescope optics fabrication process.

The linear $SF(\tau)$ have been extended to a version of two dimensional function called Area Structure Function. Liangyu He used area SF to characterize spatial frequency components of general optical surfaces in his PhD thesis in 2013. The area structure function is similarly calculated, by mathematically extending

464

the spatial separation τ from one dimension to two dimensions (τ_x, τ_y). The following figure shows a simulated surface (a), its linear SF (b) and area SF (c) respectively [4].

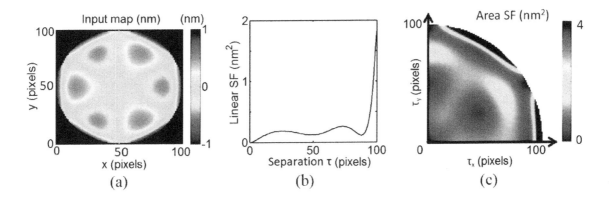

(a)   (b)   (c)

Area structure function shares some interesting relationships with other surface characteristics like the Auto Correlation Function (ACF, a special case of BE 116)

$$SF(\tau_x, \tau_y) = 2\sigma^2 \{1 - ACF(\tau_x, \tau_y)\}$$

where, σ is the two dimensional root-mean-square (BE 61) of the surface map.

Numerical features like repeatability of the spatial frequency component on the surface can be interpreted from area SF in an intuitive way. Due to some of its advantages over other characterization methods optical metrologists are now working on the potential of the SF for optical specification and conformance testing. However, the area SF is not an orthogonal representation of the spatial content of a surface compared to the Zernike Polynomials (BE 61), it is still under investigation in the optical metrology field.

References:
[1] Harrison R E W 1931 A survey of surface quality standards and tolerance costs based on 1929–1930 precision-grinding practice Trans. ASME paper no. MSP-53-12
[2] Abbott, E.J.; F.A. Firestone (1933). Specifying surface quality: a method based on accurate measurement and comparison. Mechanical Engineering 55: 569–572.
[3] Whitehouse, D.J, 1971, The properties of random surfaces of significance in their contact.
[4] He L. et al., 2013, Optical surface characterization with the structure function, CIRP Annals-Manufacturing Technology 62 (1), 539-542

KN

Equation 358: Buffon's Needle

$$\hat{\pi} = \frac{N}{n}\frac{d}{2l}$$

where π is the estimation of the value of pi, n is the number of tosses that cross a line, N is the total number of tosses, d is the distance between lines and l is the needle's length.

Buffon's needle is one of the oldest problem in geometrical probability. The problem consists in throwing a needle on an area with drawn parallel lines and estimate the probability of the needle to intersect any of the drawn lines. The probability can be shown to be related to the value of pi [1], and, therefore the experiment can be used to estimate the value of pi by Monte Carlo experiments.

The estimated value of pi is proportional to the inverse of the ratio of the needles tosses that crosses a line n over the total number of tosses N. The ratio n/N is an estimation of the probability of the dropped needle to cross a line. The value of pi is also proportional to the ratio of the distance d between the drawn lines and the length of the needle l.

It has been shown that the variance of the estimator decreases and therefore the estimation converge quicker if the lines are arranged in a double or triple-grid pattern [2].

The problem was posed and solved first by Georges-Louis Leclerc, Comte de Buffon (7 September 1707 – 16 April 1788) a French naturalist, mathematician, cosmologist and encyclopedic author.

1. G. L. Buffon, "Essai d'arithmétique morale," *Histoire naturelle, générale, et particulière*, Supplément 4, 1777 pp. 685I713.
2. Enis Siniksaran, *Throwing Buffon's Needle with Mathematica,The Mathematica Journal 11:1 ,2008 Wolfram Media, Inc.*

GM

Equation 359: Christmas Day

Merry Christmas to all our readers. No equation today but a little bit of fun that draws on some of the past equations throughout the year. Much of what follows is not original but collates the thoughts of others on the subject of Santa Claus. I hope you enjoy.

Apollo 8 arrived at the Moon on December 24, 1968 and inserted into lunar orbit.

The crew remained in orbit for 10 revolutions of the Moon. The service propulsion system engine on the service module was ignited to send the crew of Apollo 8 back toward the Earth. Upon re-establishing contact after that communications blackout,

089:32:50 Mattingly: Apollo 8, Houston. [No answer.]
089:33:38 Mattingly: Apollo 8, Houston.
089:34:16 Lovell: Houston, Apollo 8, over.
089:34:19 Mattingly: Hello, Apollo 8. Loud and clear.
089:34:25 Lovell: Roger. Please be informed there is a Santa Claus.
089:34:31 Mattingly: That's affirmative. You're the best ones to know. -NASA
So now we can confirm there is a Santa Claus, how does he visit all the houses in one day (strictly speaking he has 36 hours but more about that later).

The theme of this BE is to explain how Santa manages to visit all the houses in the world on Christmas Eve. We assume here that Santa has to obey the laws of Physics, but may have some advanced technology available to him. There are about 7 billion people in the world so let's say Santa has about one billion households to visit.

Now, in some rural places, neighbouring houses may be many kilometres apart, and in a city like London or Tokyo, may be mere metres apart. So, let's assume that each of the one billion households on Earth is 10 meters away from the previous one. This means that Santa has to travel ten million kilometres in 36 hours ($1,000,000,000 \times 0.01$).

It's that 36 hours again. If Santa starts at the beginning of Christmas Eve in places like Australia, Japan, and parts of Russia, and ends in Alaska and Hawaii just before dawn on Christmas morning, it gives him about 36 hours to play with thanks to the International Date Line.

If all he does is travel from house-to-house in this time, he needs to move at an average speed of 77 kilometres per second. Remember $speed = distance/time$ so we have $10000000/(36 \times 60 \times 60) = 77\ km/s$

For comparison the speed of light is approximately 299,792 kilometres per second and the Voyager probes are travelling at about 16 km/s). Put another way, he gets about 130 microseconds to travel to, deliver presents to, and leave each household. So although he is travelling sub-light speed he is still travelling faster than the fastest man made vehicle.

All this assumes Santa does not have to return to base to restock on presents!

So what equations will Santa need to be thinking about?

Well,

1. Travelling that fast there will be the problem of air resistance (equation 286) and maybe the need for a heat shield
2. As the sled gets hotter it will increase in size (BE 19, 196)
3. If he is going that fast could he stay in orbit (BE 84). The ISS travels at about 5 km/s. The earths escape velocity is 40,270 km/h (11 km/s)
4. Going that fast his mass will increase (BE 323)
5. But for him time will run more slowly (BE 8, 323) although not by much as he is still well below light speed
6. Going from 77 km/s to rest to 77 km/s again ($v^2 = u^2 + 2ax$, solve for a and divide by 9.81), even if we give him the whole 130 microseconds to do this, means an average acceleration of just over one billion G. If Santa weighs about 100 kg, this means that just starting and stopping his sleigh makes a force on Santa that's the equivalent of stacking three sky scrapers on top of him
7. Rudolph the redshift reindeer (BE 101)

Is there another, non-classical, way he could do it?

In 1994, Miguel Alcubierre, discovered that there is a solution of general relativity very similar Star Trek's warp drive. By artificially contracting the section of spacetime in front of the sleigh and expanding that behind, Santa and his reindeer can travel at large speed relative to the Earth while still remaining stationary within their own 'bubble' of space. This could be the solution. The trouble is this method would require an enormous amount of energy approximately equivalent to several billion times that in the entire observable universe in fact.

That brings us to our old friend quantum mechanics. Quantum mechanics also allows for things to be transported great distances in little time, and could avoid the need for Santa to take to the skies at all.

Because their position isn't a definite point but a wave spread out over space, particles can sometimes "tunnel" through barriers that, according to classical mechanics, they shouldn't be able to pass but again it would take a huge amount of energy to realize. So let's suppose that on Christmas night, Santa is in a superposition of quantum states, smeared out all the way around the planet, and each quantum state delivers presents to a single child. If just one child sees Santa, he immediately collapses into a single state, in accordance with Heisenberg explaining why it is so important that children are asleep. This would mean that no other children would receive presents that Christmas.

But maybe there is a technological solution first proposed for space exploration.

Physicist John von Neumann proposed that a spacecraft could be sent to another star system and programmed to make replicas of itself using raw materials found there. These in turn would travel to further solar systems, exponentially increasing the volume of space that can be covered.

Santa could use a similar strategy to send a delivery-sleigh to each continent, replicating itself to send one to each country, to each state, territory or county, and so on.

In fact that we have never seen a von Neumann probe is one of the arguments that extra terrestrial life does not exist (see equation 25, Drake equation). The counter argument is that any advanced civilisation would quickly realise that eventually von Neumann probes would quickly strip the galaxy of materials and would not invent them in the first place or destroy them if they found them.

I'll leave it as an exercise for the reader to calculate how many units of alcohol Santa consumes and the mass of all the mince pies eaten.

This article is based on a number of internet sources (in particular http://scienceblogs.com/startswithabang/2009/12/23/the-physics-of-santa-claus/_), that interestingly don't always come up with the same numbers.

DF

Equation 360: Vibrating Circular Membrane (The Drum)

$$w(r,\theta,t) = \sum_{n=0}^{\infty} \sum_{m=1}^{\infty} \vartheta_{mn}(r,\theta)q_{mn}(t)$$

$$= \sum_{n=0}^{\infty} \sum_{m=1}^{\infty} P_{mn}J_n(k_{mn}r)\cos(n\theta - \alpha_{mn})\sin(\omega_{mn}t - \varepsilon_{mn})$$

$$= \sum_{n=0}^{\infty} \sum_{m=1}^{\infty} J_n(k_{mn}r)(\phi_{mn}(t)\cos(n\theta) + \psi_{mn}(t)\sin(n\theta))$$

where, exhaustingly, $w(r,\theta,t)$ is the vertical dynamic displacement of the drum membrane (m), r is the radial distance from the centre of the drum (m), θ is the angular distance from an arbitrary axis drawn radially on the drum skin (radians), t is time (s), ϑ_{mn} is the two dimensional mode shape, $q_{mn}(t)$ is a generalized two dimensional modal coordinate, P_{mn} is the amplitude of a particular mode (m), J_n is a Bessel function of the first kind of order n, α_{mn} and ε_{mn} are constants that will depend on initial conditions of the membrane, $\psi_{mn}(t)$ and ψ_{mn} represent a set of normal (or modal) coordinates (m) and k_{mn} is a constant for a given mode and is determined by the boundary condition that the membrane is fixed at the outer rim of the drum at a radius R. This can be expressed by the condition that $J_n(k_{mn}R) = 0$, see table 1 below.

Finally, $\omega_{mn} = k_{mn}c$ is the natural frequency (or tone) of the mode (s^{-1}), where $c = \sqrt{T_l/\rho}$ is the velocity of wave propagation in the membrane (m·s^{-1}), T_l is the linear tension of the drum membrane (N·m^{-1}), and ρ is the area density (kg·m^{-2})

There's nothing like banging a drum for noisy festive fun, or strumming a guitar for that matter. Turns out that, being a dynamically varying surface, the mathematics for a drum requires four variables, three cylindrical coordinates and time. This is one more than was necessary for a string and substantially complicates things. Notwithstanding this, being a solution to the wave equation in two dimensions (see BE 48), the mathematics turns out to have the same structure thereby leading to similar conclusions about the nature of the sound that is produced. Analogous to a guitar string, the drum dynamics can be determined by adding up an infinite series of distinct mode shapes ϑ_{mn} each of which oscillates harmonically at frequency ω_{mn}. The amplitudes of each mode will depend on how hard and at what location the drum skin is hit.

The first six mode shapes of the membrane deformation are shown in the figure below. The first one corresponding to $m = 1$ and $n = 0$ is the lowest frequency and lowest energy mode corresponding to the fundamental tone of the drum. For the guitar string the locations at which there is no displacement (nodes) occur at points along the string. For the membrane, these nodes are lines. For the first mode there is only one node around the rim. In general there will be $n + m$ nodal lines, see if you can spot them in the figures. The maximum amplitude of this first mode is in the middle. Consequently, hitting the drum at the centre will maximally excite this modes at its frequency as well as all other $n = 0$ modes. Other modes will occur at higher frequencies in direct proportion to the values of the zeros in Table 1. However, these all have a node at the centre of the drum and therefore will not be excited. Again, just as for the guitar string that is plucked at a location at which nodes are absent for many lower order tones, a richer sound will ensue if the membrane is struck off centre. To hear these it is best to strike the drum off-centre and briefly touch the centre of the drum to damp out the $n = 0$ modes.

The last form of the equation shown expresses the generalized coordinates in a more convenient form. Exploiting the fact that the mode shapes are mathematically orthogonal functions, this las form can be used to determine the kinetic energy T and potential energy V in terms of simple quadratic functions given by

$$T = \frac{1}{4}\rho\pi R^2 \sum_{m=1}^{\infty} \sum_{n=0}^{\infty} J_n'^2(k_{mn}R)\{\dot{\phi}_{mn}^2 + \dot{\psi}_{mn}^2\}$$

$$V = \frac{1}{4}\rho\pi R^2 \sum_{m=1}^{\infty} \sum_{n=0}^{\infty} \omega_{mn}^2 J_n'^2(k_{mn}R)\{\phi_{mn}^2 + \psi_{mn}^2\}$$

Amazingly, this can then be used with Lagrange's equation (an expanded form of the Euler-Lagrange equation BE 6 and including D'Alembert's principle BE 2) to derive an infinite series of simple, single coordinate, linear differential equations that can be used to generate equations showing the motion of the drum skin for any type of force that might be applied to a drum.

The history of the development of these equations is not clear. However, it was no doubt included in the early development of Bessel's original equations and discussed as established theory by the German mathematician Hermann von Helmholtz (1821 – 1894) in his treatise on the Sensations of Tone. Probably the most comprehensive collation and extension of this work is presented by the British physicist Lord Rayleigh (1842 – 1919) in his treatise on the Theory of Sound.

References
Helmholtz H.V., 1877, On the sensations of tone, 4th edition, Dover Publications
Rayleigh J.W.S., 1894, The Theory of Sound, 2nd edition, Volume I, chapter IX, Dover publications.

Table 1: Zeros of the Bessel function of the first kind corresponding to values for $k_{mn}R$.

n \ m		m=1	m=2	m=3	m=4	m=5
n=0		2.40482556	5.5200781	8.65372791	11.7915344	14.93092
n=1		3.83170597	7.0155867	10.1734681	13.3236919	16.47063
n=2		5.1356223	8.4172441	11.6198412	14.7959518	17.95982
n=3		6.3801619	9.7610231	13.0152007	16.2234662	19.40942
n=4		7.58834243	11.064709	14.3725367	17.615966	20.82693
n=5		8.77148382	12.338604	15.7001741	18.9801339	22.2178

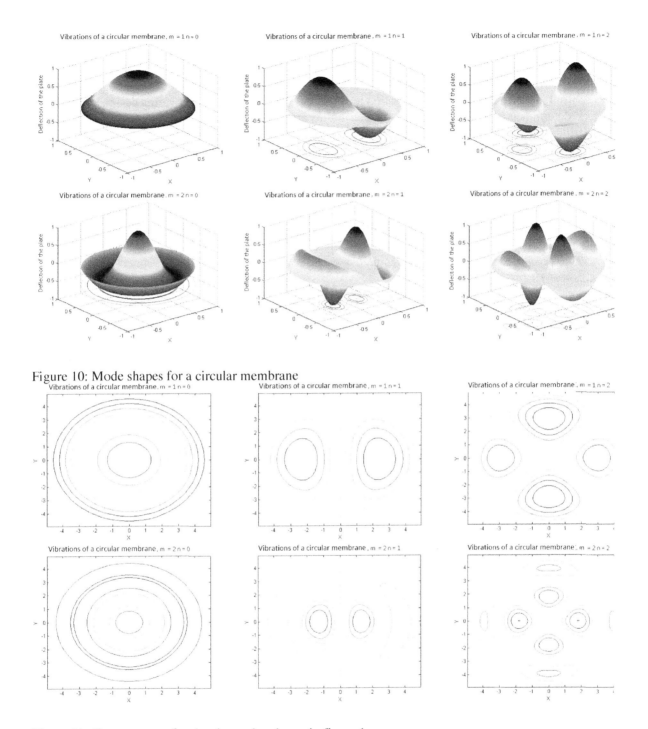

Figure 10: Mode shapes for a circular membrane

Figure 11: Contour maps for the six modes shown in figure 1.
Awesome drumming

https://www.youtube.com/watch?v=P5_G8exuIRc
https://www.youtube.com/watch?v=8ppU_XBruB4
https://www.youtube.com/watch?v=DYBTBv_h4CQ
https://www.youtube.com/watch?v=xkdzo9dYLcA

SS

471

Equation 361: Linear Hydrostatic Bearing

INTRODUCTION

A hydrostatic bearing is a type of fluid film bearing that uses a film of pressurized oil squeezed through a controlled gap to allow a very low friction load bearing surface. Hydrostatic bearing design and performance are determined by a series of trade-offs, constraints and performance requirements. The coefficients of the various performance attributes can be determined by the geometry of the bearing configuration. Discussed below are the coefficients for a typical rectangular bearing pad, however it should be noted that there are many different types of hydrostatic bearing systems and configurations each with their own specific set of coefficients and design nuances.

DESCRIPTION & APPLICATIONS

Hydrostatic bearings are typically very stiff, have high load bearing capacity and give very good damping characteristics. This makes them ideal for many types of precision machine tools and other applications where precise controlled motion is required.

In the design of hydrostatic bearing systems there are several performance considerations and trade-offs. The trade-offs are heavily influenced by the fluid characteristics, temperature and the geometry and use of the bearing surfaces themselves.

It is possible to have a stiff bearing with low flow with a viscous fluid, however this will require a lot of pumping power to push it through the system which will in turn generate a lot of heat, which is usually highly undesirable in most precision applications. More viscous fluid is good for increased damping which is a highly favourable property of hydrostatic bearings. The bearing gap has a significant influence on the flow rate of the fluid and the stiffness of the bearing.

A slight downside of hydrostatic (oil or water) bearings compared to aerostatic (air) bearings is that they typically high higher asynchronous motion.

Hydrostatic bearings should not be confused with hydrodynamic bearings which require high speed between the two bearing surfaces for the hydroplane effect to "float" the bearing. Hydrostatic bearings "float" on their bearing surfaces even when there is no relative motion between the two bearing surfaces. Hydrodynamic bearings operate in a very similar manner to a car tyre; when there is too much water on the road and the car hydroplanes, this essentially makes a hydrodynamic bearing. The power coefficient for a hydrostatic bearing can be calculated using

$$a_b = 1 - \left(\frac{b}{B}\right) - \left(\frac{l}{L}\right) + \left(\frac{2bl}{BL}\right)$$

$$q_b = \frac{1}{6a_b}\left(\frac{B-b}{l} + \frac{L-l}{b}\right)$$

$$h_b = \frac{a_b}{q_b}$$

where a_b is the load coefficient, q_b is the flow coefficient, h_b is the power coefficient, b is the bearing land width in the widest direction, B is the bearing pad width in the B direction, l is the bearing land width in the shortest direction and L is the bearing pad width in the L direction.

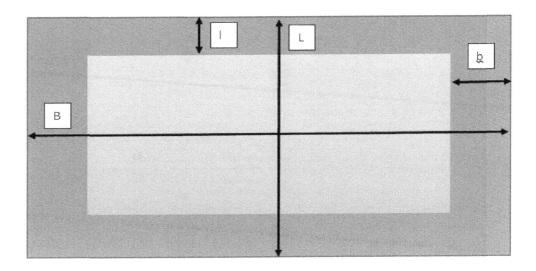

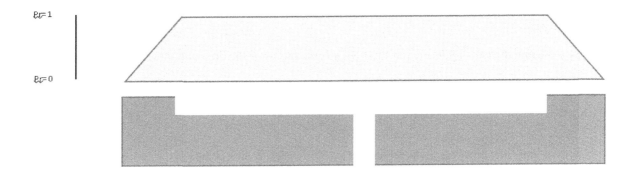

HISTORY

The history of the hydrostatic bearing principle can be traced back to the 1880s where "floating" a load on a film of fluid was considered based on the experiments of Beauchump Tower and the theoretical work of Osborne Reynolds. In 1896 Albert Kingsbury developed a pivoted shoe thrust bearing based on the research by Reynolds. In 1912 Kingdbury was contracted by the Pensilvania Water Company to produce a bearing system to replace the roller bearings which caused significant downtime. After some initial testing and development the bearing system ran for 75 years under a load of 220 tonnes with negligible wear. This lead to it being designated the 23rd International Historical Mechanical Engineering Landmark by the ASME on 22nd June 1987.

SUMMARY

Hydrostatic bearings can be complex but employ a basic physical principle to achieve a high performance bearing system. The ability to have high load capacity, long life and very low friction make it a favourable candidate for many systems.

Hydrostatic bearings are one of the more significant mechanical system developments and are used throughout the world in a wide variety of load bearing systems.

TS

Equation 362: The Mason Equation for Snow Crystal Growth

Mason's general equation for the mass growth rate of ice crystals is the following:

$$\frac{dm}{dt} = \frac{4\pi C(S-1)(1+\alpha\sqrt{Re})}{f(T)}$$

where

$$f(T) = \frac{RT}{D_v M_w p_s(T)} + \frac{L}{KT}\left(\frac{LM_w}{RT} - 1\right)$$

is a function that describes the contribution of the diffusion of water vapour towards the growth of the ice, L is latent heat of sublimation taken away from the crystal to the air, $p_s(T)$ is the saturation vapour pressure over the ice, K is the thermal conductivity of air, M_w is the molecular weight of water D_v is the diffusion coefficient of water molecules through the air, R is the universal gas constant C is the analogous electrostatic capacity of the crystal, $1 + \alpha\sqrt{Re}$ is the ventilation effect, α is a numeric factor and Re is the Reynolds number, which is expressed as

$$Re = \frac{2vr}{\nu}$$

with v being the velocity of the falling drop, r is the radius of the drop and ν the kinematic viscosity of air.

For disc-shaped crystals, the mass is expressed as:

$$m = \pi r^2 h \rho_c$$

with h being the thickness, r the radius and ρ_c the density of the crystal. Using the relationship:

$$\frac{dr}{dz} = \frac{dr}{dT}\frac{dT}{dz} = \frac{1}{v}\frac{dr}{dt}$$

with v being the terminal velocity of the crystal. Mason was able to calculate the increase in radius for different shapes of crystals falling within a certain temperature interval, with a temperature rate $dT/dz = 6.5\ K/km$. The shapes include ice spheres, thin hexagonal plates, hexagonal sector plates, prismatic columns and stellar dendrites, reported in Figure 1.

HISTORY
Snowflakes do not only attract the imagination of many children worldwide, but have been a fascination for scientists since long time. In 1611, Johannes Kepler wrote a short treatise on the possible origin of the snow crystal symmetry. In 1637, Rene' Descartes gave a detailed analysis of many different forms of natural snow crystals in his treatise on weather phenomena "Les Meteores". With the development of photography in the late 19th century, the American photographer Wilson Bentley catalogued several thousand photographs of snow crystal images, which he had acquired over several decades.

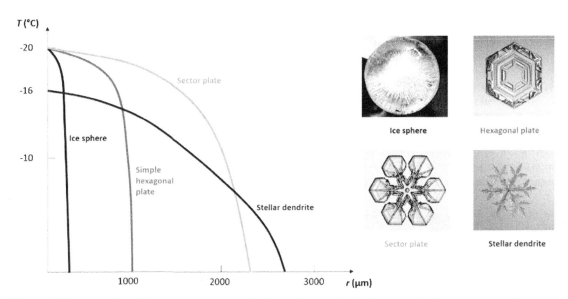

Figure 1. The growth radius vs temperature for ice spheres, simple hexagonal plates, sector paltes and stellar dendrites (Source: BJ Mason, "The Shapes of snow crystals – Fitness for purpose?" Q.J.R. Meteorol. Soc. (1994), 120, 849).

In 1930s, the Japanese physicist Ukichiro Nakaya performed the first experiments on the formation of snow crystals. He studied the morphology of snowflakes at different temperatures and supersaturation, at atmospheric pressure and categorised natural snow crystals appearing in various meteorological conditions. He combined his observations in the famous Nakaya's morphology diagram (Figure 2).

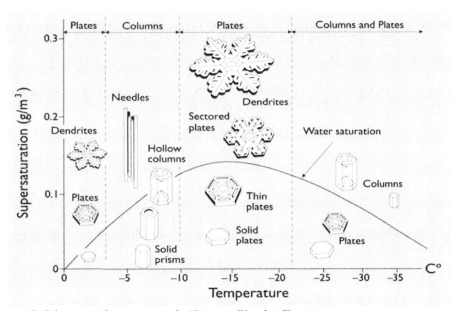

Figure 2. Nakaya's Diagram of snow crystals (Source: Physics Forme: https://physicsforme.com/2012/02/16/mathematical-model-computes-snow-flake-shapes-for-the-first-time/).

Sir Basil John Mason derived the general equation for the growth rate of an ice crystal in 1953, when he was Professor of cloud physics at Imperial College London [1]. Mason also worked as Director-General of the Meteorological Office in the UK from 1965 to 1983, where he developed models on the generation of electrical charges in thunderclouds and lightning. He also helped modernise the World Metereological Organisation during the 1960s.

The Mason equation is based on the assumption that mass diffusion towards the water drop in a supersaturated environment transports energy as latent heat, which is balanced by the diffusion of heat back across the boundary layer. The two energy transports must be nearly equal but opposite in sign, setting an interface temperature of the drop.

The term of the Mason equation associated to the ventilation effect represents the fact that vapour concentration and thermal gradients around falling snowflakes play a stronger influence on the crystal growth when compared to stationary ice crystals.

From the general equation, Mason derived equations to calculate the increase in radius experienced by a falling snow crystal within a temperature gradient at a rate of 6.5 K/km. The equations change according to the different shapes of the snowflake. Figure 1 shows the trends in the ice crystal radius as a function of the temperature.

CRYSTAL FORMATION
The formation of a snowflake begins in a cloud made of liquid water droplets, which nucleate on minute dust particles. Pure water droplets of microscopic dimensions can be supercooled up to temperatures below -40 °C before they freeze. Hence, the water vapour in the cloud is supersaturated. As the cloud temperature drops to about -10 °C, the droplets begin to freeze. However, not all of them freeze simultaneously: each droplet accumulates water molecules from its surroundings, while the vapour of the cloud remains supersaturated. Different conditions in supersaturation, air temperature and particle and heat diffusion affect the growth of the snow crystal and give rise to a great variety of shapes in snowflakes within a temperature range between -25 and 0 °C. The different shapes are illustrated in the Nakaya's diagram (Figure 2).

A transition occurs between plates and needles at -3 °C and another transition between hollow prismatic columns and plates takes place at -8 °C, within a temperature interval of less than 1 °C. A large variation in supersaturation affects the aspect ratio and growth rate of the crystal.

At temperatures corresponding to the plate shapes, an increase in supersaturation induces the transition:
very thick plate → thick plate → sector plate → dendrite.

In the columnar prism regime, an increase in supersaturation induces the transition:
short solid prism → longer column → hollow column → needle.

Therefore, shapes such as needles, sector plates and stellar dendrites appear only when vapour supersaturation exceeds values corresponding to the saturation with respect to the liquid water. In addition, when a snow crystal falls into a new environment, its sustained growth leads to new conditions and consequently the formation of hybrids or combination of forms such as columns and plates sprouting at the corners to form sector plates or dendrites.

Mason also suggests that surface diffusion of water molecules across basal and prismal faces in the early stage of the formation strongly influences the outcome in the ice crystal morphology. Once established, the shape is maintained by the orientation of the vapour water molecules and the thermal diffusion surrounding the crystal.

The transition from hexagonal plates to dendritic stellar flakes occurs when the material from the vapour phase arrives on the corners of the prism more rapidly than it can be carried away by surface diffusion along the edges of the plate. This condition occurs only when the diameter of the plate exceeds a critical value, which is proportional to the ratio between the diffusion coefficient for water molecules in air and along the prism edge.

IMPLICATION AND VARIETIES

There are precise reasons and meteorological consequences for the great variation in snowflake habits within such a narrow range of temperatures. The diagram in Figure 2 shows that the more complex shapes achieve a greater mass while falling between the same temperature intervals, as they fall more slowly. During the melting process at about 0 °C, these shapes produce larger raindrops growing by accretion of clouds droplets. Hence, the more complex shapes are more efficient in releasing precipitation, provided that the clouds last long enough – that is to say, about 1 ½ hours. Stellar dendrites are the most effective shape in terms of high precipitation.

In addition, the transition from hexagonal to dendritic structure results in a reduction of the albedo of cirrus clouds, which amounts to 1/3 of that corresponding to plates snowflakes. The reduction of albedo has significant consequences on the solar radiation balance of the atmosphere, especially in dense, extensive long-lived cirrus clouds.

Finally, the transition from plates to columns and needles shapes in the temperature range from -8 to -3 °C presents the following advantages:

- Enhanced growth rate in the clouds, which produces larger precipitations than plate crystals
- Enhanced survival rate of snowflakes to evaporation below the cloud base
- A lower concentration of primary nuclei and an increased efficiency in precipitation release, especially from supercooled convective clouds.

Ice crystals are important in cloud electrification and lightning, as they play a role in the charging mechanism in the atmosphere, involving collisions between ice particles, which ultimately depend on their surface structure.

Chemical processes in the upper atmosphere also rely on the surface properties of ice crystals to boost their reaction rates. Most meteorological phenomena are partially influenced by the structure and dynamics of the ice surface.

DEEPER UNDERSTANDING AND BETTER MODELS

In the last ten years, a better understanding of the physics of snow crystals has led to the development of more accurate simulations to model the formation of snowflakes and the transition between different shapes at different environment conditions [2].

Three main factors control the growth of a snow crystal:

- Particle diffusion, which carries water molecules to the growing crystal
- Heat diffusion, which removes the latent heat generated by solidification
- Attachment kinetics: this process determines how water molecules are incorporated into the ice crystal lattice. It is expressed by a *condensation coefficient* in the model. Water molecules adding to the flat surface of an ice crystal tend to evaporate before becoming part of the surface, due to the absence of suitable attachment sites. Therefore, the growth of snowflakes can only occur when the density of new adding molecules is high enough to form 2D islands on the surface of the ice crystals. These islands act as nucleation centres, providing steps and kinks for the attachment of other molecules. The growth of the snow crystal depends on the nucleation rate – i.e. the rate of formation of the islands – and the islands growth rate. If the nucleation rate is high, islands can form on top of other islands, leading the growth process.

In addition, surface melting, chemical impurities in the atmosphere such as alcohols, acids and hydrocarbons, and crystal imperfections such as dislocations strongly influence the crystal growth and the morphology of snowflakes.

In general, new models established that hexagonal prisms appear when the growth is dominated by the attachment kinetics, with particle and heat diffusion giving a limited contribution to the process. On the other hand, crystal branching, leading to the formation of dendritic structures, occurs when diffusion becomes the dominant force in the crystal growth dynamics. The complexity of dendritic patterns, which have a fractal structure, increases with the supersaturation.

In 2012, Harald Garcke from the University of Regensburg, Germany, and John Barrett and Robert Nurnberg from Imperial College London, UK, developed a model that successfully simulate the formation and growth of several shapes of snow crystals, including solid plates, prisms, hollow and capped columns, needles and stellar dendrites [3]. The model, which accurately reproduces the changes in the crystal habit with time, relies on the diffusion equation in the gas phase and the mass balance equation at the gas/solid interface. An important factor included in the model is the anisotropy in the surface energy of the snow crystal, which depends on the orientation of the gas/solid interface. This factor also affects the formation of different phases. The scientists successfully simulated the transition from hexagonal prisms to dendritic stellar structures with an increase in supersaturation, followed by faster evolution. Commenting on this study, physicist and snowflake maven, Ken Libbrecht, from Caltech, US, said that the scientists have solved a problem that other people have tried and failed to do [4].

Modelling the formation of snowflakes does not only give us the pride of being able to reproduce the intrinsic beauty of one of nature's fascinating creations, but has important implications in the ability of modelling structures such as dendrites, which occur in materials of technological importance such as metallic alloys.

This year, however, snowflakes have been quite elusive around the world. In London as well as in Italy, we are all still waiting them to come down from the sky. Waiting for the clouds.

References

[1] B.J. Mason, "The growth of ice crystals in a supercooled water cloud", Q.J.R.Meteorol. Soc., 79, 104, (1953).
[2] K.G.Libbrecht, "The physics of snow crystals", Rep. Prog. Phys., 68, 855, (2005)
[3] J.W.Barrett, H.Garcke, R.Nuernberg, "Numerical computations of facetted pattern formation in snow crystal growth", Phys. Rev. E, 86, 011604, (2012)
[4] R.Cowen, "Snowflake Growth Successfully Modelled from Physical Laws", Scientific American (March 2012)

AT

Equation 363: Friedmann Equations

The Friedmann equations are a set of equations derived by Alexander Friedmann in 1922. Friedmann (16th June 1888 - 16 September 1925) was a Russian physicist and mathematician who was best known for his work on the expanding universe theory.

The Friedmann equations are a set of equations in cosmology that govern the expansion of space in homogeneous and isotropic models of the universe within the context of Einstein's general relativity. In 1922, Friedmann proposed that the universe started at zero size, i.e. the Big Bang, expanded into a spherical shape, reaches a maximum size and then contracts to a Big Crunch.

The Friedmann equations are given below. Some cosmologists refer to the first equation as the Friedmann equation and the second as the Friedmann acceleration equation.

$$\frac{\dot{a}^2 + kc^2}{a^2} = \frac{8\pi G\rho + \Lambda c^2}{3}$$

$$\frac{\ddot{a}}{a} = -\frac{4\pi G}{3}\left(\rho + \frac{3p}{c^2}\right) + \frac{\Lambda c^2}{3}$$

where a is the scale factor that depends on time, $\dot{a}/a \equiv H$, where H is the Hubble constant, G is Newton's gravitational constant, Λ is the cosmological constant, c is the speed of light in a vacuum, k is a constant throughout a particular solution, ρ is the density, p is the pressure and k/a^2 is the spatial curvature of the universe.

If:

- $k = 0$, the universe is flat and has infinite volume.
- $k = -1$, the universe is open and has infinite volume with negative curvature i.e. it is saddle shaped.
- $k = +1$, the universe is closed with positive curvature negative total energy and has finite volume i.e. it is spherical.

In a Friedmann universe, the parameter Ω, known as the density parameter is the ratio of the density to the critical density ρ_c. When $k = 0$, these are equal giving a flat universe. It is known that the universe is more dense from dark matter. This dark matter, making up approximately 22% of the universes mass would lead to a contraction of the universe. However, approximately 74% of the mass of the universe is made of dark energy. To put this into context, about 3.5% of the mass of the universe is made up of gas and about 0.5 % of the universes mass is made up of stars, planets etc. Dark energy accounts for Einstein's cosmological constant and seems to accelerate the expansion of the universe. If:

- $\Omega > 1$, the universe will contract into a Big Crunch.
- $\Omega < 1$, the universe will expand forever. (BE 365)

Current measurements taken by the WMAP (Wilkinson Microwave Anisotropy Probe) spacecraft suggest the geometry of the universe is nearly flat, indicating the universe will expand forever.

A. Friedman, "On the Curvature of Space," Gen. Rel. Gravit., vol. 31, no. 12, pp. 1991-2000, 1922. Originally published in Zeitschrift fur Physik 10, 377-386 (1922), with the title Uber die Krummung des Raumes

CM

Equation 364: The Error Function

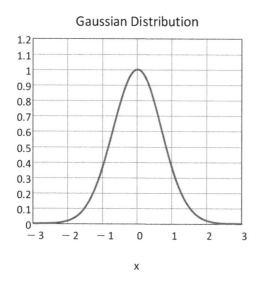

Gaussian Distribution

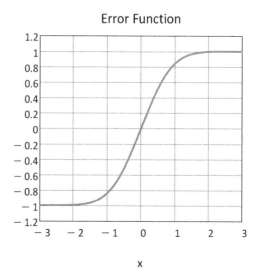

Error Function

The Gaussian function $\exp(-x^2)$ describes the beautiful bell curve shown above left, with its graceful slopes tapering towards zero on either side of a peak, although zero is never quite reached. The function extends to infinity in both directions, but the definite integral over all values of x has an analytical solution $\sqrt{\pi}$, as described in Beautiful Equation 349.

The Gaussian is of central importance to statistics, probability and random physical processes. In error analysis, the Gaussian function describes the "normal" distribution of random events, telling us the probability of observing an error of a given magnitude. In this context, it is frequently desirable to limit the integration to some portion of the Gaussian distribution. For example, we might want to know the likelihood that errors falls between specific maximum values $|x|$ in the distribution. This task is so common and important that we give it a name, the *error function*, and define it by today's beautiful equation:

$$\text{erf}(x) = \frac{2}{\sqrt{\pi}} \int_0^x \exp(-t^2)\, dt$$

The normalization $2/\sqrt{\pi}$ is such that $\text{erf}(x) = 1$ for $x = \infty$ and the factor of 2 is the result of only integrating the right half of the curve. The error function is the area under a Gaussian distribution from $-x$ to $+x$.

Unlike the integration of a Gaussian over the full range out to infinity, the error function has no analytical solution, it is an integral equation that requires a numerical approximation. The usual approach is to expand the exponent in the integrand into a Maclaurin series, which can then be integrated term by term, with the following result:

$$\text{erf}(x) = \frac{2}{\sqrt{\pi}} \left(x - \frac{x^3}{3} + \frac{x^5}{10} - \frac{x^7}{42} + \cdots \right)$$

The behaviour of the error function is as one would expect: As shown in the figure above right, the larger the value of x, the larger the value of $\text{erf}(x)$. One feature that is kind of cool is that negative values of x are allowed, even though in the context of a statistical analysis this would not seem to make much sense. It is a clue that there are many other applications for $\text{erf}(x)$.

The error function has a long history given its practical importance. J.W.L. Glaisher provides an early description of a close cousin to the error function in paper entitled "On a class of definite integrals," appearing in Philosophical Magazine (Series 4, Vol.42, Issue 280, 1871). Applications of the function include characterizing spectral distributions and apodised beam profiles, solving differential equations, describing diffusion and dissipation, and analysing uncertainty in metrology. One of the things that I find most charming about this equation is the name *error* function: It is one of the few occasions for which error is the goal, not the accidental outcome, of a mathematical analysis.

PdG

Equation 365: Big Rip – Time to the End of the Universe

We have now come to the end of the Beautiful Equations adventure and the end of 2015. So today's equation is about a scenario for the end of the universe. First predicted in 2003 by Robert Caldwell, Marc Kamionkowski and Nevin Weinberg (all from USA), The Big Rip is a theory of how the universe will end by being ripped apart by the expansion of the universe some time in the future. Many folk have heard us physicists talking about dark energy (sometimes called phantom energy or quintessence) and dark matter. We need to add these dark concepts to get our current theories of physics (mainly General Relativity – see BE 337) to match astronomical observations. We've known that the universe is expanding since Edwin Hubble's observations, but we also recently observed that the expansion rate seems to be accelerating. These observations, plus the observed motion of galaxies, means that we need to evoke some kind of dark matter and dark energy, which is prevalent throughout the universe. We have no idea what these dark things are (but that's another story…). The ratio of dark stuff to regular stuff (i.e. stars, atoms, you) and the ratio of dark energy pressure to energy density (w, known as the equation of state parameter) will determine the ultimate fate of the universe. At a certain ratio (when $w < -1$), the acceleration of the universe will continue until it becomes faster than the speed of light; at which point there can be no interactions between anything, i.e. all the force carrying particles cannot mediate forces between matter particles, and everything is essentially ripped apart. Note that relativity prohibits any matter or energy travelling faster than the speed of light, but it doesn't rule out spacetime exceeding this limit.

Today's equation is an estimate of the time to the Big Rip, t_{rip}; H_0 is the Hubble constant (see BE 166), Ω_m is the current value of the density of matter in the universe and t_0 is the current age of the universe. Plugging in the values predicted in the 2003 paper, we get a prediction that the universe will be ripped apart in around 22 billion years.

About 60 million years before the Big Rip, gravity would be too weak to allow galaxies to stay together. Approximately three months before the end, any planetary systems would become gravitationally unbound, and in the last few minutes, stars and planets would be ripped apart. Finally, just before the "end", atoms would get it.

So, no need to lose any sleep just yet and, of course, this is just a hypothesis, which could be wildly incorrect, depending on what we ultimately find out about the dark things in the universe.

Happy New Year everyone!

$$t_{rip} - t_0 \approx \frac{2}{3|1 + w|H_0\sqrt{1 - \Omega_m}}$$

RL

The Contributors

Vadim Baines-Jones is a Product Conformance Engineer specialising in usability engineering and risk management of safety-critical hospital equipment. His PhD was awarded by Liverpool John Moore's University and covered fluid application and CFD in the grinding environment. Previous to his current role he worked as a Process Development Engineer focusing on Manufacturing Engineering – specifically the mechanisms when grinding cam- and crank-shafts. He is an active STEM (Science, Technology, Engineering and Mathematics) ambassador, Chairman of a local volunteer-led male carers group and an active enlightened agent in augmented reality game, Ingress.

Patrick Baird is a R&D scientist at Gnosys in the Surrey Research Park in Guildford, where he works primarily in spectroscopy and software development for multivariate analysis. His background is in physics and he received his PhD from Brunel University in 1996 for work in metrology at the National Physical Laboratory. For some years he carried out postdoctoral research in scanning microscopy at The University of Surrey and Imperial College. In addition to physics and mathematics, his other interests are music (particularly the piano), astronomy, history and palaeontology. Patrick has a love for equations and their solutions, and has developed an interest in their history and original derivations.

Harish Cherukuri is a professor in the Department of Mechanical Engineering and Engineering Science at the University of North Carolina at Charlotte. His research interests include mechanics of deformable bodies, modelling manufacturing processes, and computational methods with emphasis on the finite element method and smoothed particle hydrodynamics. His other interests include nutrition, biochemistry, and coding.

Peter de Groot (PdG) is the Executive Director of R&D at Zygo Corporation, in Middlefield, Connecticut. ZYGO designs, develops, and manufactures optical metrology solutions that provide process control for surface shape, roughness, material characteristics, film thickness and stage positioning, largely based on ZYGO's unique expertise in optical interferometry. Peter heads a R&D Team comprised of 7 PhD scientists, focused on the invention and concept demonstration of new optical instrument products and product enhancement. As R&D Group Leader and as a Principal Scientist, Peter has contributed to nearly every ZYGO metrology product since 1992, is an inventor for 130 US patents, and has published 145 technical papers, book chapters and review articles. Peter has additional experience as an adjunct professor of optics and optical metrology, and served as a Peace Corps volunteer high school physics teacher and teacher trainer in Africa. See www.zygo.com

Massi Ferrucci is a PhD student at KU Leuven and is performing his doctoral research at the National Physical Laboratory in Teddington, United Kingdom. Massi received a B.Sc. in Physics and a B.A. in Russian Language and Literature from the University of Maryland in College Park. While working at the National Institute of Standards and Technology in Gaithersburg, Maryland, Massi pursued and received a M.Sc. in Applied Physics from the Johns Hopkins University in Baltimore, Maryland. Massi has a passion for dimensional metrology and, in particular, for the development of geometrical calibration procedures for coordinate measuring systems.

David Flack has been a dimensional metrologist for thirty-five years. He specialises in CMM, diameter, roundness and angle measurements. He is responsible for dimensional measurement services at a world leading NMI.

Han Haitjema is director of Mitutoyo Research Center Europe in Best, The Netherlands. He was previously at the Eindhoven University of Technology and the Van Swinden Laboratory, the Dutch national laboratory in Delft, The Netherlands. His research interest has been dimensional metrology in all its aspects in the last 25 years. His hobbies include photography, piano- and harpsichord playing and classical music.

Mike Goldsmith is a freelance acoustician and science-writer. He has a PhD in astrophysics from the University of Keele and continues to study the dwarf planet Pluto, but has worked for the last thirty years primarily in acoustics; formerly being Head of Acoustics at the National Physical Laboratory. He is particularly interested in problems related to noise, including underwater and high-frequency phenomena. Other interests include the history of science, running, cooking, and science fiction.

Olaf Kievit is a teacher of IB Chemistry and Physics at Sotogrande International School in Spain. He received his PhD in Chemistry from Yale University in 2000 for research on electrochemistry of proteins involved in photosynthesis. He has previously taught at schools in the UK and the US. He has an interest in physical science, with a focus on the big conceptual ideas, and their expression in simple but elegant equations. In addition to this, his interests include building scale models, photography, philosophy, meditation and golf.

Richard Leach is a professor of metrology at the University of Nottingham. He was previously at the UK's National Physical Laboratory for 25 years and is an expert on surface measurement for advanced manufacturing, especially using optical techniques. His hobbies include writing (text books and other non-fiction), fitness and nutrition and working too hard. Richard has always loved equations, and once won a competition for saying the solution to a quadratic equation the fastest in his school and never ceases to be amazed at how nature can follow simple, yet elegant rules.

Ian Lee-Bennett is an Executive Director of Taylor Hobson Ltd, a leading global provider of metrology solutions (www.taylor-hobson.com). One of his focuses there is the strengthening of the company's IP portfolio, and recently his team has been working on new optimised filtering routines. He holds a number of patents in the fields of novel mechanical systems and computational algorithms. Ian was originally a machine code software engineer and specialised in algorithms, many based on old adding machine techniques. He now loves all areas of science and maths, and has a particular interest in machine consciousness and quantum physics and an everlasting fascination with prime numbers. Ian has always loved solving problems and employing mathematical equations is a fantastic way of doing that - and furthermore, equations and the mathematics they are built on seem to transcend the physical universe and allow us to experience a more fundamental truth than even the universe itself.... And they are cool! In addition to his scientific interests, he is also a song writer, actor and photographer. He lives in Leicester.

Campbell McKee Campbell McKee is a structural health monitoring engineer at James Fisher Testing Services (Strainstall) focusing on large structure applications. He obtained both his degree in Applied Physics and PhD in Electrical and Electronic Engineering from the University of Strathclyde in Glasgow. His research interests include non-contact techniques for structural monitoring and NDT, digital signal processing, computer networks and instrumentation. His interests include photography, rugby, walking and reading.

Giuseppe Moschetti is a final year PhD student enrolled at Huddersfield University (UK), but is currently doing his research at the UK's National Physical Laboratory. Before starting his PhD, he was educated at the University of Bologna (Italy) in electronic engineering. Giuseppe's research interests are various and as such they found full expression in the multidisciplinary field of precision engineering: optical design, digital signal processing, error propagation and high performance computing. In his free time Giuseppe enjoys cycling, juggling and team sports (basketball, football and lately rugby).

Kang Ni is a PhD student at the University of North Carolina at Charlotte. He was educated in the Hefei University of Technology in China before he came to the U.S. His encounter with precision engineering started with his master degree work of carrying out dimensional and machine tool metrology on a lithography machine. He has been trying to appreciate the art of metrology driven by his passion for precision ever since. Kang loves swimming, photography and sending post cards to himself and friends while travelling. He is fascinated with the way Mother Nature exists and the stories behind beautiful equations.

Robert Oates is a software intensive systems engineer specialising in the cyber security of safety-critical systems. His PhD was awarded by The University of Nottingham and covered security, robotics and artificial immune systems. After working for several years as a post-doctoral research fellow in both Computer Science and Manufacturing Engineering, he became a consulting engineer working in the aerospace, marine and nuclear sectors. He is an active STEM (Science, Technology, Engineering and Mathematics) ambassador, and enthusiastic about encouraging more people to discover the incredible things that those disciplines have to offer: the awe to be found in science; the humanity-changing effects of technology and engineering; and the beauty of equations.

Frank Roberts is a private tutor of physics and maths, as well as working in colleges with students in the 16+ age group. He is currently engaged by the BTEC engineering and A-Level Maths departments at a local further education college. His teaching career was interspersed by over a decade involved with the software side of 'real time' computer systems. Pastimes include swimming, reading and chess (he runs a popular chess club for juniors).

Paul Rubert is the director of a small UK metrology company, which manufactures surface reference standards and also operates a calibration laboratory. He graduated in Physics at Imperial College, London, and then studied engineering at the University of Salford, Manchester. He has made a few small and insignificant contributions to surface metrology at international conferences and at ISO, where he is a member of the committee responsible for surface measurement. His hobbies include flying radio-controlled planes, playing the violin, and arguing about controversial subjects with anyone who will listen.

Stuart Smith is a professor at the University of North Carolina at Charlotte and works in the field of scientific instrument and machine design. This work is a continuing struggle to harness technology for the purpose of utilizing or squeezing out information about natures laws that we are so privileged to able to comprehend. Often the effects that form the focus of any particular study can be quantifiably described using the tools of mathematics that, in turn, inform an understanding of experimental results. Of course in all experiments, all of the laws of physics are operating all of the time and it becomes necessary, as an experimentalist, to assimilate a knowledge of the mathematical structure of physical laws that, in turn, serve to illuminate the beauty of nature.

Trevor Stolber is a precision engineer at 3M Company working with the structured surfaces group making microstructure on a large scale. The main focus of this work is to develop and advance the state of the art of high precision machine tools and precision processes. Trevor worked at Cranfield Precision for 13 years and studied at Cranfield University undertaking an MSc in Ultra Precision Technologies. He is fascinated by the fabric of material and the chase for nanometres in the search for ever higher accuracy.

Antonio Torrisi studied a Degree in Materials Science at the University of Turin (Italy). After the degree, he moved to London (UK) in 2003 to study a PhD in Computational Modelling of the Organic Solid State at University College London (UCL), where he continued to work as Post-Doctorate Research Associate until September 2011; his research interests focused on modelling Metal-Organic Framework materials and advanced inorganic electrides for carbon capture and catalysis and hydrogen storage. In 2011, he studied a MSc in Science Communication at Imperial College London, with a dissertation on the impact of Citizen Science and Street Science projects in science education, with interests in multimedia production and science writing. Following a short project worked as Research Scientist at Darlow Smithson Production for the production of the documentary "Earth From Space", in 2013, Antonio worked as Deputy Editor of the Imperial College-based publication *A Global Village* (now called *ANGLE*) on science and society, and in 2014 he worked as Reporter for the B2B publication Industrial Minerals. He joined the National Physical Laboratory in 2015 as Assistant Instructional Designer and since 2016 he has been moving his career towards science teaching and education.

Acknowledgements

The authors would like to thank Adam Thompson, Isam Bitar, Patrick Bointon, Waiel Elmadih, Lars Körner, Lewis Newton, Danny Sims-Waterhouse, Nicholas Southon, Matthew Thomas and Luke Todhunter (all PhD students at The University of Nottingham) for their contributions in editing this book. Robert Oates would also like to thank Victor Malysz and David Banham for fruitful discussions.

Made in the USA
Middletown, DE
01 July 2022